METEOROLOGICAL MONOGRAPHS

VOLUME 30   AUGUST 2003   NUMBER 52

RADAR AND ATMOSPHERIC SCIENCE: A COLLECTION OF ESSAYS IN HONOR OF DAVID ATLAS

Edited by

Dr. Roger M. Wakimoto
Dr. Ramesh Srivastava

American Meteorological Society
45 Beacon Street, Boston, Massachusetts 02108

ISBN 1-878220-57-8
ISSN 0065-9401

Published by the American Meteorological Society
45 Beacon St., Boston, MA 02108

Printed in the United States of America
by Allen Press, Inc., Lawrence, KS

TABLE OF CONTENTS

PREFACE

This monograph pays tribute to one of the leading scientists in meteorology, Dr. David Atlas. Dave has contributed in a number of ways, as eloquently summarized in chapter 1 of this monograph, during a span of time that covers nearly six decades. He has made important contributions as a researcher, inventor, laboratory leader, and educator. In addition, Dave has compiled an impressive record of service to the community, in general, and the American Meteorological Society (AMS), in particular. Dave's name is synonymous with radar meteorology and he has mentored or extensively interacted with almost every prominent scientist in this field.

A named symposium was organized on 14 January 2002 at the AMS's Annual Meeting in Orlando, Florida, in order to honor and celebrate his distinguished career. A program committee composed of Robert Serafin, Richard (Rit) Carbone, Roddy Rogers, and Roger Wakimoto was assembled (with Bob and Rit acting as cochairs) to identify topics and compile a list of invited speakers. Narrowing the number of subject areas was a difficult task and, after extensive discussion, 10 topics were selected for oral presentations (including one on radar's impact on education). The committee felt that it was important to select young, rising stars in the field of radar meteorology as speakers in addition to the more established senior scientists. A panel discussion was organized for the last session of the symposium to debate the future of radar applications. The end of the day was highlighted by an evening banquet. This special dinner provided an opportunity for friends and colleagues to pay tribute to Dave and his wife, Lucille. The gala event closed with Dave providing an emotional address to the audience.

The work was not over at the end of the symposium. All of the speakers were asked to write papers on the material they had presented for inclusion in this monograph. The common theme of all of the talks was "radar." Although we requested that the authors present the state of the science, we also emphasized that it was appropriate to note Dave Atlas's contribution to the subject area whenever appropriate. Rit and Bob kindly agreed to begin the monograph with the aforementioned chapter that provides an overview of Dave's career and pays tribute to the numerous and pioneering contributions that he has made over the years. A special effort was made to document the symposium. Numerous photographs were taken of the speakers, audience, and the events that unfolded at the evening banquet. A subset of these photos is included in this monograph.

Dave provided editorial leadership in overseeing *Radar in Meteorology*, another AMS monograph published in 1990 to both honor Professor Louis J. Battan and to commemorate the anniversary of the first radar conference in 1947. The book, among other goals, provides an excellent treatise on radar meteorology. A quick perusal of the table of contents of that book will reveal that several of the titles overlap with those presented in the current monograph. Our instructions to the authors were that they should not duplicate the material presented back then unless it was necessary to set the background for their articles. Instead, the emphasis was to be on the advances made on their assigned topic since the publication of *Radar in Meteorology*. In light of these instructions, the amount of new material that has accrued over the past decade as documented in several of the chapters is truly impressive. This provides further evidence of the dynamic nature of our field.

It is difficult to assess this monograph's ability to withstand the test of time. While many of the articles provide excellent summaries on a variety of subjects within our discipline, radars are constantly undergoing a never-ending list of modifications and improvements in order to remotely probe the atmosphere in novel ways. Accordingly, new radar techniques will always be just around the corner and will invariably result in a steady stream of new datasets. The future is bright, and we all owe a great deal of gratitude to Dave for his incredible insight, ingenuity, and leadership during the formative years of radar meteorology. Indeed, he continues to be a guiding light to many of us today.

It is our sincere hope that this monograph will help stimulate the next generation of radar meteorologists, provide an excellent resource book for scientists and educators, and serve as a historical record of the gathering of many of our distinguished scientists in our field who converged one January day in Orlando to honor one of the giants in our discipline.

Photographs of the presenters at the Atlas Symposium are included with the respective chapters.

Roger M. Wakimoto and Ramesh Srivastava
Editors

CONTRIBUTORS

HOWARD B. BLUESTEIN
School of Meteorology
University of Oklahoma
100 E. Boyd, Rm. 1310
Norman, OK 73019
E-mail: hblue@ou.edu

KEITH A. BROWNING
Department of Meteorology
University of Reading
P.O. Box 243, Reading RG6 6BB
United Kingdom
E-mail: k.a.browning@reading.ac.uk

RICHARD E. CARBONE
Mesoscale and Microscale Meteorology Division
NCAR
P.O. Box 3000
Boulder, CO 80307-3000
E-mail: carbone@ucar.edu

V. CHANDRASEKAR
Colorado State University
1373 Campus Delivery
Fort Collins, CO 80523-1373
Email: chandra@colostate.edu

FRÉDÉRIC FABRY
Department of Atmospheric and Oceanic Sciences
McGill University
805 Sherbrooke St. W.
Montreal, QC H3A 2K6
Canada
E-mail: frederic@radar.mcgill.ca

KENNETH S. GAGE
NOAA Aeronomy Laboratory
325 Broadway
Boulder, CO 80305-3328
E-mail: Kenneth.S.Gage@noaa.gov

EARL E. GOSSARD
NOAA/Environmental Technology Laboratory
325 Broadway
Boulder, CO 80305-3328
E-mail: egossard@indra.com

DAVID P. JORGENSEN
NOAA/National Severe Storms Laboratory
Mesoscale Research Division—Boulder
325 Broadway
Boulder, CO 80305-3328
E-mail: David.P.Jorgensen@noaa.gov

R. JEFFREY KEELER
Research Engineer
Atmospheric Technology Division
NCAR
P.O. Box 3000
Boulder, CO 80307-3000
E-mail: keeler@ucar.edu

FRANK D. MARKS JR.
NOAA/AOML
Hurricane Research Division
4301 Rickenbacker Causeway
Miami, FL 33149-1097
E-mail: Frank.Marks@noaa.gov

ROBERT MENEGHINI
Code 975
NASA/Goddard Space Flight Center
Greenbelt, MD 20771
E-mail: Robert.Meneghini-1@nasa.gov

DANIEL ROSENFELD
Institute of Earth Sciences
The Hebrew University of Jerusalem
Jerusalem 91904, Israel
E-mail: Daniel.Rosenfeld@huji.ac.il

STEVEN A. RUTLEDGE
Department of Atmospheric Science
Colorado State University
Ft. Collins, CO 80525
E-mail: rutledge@atmos.colostate.edu

ROBERT J. SERAFIN
Director Emeritus
NCAR
P.O. Box 3000
Boulder, CO 80307-3000
E-mail: serafin@ucar.edu

JUANZHEN SUN
NCAR
P.O. Box 3000
Boulder, CO 80307-3000
E-mail: sunj@ucar.edu

CARLTON W. ULBRICH
Department of Physics and Astronomy
Clemson University
Clemson, SC 29634
E-mail: cwu@bellsouth.net

Roger M. Wakimoto
Department of Atmospheric Sciences
UCLA
405 Hilgard Ave.
Los Angeles, CA 90095-1565
E-mail: roger@atmos.ucla.edu

Tammy M. Weckwerth
Atmospheric Technology Division
NCAR
P.O. Box 3000
Boulder, CO 80307-3000
E-mail: tammy@ucar.edu

James W. Wilson
NCAR
P.O. Box 3000
Boulder, CO 80307-3000
E-mail: jwilson@ucar.edu

Isztar Zawadzki
Department of Atmospheric and Oceanic Sciences
McGill University
805 Sherbrooke St. W.
Montreal, QC H3A 2K6
Canada
E-mail: isztar@radar.mcgill.ca

Chapter 1

David Atlas: The Man and His Works

Robert J. Serafin and Richard E. Carbone

National Center for Atmospheric Research, Boulder, Colorado

Serafin   Carbone

1. Introduction

During his 57-yr career, David Atlas has been among the more influential people in the field of meteorology and a leading figure in the subdiscipline of radar meteorology. Researcher, inventor, laboratory leader, and educator, the contributions made by Atlas have been both broad and deep. Recognition from the American Meteorological Society (AMS) includes the Meisinger Award and Rossby Medal for research, the Cleveland Abbe Award for service, and Honorary Membership for the totality of his contributions. He also received the Symons Memorial Medal from the Royal Meteorological Society, was elected to the National Academy of Engineering and to the Presidency of the American Meteorological Society, and he has served science and society in many other capacities.

Atlas's achievements are the consequence of his many qualities, which include his persistence, intellect, creativity, enthusiasm, and love for science. A hard-driving man, his influence and impact on those who surrounded him are well known. Dave's determination to succeed continually forced a creative tension and occasionally tested one's capacity to cope. As a taskmaster, there is little doubt that he elevated the accomplishments of others to levels they might not otherwise have achieved.

There are other factors central to shaping his career, perhaps best described as "serendipitous," a term often used by Dave. In his memoir (Atlas 2001) Dave comments "... I also began to realize that one had to be opportunistic and flexible to exploit events when they occurred." This comment was in reference to an event in September 1945, after Atlas had photographically recorded radar echoes from a hurricane that, unexpectedly, had passed directly overhead. *Carpe diem* has been a guiding principle in his life, as with many scientists and exceptional achievers in all walks of life.

Still other events were of an accidental or circumstantial nature. He was born in 1924, which placed him in college during September 1941. He planned to be an electrical engineer when he entered the City College of New York (CCNY). However, the events of 7 December 1941 changed his plans and those of millions of people around the world. Atlas accelerated his education with a heavy course load and summer school, expecting to join the military in some capacity. Soon thereafter, he was accepted by the Army Air Corps in a premeteorology training program. Prior to this he apparently had little, if any, interest in meteorology. An event of which he had no knowledge was the invention of radar in the mid-1930s (Serafin 1996), in addition to the early detection of weather with military radars in 1941 (Rogers and Smith 1996). In the military Dave met Louis Battan. They would become roommates at New York University (NYU), lifelong friends, and professional colleagues, along with others who would lead the field in years to come.

Another factor that was responsible for setting the

stage for his career was the GI Bill, which provided financial assistance that led to his earning his doctorate of science in 1955. A mere two years later he received the Meisinger Award, his reputation having grown by leaps and bounds in what was now broadly recognized as an exciting new branch of meteorology. A man who set out to become an electrical engineer was drawn into meteorology by his assignment to service in war. As it turned out, the multidisciplinary education and training that he received had prepared him ideally for the emerging field of radar meteorology. Dave used his innate talents and a lot of hard work to accomplish the rest.

Among Atlas's early and very significant accomplishments was his invention of the isoecho contour mapping concept. This occurred in 1947, while he was a member of the All Weather Flying Division at Clinton Air Force Base in Wilmington, Ohio. (This was just one year after he had received his baccalaureate degree based upon his petition that the many courses he had taken at CCNY, NYU, MIT, and Harvard, and those given by the military were equivalent to the requirements for the degree, which was granted by NYU.) A patent was granted in 1953. The isoecho contour method was the first to quantize and therefore quantify weather radar reflectivity information on cathode ray tubes. This relatively simple concept received widespread use for decades on commercial aircraft, ground-based operational weather radars, and by the research community. It was not until the advent of color displays in the early 1970s that isoecho displays began to be replaced. Indeed, there were many airline pilots who objected strongly to the loss of the traditional CRT displays when color technology became available. During this period Dave also became associated with the Thunderstorm Project (Byers and Braham 1949). By the age of 24, Atlas was hooked on science.

2. In command at the Air Force Cambridge Research Laboratory (AFCRL)

In 1948 Dave married Lucille Rosen and joined the AFCRL as chief of the Weather Radar Branch, a lofty position for one so young. In his 18 years at AFCRL, Atlas forged strong collaborative relationships with colleagues at the lab, in Canada, in universities, and in private sector companies. Some of the luminaries with whom Atlas interacted included Stewart Marshall and Walter Hitschfeld with the Stormy Weather Group at McGill University, Herb Groginsky with Raytheon Company, and Roger Lhermitte from France. During this period there were many exciting new discoveries made in relation to the generation of and estimation of precipitation, clear air echoes and their causes, and the initial studies with Doppler radar. Dave was able to attract Roger Lhermitte to AFCRL and, in doing so, brought to the United States from France one of its more creative scientists and engineers.

On the occasion of the first Doppler velocity-azimuth display (VAD) measurements, when Lhermitte attached an audio speaker to the Doppler output, Dave said, "To our astonishment and exquisite pleasure on 2 Dec 1957, we heard and tape recorded the Doppler shift as it varied in pitch from near zero frequency when it was pointed crosswind, to high frequencies when it was pointing either up- or downwind" (Atlas 2001). This set the stage for several decades of Doppler radar research and, ultimately, operational applications that include similar VAD techniques (Browning and Wexler 1968). Somewhat to Dave's dismay, the United States was unable to install a national Doppler radar network until the 1990s.

Multiple Doppler radar networks have been used by researchers since 1970 (Lhermitte and Miller 1970) for measurements of three-dimensional kinematic fields in precipitation and later in clear air. A multi-Doppler study by Carbone (1982, 1983) observed cold frontal scale collapse and the ensuing generation of tornadoes in a conditionally neutral atmosphere. Application of the equations of motion to Doppler-derived kinematic fields eventually led to the retrieval of thermodynamic fields such as pressure perturbations, buoyancy (e.g., Gal-Chen 1978; Hane and Ray 1985), and microphysical retrievals (e.g., Ziegler 1985). Airborne Doppler research radars have been used effectively since the early 1980s, including the first airborne rapid scan system, developed at the National Center for Atmospheric Research (NCAR; Hildebrand et al. 1994). Roger Lhermitte and Dave Atlas laid the groundwork for these advances that have revolutionized mesoscale meteorology and have saved many lives and billions of dollars in property damage in the aviation industry alone.

3. The lure of academia

After 18 years at AFCRL, Atlas received an offer he could not refuse. He accepted a position as professor at the University of Chicago and set out to establish a world-class research activity in radar meteorology. He brought Ramesh Srivastava to Chicago from India and attracted a number of excellent graduate students. It was during this time that he established a relationship with Juergen Richter at the Naval Electronics Laboratory in San Diego, California. Dave became very excited about their frequency-modulated continuous-wave (FM–CW) radar and made a short visit to San Diego, California. It was during this visit that he obtained some exciting observations of very thin layers in the boundary layer, wave motions in these layers, and their eventual breakdown into turbulence.

While at Chicago, he forged a relationship with The Illinois Institute of Technology and created the joint Laboratory for Atmospheric Probing. He also received an National Science Foundation (NSF) grant of $650,000 to create a Doppler research radar facility for regional and national field programs. The radar would be the first of its kind. It would be transportable, would have high resolution, and would have a high-speed digital signal processor. It was the signal processor that was most challenging. Control Data Corporation won the bid to build the signal processor, which would use dynamic gain control to achieve high dynamic range in all range gates. It would provide for full Doppler spectral processing since the pulse pair processor was not yet well established as a research tool. The Illinois State Water Survey joined the team and the new radar was coined CHILL, for CHicagoILLinois. The radar was employed extensively for many years in a wide range of experiments around the country and gathered a fine reputation. The signal processor was fraught with a variety of nagging problems that required students to manually decipher hexadecimal dumps of processor output. Despite these shortcomings, CHILL produced many excellent datasets. It was a sophisticated design, well ahead of its time, which set the stage for the development of other transportable research radars at NCAR and elsewhere a few years later. Today the CHILL still operates as a research and educational facility at Colorado State University (Rutledge et al. 1993; also see chapter 11 in this monograph). The processor and antenna have been upgraded and polarization diversity has been a feature for many years. Dave's dream of a first-class radar that would reside at a university and serve to educate students in the discipline while also serving as a state-of-the-art research facility has been realized.

4. The NCAR challenge

In 1972, Atlas joined NCAR. At NCAR he would have the opportunity to transform and lead the Facilities Laboratory, which provided a broad range of observational and computational facilities for use by scientists. This was a difficult decision for Dave since he was quite attached to his students and university life in Chicago. But the NCAR opportunity presented a special challenge to Dave because it was judged at that time that NCAR's facility support had become obsolete and that modernization was in order. Atlas brought a vision to NCAR that included a state-of-the-art array of next-generation observing facilities. There would be Doppler radars, automated surface stations, next-generation sounding systems, lidars, acoustic sounders, and new airborne instruments, including an airborne Doppler radar. All of this came to pass in the next 20 years or so. Dave's major contributions were to get the ball rolling by articulating this vision and then by hiring several key staff, including Robert Serafin, for its implementation. Dave also felt that the Facilities Laboratory needed a new name. Vin Lally jokingly suggested that the name should be the Advanced Technology Laboratory for Atmospheric Science (ATLAS). Dave warmed to this suggestion, tongue in cheek, but after some deliberation the administration and Dave agreed that the new name should be the Atmospheric Technology Division (ATD).

Within two years there were two transportable C-Band Doppler radars that became a mainstay of university research for about two decades. The Portable Automated Mesonetwork (PAM) was the first fully automated mesonet reporting its data via radio telemetry and later via satellite. The new radars and PAM helped to transform the way in which field experiments were conducted because the real-time availability of data and displays greatly facilitated the knowledge of "present weather" and allocation of discretionary observing resources. These research systems led to much improved understanding of the processes leading to the initiation, growth and decay of convective storms, extratropical cyclones, tropical rainfall, and other phenomena. The study and detection of hazardous wind shear and microbursts using these radars led to the eventual deployment of Terminal Doppler Weather Radars for aviation safety at major airports in the United States and in many locations internationally. Exciting research was conducted with NCAR's facilities by many investigators in the 1970s, 1980s, and 1990s when ATD was internationally recognized as a leader in the field of atmospheric observations. Countless students were introduced to Doppler radar and field projects more generally through the use of these radars in pursuit of their advanced degrees.

Two years after joining NCAR, Dave was asked to assume leadership of the National hail Research Experiment (NHRE). Its director, Dr. Bill Swinbank, had suddenly and unexpectedly passed away and a successor was needed. Dave accepted this new challenge. NHRE was a weather modification program aimed at demonstrating that it would be possible to suppress hail in the high plains of northeastern Colorado. NHRE set out to test a Soviet hypothesis that accumulation zones of supercooled water drops could be seeded early enough to deplete liquid water supply in hail-producing storms and therefore reduce the incidence of large and damaging hail. NHRE had high national and international visibility; many U.S. universities were involved as were international scientists from Canada, Europe, and Africa. An impressive array of observational facilities was brought to bear.

Before long, it appeared to Atlas that no positive effect on the suppression of hail would be detectable, if only because of the substantial natural variability of hailstorms. Moreover, investigations of the microphysics of the northeast Colorado storms indicated that precipitation occurred primarily through the ice phase and that few, if any, large supercooled drops existed in the cold-base clouds there. Thus, the Russian hypothesis did not apply to these storms. In addition, Browning and Foote (1976) had suggested that seeding could even serve to increase hail. Dave felt strongly that this collective evidence was sufficient to stand down from cloud seeding for a period and to focus solely on research that would re-examine the basic hypothesis for cloud seeding. To some, NHRE was considered a failure because the ability to suppress hail was not demonstrated but, from a research perspective, NHRE was a success because substantial new understanding was gained about precipitation processes in high plains environments, which benefited the understanding of deep convection more generally. NHRE also established a paradigm for modern observational infrastructure, which served as a model for the conduct of many subsequent field experiments in the years to come.

During his tenure at NCAR, Dave was elected to the AMS presidency. His term was marked by a focus on atmospheric science and public policy. This included a first of its kind special symposium for a subcommittee of the U.S. Congress at the 1975 AMS Conference on Severe Storms and a general symposium on public policy at the annual meeting in 1976. As president elect in 1974, Dave and Lucille were included in the first scientific delegation to visit the Peoples Republic of China since before the cultural revolution. This historical visit was the forerunner of decades of scientific collaboration between the two countries and today remains as one of the areas in which the two countries find common ground.

5. NASA calls

Dave left NCAR in January 1977 to join the National Aeronautics and Space Administration/Goddard Space Flight Center (NASA/GSFC). The opportunity at GSFC was irresistible, since he was given carte blanche to build a new laboratory. He promptly established the Goddard Laboratory for Atmospheric Sciences (GLAS) and set out a new vision for atmospheric research programs there. As in the past, Dave placed scientific excellence at the top of his priorities and proceeded to attract some excellent people to GLAS, borrowing on his experience at the University of Chicago and at NCAR. These people distinguished themselves as researchers in their own rights and also helped to establish GLAS as a first-rate center for atmospheric research. While still very much interested in radar meteorology, Dave's interests quickly broadened to encompass the full spectrum of active and passive remote sensing of the atmosphere, oceans, and earth's surface. Dave played a prominent role in defining the Tropical Rainfall Measuring Mission (TRMM), working closely with Joanne Simpson and colleagues from Japan, to implement the first meteorological radar in space. This turned out to be an enormous success. TRMM has provided unprecedented detail on the structure and distribution of rainfall and improved estimation of rainfall amount over tropical oceans. This information is essential to understanding the earth's energy budget and water cycle, which is crucial to understanding the climate system. While not mentioned earlier, Dave has always sought and established effective collaborations with colleagues throughout the world, TRMM being just one prominent example.

6. The written legacy

Among Dave's principal written legacies is *Radar in Meteorology* (Atlas 1990), produced and edited from proceedings at the Battan Memorial and 40th anniversary Radar Meteorology Conference. Lou Battan, Dave's close friend and colleague dating back to WWII, died in 1986. Lou had been an outstanding educator, researcher, and leader in the field and Dave decided to organize a special radar meteorology conference that would commemorate Lou and 40 years of AMS conferences on radar meteorology. The conference format was designed by Dave, working closely with the AMS Committee on Radar Meteorology, and tutorial papers were written and delivered by the foremost experts in the field. Owing to Dave's dogged determination and prodding of authors, *Radar in Meteorology* contains the most comprehensive collection of contributions that has ever been produced under one cover in radar meteorology. It remains a reasonably current and superb reference for students and researchers to this day.

It is informative to examine Atlas's extensive publication record from the viewpoint of peer interest in his work. Among more than 230 papers, a few were cited more frequently than others. The most highly cited works span a period of 40 years, from 1953 to 1993. These papers originated in similar proportion at each of Atlas's principal "venues"—AFCRL, University of Chicago, NCAR, and GSFL. The most frequently cited publication overall is *Advances in Radar Meteorology* (Atlas 1964), the first textbook type of publication that reviewed Doppler signal theory in depth. *Advances in Radar Meteorology* served as a treasure chest of empirical relationships among reflectivity factor, attenuation, water content, and rainfall rate and presented some novel interpretations of the radar equation. In addition, it treated the complications of Mie scattering and multiple wavelength responses to hail and other hydrometeors.

Rounding out the "top 10" research publications, as determined by total citations, are papers that fall into three broad categories:

- scattering and attenuation properties of hydrometeors,
- techniques for reduction of bias and uncertainty in rainfall estimation, and
- studies related to turbulence and mesoscale organization.

The most frequently cited piece of original research is entitled "Doppler characteristics of precipitation at vertical incidence" (Atlas et al. 1973). This work was conducted during the heart of Dave's career, in times of remarkable productivity and creativity at the University of Chicago. It is a principal legacy of the Laboratory for Atmospheric Probing FPS-18 radar, which was limited to observations at vertical incidence from a rooftop antenna at the Illinois Institute of Technology. While the original research presented in this paper was nearly all theoretical, many of the ideas had their foundation in diagnostic studies using the FPS-18 radar. The creativity of Atlas and the brilliance of Srivastava combined to improve understanding of Doppler spectral techniques and to quantify the evolution of precipitation from the ice phase aloft to quasi-equilibrium rainfall below. Perhaps the most important findings are related to improved understanding of the limitations of such techniques in the presence of vertical air motions, turbulence, crosswinds, and uncertainties in particle terminal fall speeds. In chronological order, other works related to the electromagnetic properties of hydrometeors, precipitation estimation, and hydrometeor discrimination, include

"Scattering and attenuation by non-spherical particles" (Atlas et al. 1953);

"Multi-wavelength radar reflectivity of hailstorms" (Atlas and Ludlam 1961);

"The physical basis for attenuation–rainfall relationships and measurements of rainfall parameters" (Atlas and Ulbrich 1974);

"Path- and area-integrated rainfall measurement by microwave attenuation" (Atlas and Ulbrich 1977);

"Estimation of convective rainfall by area integrals, theory and empirical basis" (Atlas et al. 1990).

The remaining highly cited works engage issues related to turbulent flow and mesoscale organization. Notably, each of these papers has significance beyond the immediate realm of radar or remote sensing techniques.

The earliest of these addressed precipitation-induced wind perturbations in the melting layer (Atlas et al. 1969). Until the publication of this paper, the dynamical role of the melting level was vastly under appreciated and sometimes unrecognized, since evaporation of precipitation was considered to be the more important diabatic cooling process. Since then both the stabilizing and destabilizing effects of diabatic cooling due to melting have been extensively explored in extratropical and tropical cyclones and in the stratiform regions of mesoscale convective systems.

Atlas spent a significant fraction of his career fascinated by the occurrence of "angels," that is, radar echoes of various types with origin in the optically clear atmosphere. These investigations included a wide range of conditions and phenomena and with probes that spanned the microwave and UHF spectrum. His study on the birth of clear air turbulence in the marine boundary layer of southern California (Atlas and Metcalf 1970) is among those papers most frequently cited. The excitement associated with FM/CW radar images of breaking Kelvin–Helmholtz waves is surely a high point of Dave's career. The applicability of gradient Richardson number theory to observations in the atmosphere was both straightforward and persuasive, thus charting the evolution of strongly sheared flow in a statically stable fluid to the eventual cascade of turbulent kinetic energy in the inertial subrange.

Dave switched oceans and gears between 1970 and 1986 when he investigated the structure of a convectively unstable marine boundary layer as viewed by lidar and aircraft observations (Atlas et al. 1986). Cold air outbreaks from North America in winter lead to extreme interfacial fluxes of latent and sensible heat over the ocean, resulting in the moistening and destabilization of air masses over coastal waters. This work graphically illustrated and quantified the growth of the convective marine boundary layer and the mesoscale organization associated with shallow convection in the presence of vertical shear and strong surface fluxes.

7. Old soldiers never die

Dave retired from NASA in 1984, 17 years ago. Although formally retired from government service, he remains professionally active to this day. After GLAS, Dave formed his own small company that would serve as a base for him to continue to do research. He has contributed substantially to the understanding of tropical rainfall processes through his many collaborative papers on TRMM-related topics. Dave became interested in microburst and wind shear detection for aviation safety and invented and patented a technique through which low-level wind shear could be detected with fan beam, air traffic control radars at airports.

Dave's accomplishments continued to be recognized after leaving GSFC. In 1986, he was elected to the National Academy of Engineering. In 1996 he received the Carl-Gustaf Rossby Research Medal and in 2001 became an Honorary Member of the AMS. These latter two awards are the highest forms of recognition awarded by the Society.

Until the 2001 meeting in Munich, Dave had contributed to all 29 AMS Conferences on Radar Meteorology. This is an amazing record in itself given the sheer span of time and other responsibilities.

While still threatening to retire, Dave's creative energy stands in the way. He still sees ways to measure rainfall more accurately and ways to build better radars. The young man who set out to become an electrical engineer has become among the most influential meteorologists of the past half-century. His enthusiasm for science and discovery remains as strong today as when he first conceived of the isoecho contour device almost 50 years ago. A man who set very high professional standards is also a man with great compassion for the lives of others, a man who would do almost anything to help a friend. He has touched the lives of hundreds, perhaps thousands of people worldwide. Advances in the field, which he influenced directly and indirectly, have contributed to a better life for all of society. Those of us who can make the claim to be a friend and colleague of Dave Atlas share a privilege and honor that we cherish greatly.

REFERENCES

Atlas, D., 1964: Advances in radar meteorology. *Advances in Geophysics,* Vol. 10. Academic Press, 317–348. (Reprinted in Russian by Hydrometeorological Publications, Leningrad, Russia.)

——, Ed., 1990: *Radar in Meteorology.* Amer. Meteor. Soc., 806 pp.

——, 2001: *Reflections: A Memoir.* Amer. Meteor. Soc., 142 pp.

——, and F. H. Ludlam, 1961: Multi-wavelength radar reflectivity of hailstorms. *Quart. J. Roy. Meteor. Soc.,* **87,** 523–534.

——, and J. I. Metcalf, 1970: The birth of "CAT" and microscale turbulence. *J. Atmos. Sci.,* **27,** 903–913.

——, and C. W. Ulbrich, 1974: The physical basis for attenuation–rainfall relationships and the measurement of rainfall parameters by combined attenuation and radar methods. *J. Rech. Atmos.,* **8,** 275–298.

——, and ——, 1977: Path- and area-integrated rainfall measurement by microwave attenuation in the 1–3 cm band. *J. Appl. Meteor.,* **16,** 1322–1331.

——, M. Kerker, and W. Hitschfeld, 1953: Scattering and attenuation by non-spherical atmospheric particles. *J. Atmos. Terr. Phys.,* **3,** 108–119.

——, R. Tatehira, R. C. Srivastava, W. Marker, and R. E. Carbone, 1969: Precipitation-induced mesoscale wind perturbations in the melting layer. *Quart. J. Roy. Meteor. Soc.,* **95,** 544–560.

——, R. C. Srivastava, and R. S. Sekhon, 1973: Doppler radar characteristics of precipitation at vertical incidence. *Rev. Geophys. Space Phys.,* **11,** 1–35.

——, S. H. Chou, B. Walter, and P. J. Sheu, 1986: The structure of the unstable marine boundary layer viewed by lidar and aircraft observations. *J. Atmos. Sci.,* **43,** 1301–1318.

——, D. Rosenfeld, and D. A. Short, 1990: The estimation of convective rainfall by area integrals. Part 1: Theoretical and empirical basis. *J. Geophys. Res.,* **95,** 2153–2160.

Browning, K. A., and R. Wexler, 1968: A determination of kinematic properties of a wind field using Doppler radar. *J. Appl. Meteor.,* **7,** 105–113.

——, and G. B. Foote, 1976: Airflow and hail growth in supercell storms and some implications for hail suppression. *Quart. J. Roy. Meteor. Soc.,* **102,** 499–533.

Byers, H. R., and R. R. Braham Jr., 1949: *The Thunderstorm.* U. S. Government Printing Office, Washington, D.C., 287 pp.

Carbone, R., 1982: A severe frontal rainband. Part I: Stormwide hydrodynamic structure. *J. Atmos. Sci.,* **39,** 258–279.

——, 1983: A severe frontal rainband. Part II: Tornado parent vortex circulation. *J. Atmos. Sci.,* **40,** 2639–2654.

Gal-Chen, T., 1978: A method for the initialization of the aneleastic equations: Implications for matching models with observations. *Mon. Wea. Rev.,* **106,** 587–606.

Hane, C. E., and P. S. Ray, 1985: Pressure and buoyancy fields dervided from Doppler radar data in a tornadic thunderstorm. *J. Atmos. Sci.,* **42,** 18–35.

Hildebrand, P. H., C. A. Walther, C. L. Frush, J. Testud, and F. Baudin, 1994: The ELDORA/ASTRAIA airborne Doppler weather radar—Goals, design, and first field tests. *Proc. IEEE,* **82,** 1873–1890.

Lhermitte, R. M., and L. J. Miller, 1970: Doppler radar methodology for the observation of convective storms. Preprints, *14th Radar Meteorology Conf.,* Tuscon, AZ, Amer. Meteor. Soc., 133–138.

Rogers, R. R., and P. L. Smith, 1996: A short history of radar meteorology. *Historical Essays on Meteorology, 1919–1995,* J. R. Fleming, Ed., Amer. Meteor. Soc., 57–98.

Rutledge, S. A., P. C. Kennedy, and D. A. Brunkow, 1993: Use of the CSU–CHILL radar in radar meteorology education at Colorado State University. *Bull. Amer. Meteor. Soc.,* **74,** 25–32.

Serafin, R. J., 1996: Evolution of atmospheric measurement systems. *Historical Essays on Meteorology, 1919–1995,* J. R. Fleming, Ed., Amer. Meteor. Soc., 43–56.

Ziegler, C. L., 1985: Retrieval of thermal and microphysical variables in observed convective storms. Part 1: Model development and preliminary testing. *J. Atmos. Sci.,* **42,** 1487–1509.

Chapter 2

Mesoscale Substructure of Extratropical Cyclones Observed by Radar

KEITH A. BROWNING

Joint Centre for Mesoscale Meteorology, University of Reading, Reading, United Kingdom

Browning

1. Introduction

This is the fourth in a sequence of review articles in American Meteorological Society (AMS) monographs in which we deal with the observed structure of extratropical cyclones. The first of these articles (Browning 1990b), following the 1988 Palmén Memorial Symposium in Helsinki, classified cyclones into comma-cloud systems, instant-occlusion systems, and frontal wave cyclones according to the location of an upper-level trough/vortex in relation to the main baroclinic zone (see also Zillman and Price 1972). It presented the system-relative airflow concepts surrounding warm and cold conveyor belts (Green et al. 1966; Harrold 1973; Carlson 1980) and described two contrasting types of frontal system according to whether the warm conveyor belt is characterized by rearward-sloping or forward-sloping ascent anabatic (ana-) and katabatic (kata-) cold fronts, respectively. A radar-oriented version of this material appeared as a chapter (Browning 1990a) in the Battan Memorial Volume, an AMS monograph edited by Dave Atlas. These articles also drew attention, following Reed and Danielsen (1959) and Danielsen (1966), to the importance in developing cyclones of air descended from the upper troposphere and lower stratosphere, which forms a third airstream referred to as the dry intrusion. Although these airstream concepts suffer from being somewhat subjective in their application, they provide a useful framework within which to in-

terpret the mesoscale substructure of cyclones, and in the 1990 reviews a start was made in setting mesoscale rainband structures within such a context. In the last several years progress has been made in giving Lagrangian-airstream descriptions an objective basis (Wernli 1997; Davies and Wernli 1997; Rossa et al. 2000).

The third AMS monograph review article (Browning 1999a), following the 1994 Bergen Symposium on the Life Cycles of Extratropical Cyclones, put these conceptual ingredients into the framework of the frontal-fracture cyclone model that Shapiro and Keyser (1990) had presented earlier at the Palmén symposium. This model was reconciled with the conveyor-belt paradigm, and prominence was given to an extra feature of the flow in a developing cyclone, referred to as a secondary warm conveyor belt. The notion of such a flow peeling off from the base of the primary warm conveyor belt was introduced by Young et al. (1987) and has been developed by Young (1989), Browning and Roberts (1994), Bader et al. (1995), and Lemaître et al. (1999). This flow, together with the cold conveyor belt, produces a characteristic cloud feature known as a cloud head, first described by Böttger et al. (1975). The relationship of the dry intrusion to the cloud head was discussed in the 1999 AMS monograph review article and a start was also made in addressing the mesoscale substructure of cold fronts and convection beneath the dry intrusion. Further details about the dry intrusion, the associated tropopause fold, and their involvement in the generation of the different kinds of frontal structure were presented in a special issue of *Meteorological Applications* that was devoted to extratropical cyclones (Browning 1997).

In this, the fourth of our AMS monograph reviews on extratropical cyclones, we focus more on their detailed mesoscale substructure. First, however, in section 2 we set the context by providing a radar-oriented overview of the conveyor-belt concepts that were presented in the earlier reviews. We then devote sections to slantwise and upright convection at cold fronts (section 3a), the role of evaporation in slantwise convection (section 3b), and shearing instability in frontal zones (section 3c). These are followed by sections on the more detailed substructure of the upright convection (line convection) along the surface cold front (section 3d) and bands of upright convection beneath the dry intrusion (section 3e). The review is highly selective: in order to present a coherent view of a subset of interrelated issues, it concentrates on a particular kind of cyclone—cold-season maritime cyclones in northwestern Europe. The emphasis throughout this article is on the use of radar observations to address these topics. This is appropriate since it is through radar that one is best able to observe many of the smaller-scale phenomena and processes that are the focus of this article. It is doubly appropriate here because this monograph has been prepared to honor Dave Atlas who has inspired so many, including this author, to make full use of radar in meteorology.

2. Overview of cyclone structure at the time of frontal fracture

a. Conceptual model

Intense cyclonic development occurs when an upper-level anomaly in potential vorticity (PV) associated with a local tropopause depression overruns a frontal zone in the lower troposphere. The accompanying circulation leads to a fracturing of the frontal zone (Shapiro and Keyser 1990) and, as the cyclone begins to mature, it leads to characteristic patterns of airflow, cloud, and precipitation as shown in Fig. 2.1. In fact there is considerable generality in the process of frontal fracture: Schultz et al. (1998) have shown that frontal fracture can be explained simply by the rotation of the isentropes around a vortex, even in the absence of interaction between upper and lower levels. The frontal-fracture model is applicable to many cyclones in northwestern Europe and northwestern America as well as to those off the east coast of north America for which the model was originally derived.

The four panels in Fig. 2.1 show different aspects of the cyclone structure, all for the same time: each panel covers the same area and it is necessary to visualize the superposition of these panels. The surface frontal analysis is shown in all panels to assist in relating the panels

to one another. As shown by the dashed isopleths of wet-bulb potential temperature (θ_w)[1] in Fig. 2.1a, the marked fronts are sharply defined. The two parts of the fractured surface cold front (SCF) are separated by a diffuse cold frontal zone. The part of the cold front that goes through the low pressure center (L) is continuous with the warm front and is sometimes referred to as a bent-back warm front. Here we shall call it simply a bent-back front and restrict the term cold front to the other part of the cold front.

Figure 2.1b shows the distribution of cloud as seen by satellite and the broad distribution of surface precipitation (stippled) as seen by radar. Between the two cold fronts there is a region of no, or reduced, cloudiness separating the cloud head from the main polar front cloud band. This is referred to as the dry slot.

Figure 2.1c shows the conveyor-belt analysis, the arrows showing motion relative to the cyclone system. The flow W1 is the primary warm conveyor belt, responsible for the polar front cloud band. It consists of a strong flow of warm moist air characterized by high θ_w. To a first approximation W1 travels parallel to the cold front and it eventually ascends above the warm front. In the region of the sharp surface cold front, there is a superimposed transverse circulation within the W1 flow that leads to some of the W1 air rising above and behind the surface position of the cold front as it travels along the front: this is why the W1 arrow is drawn straddling the cold front.

The low-level flow ahead of the warm front and bent-back front consists of air with low θ_w and is known as the cold conveyor belt (Fig. 2.1c). Part of this flow ascends within the cloud head and contributes to the generation of precipitation in this region, although according to Schultz (2001) this contribution is usually relatively small. The other contribution to the growth of precipitation in the cloud head is produced by the secondary warm conveyor belt, represented in Fig. 2.1c by the set of arrows labeled W2. This flow consists of high-θ_w air originating from near the edge of the primary warm conveyor belt W1. The W2 air travels at low levels as a "shallow moist zone" (SMZ) within the dry slot and then ascends over the cold conveyor belt within the upper part of the cloud head (Browning and Roberts 1994).

The bold arrow in Fig. 2.1d shows the midtropospheric flow that accounts for the dry slot. This is the dry-intrusion flow, which descends from upper levels upstream in the region of a tropopause fold. The dry intrusion (DI) is associated with the upper-level PV anomaly (maximum) that contributes to the process of cyclogenesis (Hoskins et al. 1985). The same PV anomaly is also responsible in part for inducing the W2 flow

---

[1] This article emphasizes airstreams (conveyor belts) and airstream boundaries, and so fronts tend to be described in terms of θ_w fronts, which is appropriate when there is moist ascent leading to rainbands along the θ_w fronts.

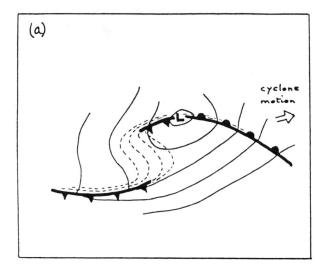

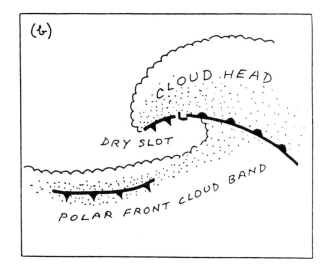

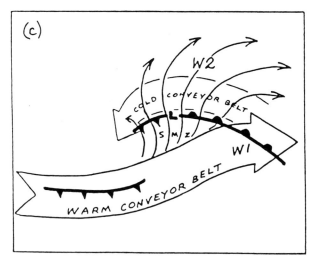

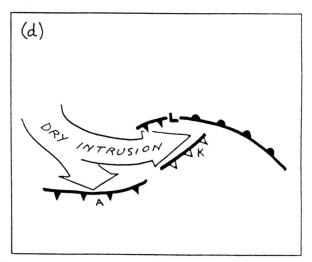

FIG. 2.1. Frontal-fracture stage in the evolution of a maritime extratropical cyclone: a conveyor-belt interpretation of the Shapiro–Keyser model (from Browning 1999b). (a) Surface analysis, showing isobars (continuous lines), isopleths of wet-bulb potential temperature (θ_w) (dashed lines) and surface fronts (analyzed only where sharp); (b) cloud (cold-side boundaries delineated) and precipitation (shown stippled), superimposed on surface fronts; (c) conveyor-belt analysis superimposed on surface fronts (see text for explanation); (d) dry intrusion, with an upper cold (θ_w) front at its leading edge (open cold front symbols), superimposed on surface fronts (solid symbols); the labels K and A denote katabatic cold front and anabatic cold front, respectively.

that undercuts the leading edge of the DI as an SMZ before rising within the cloud head. The DI air entering the dry slot has a lower value of θ_w than that in the W2 flow in the SMZ, but the resulting potential instability may not be realized unless the cyclone deepens rapidly or there are orographic effects leading to strong ascent in the dry slot.

The bifurcation of the dry-intrusion arrow, as drawn in Fig. 2.1d, is meant to indicate diffluence of the flow rather than a splitting into two distinct flows. Nevertheless, it properly draws attention to two distinct portions of the cold front. Air at the right-hand edge of the DI (near A in Fig. 2.1d) tends to undercut the warm-conveyor-belt flow, W1, behind the main cold front as part of a thermally direct transverse circulation (anabatic

cold front), as shown later in Fig. 2.4. On the other hand, at K, closer to the leading edge of the DI, air in the DI overruns the warm airflow, W2, at middle levels. Here its leading edge behaves as an upper cold front (denoted in Fig. 2.1d by open cold frontal symbols). Usually this is better defined as a moisture front than as a temperature front and, strictly, it is better to refer to it as an upper cold θ_w front. Whereas an observer on the ground at location A in Fig. 2.1d would tend to experience the passage of the sharp cold front as a sudden drop in temperature, an observer encountering the overrunning DI at K would find a gradual decrease in temperature occurring at the surface long after the passage of the upper cold θ_w front. This is the kind of cold frontal structure that is referred to as a kata-cold front

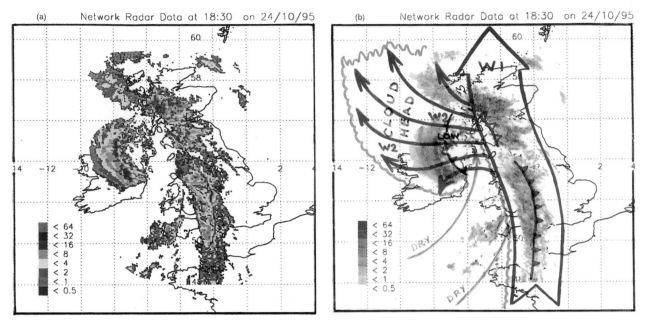

FIG. 2.2. Structure of a cyclone during the frontal-fracture stage, as observed at 1830 UTC 24 Oct 1995 (from Browning 1999b). (a) Radar-network picture. Colors represent approximate rainfall intensity in mm h^{-1} according to the key. (b) Same as (a) but rainfall intensity represented by a grayscale, with the following superimposed in color: satellite-detected edges of the polar front cloud band and the main cloud head (green), primary warm conveyor belt (broad red arrow labeled W1), secondary warm conveyor belt (set of orange arrows labeled W2), and dry intrusion (two yellow arrows approaching from southwest). For simplicity the cold conveyor belt is omitted. Cold fronts (including the bent-back front) are shown blue with solid symbols for surface fronts and open symbols for upper cold θ_w front.

(Sansom 1951), or split cold front (Browning and Monk 1982). Midlevel or upper-level cold fronts are also a common feature of North American cyclones (e.g., Martin et al. 1990; Locatelli et al. 1995), although their detailed behavior differs from the split cold front model.

The main areas of precipitation in Fig. 2.1b are characterized as follows.

1) *Within the polar front cloud band.* A broad band of rain is associated with the W1 flow and polar front cloud band. This tends to be subdivided into mesoscale rainbands parallel to the SCF and either behind or ahead of it [referred to as wide cold frontal rainbands and warm sector rainbands by Matejka et al. (1980)]. The precipitation is mainly stratiform but there is some embedded cellular convection near the cloud top. There may be stronger and more organized cellular convection along the upper cold θ_w front. Along the sharp SCF itself there may be "line convection" (Browning and Harrold 1970), giving a narrow cold frontal rainband (Matejka et al. 1980).

2) *Within the cloud head and warm frontal region.* An often extensive region of mainly stratiform precipitation occurs on the cold side of the warm front and bent-back front where the cold conveyor belt is overrun by the W1 and W2 flows, respectively. Embedded convective precipitation occurs, especially close to the surface front where the W2 flow is itself overrun by dry-intrusion air. There is often a narrow cold frontal rainband associated with line convection at the bent-back front.

Convective showers and sometimes squall lines can also occur within the dry-slot region itself and these are discussed later, in section 3d(3).

b. Example of application of conveyor-belt model

Airborne Doppler radar observations of a very rapidly deepening cyclone undergoing frontal fracture over the Atlantic have been described by Wakimoto et al. (1992). A comprehensive mesoscale analysis of the same case was presented by Neiman et al. (1993). We now present another example of the frontal-fracture stage, as seen by ground-based radar during the development of a less extreme cyclone that nevertheless brought heavy rain and strong winds to the British Isles (Browning et al. 1997b). Figure 2.2 shows a radar-network picture of the rainfall distribution (a) by itself and (b) with an analysis superimposed, based on satellite and radar imagery and on operational mesoscale model products interpreted in the light of the conceptual model in Fig. 2.1. (Cyclones tend to travel from west to east (left to right in Fig. 2.1) roughly parallel to the broad zones of north–south temperature gradient. However, the cyclone shown in Fig. 2.2 happened to be associated with the west–east temperature gradient ahead of a deep trough and so Fig. 2.1 should be viewed with the page rotated through 90° to achieve a proper comparison with Fig. 2.2).

Figure 2.2b shows the primary and secondary warm-conveyor-belt flows, W1 (red) and W2 (orange), together with the cold-side boundaries of the polar front

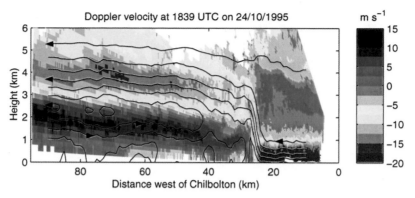

FIG. 2.3. Flow in a cross section in southern England at right angles to the anabatic cold front in Fig. 2.2, obtained by Doppler radar at 1839 UTC 24 Oct 1995. West is to the left and the front is traveling from left to right. Colors show the transverse wind component relative to the front according to the scale (m s⁻¹). Solid curves are inferred streamlines (revised version provided by Danny Chapman, University of Reading).

cloud band and cloud head (green) as determined from satellite imagery. The 150-km-wide band of cold frontal rain due to the W1 flow was associated with part of a north–south belt of moist air with θ_w in excess of 14°C. The equally wide band of cloud-head precipitation over and to the north of Ireland is associated with the W2 flow (orange arrows) and with the cold conveyor belt (not shown) that it was overrunning. The apparent wedge-shaped gap in the cloud-head precipitation just north of Ireland between 7° and 9°W is an artifact owing to the lack of good radar coverage in that region; there was no corresponding gap in the satellite imagery. The large canopy of cloud extending between 9° and 13°W in the region of the label "cloud head" was not associated with surface precipitation because of evaporation within undercutting dry air. According to the mesoscale model (not shown), the W2 flow extended westward toward the cloud head within a shallow moist zone (cf. SMZ in Fig. 2.1c) over the northern Irish Sea. This was seen in the model as a region with $\theta_w > 14°C$ and RH > 95%, which was underlying dry-intrusion air with lower θ_w.

Also drawn in Fig. 2.2b are the cold fronts (shown in blue). The positions of the surface fronts are consistent with the leading edges of strong θ_w gradients in the mesoscale model. The precise positions of the surface fronts over England and Ireland are drawn in Fig. 2.2b to be consistent with the lines of heaviest rain corresponding to the narrow cold frontal rainbands seen more clearly in Fig. 2.2a.

Finally, in Fig. 2.2b the two yellow arrows from the southwest represent dry-intrusion air consistent with the mesoscale model's relative-humidity pattern [see Figs. 10c, and d in Browning et al. (1997b)]. The 650-hPa flow associated with the northern arrow, which overruns the W2 flow, terminates at the upper cold θ_w front (open symbols): there was no sharp surface cold front at this position. The southern (yellow) arrow descends from 650 to 850 hPa behind the sharp surface cold front and

was seen in the model [Fig. 11d of Browning et al. (1997b)] as a well-defined dry finger intruding at lower levels (850 hPa): this feature corresponds to what is sometimes referred to as a rear-to-front, or rear-inflow, jet in the context of mesoscale convective systems (e.g., Smull and Houze 1985; Johnson 2001).

Between the two branches of the dry intrusion, Fig. 2.2 shows a small area of stratiform rain and line convection forming over the southwestern tip of England in association with a newly developing mini–cloud head. The corresponding analysis is not drawn in Fig. 2.2b because of the danger of overcomplicating the diagram. However, we mention it in passing because small features like this are often encountered due to mesoscale upper-level PV maxima overrunning low-level cold frontal zones. Another example of this behavior is given by Browning and Golding (1995) who show that the pockets of dry air associated with individual mesoscale PV maxima can be detected within satellite water-vapor imagery. One of the mesoscale PV maxima in their study led to localized intense line convection with a tornado. Such mesoscale systems are a common occurrence but are difficult to forecast because operational NWP models tend not to resolve them adequately.

Figure 2.3 shows a Doppler radar cross section through and at right angles to the cold front, close to the south coast of England. The flow is rather two dimensional in this region and it has been possible to derive realistic streamlines assuming approximately 2D continuity. Although the main W1 flow is south–north, into the plane of this cross section, Fig. 2.3 shows that there is nevertheless also an intense west–east circulation transverse to the front. Strong rearward flow relative to the system (shown red in Fig. 2.3) corresponds in part to warm boundary layer air within the W1 flow ascending abruptly as line convection at the surface cold front before ascending slantwise behind it. Air in the rear-inflow jet (shown blue) can be seen entering from the rear beneath the W1 flow before some of it, too, ascends (see the streamline changing direction at $x = 41$ km) and then returns as a rearward flow at the base of the (yellow and red) rearward-directed slantwise ascent. A rather similar transverse circulation occurs within cloud heads.

A notable feature of the line convection occurring at cold-season surface cold fronts is the combination of the vigor and shallowness of the convection. Although line convection in the United Kingdom is occasionally as deep as 5 km [e.g., Fig. 8b of Browning et al. (2001b)], it is more often confined to the lowest 3 km of the troposphere. Yet despite the shallowness of the convection, low-level convergence is typically of order 10^{-2} s^{-1} and leads to peak updrafts of up to about 10 m s^{-1}. As a result, intense radar reflectivities (in excess of 50 dBZ, for the case in Fig. 2.3) are not uncommon. The line convection in Fig. 2.3 is to some extent forced convection since it occurs in a boundary layer characterized by strong cyclonic shear but relatively weak convective available potential energy (CAPE) the potential instability is restricted to the lowest 3 km, with θ_w decreasing by about 1°C over this layer. Thermodynamic retrievals from dual-Doppler observations, together with radiosondes released just ahead of a similar occurrence of line convection in northwest France, demonstrated even more convincingly the small magnitude of the CAPE in these circumstances (Roux et al. 1993). In warm seasons over land, and sometimes in cold seasons over the sea (Braun et al. 1997), the existence of stronger convective instability gives rise to more three-dimensional cellular convection at cold fronts and it typically extends to much greater heights.

We shall now examine the mesoscale processes that determine the detailed structure of some of the features outlined above.

3. Mesoscale substructure

a. Slantwise and upright convection in cold frontal zones

1) THEORETICAL BACKGROUND

It is well known that classic (anabatic) cold fronts tend to be characterized by mesoscale circulations in which upright line convection feeds a layer of rearward slantwise ascent as in Fig. 2.3. The transverse circulation depicted in Fig. 2.3 is relatively simple in that it consists of a *single* rear-inflow jet of cold air descending slantwise beneath a *single* layer of slantwise ascent, which is being fed by boundary layer air lifted within a *single* line of intense upright convection. Often, however, the transverse circulations at cold fronts are much more complicated than this, and *multiple* layers of slantwise ascent and descent coexist with *multiple* regions of upright convection. It is often suspected that the frontogenetically forced circulations at anabatic cold fronts are enhanced by mesoscale processes such as conditional symmetric instability (CSI) and geostrophic mo-

mentum adjustment (explained below), but it is notoriously difficult to discriminate between these circulations and the larger-scale transverse circulation within which they are embedded. However, the common occurrence of multiple circulations with small vertical scale helps to distinguish the mesoscale processes from the large-scale circulation and this is what motivates our detailed examination of observed cases of multiple layering. We present the observations in section 3a(2), but first we summarize the theoretical background to CSI and geostrophic momentum adjustment.

In a baroclinic atmosphere, moist convection may take a variety of forms depending on the relative gravitational and inertial stability to vertical and horizontal motions. Pure gravitational instability may be released by upright motions and pure inertial instability by horizontal motions. Conditional symmetric instability arises from the combination of gravitational and inertial accelerations in a baroclinic atmosphere and this may be released by slantwise motions (Bennetts and Hoskins 1979). Such motions are referred to as slantwise convection.

As described in the review of CSI by Schultz and Schumacher (1999), slantwise and upright convection may coexist in regions that are unstable to both. Emanuel (1994) noted that this is especially likely in the warm sectors of baroclinic cyclones as this air enters the frontal circulation. The slantwise convection extends transverse to the front over a horizontal scale of about 100 km, which is more than an order of magnitude greater than that for upright convection. It yields vertical velocities of tens of centimeters per second, more than an order of magnitude smaller than the updrafts within upright convection. If an initially gravitationally and inertially stable baroclinic atmosphere is destabilized, then symmetric instability will appear prior to pure gravitational instability (Emanuel 1994). Even so, the larger growth rate of gravitational instability might be expected to imply that if both instabilities are present, then the release of pure gravitational instability (leading to upright convection) would tend to dominate in time. In practice, however, these instabilities appear to interact such that the release of one type of instability preconditions the atmosphere for the other type of instability.

Two mechanisms for the interaction of symmetric and pure gravitational instability in frontal rainband development are discussed by Xu (1986). He refers to the process of rainband development suggested by Bennetts and Hoskins (1979) as "downscale" development. In this scenario, the release of CSI as slantwise convection causes the midtroposphere to become unstable to smaller-scale upright gravitational instability. In the alternative, "upscale" development, small-scale upright convection of boundary layer air occurs first; this stabilizes the atmosphere to upright gravitational instability and meso-β-scale slantwise convection then occurs. Even in an atmosphere stable to CSI it is still possible for the upright convection to trigger slantwise convec-

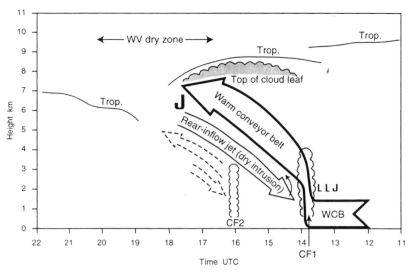

FIG. 2.4. Synthesis of features in a time–height section derived from a combination of UHF and VHF wind-profiler radars, showing double structure during passage of an ana-cold front on 22 Feb 1995. Solid arrows denote primary slantwise circulation and dashed arrows denote secondary slantwise circulation: CF, cold front; WCB, warm conveyor belt; LLJ, low-level jet; J, upper-level jet; WV dry zone, dry zone as seen in satellite water-vapor channel; Trop., tropopause (from Browning et al. 1998).

tion, but through an inertially stable process known as geostrophic momentum adjustment (or ΔM adjustment). In this process, upright convection lifts relatively low-momentum air upward, leading to subgeostrophic momentum anomalies in the midtroposphere, which are ultimately removed by slantwise convection (Holt and Thorpe 1991; Fischer and Lalaurette 1995).

It is widely recognized that the representation of moist convection within numerical models is one of the great challenges standing in the way of the improvement of weather and climate prediction. The scales of convective motions sit uneasily between being partially resolved and being subgrid-scale processes whose effects need to be parameterized: there is therefore a danger of double accounting. An important ingredient for progress is the development of a better understanding of the nature of both upright and slantwise circulations and the ways in which they interact. Radar observations, presented next, are beginning to shed light on these complex issues.

2) RADAR OBSERVATIONS

A common structure observed for ana-cold fronts in the United Kingdom is shown in Fig. 2.4. This is a synthesis of features identified in the plots of velocity and received power obtained using collocated UHF and VHF wind-profiling radars [see Browning et al. (1998) for the full set of plots]. The front was travelling at 15 m s^{-1} so that the 11-h time–height record in Fig. 2.4 corresponds to a distance of about 600 km resolved perpendicular to the front. Satellite imagery showed a discrete cloud band associated with a warm conveyor

belt oriented into the plane of the diagram. The radar data summarized in Fig. 2.4 show a pattern of flow transverse to the front (i.e., normal to the axis of the warm conveyor belt) that is broken into two pairs of circulations: each circulation resembles that in Fig. 2.3 although one is more intense than the other. The ascending limb of the primary circulation (solid arrow) is fed by warm-conveyor-belt air originating in a low-level jet (LLJ), which makes its initial ascent within line convection at the main surface cold front (CF1). The descending branch of the primary circulation corresponds to dry-intrusion air associated with a tropopause fold beneath the main upper-level jet. The weaker secondary circulation (dashed arrows) is situated on the cold side of the frontal zone, 100 km behind and 3 km beneath the primary circulation.

Several other recent studies (Browning et al. 2001a,b; Browning and Wang 2002) have revealed a mesoscale substructure similar to Fig. 2.4. Detailed plots from one of these studies are shown in Fig. 2.5. The time–height sections in Fig. 2.5 show the (a) raw front-normal and (b) front-parallel components of the wind obtained from a UHF wind-profiling radar. Again there are two regions of line convection (L1 and L2). Each is associated with strong cyclonic shear and an upward penetration of low-momentum air just on the cold-air side of the line-convection features (Fig. 2.5b). The primary line convection (L1) feeds the slantwise ascent S1 while the weak and intermittent line convection L2 is associated with either or both of the layers of slantwise ascent, S2a and S2b. The vertical spacing between adjacent slantwise ascending flows is typically 2 km and their horizontal spacing about 100 km. Vestiges of other slantwise cir-

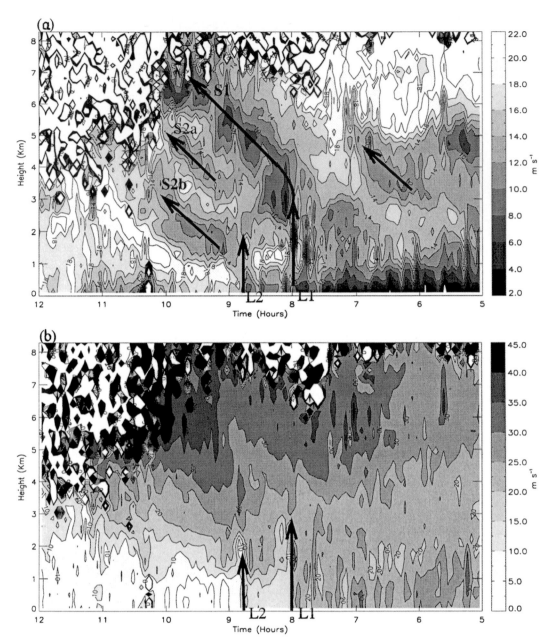

FIG. 2.5. Multiple convective circulations associated with the passage of an ana-cold front on 5 Nov 1999 as shown by time–height plots at high temporal resolution (3½ min) of (a) front-perpendicular and (b) front-parallel wind components (m s⁻¹) as measured by a UHF wind-profiler radar located in southwest England. (a) The isotachs are relative to the ground: the superimposed inclined arrows denote system-relative front-to-rear slantwise flows. The vertical arrows, L1 and L2, represent line-convection features: L1 is related to slantwise ascent S1, and L2 is related to S2a and/or S2b (from Browning and Wang 2002).

culations also exist ahead of the primary one. Similar multiple circulations have been documented within cloud heads by Browning et al. (1995) and Roberts and Forbes (2002).

The plan distribution of near-surface rainfall intensity at the time when the line convection L1 in Fig. 2.5 was passing over the wind profiler is shown in Fig. 2.6a. The narrow cold frontal rainband associated with L1 shows up clearly as a thin red line embedded within the broad area of cold frontal precipitation. Three hours later (Fig. 2.6b), L1 can be seen close to the leading edge of the broad cold frontal rain area while, behind it, the broader band of heavy rain in Fig. 2.6a has acquired a somewhat narrower multibanded structure (Fig. 2.6b), consistent with the existence of the intermittent secondary line convection L2.

An issue discussed by Browning et al. (2001a) is whether the slantwise ascent is caused by or is respon-

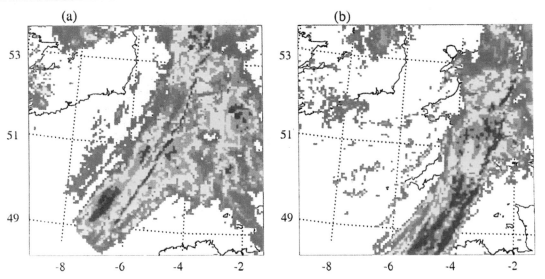

FIG. 2.6. Weather radar network displays showing multiple rainbands associated with an ana-cold front at (a) 0800 and (b) 1100 UTC 5 Nov 1999. (Blue > 0.5, green > 1, pale blue > 2, yellow > 4, orange > 8, red > 16, maroon > 32 mm h^{-1}).

sible for the line convection. This question remains unresolved although it is possible that both processes may be occurring; for example, the primary slantwise circulation might be an upscale development due to geostrophic momentum adjustment (ΔM adjustment) of relatively low-momentum air lifted by the primary line convection, while the weaker secondary circulation might be due to CSI, which in turn triggers the secondary line convection, that is, downscale development.

Another form of downscale development is the triggering of upper-level upright convection due to the buckling of θ_w surfaces by the slantwise circulation as first proposed by Bennetts and Hoskins (1979). Upper-level convective cells occurred at the top of the S1 slantwise ascent on the occasion shown in Figs. 2.5 and 2.6, and these were observed by a microwave Doppler radar as shown in Fig. 2.7. Figure 2.7d shows the reflectivity pattern associated with the upper-level convective cells; the other panels in Fig. 2.7 establish the context for these cells. The top two panels are Doppler velocity and reflectivity plots for an RHI scan perpendicular to the front when the main line convection L1 was 36 km upstream (west) of the radar. The bottom two panels are corresponding plots for a later scan *parallel* to the front when L1 was 70 km *downstream* of the radar. Fig. 2.7a shows the (red) boundary layer air being overtaken by the primary line convection L1 at 36 km, which lifts this air near vertically to around 3 km where it feeds the (red and yellow) layer of primary slantwise ascent S1. Figure 2.7b shows the shallow reflectivity maximum (red) due to the narrow cold frontal rainband associated with L1. The Doppler velocities in the front-parallel RHI in Fig. 2.7c show the wind increasing strongly with height from the south (the thermal wind) up to the level where the uniform precipitation echo gives way to cel-

lular echoes. Returning to the reflectivity pattern in Fig. 2.7d we see that the precipitation generator cells between 7 and 8½ km altitude are due to upright convection, which is at the top of the layer of primary slantwise ascent S1 shown in Fig. 2.7a.

Convective cells are often observed above slantwise ascent. When the slantwise ascent is vigorous the cloud-top convection can become organized into bands oriented roughly perpendicular to the frontal zone. This sometimes leads to cloud-top striations detectable by satellite. Feren (1995) refers to them as "striated delta" cloud systems and he found that they were usually associated with explosively deepening cyclones and gale-force winds. The wavelength of the striations may be as large as 30–55 km (Dixon et al. 2000; Browning and Wang 2002) but we have no good dynamical explanation yet to account for the large wavelength.

b. Evaporation in frontal systems

1) DYNAMICAL ROLE OF EVAPORATION IN SLANTWISE CONVECTION

The importance of latent heat release from condensation in enhancing the development of extratropical cyclones has long been recognized (e.g., Manabe 1956). However, the role of evaporation in these situations has received detailed attention only over the last decade or so and it is now attracting increased interest. Huang and Emanuel (1991), in studying the effect of the evaporation of rain in a semigeostrophic model of frontogenesis, demonstrated an important effect. Parker and Thorpe (1995) studied the effect of evaporation (sublimation) of snow on the development of a baroclinic wave. They found only a weak effect in accelerating

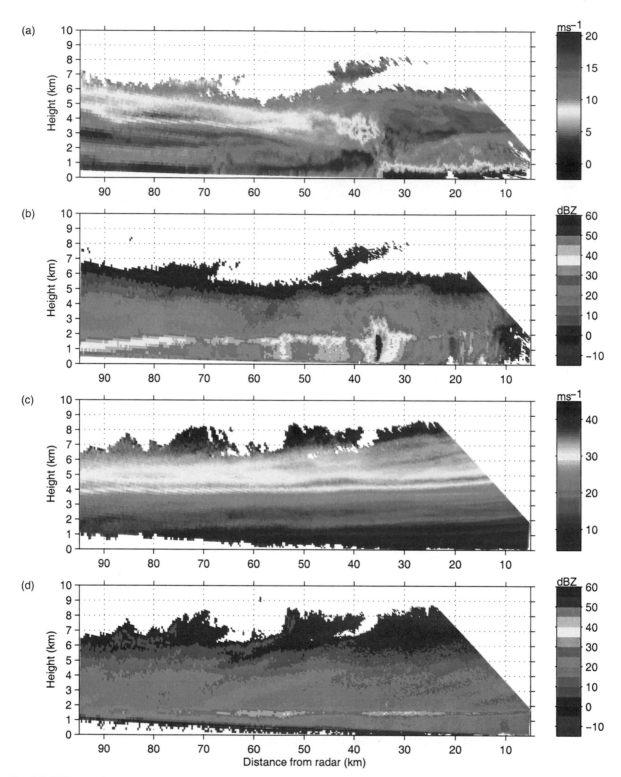

Fig. 2.7. RHI scans from a microwave Doppler radar showing convective generator cells above 7 km at the top of a strongly sheared cold anafrontal zone at (a),(b) 1214 and (c),(d) 1401 UTC on 5 Nov 1999. (a),(c) Doppler velocity in m s^{-1}; the strongest flows from left to right are shown in (a) blue and (c) red; (b),(d) reflectivity in dBZ. (a),(b) The scan was obtained roughly at right angles to the surface cold front, approaching radar from the left (west). (c),(d) The scan was obtained roughly parallel to the SCF (south is to the left) (after Browning and Wang 2002).

frontogenesis but suggested it might have a significant impact on the smaller frontal scale by enhancing the descending limb of the slantwise circulation. Yang and Houze (1995) found that evaporative cooling was important in maintaining rear-to-front flow all the way to the position of a surface gust front. Sensitivity studies by Forbes et al. (2000) using a mesoscale numerical weather prediction model indicated a strong sensitivity of the development of cold frontal troughs and associated cloud heads to the evaporation of ice, which affects the mesoscale patterns of cloud and precipitation. Rutledge (1989), who investigated the role of diabatic processes in a case study of an intense cold frontal rainband, found that it is the combination of condensation heating in the ascent ahead of the frontal zone and evaporative cooling in the descent behind that is important in maintaining the density contrast across the front.

Clough and Franks (1991) have used simple kinetic-thermodynamic models to study the evaporation of ice beneath the ascending limb of a slantwise frontal circulation. They pointed out that the evaporative cooling from falling ice particles tends to be dynamically much more important than the cooling from evaporating raindrops because the cooling is localized in the vertical owing to the low fall speeds of ice particles compared with raindrops. They estimated that, in stable conditions, evaporation of ice could maintain a moist slantwise descending flow of up to 20 to 30 cm s^{-1}. Using a 2D microphysical retrieval model, Marécal and Lemaître (1995) also demonstrated the crucial role of snow evaporation in maintaining the slantwise descending part of a CSI-induced circulation in a nearly saturated state. Model simulations by Clough et al. (2000) of a Fronts and Atlantic Storm-Track Experiment (FASTEX) cyclone observed by radar and dropsondes showed that the evaporative cooling appreciably enhanced the amount of negative potential vorticity; they concluded that the evaporative cooling actually generated the CSI rather than merely helping to sustain CSI-induced circulations.

The evaporatively cooled air descending beneath the rearward-sloping ascent in an ana-cold front can reach the ground as a cold pool. As long ago as 1953, Oliver and Holzworth pointed out that evaporatively cooled air behind the surface cold front is associated with a mesoscale ridge of high pressure and they suggested that the head of cold air could lead to an acceleration of the surface front toward the warm air. Cooling due to melting also has an important effect. The cold air behaves in some respects like a density current (Carbone 1982; Hobbs and Persson 1982; Parsons 1992). However, as explained by Smith and Reeder (1988), there remains some question about the closeness of the analogy to a density current.

2) RADAR OBSERVATIONS

Evidence of evaporation of precipitation within the descending limb of an ana-cold frontal circulation can be seen in part of the RHI section in Fig. 2.7b. It is seen most clearly in the decrease in reflectivity of several dBZ (pale green to pale blue) between 3½ and 2½ km at 90-km range. The possibility cannot be ruled out that this decrease is due to horizontal nonuniformity in the precipitation, with streamers entering the plane of the section owing to the wind direction changing with height; evaporation is, however, a likely explanation since the decrease is collocated with the primary rear-inflow jet (blue) in Fig. 2.7a. Occasionally the effect of evaporation is much more obvious, as in the case depicted in the time–height plots in Fig. 2.8, obtained from a UHF radar wind profiler.

The pattern of received power in Fig. 2.8a depicts a region of pronounced evaporation below a sloping reflectivity maximum highlighted by the dashed line. Between 0440 and 0530 UTC there is total evaporation over a depth of 1 km. The accompanying plot in Fig. 2.8b shows regions of inferred slantwise ascent and descent superimposed on the pattern of the front-normal wind component represented by the shading. The most pronounced area of forward-sloping descent is highlighted by the long dashed arrow extending down to the boundary layer after the passage of the main surface cold front C1. Comparison between Figs. 2.8a and 2.8b shows that this descending flow corresponds closely to the evaporation layer. Output from the mesoscale model (not shown) makes it clear that this was a layer of high PV connected to a tropopause fold. High-PV air descending in a tropopause fold plays a role in promoting cyclogenesis where it overruns a baroclinic zone at lower levels (Hoskins et al. 1985).

A feature of Fig. 2.8 is the shallowness of the evaporating layer and the associated descending flow. R. M. Forbes (2002, personal communication) has carried out a statistical analysis of a long time series of reflectivity data from a vertically pointing 94-GHz cloud radar with high (60 m) vertical resolution to determine the typical range of ice evaporation depth scales for all kinds of frontal cloud systems. Using a three-month dataset, he found that the depth from the level of maximum ice content to where it drops to a tenth of the maximum was usually as little as 500 m. A typical time–height record from the cloud radar is shown in Fig. 2.9: this example is of the passage of a warm frontal cloud system. Although the evaporation depth scale in Fig. 2.9 is only about 500 m, there is some evidence of downward extensions of reflectivity, so-called "stalactites", at the base of the echo layer between 1300 and 1400 UTC. A Doppler radar study of stalactites by Harris (1977) revealed updrafts and downdrafts of ±1–3 m s^{-1}, which he attributed to evaporatively induced convective cells (mammatus clouds).

According to R. M. Forbes (2002, personal communication), numerical weather prediction models significantly overestimate ice evaporation depth scales. This was true even for special mesoscale model runs

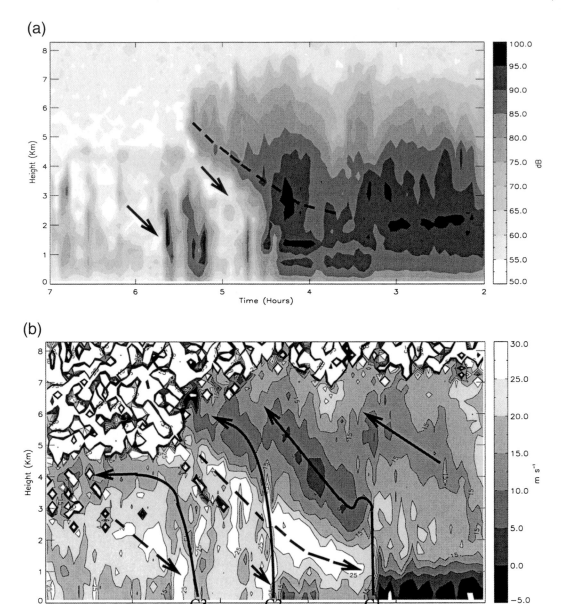

FIG. 2.8. An occasion of strong evaporation just below the interface between rearward slantwise ascent and forward slantwise descent, as shown by time–height plots at high temporal resolution (3½ min) of data from a UHF wind-profiler radar during the passage of an anabatic cold front on 30 Oct 2000. (a) Received power (dB); the dashed line is drawn through the level of maximum power below which evaporation was reducing the size of the ice-particle targets, and the arrows draw attention to regions of weak power associated with dry slantwise descent. (b) The front-perpendicular component of the wind (m s^{-1}, shaded according to the key); it also shows the regions of slantwise ascent (solid arrows) fed by lines of convection (C1, C2, and C3), and slantwise descent (inclined dashed arrows), as inferred from front-perpendicular wind components that are, respectively, less and more than the frontal velocity of 23 m s^{-1} (from Browning et al. 2001b).

made at high resolution (150 m in the vertical). There will be a tendency for very shallow layers of cooling to act, along with appropriate larger-scale deformation fields, to sharpen gradients of temperature and velocity within frontal zones. This is counteracted by the effect of shearing instability, another process that is inadequately represented in present-day NWP models.

c. Shearing instability in frontal zones

The possibility that frontal zones can be in a state of dynamical equilibrium between processes tending to sharpen them and mixing tending to diffuse them was discussed by Browning et al. (1970). They supported the idea using Doppler radar measurements. The idea

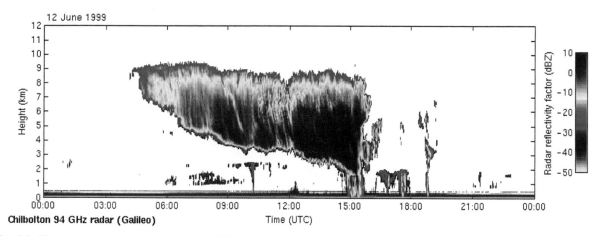

FIG. 2.9. Time–height record of received power (dB) as measured by a high-resolution vertically pointing 94-GHz cloud radar, showing evaporation beneath an ice crystal "cloud" during the passage of a warm front on 12 Jun 1999 (courtesy of Robin Hogan, University of Reading).

was further supported by dropsonde measurements from the FRONTS 92 project [Browning 1995; see also Fig. 17d in the AMS review article by Browning (1999a)], which revealed a coherent shallow layer with Richardson number less than ¼ just beneath the ascending limb of an ana-cold frontal circulation. The Richardson number Ri, defined by $N^2 S^{-2}$, where N is the Brunt–Väisälä frequency and S is the vertical wind shear, is a measure of the stability of a statically stable sheared flow. A necessary condition for shearing instability is that Ri < 0.25 somewhere in the flow (Miles and Howard 1964). The resulting instability, known as the Kelvin–Helmholtz (KH) instability, takes the form of billows, which amplify and break to give turbulent mixing.

Doppler radar can detect individual KH billows in different ways. Provided a radar has sufficient power and resolution, and a wavelength of 10 cm or more, it can be used in the clear air to detect temperature and humidity inhomogeneities within statically stable parts of the billows (Atlas et al. 1966). In the case of large billows, this gives rise to "cat's-eye" echo patterns where the billows distort the original stable layer (Hicks and Angell 1968; Browning 1971). For billows forming within precipitating frontal zones, the echo due to clear air returns at 10-cm wavelength is obscured by the return from the precipitation; however, it is possible to use Doppler velocity information from the precipitation tracers to observe instead the vertical redistribution of shear by the billows. The distributions of vertical shear obtained in Doppler studies by James and Browning (1981) and Chapman and Browning (1997), and derived from the study by Takahashi et al. (1993), show similar structures to the distribution of thermal stratification implied by the clear-air-echo studies. This parallel behavior between the shear and thermal stratification in KH billows is supported by numerical simulations by Scinocca (1995).

Doppler radar observations of wind shear in a variety of frontal precipitation events (Chapman and Browning 1998) have shown KH billows to be common in both warm and cold frontal zones. However, the largest billows were observed in warm fronts where their crest-to-trough amplitude can be as great as 1 km. Figure 2.10 shows an example of a west–east RHI scan, parallel to a warm front and at right angles to the axes of the billows for one such event. Figure 2.10a shows westerly wind components (red) overlying easterlies (blue); the white lines show the mass streamfunction derived assuming 2D flow. Figure 2.10b shows the pattern of vertical velocity implied by Fig. 2.10a, with areas of ascent red and descent blue. Figure 2.10c, depicting the pattern of vertical shear derived from Fig. 2.10a, shows the characteristic cat's-eye pattern associated with KH billows. Figure 2.10d depicts the pattern of vertical shear 20 min earlier.

The location of the billow event in Fig. 2.10 with respect to the frontal zone is shown in Fig. 2.11. The billows were situated between 50 and 100 km ahead of the surface warm front near an altitude of 1 km where a high-θ_w warm-conveyor-belt flow was ascending above a strong warm frontal zone. As shown in Fig. 2.11, the 10–m s^{-1} thermal wind shear across the 1-km depth of the billows was oriented at right angles to the mean velocity of the air over their depth, which, like the axes of the billows, was oriented approximately at right angles to the front. The cat's-eye pattern of overturning was detected by scanning at right angles to the axes of the billows. In cases such as this, in which the direction of the billow axes corresponds to the direction of mean flow, the billows at the upwind end are young billows while the billows at the downwind end are old broken billows. The along-axis length is therefore limited by the natural lifetime of the evolving billows (although presumably the billows may persist for longer if the larger-scale deformation field is tending to reduce the Richardson number).

Figure 2.12 depicts the structure of the warm frontal zone in a south–north section through the location of

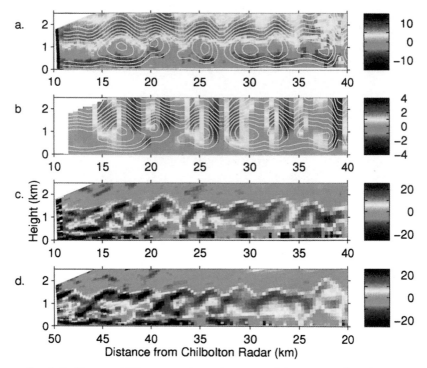

FIG. 2.10. West–east RHI cross sections (almost) parallel to a warm frontal zone exhibiting large-amplitude Kelvin–Helmholtz billows on 6 Sep 1995. (a) Doppler velocity, effectively giving westerly wind components (m s^{-1}, blue to the left and red toward the right) at 2220 UTC; superimposed are contours of a mass streamfunction (intervals of 1000 kg m^{-1} s^{-1}) projected onto the plane of the RHI. (b) As in (a) except shading shows vertical velocity (m s^{-1}) at 2220 UTC. (c),(d) Vertical west–east cross sections of shear (m s^{-1} km^{-1}) derived from Doppler velocity RHIs at (c) 2220 and (d) 2200 UTC (from Chapman and Browning 1997).

the radar as derived from a mesoscale model. As shown by the shaded area in Figs. 2.12a,b, the billows occurred where the warm-conveyor-belt air riding up the frontal zone (bold arrows) became saturated (bold contour). Because of the discontinuous drop in its static stability when an air mass becomes saturated, the Richardson number drops abruptly with the onset of saturation as shown in Fig. 2.12b. Prior to saturation the lowest values of Ri were close to 1. Following saturation Ri was of order 0.1. Subcritical values of Ri developed suddenly over a substantial depth, thereby triggering the development of KH instability over a deep layer. This sudden development and release of shearing instability is analogous to the production and release of convective instability when a potentially unstable air mass is lifted to saturation, and an airmass where this is possible is referred to as possessing potential shearing instability (Chapman and Browning 1999). The billows also appeared to induce small drizzle shower clouds in a convectively unstable region that existed just above the frontal layer containing the billows (see Fig. 2.12a for model evidence of potential instability). These showers then produced weak reflectivity cores oriented parallel to the billows, but traveling at the velocity of air above the shear layer, so moving out of phase with the billows.

Wakimoto and Bosart (2001) observed what was in all probability a similar occurrence of KH billows during an aircraft flight through a warm frontal zone in the FASTEX project. Figure 2.13 shows a plan section at midbillow level and Fig. 2.14 shows a vertical section across the axes of the billows, as derived from the airborne Doppler radar. The solid black contours in Fig. 2.13a show a series of elongated vertical-vorticity maxima corresponding to the ascent maxima along the axes of the train of billows. The collocated ridges of maximum wind speed and direction in Fig. 2.13b have an orientation similar to the axes of maximum ascent and are a result of higher-momentum air being brought down to a lower level by the billows. The wavelength of the billow train is about 10 km, twice that observed in Fig. 2.10. Linear theory predicts that KH billows should exhibit a wavelength between 4.4 and 7.5 times the vertical depth of the frontal shear (Miles and Howard 1964) and this is consistent with the observed billow depth of about 2 km in Fig. 2.14, compared with 1 km for the billows in Fig. 2.10.

What is not entirely clear is whether the magnitude of dissipation rate due to billow-induced turbulence is large enough to have a major effect on the evolution of the larger-scale frontal structure. The dissipation rate

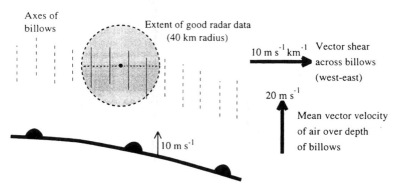

FIG. 2.11. Diagram depicting in plan view the location and context of the Kelvin–Helmholtz billows depicted in Fig. 2.10. It shows the extent of radar coverage (shaded circle), the orientation of the RHIs in Fig. 2.10 (west–east dotted line), the observed orientation of billow axes (solid lines), and the likely orientation of other billows (dashed lines) in relation to the surface warm front at 2200 UTC 6 Sep 1995. The sketch of the billow axes is schematic insofar as the actual wavelength was about 5 km rather than the 20 km shown in this diagram.

due to turbulence can in principle be estimated from the width of the Doppler spectrum. Although in practice it is difficult to do this reliably in frontal zones, Chapman and Browning (2001) show that if certain precautions are followed, it is possible to make rough estimates using a high-resolution radar. Large-amplitude billows of the kind described above occur only occasionally. More generally, the dissipation is likely to be due to small billows occurring extensively within shallower layers. Figure 2.15b shows an example of the radar-derived pattern of dissipation rate associated with an ana-cold front consisting of vertically stacked mesoscale slantwise circulations with multiple shear layers as shown in Fig. 2.15a. The origin of the multiple shear layers was discussed above in section 3a. Further splitting of the shear layers may occur as a result of the billow-induced mixing itself. The interaction between processes leading to mesoscale slantwise circulations and Kelvin–Helmholtz instability determines how mixing takes place in the stratified free atmosphere. If the

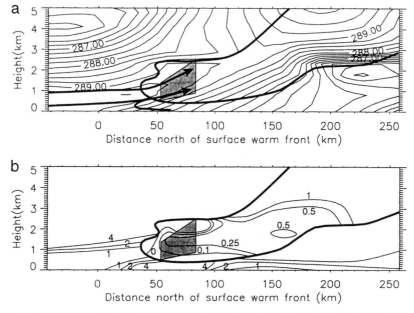

FIG. 2.12. South–north vertical cross section passing through the radar location, at 2200 UTC 6 Sep 1995, as derived from the operational Met Office mesoscale model: (a) wet-bulb potential temperature (thin lines labeled in K) and front-relative streamlines (bold lines with arrow heads) entering the region containing the billows; (b) Richardson Number (thin contours). The region of saturation is delineated by a bold contour, and the location of the radar-detected billows is shown shaded, in both (a) and (b) (from Chapman and Browning 1999).

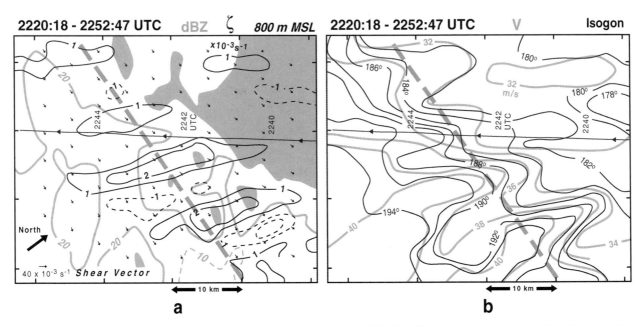

FIG. 2.13. Analyses within a plan section at 800 m showing the structure of KH billowlike features within a warm frontal zone derived from airborne Doppler radar measurements at 2220–2253 UTC 5 Feb 1997. (a) Radar reflectivity (gray lines with values greater than 30 dBZ shaded gray), vertical vorticity (solid and dashed black lines at intervals of 1×10^{-3} s^{-1} representing positive and negative values, respectively), and shear vectors for the layer from 0.4 to 2.0 km. (b) Wind direction (black lines at 2° intervals) and speed (gray lines at 2 m s^{-1} intervals). The aircraft flight track is shown by the thin black line. The dashed gray line represents the orientation of the warm frontal zone (from Wakimoto and Bosart 2001).

mixing in frontal zones is to be represented physically in NWP models, both of these processes will need to be taken into account.

d. Mesoscale organization of upright convection in cyclones

As discussed in section 3a, upright convection can be in the form of upper-level cellular convection at the top of a layer of slantwise ascent or it can be associated with air convecting from the boundary layer. The latter is of course responsible for the highest rainfall intensities encountered in cyclones. In this section we shall take a closer look at the latter and divide it into two categories that are common in the United Kingdom: the line convection responsible for narrow cold frontal rainbands [section 3d(1)] and convective squall lines within the dry-slot region [section 3d(3)]. We shall also examine the transition between the convection occurring in these two regions [section 3d(2)].

1) STRUCTURE OF LINE CONVECTION

Line convection is the shallow but intense line of convection that occurs at the primary surface cold front (and also sometimes, in a weaker form, up to about 100 km behind) in association with cold-season anabatic cold fronts. The context and broad attributes of line convection have been discussed in section 3a. It was depicted in Figs. 2.4 and 2.5 as the upright ascent up

to 3–4 km labeled CF1 and L1, respectively. It is seen in the RHI in Fig. 2.3 as the abrupt ascent at ranges between 22 and 23 km, and in the RHI in Figs. 2.7a,b as the velocity discontinuity and shallow reflectivity maximum at 35 km. In plan view it can be seen as the thin line of intense (red) echo in Fig. 2.6. The vertical and horizontal structure of line convection was reviewed in the 1990 AMS monograph article (Browning 1990b, chapter 8.9) and so we shall restrict ourselves here to some recent refinements in understanding, particularly in relation to the way in which the line convection is broken into line elements. These line elements were first discussed in detail by James and Browning (1979) and Hobbs and Biswas (1979). Moore (1985) suggested that their formation is due to horizontal shearing instability at the strongly sheared edge of the prefrontal low-level jet. Thorpe and Emanuel (1985) suggested that they are due to an internal baroclinic/barotropic instability.

A unique dataset describing the 3D structure of cold frontal line-convection elements was acquired during the FASTEX project using airborne Doppler radar (Wakimoto and Bosart 2000). The line convection was very shallow, with tops below 2 km. As is usual for line convection, however, there was strong convergence in the boundary layer, leading to peak updraft velocities of several meters per second at 400 m. Figure 2.16 shows a plan section at 400 m through three of the line elements. The black contours in Fig. 2.16b show the elongated updraft cores and the shaded areas show the corresponding precipitation cores extending several ki-

Cross Section Parallel to Front

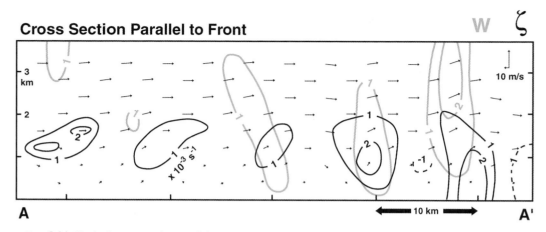

FIG. 2.14. Vertical cross section parallel to the warm front roughly along the dashed gray line in Figs. 2.13a,b, showing vertical velocity (gray contours for 1 and 2 m s^{-1}) superimposed on vertical vorticity (black lines for 1 and 2 × 10^{-3} s^{-1}) and wind vectors in the plane of the cross section (from Wakimoto and Bosart 2001).

lometers downwind, owing to the displacement of the precipitation falling through the strong environmental flow. Figure 2.16 shows that the updrafts forming the line-convection elements are within a 4-km-wide strip of concentrated vertical vorticity (Fig. 2.16a), wind veer (Fig. 2.16c), and wind speed gradient (Fig. 2.16d). Peak values of vertical cyclonic vorticity (horizontal shear) exceed 6 × 10^{-3} s^{-1}. This is similar to values derived from the early ground-based Doppler measurements of Browning and Harrold (1970) who also showed that the observed mass convergence in the friction layer was related to the cyclonic shear at the top of the friction layer according to a relationship given by Eliassen (1959).

The orientation of individual precipitation cores in Fig. 2.16 is rotated by 10°–15° from the mean orientation of the overall frontal shear zone: the updraft cores are rotated by twice this amount. The 5–10-km-wide

gaps between the convective cores are characterized by regions of much weaker vertical motion. Flights across the cold front at 600 m show weak gradients within the gap region (Fig. 2.17b) compared with the precipitation core (Fig. 2.17a): the main drop in equivalent potential temperature occurs over a distance of 12 km in the gap compared with 4 km in the core. The gap regions in line convection are favored locations for the development of small tornadoes (e.g., Carbone 1982; Browning and Golding 1995).

2) TRANSITION FROM CLASSIC LINE CONVECTION ALONG TRAILING COLD FRONT TO CONVECTIVE ELEMENTS NEAR DRY SLOT

Browning and Roberts (1996), in their study of the variations in structure along a cold front, drew attention to the way in which the narrow cold frontal rainband

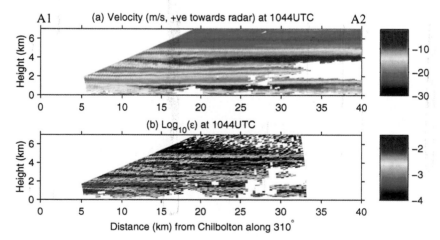

FIG. 2.15. Radar cross section (RHI) through an ana-cold front perpendicular to and looking away from the surface cold front at 1044 UTC 10 Feb 2000; (a) Doppler velocity in m s^{-1}; (b) logarithm of the dissipation rate, where dissipation rate is in m^2 s^{-3} (from Chapman and Browning 2001).

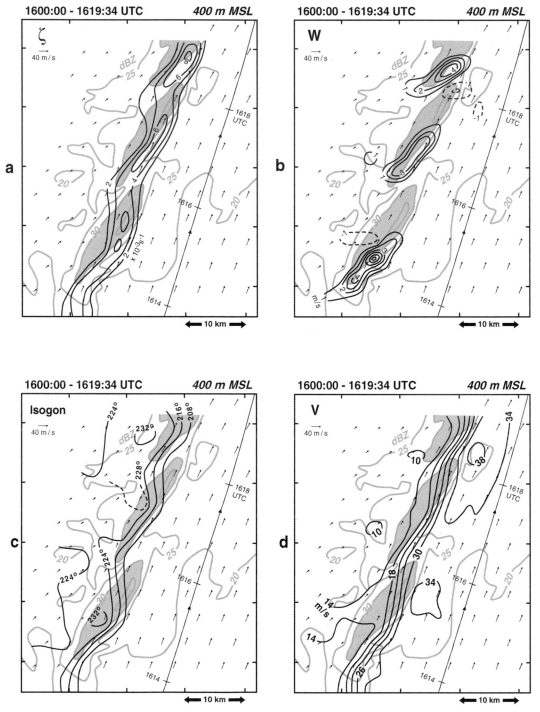

FIG. 2.16. Analyses within a plan section at 400 m showing the dynamical structure in the vicinity of three precipitation cores (shaded) within a narrow cold frontal rainband derived from airborne Doppler radar measurements at 1600–1620 UTC 12 Jan 1997. Radar reflectivity (gray contours, with shading greater than 30 dBZ) and ground-relative wind velocities (arrows) superimposed on (a) vertical vorticity (black contours at intervals of 2×10^{-3} s^{-1}), (b) vertical velocity (black contours every 1 m s^{-1}), (c) isogons (solid black contours every 8°), and (d) magnitude of the horizontal velocity (black contours every 4 m s^{-1}). The aircraft flight track is shown by the thin black line (from Wakimoto and Bosart 2000).

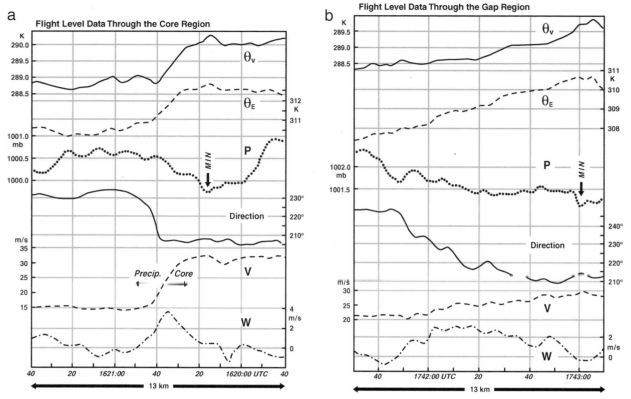

FIG. 2.17. Time series of in situ measurements during a flight at 600 m through (a) a precipitation core and (b) a gap between precipitation cores for the narrow cold frontal rainband depicted in Fig. 2.16. (a) The approximate location of the precipitation core is labeled. (a),(b) The dotted curves show the derived surface pressure (from Wakimoto and Bosart 2000).

may change in character along the length of the front. In particular they showed how the line-convection elements become more widely spaced and oriented at a larger angle to the overall chain of line elements as one goes from the trailing cold anafrontal region toward the dry slot. Their findings are exemplified by the more recent case depicted in Fig. 2.18a, which shows a radar-network picture of the precipitation accompanying an intense cyclone centered over the west coast of Wales. The figure is interpreted in Fig. 2.18b within the context of the conveyor-belt model in Fig. 2.1 (in much the same way as Fig. 2.2 was). Figure 2.18b draws attention to the narrow cold frontal precipitation cores due to intense line-convection elements embedded within the overall region of precipitation associated with the primary warm conveyor belt (W1). In the trailing ana-cold frontal part of the front, where the surface cold front is sharp, the line elements are indeed seen to be aligned at a small angle to the front (as in Fig. 2.16). Farther north, however, the size of the gaps between the line elements is seen to increase and there is a tendency for the convection in them to be less line-like. An analysis of surface reports showed that the warm-sector air in this region extended westward as a W2 flow toward the low center L (see yellow contour in Fig. 2.18b); that is, high-θ_w air lags behind the convective elements at low levels despite the encroachment of low-θ_w air aloft. As

pointed out in section 2, the cold front in this region is a kata-cold front and the convective elements under discussion are associated with the upper cold (θ_w) front rather than a surface cold front. The reason for the convection taking on the structure that is observed needs further study but it is probably related to the particular combination of vertical wind shear and convective instability in the kata-cold frontal region. The kata-cold front is characterized by significant (though rather shallow) potential instability due to the overrunning dry intrusion. The way in which the potential instability develops in this region is illustrated in Fig. 2.19.

The three panels in Fig. 2.19 show three snapshots from an operational mesoscale model run depicting the evolution of the θ_w field during the deepening of a surface low (at location L). For simplicity a single θ_w surface, corresponding to the middle of the frontal zone, is analyzed where it intersects the 950- and 700-hPa levels. The sequence shows how the surface winds up with time. This is associated with the field of rotation shown in the larger-area analysis on the right of Fig. 2.19. The system-relative winds shown are for the 700-hPa level where the rotation tends to be greatest. Lower down, the θ_w surface rotates less, although it remains to be established to what extent this is due to differential rotation or to diabatic warming modifying θ_w closer to the ground. In any case, this kind of effective differential

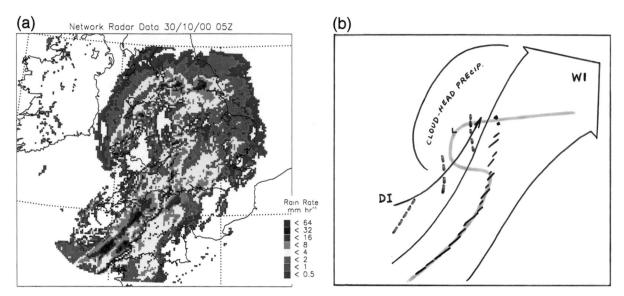

FIG. 2.18. Precipitation distribution accompanying a rapidly deepening cyclone at 0500 UTC 30 Oct 2000: (a) radar-network display, with rainfall intensity as shown by the color bar; (b) analysis covering exactly the same area as (a), with precipitation cores along the narrow cold frontal rainband shown by short black lines and other convective rainbands beneath the dry intrusion/dry slot shown dashed in dark green. The primary warm-conveyor-belt and dry-intrusion flows are depicted schematically by the broad and narrow arrows, labeled W1 and D1, respectively. The yellow line delineates the observed westernmost extent of warm air at the surface (wet-bulb temperature exceeding 11°C); in the north it corresponds to the surface warm front and bent-back front, in the south it corresponds to the surface cold front, and in between it protrudes westward toward the cyclone center (L) as a high-θ_w SMZ beneath the D1.

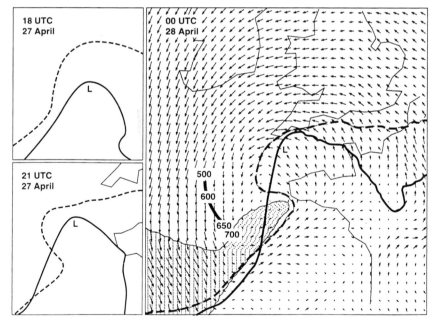

FIG. 2.19. Output from the operational Met Office mesoscale model, for the rapidly deepening cyclone observed in the FRONTS 92 project on 27–28 Apr 1992, depicting the winding up of the $\theta_w = 9°C$ surface under the influence of differential cyclonic rotation; three-hourly sequence from 1800 UTC 27 Apr (T + 0) to 0000 UTC 28 Apr (T + 6). The two panels on the left side show the 9°C surface at 700 hPa (dashed) and 950 hPa (solid). The right-hand panel shows system-relative flow and extent of dry air (RH < 30% shaded) at 700 hPa superimposed on the $\theta_w = 9°C$ surface at 700 and 950 hPa for 0000 UTC 28 Apr. The kata-cold frontal region is where the dry air overruns and the θ_w surface locally slopes forward. Numbers show positions of the axis of the maximum PV corresponding to the descending tropopause fold between the 500- and 700-hPa levels. The right-hand diagram is plotted to the same scale as those on the left-hand side but it covers a larger area (after Browning et al. 1997a).

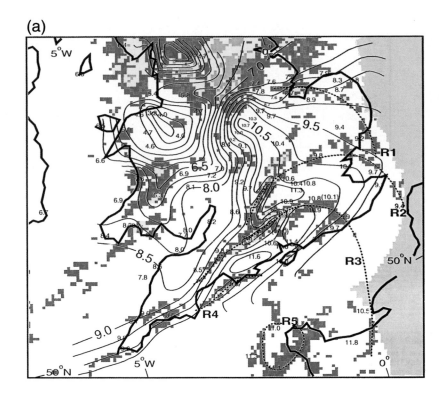

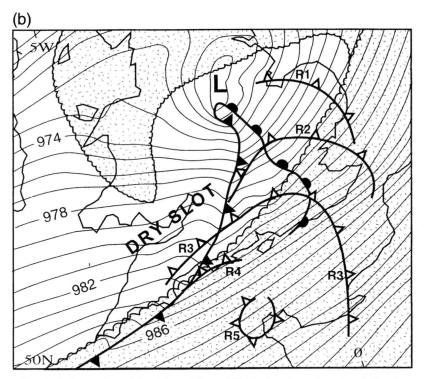

FIG. 2.20. Convective rainbands/squall lines within the dry-slot region of a small cyclone observed at 2100 UTC 24 Feb 1997 during the FASTEX project. (a) Radar rainfall pattern superimposed on isopleths of surface wet-bulb potential temperature, with axes of convective rainbands R1–R5 drawn dotted (including over newly formed gaps in the rainbands); the θ_w analysis is based on individual observations shown by the small numbers. (b) The corresponding surface pressure and frontal analysis, with the axes of convective rainbands depicted by fronts with open cold frontal symbols. The stippled area is added to show the extent of cloud with tops above the $-15°C$ level (from Browning and Roberts 1999).

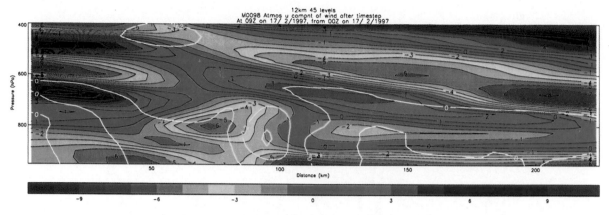

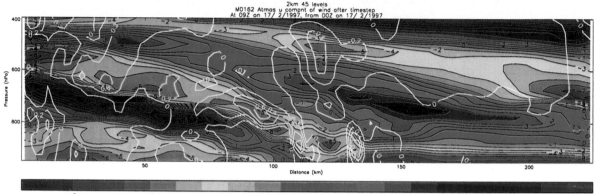

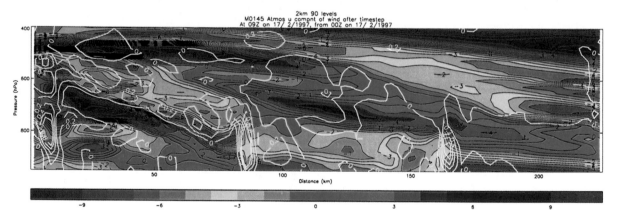

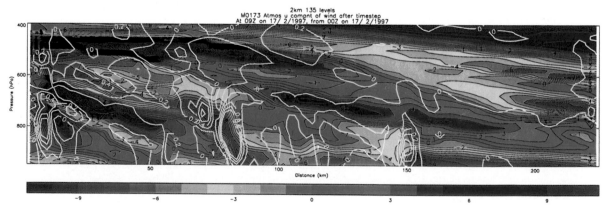

rotation with height is commonly observed during frontal fracture and it leads to a localized region of overrunning (low θ_w above high θ_w) just south of the cyclone center. Figure 2.19 shows that the region of maximum overrunning coincides with the driest part of the dry intrusion (stippled) and this of course inhibits the release of the potential instability in the absence of other factors such as enhanced ascent or evaporative cooling.

The dry intrusion in Fig. 2.19 was associated with a tropopause fold, the axis of which is shown extending down from 500 hPa. The fold at low levels was a thin filament and thus probably not dynamically very active; however, at higher levels it was connected to a substantial tropopause depression, which, as demonstrated by Griffiths et al. (2000), may have played an important role in inducing the circulations that led to both the development of the potential instability and its eventual release through locally enhanced lifting.

3) Convective squall lines within the dry-slot region

The northernmost convective elements discussed in section 3d(2) were located toward the leading edge of (rather than within) the dry slot. According to Carr and Millard (1985), convection-related severe weather events tend to be associated with DIs and they found that 93% of such events occur at the boundaries of the DIs. Within the main body of the dry slot, where the DI air systematically overruns the SMZ (in Fig. 2.1), there may be an almost total absence of precipitation or there may be a few bands of convective precipitation spanning it as shown in Fig. 2.18 (see green dashed lines in Fig. 2.18b). In this case, the convective rainbands were associated with a succession of surface convergence lines, roughly parallel to the primary surface cold front, which advanced through the system. On other occasions the organization of rainbands within a dry slot may be more complicated; we now consider such a case.

Figure 2.20 shows analyses of convective rainbands observed in the dry slot of a small but vigorous cyclone that crossed the United Kingdom during the FASTEX project. Figure 2.20a depicts the radar rainfall distribution superimposed on contours of the surface θ_w. The rainbands in question are highlighted by dotted lines labeled R1, R2, R3, R4, and R5: some of these lines

span regions with no detectable precipitation at the analysis time, but continuity with the rainband structure observed at other times provides credibility in the analysis. Parts of rainbands R2, R3, and R4 occurred within a warm sector that was characterized by surface values of θ_w greater than 10.5°C. Here the rainbands had squall-line characteristics, with convective downdrafts bringing lower-θ_w air and very strong winds to the surface. Other parts of the rainbands, notably parts of R2 and all of R1, were situated ahead of the surface warm front and, being cut off from the source of high-θ_w air near the surface, were characterized by only rather weak mid-level convection.

Figure 2.20b shows a mesoanalysis for the time corresponding to Fig. 2.20a. The cyclone center is located in central England within a large dry slot seen in satellite imagery. The stippled cloud area to the north of the cyclone is the cloud head and that in the southeastern half of the figure is the polar front cloud band (cf. Fig. 2.1). As often happens, the dry intrusion was much more extensive than was evident from the satellite imagery, and in this case it extended beneath most parts of the polar front cloud band shown in this figure. The fronts drawn with open cold front symbols are believed to correspond to successive pulses of increasingly low-θ_w air aloft within the dry intrusion, that is, cold upper (θ_w) fronts, although this remains conjecture because of the difficulty in obtaining sufficiently well-resolved observations of thermal structure. They may resemble in some respects the hyperfine baroclinic zones observed by Kreitzberg and Brown (1970). These upper cold fronts are shown in Fig. 2.20b to be overrunning the warm sector where they trigger the convective squall lines; they traveled at a high speed (39 m s^{-1}), thereby overtaking the lower-level features and advancing ahead of the warm front where they began to decay. The almost circular rainband R5, at the apparent "epicenter" of the arc rainbands, is intriguing. The writer has observed similar features on one or two other occasions. Their origin is obscure but we speculate that they may radiate outward from where downward-propagating pulses of low-θ_w air first impinge on the top of the boundary layer.

An important feature of dry-slot convection is the tendency for the production of evaporatively chilled downdrafts in a region where, especially in the case of intense cyclones, the winds at about 800 hPa can be very strong. The mixing down of high-momentum air

←

FIG. 2.21. Vertical section across an ana-cold frontal zone with multiple slantwise circulations, at 0900 UTC 17 Feb 1997 (FASTEX intensive observation period 16), as derived from 9-h forecasts for the same location, from four versions of the Met Office mesoscale model operating with the so-called new dynamics code but at different resolutions: (a) 12-km grid, 45 levels; (b) 2-km grid, 45 levels; (c) 2-km grid, 90 levels; (d) 2-km grid, 135 levels. Black contours give front-perpendicular wind component (m s^{-1}), with the regions of strongest front-to-rear flow (negative) shown yellow/orange/red and the regions of strongest rear-to-front flow shown blue/maroon. White contours show vertical velocity (m s^{-1}); (c),(d) one updraft core at low levels at 85 km is associated with line convection that feeds some rearward slantwise ascent, but another updraft core at or beyond 150 km is due to a chance encounter with an isolated convective cell not associated with organized slantwise ascent. The uppermost layer(s) of front-to-rear slantwise flow is (are) fed by a region of line convection located beyond the right-hand side of the section (courtesy of Humphrey Lean, Joint Centre for Mesoscale Meteorology, University of Reading).

by the downdrafts often leads to damaging gusts at the surface. Although the convection is frequently very shallow, the strength of the gusts is increased somewhat by the kinetic energy of the downdraft itself (Nakamura et al. 1996). Damaging gusts due to dry-slot convection occurred in the case described in this section. Surface gusts in excess of 40 m s^{-1} also occurred in the case discussed in section 3d(2) in association with the dry-slot convective bands shown by the dashed green lines in Fig. 2.18b. The damaging winds that led to the loss of life in the 1979 Fastnet yachting race were also attributed to gusts associated with shallow dry-slot convection (Bond et al. 1981).

4. Concluding remarks

This article has focused on the meso-beta-scale and smaller-scale substructure of cyclones, although larger-scale processes have been touched upon to provide the appropriate context. Particular features discussed include layers of slantwise convection and associated shallow layers of evaporative cooling and shearing instability, as well as lines of convection at major cold fronts and within the dry-slot region of cyclones. Radar is the best tool for observing these features and results have been presented from a variety of different kinds of radar systems.

We are low down on the learning curve where these mesoscale phenomena are concerned: we are only just beginning to see in the most qualitative way how they are organized, where in a cyclone system they are most likely to be encountered and why. In many cases these phenomena are important in themselves because they are closely associated with significant weather events such as heavy rain or wind gusts. What is not yet clear in some cases is the extent to which these phenomena and the details of the related processes have an important impact on the larger-scale dynamics.

Many of the features discussed here are at or below the limit of resolution of present-day operational mesoscale NWP models. Line-convection elements, for example, require a horizontal grid spacing of about 1 km. Slantwise convection is particularly sensitive to the vertical grid spacing and, according to Persson and Warner (1993), a vertical spacing of as little as 200 m is necessary if it is to be explicitly resolved. Alternatively it could be parameterized (Lindstrom and Nordeng 1992), but then there is the danger of double accounting as model resolution is improved. Other processes, such as ice microphysics, need to be parameterized better if, for example, the dynamical effects of evaporation are to be properly represented. Then there are further processes that are neither resolved nor parameterized in most models. These include shearing instability in the free troposphere, and also the vertical transport of momentum by line convection, which is thought to trigger slantwise convection through the process of geostrophic momentum adjustment. There remains much to be done, there-

fore, in order to develop, and observationally validate, a new generation of NWP models running at very high resolution.

As an example of the promising development of a mesoscale model, we end this article by showing the improvement in the representation of stacked slantwise convective circulations achievable by increasing the resolution, particularly the vertical resolution (H. Lean 2001, personal communication). The panels in Fig. 2.21 show the modeled transverse circulations for a particular frontal cross section at four different resolutions of the so-called new dynamics (nonhydrostatic) version of the Met Office Unified Model. The red and yellow shading shows flows with a system-relative right-to-left component. When a 45-level version of the model is used, there are two main layers of rearward (right to left) slantwise ascent, with an increase in the coherence as the horizontal grid length decreases from 12 km (Fig. 2.21a) to 2 km (Fig. 2.21b). When the number of model levels is increased to 90, the number of resolved layers of slantwise ascent increases to 4 and their wavelength is roughly halved (Fig. 2.21c); this is close to the values actually observed in this and many other ana-cold frontal situations. Reassuringly, the number of resolved layers does not continue to increase as the number of model levels is increased to 135 (Fig. 2.21d). This supports the view that about 100 levels may be adequate to resolve the release of the mesoscale instabilities responsible for the slantwise convection.

Parallel observational and modeling studies will need to continue for many years (a) to investigate further the nature of the mesoscale processes, (b) to ascertain whether they have any upscale impact, and (c) to incorporate them where necessary into operational mesoscale NWP models in order to provide better very-short-range forecasts of the locally severe weather that can accompany extratropical cyclones. Radar is uniquely suited to obtaining many of the required observations and it will continue to play a central role in both research and operational practice. We owe a debt of gratitude to Dave Atlas for his foresight and leadership in promoting the application of both ground-based and airborne radar technology.

Acknowledgments. I thank colleagues in the Joint Centre for Mesoscale Meteorology, especially Danny Chapman, Pete Clark, Richard Dixon, Richard Forbes, Sue Gray, Robin Hogan, Anthony Illingworth, Humphrey Lean, Peter Panagi, Nigel Roberts, and Chang Wang; and also Catherine Gaffard, John Nash, and Tim Oakley in the Met Office at Beaufort Park, for contributing ideas and data. I am also grateful to Brad Smull for helpful comments on this paper.

REFERENCES

Atlas, D., K. R. Hardy, and K. Naito, 1966: Optimizing the radar detection of clear air turbulence. *J. Appl. Meteor.,* **5,** 450–460.

Bader, M. J., G. S. Forbes, J. R. Grant, R. B. E. Lilley, and A. J. Waters, Eds., 1995: *Images in Weather Forecasting.* Cambridge University Press, 499 pp.

Bennetts, D. A., and B. J. Hoskins, 1979: Conditional symmetric instability—a possible explanation for frontal rainbands. *Quart. J. Roy. Meteor. Soc.,* **105,** 945–962.

Bond, J. E., K. A. Browning, and C. G. Collier, 1981: Estimates of surface gust speeds using radar observations of showers. *Meteor. Mag.,* **110,** 29–40.

Böttger, H., M. Eckardt, and U. Katergiannakis, 1975: Forecasting extratropical storms with hurricane intensity using satellite information. *J. Appl. Meteor.,* **14,** 1259–1265.

Braun, S. A., R. A. Houze Jr., and B. F. Smull, 1997: Airborne dual-Doppler observations of an intense frontal system approaching the Pacific Northwest coast. *Mon. Wea. Rev.,* **125,** 3131–3156.

Browning, K. A., 1971: Structure of the atmosphere in the vicinity of large-amplitude Kelvin–Helmholtz billows. *Quart. J. Roy. Meteor. Soc.,* **97,** 283–299.

——, 1990a: Organization and internal structure of synoptic and mesoscale precipitation systems in midlatitudes. *Radar in Meteorology,* D. Atlas, Ed., Amer. Meteor. Soc., 433–460.

——, 1990b: Organization of clouds and precipitation in extratropical cyclones. *Extratropical Cyclones: Erik Palmén Memorial Volume,* C. W. Newton and E. O. Holopainen, Eds., Amer. Meteor. Soc., 129–153.

——, 1995: On the nature of the mesoscale circulations at a kata-cold front. *Tellus,* **47A,** 911–919.

——, 1997: The dry intrusion perspective of extra-tropical cyclone development. *Meteor. Appl.,* **4,** 317–324.

——, 1999a: Mesoscale aspects of extratropical cyclones: An observational perspective. *The Life Cycles of Extratropical Cyclones,* M. A. Shapiro and S. Grønås, Eds., Amer. Meteor. Soc., 265–283.

——, 1999b: Use of radar in mesoscale analysis. *Proc. COST-75 Final Int. Seminar on Advanced Weather Radar Systems,* Lacarno, Switzerland, European Commission, 239–256.

——, and T. W. Harrold, 1970: Air motion and precipitation growth at a cold front. *Quart. J. Roy. Meteor. Soc.,* **96,** 369–389.

——, and G. A. Monk, 1982: A simple model for the synoptic analysis of cold fronts. *Quart. J. Roy. Meteor. Soc.,* **108,** 435–452.

——, and N. M. Roberts, 1994: Structure of a frontal cyclone. *Quart. J. Roy. Meteor. Soc.,* **120,** 1535–1557.

——, and B. W. Golding, 1995: Mesoscale aspects of a dry intrusion within a vigorous cyclone. *Quart. J. Roy. Meteor. Soc.,* **121,** 463–493.

——, and N. M. Roberts, 1996: Variation of frontal and precipitation structure along a cold front. *Quart. J. Roy. Meteor. Soc.,* **122,** 1845–1872.

——, and ——, 1999: Mesoscale analysis of arc rainbands in a dry slot. *Quart. J. Roy. Meteor. Soc.,* **125,** 3495–3511.

——, and C.-G. Wang, 2002: Cloud-top striations above ana-cold frontal circulations. *Quart. J. Roy. Meteor. Soc.,* **128,** 477–499.

——, T. W. Harrold, and J. R. Starr, 1970: Richardson number limited shear zones in the free atmosphere. *Quart. J. Roy. Meteor. Soc.,* **96,** 40–49.

——, S. A. Clough, C. S. A. Davitt, N. M. Roberts, T. D. Hewson, and P. G. W. Healey, 1995: Observations of the mesoscale substructure in the cold air of a developing frontal cyclone. *Quart. J. Roy. Meteor. Soc.,* **121,** 1229–1254.

——, S. P. Ballard, and C. S. A. Davitt, 1997a: High-resolution analysis of frontal fracture. *Mon. Wea. Rev.,* **125,** 1212–1230.

——, N. M. Roberts, and A. J. Illingworth, 1997b: Mesoscale analysis of the activation of a cold front during cyclogenesis. *Quart. J. Roy. Meteor. Soc.,* **123,** 2349–2375.

——, D. Jerrett, J. Nash, T. Oakley, and N. M. Roberts, 1998: Cold frontal structure derived from radar wind profilers. *Meteor. Appl.,* **5,** 67–74.

——, D. Chapman, and R. S. Dixon, 2001a: Stacked slantwise convection circulations. *Quart. J. Roy. Meteor. Soc.,* **127,** 2513–2536.

——, R. S. Dixon, C. Gaffard, and C.-G. Wang, 2001b: Wind profiler measurements in the storm of 30 October 2000. *Weather,* **56,** 367–373.

——, P. Panagi, and E. M. Dicks, 2002: Multi-sensor synthesis of the mesoscale structure of a cold-air comma cloud system. *Meteor. Appl.,* **9,** 155–175.

Carbone, R. E., 1982: A severe frontal rainband. Part I: Stormwide hydrodynamic structure. *J. Atmos. Sci.,* **39,** 258–279.

Carlson, T. N., 1980: Airflow through midlatitude cyclones and the comma cloud pattern. *Mon. Wea. Rev.,* **108,** 1498–1509.

Carr, F. H., and J. P. Millard, 1985: A composite study of comma clouds and their association with severe weather over the Great Plains. *Mon. Wea. Rev.,* **113,** 370–387.

Chapman, D., and K. A. Browning, 1997: Radar observations of wind-shear splitting within evolving atmospheric Kelvin–Helmholtz billows. *Quart. J. Roy. Meteor. Soc.,* **123,** 1443–1439.

——, and ——, 1998: Use of wind-shear displays for Doppler radar data. *Bull. Amer. Meteor. Soc.,* **79,** 2685–2691.

——, and ——, 1999: Release of potential shearing instability in warm frontal zones. *Quart. J. Roy. Meteor. Soc.,* **125,** 2265–2289.

——, and ——, 2001: Measurements of dissipation rate in frontal zones. *Quart. J. Roy. Meteor. Soc.,* **127,** 1939–1959.

Clough, S. A., and R. A. A. Franks, 1991: The evaporation of frontal and other stratiform precipitation. *Quart. J. Roy. Meteor. Soc.,* **117,** 1057–1080.

——, H. W. Lean, N. M. Roberts, and R. M. Forbes, 2000: Dynamical effects of ice sublimation in a frontal wave. *Quart. J. Roy. Meteor. Soc.,* **126,** 2405–2434.

Danielsen, E. F., 1966: Research in four-dimensional diagnosis of cyclonic storm cloud systems. Scientific Rep. 1, Contract AF 19(628)-4762, The Pennsylvania State University, 53 pp.

Davies, H. C., and H. Wernli, 1997: On studying the structure of synoptic systems. *Meteor. Appl.,* **4,** 365–374.

Dixon, R. S., K. A. Browning, and G. J. Shutts, 2000: The mystery of striated cloud heads in satellite imagery. *Atmos. Sci. Letters,* **1,** 1–13. [Available online at http://dx.doi.org./doi:10.1006/asle.2000.0001.]

Eliassen, A., 1959: On the formation of fronts in the atmosphere. *The Atmosphere and the Sea in Motion,* B. Bolin, Ed. Rockefeller Institute Press, 277–287.

Emanuel, K. A., 1994: *Atmospheric Convection.* Oxford University Press, 580 pp.

Feren, G., 1995: The striated-delta cloud system—A satellite imagery precursor to major cyclogenesis in the eastern Australian–western Tasman Sea region. *Wea. Forecasting,* **10,** 286–309.

Fischer, C., and F. Lalaurette, 1995: Meso-β-scale circulations in realistic fronts.II: Frontogenetically forced basic states. *Quart. J. Roy. Meteor. Soc.,* **121,** 1285–1321.

Forbes, R. M., H. W. Lean, N. M. Roberts, and P. A. Clark, 2000: A study of the FASTEX IOP-16 mid-latitude cyclone: Implications for mesoscale modelling. JCMM Internal Rep. 107, Met Office, United Kingdom, 43 pp. + 55 pp. of figures.

Green, J. S. A., F. H. Ludlam, and J. R. F. McIlveen, 1966: Isentropic relative-flow analysis and the parcel theory. *Quart. J. Roy. Meteor. Soc.,* **92,** 210–219.

Griffiths, M., A. J. Thorpe, and K. A. Browning, 2000: Convective destabilisation by a tropopause fold diagnosed using potential-vorticity inversion. *Quart. J. Roy. Meteor. Soc.,* **126,** 125–144.

Harris, F. I., 1977: The effects of evaporation at the base of ice precipitation layers: Theory and radar observations. *J. Atmos. Sci.,* **34,** 651–672.

Harrold, T. W., 1973: Mechanisms influencing the distribution of precipitation within baroclinic disturbances. *Quart. J. Roy. Meteor. Soc.,* **99,** 232–251.

Hicks, J. J., and J. K. Angell, 1968: Radar observations of breaking gravitational waves in the visually clear atmosphere. *J. Appl. Meteor.,* **7,** 114–121.

Hobbs, P. V., and K. R. Biswas, 1979: The cellular structure of narrow cold-frontal rainbands. *Quart. J. Roy. Meteor. Soc.,* **105,** 723–727.

——, and P. O. G. Persson, 1982: The mesoscale and microscale structure and organization of clouds and precipitation in midlatitude cyclones. Part V: The substructure of narrow cold-frontal rainbands. *J. Atmos. Sci.,* **39,** 280–295.

Holt, M. W., and A. J. Thorpe, 1991: Localized forcing of slantwise motion at fronts. *Quart. J. Roy. Meteor. Soc.,* **117,** 943–963.

Hoskins, B. J., M. E. McIntyre, and A. W. Robertson, 1985: On the use and significance of isentropic potential vorticity maps. *Quart. J. Roy. Meteor. Soc.,* **111,** 877–946.

Huang, H.-C., and K. A. Emanuel, 1991: The effects of evaporation on frontal circulation. *J. Atmos. Sci.,* **48,** 619–628.

James, P. K., and K. A. Browning, 1979: Mesoscale structure of line convection at surface cold fronts. *Quart. J. Roy. Meteor. Soc.,* **105,** 371–382.

——, and ——, 1981: An observational study of primary and secondary billows in the free atmosphere. *Quart. J. Roy. Meteor. Soc.,* **107,** 351–365.

Johnson, R. H., 2001: Surface mesohighs and mesolows. *Bull. Amer. Meteor. Soc.,* **82,** 13–32.

Kreitzberg, C. W., and H. A. Brown, 1970: Mesoscale weather systems within an occlusion. *J. Appl. Meteor.,* **9,** 417–432.

Lemaître, Y., A. Protat, and D. Bouniol, 1999: Pacific and Atlantic "bomb-like" deepening in mature phase: A comparative study. *Quart. J. Roy. Meteor. Soc.,* **125,** 3513–3534.

Lindstrom, S. S., and T. E. Nordeng, 1992: Parameterized slantwise convection in a numerical model. *Mon. Wea. Rev.,* **120,** 742–756.

Locatelli, J. D., J. E. Martin, J. A. Castle, and P. V. Hobbs, 1995: Structure and evolution of winter cyclones in the central United States and their effects on the distribution of precipitation. Part III: The development of a squall line associated with weak cold frontogenesis aloft. *Mon. Wea. Rev.,* **123,** 2641–2662.

Manabe, S., 1956: On the contribution of heat released by condensation to the change in pressure pattern. *J. Meteor. Soc. Japan,* **34,** 308–320.

Marécal, V., and Y. Lemaître, 1995: Importance of microphysical processes in the dynamics of a CSI mesoscale frontal cloud band. *Quart. J. Roy. Meteor. Soc.,* **121,** 301–318.

Martin, J. E., J. D. Locatelli, and P. V. Hobbs, 1990: Organization and structure of clouds and precipitation on the mid-Atlantic coast of the United States. Part III: The role of a middle-tropospheric cold front in the production of heavy precipitation. *Mon. Wea. Rev.,* **118,** 195–217.

Matejka, T. J., R. A. Houze, and P. V. Hobbs, 1980: Microphysics and dynamics of clouds associated with mesoscale rainbands in extratropical cyclones. *Quart. J. Roy. Meteor. Soc.,* **106,** 29–56.

Miles, J. W., and L. N. Howard, 1964: Note on a heterogeneous shear flow. *J. Fluid Mech.,* **20,** 311–336.

Moore, G. W. K., 1985: The organization of convection in narrow cold-frontal rainbands. *J. Atmos. Sci.,* **42,** 1777–1791.

Nakamura, K., R. Kershaw, and N. Gait, 1996: Prediction of near-surface gusts generated by deep convection. *Meteor. Appl.,* **3,** 157–167.

Neiman, P. J., M. A. Shapiro, and L. S. Fedor, 1993: The life cycle of an extratropical marine cyclone. Part II: Mesoscale structure and diagnostics. *Mon. Wea. Rev.,* **121,** 2177–2199.

Oliver, V. J., and G. C. Holzworth, 1953: Some effects of the evaporation of widespread precipitation on the production of fronts and on changes in frontal slopes and motions. *Mon. Wea. Rev.,* **81,** 141–151.

Parker, D. J., and A. J. Thorpe, 1995: The role of snow sublimation in frontogenesis. *Quart. J. Roy. Meteor. Soc.,* **121,** 763–782.

Parsons, D. B., 1992: An explanation for intense frontal updrafts and narrow cold-frontal rainbands. *J. Atmos. Sci.,* **49,** 1810–1825.

Persson, P. O. G., and T. T. Warner, 1993: Nonlinear hydrostatic conditional symmetric instability: Implications for numerical weather prediction. *Mon. Wea. Rev.,* **121,** 1821–1833.

Reed, R. J., and E. F. Danielsen, 1959: Fronts in the vicinity of the tropopause. *Arch. Meteor. Geophys. Bioklimatol.,* **A11,** 1–17.

Roberts, N. M., and R. M. Forbes, 2002: An observational study of multiple cloud head structure in the FASTEX IOP-16 cyclone. *Atmos. Sci. Lett.,* **3,** 59–70.

Rossa, A. M., H. Wemli, and H. C. Davies, 2000: Growth and decay of an extra-tropical cyclone's PV tower. *Meteor. Atmos. Phys.,* **73,** 139–156.

Roux, F., V. Marécal, and D. Hauser, 1993: The 12/13 January 1988 narrow cold-frontal rainband observed during MFDP/FRONTS 87. Part I: Kinematics and thermodynamics. *J. Atmos. Sci.,* **50,** 951–974.

Rutledge, S. A., 1989: A severe frontal rainband. Part IV: Precipitation mechanisms, diabatic processes, and rainband maintenance. *J. Atmos. Sci.,* **46,** 3570–3594.

Sansom, H. W., 1951: A study of cold fronts over the British Isles. *Quart. J. Roy. Meteor. Soc.,* **77,** 96–120.

Schultz, D. M., 2001: Reexamining the cold conveyor belt. *Mon. Wea. Rev.,* **129,** 2205–2225.

——, and P. N. Schumacher, 1999: The use and misuse of conditional symmetric instability. *Mon. Wea. Rev.,* **127,** 2709–2732; corrigendum, **128,** 1573.

——, D. Keyser, and L. F. Bosart, 1998: The effect of large-scale flow on low-level frontal structure and evolution in midlatitude cyclones. *Mon. Wea. Rev.,* **126,** 1767–1791.

Scinocca, G. P., 1995: The mixing of mass and momentum by Kelvin–Helmholtz billows. *J. Atmos. Sci.,* **52,** 2509–2530.

Shapiro, M. A., and D. Keyser, 1990: Fronts, jet streams, and the tropopause. *Extratropical Cyclones: The Erik Palmén Memorial Volume,* C. W. Newton and E. O. Holopainen, Eds., Amer. Meteor. Soc., 167–191.

Smith, R. K., and M. J. Reeder, 1988: On the movement and low-level structure of cold fronts. *Mon. Wea. Rev.,* **116,** 1927–1944.

Smull, B. F., and R. A. Houze Jr., 1985: A midlatitude squall line with a trailing region of stratiform rain: Radar and satellite observations. *Mon. Wea. Rev.,* **113,** 117–133.

Takahashi, N., H. Uyeda, and K. Kikuchi, 1993: A Doppler radar observation on wave-like echoes generated in a strong vertical shear. *J. Meteor. Soc. Japan,* **71,** 357–365.

Thorpe, A. J., and K. A. Emanuel, 1985: Frontogenesis in the presence of small stability to slantwise convection. *J. Atmos. Sci.,* **42,** 1809–1824.

Wakimoto, R. M., and B. L. Bosart, 2000: Airborne radar observations of a cold front during FASTEX. *Mon. Wea. Rev.,* **128,** 2447–2470.

——, and ——, 2001: Airborne radar observations of a warm front during FASTEX. *Mon. Wea. Rev.,* **129,** 254–274.

——, W. Blier, and C. H. Liu, 1992: The frontal structure of an explosive oceanic cyclone: Airborne radar observations of ERICA IOP 4. *Mon. Wea. Rev.,* **120,** 1135–1155.

Wernli, H., 1997: A Lagrangian-based analysis of extratropical cyclones: A detailed case-study. *Quart. J. Roy. Meteor. Soc.,* **123,** 1677–1706.

Xu, Q., 1986: Conditional symmetric instability and mesoscale rainbands. *Quart. J. Roy. Meteor. Soc.,* **112,** 315–334.

Yang, M.-J., and R. A. Houze Jr., 1995: Sensitivity of squall-line rear inflow to ice microphysics and environmental humidity. *Mon. Wea. Rev.,* **123,** 3175–3193.

Young, M. V., 1989: Investigation of a cyclogenesis event, 26–29 July 1988, using satellite imagery and numerical model diagnostics. *Meteor. Mag.,* **118,** 185–196.

——, G. A. Monk, and K. A. Browning, 1987: Interpretation of satellite imagery of a rapidly deepening cyclone. *Quart. J. Roy. Meteor. Soc.,* **113,** 1089–1115.

Zillman, J. W., and P. G. Price, 1972: On the thermal structure of mature Southern Ocean cyclones. *Aust. Meteor. Mag.,* **20,** 34–48.

Chapter 3

State of the Science: Radar View of Tropical Cyclones

FRANK D. MARKS JR.

Hurricane Research Division, NOAA/AOML, Miami, Florida

Marks

1. Introduction

Radar played an important role in studies of tropical cyclones since it was developed in the 1940s. The spiral echo pattern that is characteristic of tropical cyclones was one of the first meteorological features described by radar. The pioneering studies of Dave Atlas and colleagues established radar as an ideal remote sensing tool for studies of the structure and evolution of tropical cyclones (e.g., Kessler and Atlas 1956; Kessler 1957, 1958; Atlas et al. 1963). Since then our understanding of tropical cyclones and their characteristics improved dramatically, in part due to the use of radars, both non-coherent and coherent. Advances in understanding closely parallel advances in radar technology and data processing. These advances rapidly accelerated over the last 15 years with the advent of operational ground-based Doppler radar networks in the United States, Taiwan, Japan, Hong Kong, and the French islands of La Réunion and Martinique, and the use of airborne Doppler radars on the two National Oceanic and Atmospheric Administration (NOAA) WP-3D aircraft, which regularly fly into tropical cyclones each hurricane season.

Airborne Doppler radar datasets collected in the 1980s represented our best information on the kinematic structure of tropical cyclones (e.g., Marks and Houze 1984, 1987; Marks et al. 1992; Houze et al. 1992; Gamache et al. 1993), while airborne and ground-based Doppler radars were instrumental in determining the kinematic structure of the outer rainbands (e.g., Ishihara et al. 1986; Powell 1990a,b; Barnes et al. 1991; Tabata et al. 1992; Ryan et al. 1992; Barnes and Powell 1995). These datasets revolutionized our perception of the kinematic structure of a mature tropical cyclone from an axisymmetric vortex to one that recognizes the significance of asymmetric motions to track and intensity changes.

While these datasets enabled the documentation of the axisymmetric primary (rotational) and secondary[1] circulations and the major asymmetries in a few cases, additional airborne and ground-based Doppler radar datasets were crucial to the documentation of these characteristics over a broad range of tropical cyclones. Also, while the airborne Doppler datasets were instrumental in describing the major circulation features, they were not very useful for describing the temporal evolution of these features. Unfortunately, prior to 1989 high–temporal resolution ground-based Doppler radar datasets in mature tropical cyclones in the Tropics were limited to that collected in Taiwan (e.g., Typhoon Alex in 1987; Jou et al. 1994; Lee et al. 1999). Subsequently, there was an explosion in the number of Doppler radar observations of tropical cyclones through the deployment of operational Doppler radar networks in the United States and Taiwan, and the addition of numerous re-

---

[1] Characterized by flow radially inward at low levels, upward in the eyewall, and radially outward at high levels.

search Doppler radars (both fixed and mobile) in the United States, Australia, and Japan (see appendix for summary of these technological developments).

This review covers the state of the science of tropical cyclones since the review presented at the Battan Memorial, 40th Conference on Radar Meteorology (Marks 1989).[2] The major change since the last review is the addition of numerous high-quality Doppler radar datasets in tropical cyclones. In the last 15 years, the instrumentation on NOAA WP-3D research aircraft, in particular the airborne tail Doppler radar and other remote and in situ sensors, improved sufficiently to produce a new generation of tropical cyclone data whose analysis and interpretation is rapidly expanding our understanding of these storms. The advent of the operational Weather Service Radar 1988-Doppler (WSR-88D) radar network, and the construction of portable Doppler radars and profilers that can be moved to the a location near tropical cyclone landfall, has also generated new and unique high–temporal resolution datasets that are enabling improved understanding of storm wind structure as the storm moves from over the sea to over land. Thanks to the addition of these new Doppler radar systems, the number of datasets in the 1990s increased by an order of magnitude over that collected in the 1980s. These new datasets have given us an unprecedented opportunity to document the dynamics of tropical cyclones and have led to improved understanding of the symmetric vortex and the major asymmetries. The ability to map the complete three-dimensional circulation of the storm in a short time (<1 h) enabled the partitioning of the wind to examine the roles of asymmetries in studies of storm motion and intensity change. This review will focus on the results from these studies.

2. Tropical cyclone structure and dynamics as revealed by radar observations

a. Basic structure

Many of the common features of tropical cyclones, such as the eyewall and the rainbands, were first described in early radar studies (e.g., Maynard 1945; Wexler 1947; Kessler and Atlas 1956; Kessler 1957, 1958; Jordan et al. 1960; Atlas et al. 1963). In these studies, radar observations often served as a context for describing tropical cyclone structure. In the 1980s researchers started combining radar observations from the NOAA airborne radars together with the flight-level wind information to relate the observed radar structure in a number of tropical cyclones to the wind field in a consistent manner (e.g., Jorgensen 1984a,b; Willoughby et al. 1982, 1984). This approach indicated that the storm structure can be described in terms of a "stationary band complex" (SBC) that consisted of an eye-

wall, a principal rainband, connecting bands, and several secondary rainbands outside the eyewall (Figs. 3 and 3.2, Willoughby et al. 1984).

As seen in Fig. 3.1, radar reflectivity patterns in tropical cyclones provide a good means for flow visualization although they represent precipitation, not winds. Descending motion occupies precipitation-free areas, such as the eye. The axis of the cyclone's rotation lies near the center of the eye, which is surrounded by the eyewall. In intense tropical cyclones, it may contain reflectivities as high as 50 dBZ,[3] equivalent to rainfall rates of 74 mm h^{-1}.[4] Less extreme reflectivities, 40 dBZ (13 mm h^{-1}), characterize most convective rainfall in the eyewall and spiral bands.

In Fig. 3.1a the principal band marks the outer boundary of the inner core of the vortex. Outside this rainband, streamlines converge into the band or are deflected around the inner core. A secondary wind maximum is typically observed in the principal band (e.g., Marks and Houze 1984; Barnes and Powell 1995; Samsury and Zipser 1995). The primary circulation of the vortex dominates the flow inside the principal band, where the eyewall is the focus of the upward branch of the secondary circulation of the vortex (e.g., Willoughby et al. 1992; Jorgensen 1984a,b). The eyewall may vary in diameter from 12 to 80 km (e.g., Marks 1985).

Secondary rainbands are smaller than the principal rainband and tend to be embedded in regions of stratiform precipitation surrounding the eyewall. Outside convection, reflectivities are still weaker, 30 dBZ, equivalent to a 2.4–mm h^{-1} rain rate. This "stratiform rain," denoted by a distinct reflectivity maximum or "bright band", a horizontal reflectivity enhancement near 5-km altitude, the altitude of the 0°C isotherm (cf. Fig. 3.2), falls out of the anvil cloud that grows from the convection. No wind maxima are found in these rainbands and the vertical velocity is weak. The connecting band is a special form of the secondary bands. It is characterized by weak reflectivity and little vertical motion. One or more connecting bands are found between the downwind portion of the principal band and the eyewall (cf. Fig. 3.1a).

Willoughby et al. (1984) noted that concentric eyewalls may be found inside of the principal rainband in strong storms (maximum winds >50 m s^{-1}), with large inner core circulations. Figure 3.1b shows the concentric eyewalls in Hurricane Gilbert (1988) described by Black and Willoughby (1992) and Dodge et al. (1999). The outer eyewall contained a secondary wind maximum and had the same characteristic circulation and structure as the inner eyewall. The outer eyewall often contracts inward with time, eventually replacing the inner eyewall. Airborne radar composites over a 6-day period in Hurricane Allen (1980) showed that the eyewall di-

---

[2] There is another excellent review in chapter 10 of the book by Houze (1993).

[3] $10 \log_{10}(Z)$, where Z is equivalent radar reflectivity factor (mm^6 m^{-3}).

[4] Using the relation $Z = 300R^{14}$ from Jorgensen and Willis (1982).

(a)   (b)

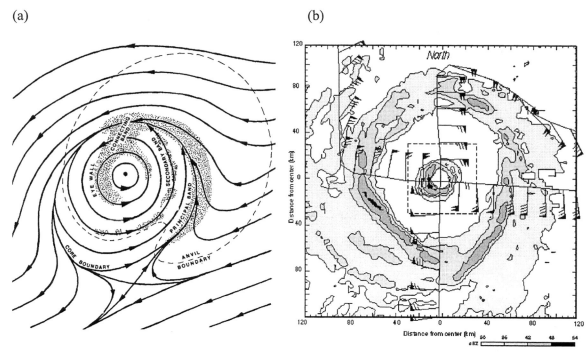

Fig. 3.1. (a) Schematic representation of the stationary band complex, the entities that compose it, and the flow in which it is embedded (Willoughby et al. 1984); (b) lower fuselage radar image from data collected 1008–1015 UTC 14 Sep 1988 (Dodge et al. 1999). The domain is 240 × 240 km, and the major tick marks are at 20-km intervals. Successively darker shading indicates radar reflectivities 30, 36, 42, and 48 dBZ. The small dashed box encloses the eye Doppler analysis domain. The line with the wind barbs is the flight track from 0858 to 1016 UTC. Wind barbs are plotted every 2 min, using the standard convention: a pennant is 25 m s^{-1}, a barb represents 5 m s^{-1}, and a half-barb is 2.5 m s^{-1}.

ameter changes as the tropical cyclone intensity changes (Marks 1985). As the vortex strengthened, the eyewall diameter contracted.

The radial variation of the vertical reflectivity structure of the inner 200 km of a tropical cyclone (Fig. 3.2a) is very similar to that of a mature mesoscale convective system (MCS; e.g., Jorgensen 1984a; Marks 1985; Black and Hallett 1986). The eyewall (located at radii 12–25 km in Fig. 3.2a) is characterized by large horizontal reflectivity gradients and the large vertical extent

(a)   (b)

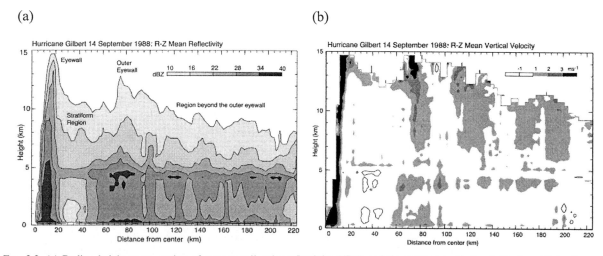

Fig. 3.2. (a) Radius–height cross section of average tail radar reflectivity (dBZ) and (b) average vertical velocity (m s^{-1}) for the region within 225 km of the center of Hurricane Gilbert on 14 Sep 1988; (a) successively darker shades of gray enclose areas with average reflectivity 10, 16, 22, 28, 34, and 40 dBZ; (b) upward vertical velocities are indicated by successively darker shades of gray, with black enclosing areas ≥3 m s^{-1}. The white area comprises weak vertical velocities from −1 to 1 m s^{-1}, and the stippled regions enclosed by the black contour line have vertical velocities <−1 m s^{-1} (from Dodge et al. 1999).

of the reflectivity maximum. The eyewall is surrounded by an extensive region of stratiform precipitation (10–40 km outside the eyewall in Fig. 3.2a). The reflectivity maxima in the eyewall and the rainbands tend to slope downwind with decreasing altitude, as first shown by Atlas et al. (1963).

The radial variation of the vertical structure of the vertical velocity from vertical incidence Doppler observations within the inner 200 km of a tropical cyclone is also similar to that of a mature MCS. The inner edge of the eyewall (5–10-km radius in Fig. 3.2b), at altitudes from near the surface to the storm top, is characterized by upward motion, with peak values as high as 20 m s^{-1}, and median values of 3–4 m s^{-1} (e.g., Marks and Houze 1987; Black et al. 1994; Black et al. 1996). Weak downdrafts are found generally below 6-km altitude coincident with the maximum eyewall reflectivity, while the strongest downdrafts, as high as -15 m s^{-1}, generally occur at altitudes >8 km (Black et al. 1994; Black et al. 1996). Outside the eyewall (radii >30 km in Fig. 3.2b) the vertical velocities are much weaker, with weak ascent found above the height of the bright band, and weak descent below the bright band (e.g., Marks and Houze 1987; Black et al. 1996). Stronger vertical velocities can be found in strong rainbands, like the secondary eyewall between 60- and 80-km radius in Fig. 3.2b (e.g., Black et al. 1996).

b. Inner core

Before the advent of the airborne Doppler, details of the kinematic structure of the inner core of tropical cyclones were determined by combining aircraft observations from one or more aircraft at different flight levels over a 4–6-h period (e.g., LaSeur and Hawkins 1963; Hawkins and Rubsam 1968; Hawkins and Imbembo 1976; Jorgensen 1984b; Frank 1984), from a composite of rawindsonde observations from many different stations over a 12–24-h period (e.g., Frank 1977), or from a composite of flight-level data obtained in many storms (Shea and Gray 1973; Gray and Shea 1973; Willoughby et al. 1982; Jorgensen 1984a; and Willoughby 1990). These studies provided extensive information on the axisymmetric structure of the vortex and a limited description of the asymmetric structure of the wind field. Since 1982, the Hurricane Research Division used the NOAA airborne Doppler radars to map the three-dimensional wind and reflectivity data in the inner core of tropical cyclones (e.g., Marks and Houze 1984, 1987; Marks et al. 1992; Black et al. 1994; Lee et al. 1994; Roux and Marks 1996; Roux and Viltard 1995; Dodge et al. 1999; Reasor et al. 2000). These observations have shown the utility of the airborne Doppler radar to sample the wind field in three dimensions with high spatial (~1 km) and temporal (~0.5–2.0 h) resolution.

Airborne Doppler data, collected from two nearly orthogonal flight legs, can be combined in a "pseudo-" (using data from the same radar at two different times)

or true dual-Doppler analysis of the three-dimensional wind field. The derived Doppler wind fields cover horizontal and vertical dimensions of ~75–100 and 15 km, respectively. The three-dimensional wind analyses cover 0.5–1.0-h intervals (equivalent to a one-dimensional sample of flight-level winds 200–400 km in length), compared with the 4–6 h needed for the composite analyses of flight-level data. Hence, storm structure on scales of 20 km and 0.5–1.0 h, particularly asymmetries that were difficult to infer from the composite analyses, can now be deduced.

Marks et al. (1992) provided a framework by which the asymmetric nature of the inner core of a tropical cyclone can be described. The Doppler analyses of two tropical cyclones—Hurricane Gilbert on 14 September 1988 (Dodge et al. 1999) and Hurricane Hugo on 17 September 1989 (Roux and Marks 1996)—are used to illustrate this framework. The two storms provide a representative cross section of intensities and intensity change, where Hugo was a rapidly intensifying storm (central pressure falling at >2 mb h^{-1}) with a central pressure of 952 hPa, while Gilbert was near peak intensity with a central pressure of 890 hPa for 12–18 h.

The three-dimensional wind field within each storm was derived from a pseudo-dual-Doppler analysis using the NOAA WP-3D airborne Doppler radars. The flight legs used in the dual-Doppler analysis varied from 75 to 110 km in length (10–15 min). The Doppler-derived wind fields for each storm consist of a storm-composite wind field that was centered on the storm circulation center covering a domain 150 km × 150 km on a side for Hurricane Hugo, and 60 km × 60 km for Hurricane Gilbert. The wind field for both storms extended from 0.5 to 16.0 km in altitude. The average time separation between radial velocity estimates from each flight leg in the analysis was 27 min, implying that there was no possibility of deriving estimates of convective-scale motions. Because of limitations and differences in azimuthal and radial resolution, these analyses concentrate on the mesoscale motions within the inner circulation (<100-km radius) with timescales of 45–60 min and space scales of wavenumber 2 in azimuth and 2–5 km in radius. Hence, these analyses offer an ideal opportunity to describe all of the components of the tropical cyclone circulation, including asymmetries, and to investigate the interaction of the vortex with its environment.

1) WIND FIELD DECOMPOSITION

The Doppler analysis renders the "total" wind field (relative to the ground) in a storm-centered coordinate system. Historically, observational studies of the kinematic structure of tropical cyclones have discussed the "relative" wind, that is, the circulation relative to the moving storm. In a cylindrical coordinate system centered on the storm, the horizontal relative wind vector $\mathbf{v}_r$ can be expressed as

(a)   (b)

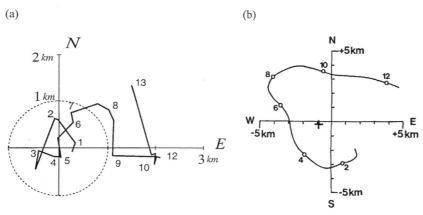

FIG. 3.3. Circulation centers at each level in the three-dimensional Doppler analyses for Hurricanes (a) Gilbert on 14 Sep 1988 (Dodge et al. 1999) and (b) Hugo on 17 Sep 1989 (Roux and Marks 1996). Positions are plotted relative to (a) the mean center of circulation in the first 6 km, which is 1.5 km west and 3.0 km north of the reflectivity center, and (b) the flight-level circulation center. The centers are plotted and labeled. (a) The reference circle is 1 km in radius. (b) The +-sign represents the density-weighted mean position.

$$v_r(r,\ \lambda,\ z) = \mathbf{v}(r,\ \lambda,\ z) - \mathbf{v}_s, \qquad (3.1)$$

where $\mathbf{v}$ is the total wind, $\mathbf{v}_s$ is the horizontal storm-motion vector (a constant defined by the objective fit to the centers from the flight-level data; e.g., Willoughby and Chelmow 1982), r is radius, λ is azimuth, and z is height.

The wind field is partitioned in a cylindrical coordinate system, where the circulation center varies with altitude as defined by Marks et al. (1992). The storm-relative horizontal wind vector in the cylindrical coordinate system is defined as

$$\mathbf{v}_r(r,\ \lambda,\ z) = \overline{\mathbf{v}_r}(z) + \mathbf{v}'(r,\ \lambda,\ z), \qquad (3.2)$$

where $\mathbf{v}'$ is the horizontal storm wind field and $\overline{\mathbf{v}_r}(z)$ is the horizontally averaged wind vector over a cylindrical portion of the Doppler analysis domain (πr_{max}^2) centered on the storm,

$$\overline{\mathbf{v}_r}(z) \equiv \frac{1}{2\pi} \int_0^{2\pi} \int_0^{r_{max}} \mathbf{v}_r(r,\ \lambda,\ z)\ dr\ d\lambda. \qquad (3.3)$$

Here r_{max} is defined as the largest radius that surrounds the center within the Doppler analysis domain at every level in the vertical (for Hugo $r_{max} = 75$ km and for the Gilbert $r_{max} = 40$ km).

Assuming that circular symmetry of the horizontal storm wind field $\mathbf{v}'$ can be decomposed into two components,

$$\mathbf{v}'(r,\ \lambda,\ z) = \mathbf{v}_0'(r,\ z) + \mathbf{v}_a'(r,\ \lambda,\ z), \qquad (3.4)$$

where $\mathbf{v}_0'(r,\ z)$ is the axisymmetric component of the tropical cyclone vortex, given by

$$\mathbf{v}_0'(r,\ z) \equiv \frac{1}{2\pi} \int_0^{2\pi} \mathbf{v}'(r,\ \lambda,\ z)\ d\lambda, \qquad (3.5)$$

and $\mathbf{v}_a'$ is the deviation from $\mathbf{v}_0'(r,\ z)$. By definition $\mathbf{v}_0'(r,\ z)$ includes the axisymmetric mean tangential (primary

circulation) and radial (secondary circulation) components. In terms of a Fourier decomposition, $\mathbf{v}_0'$ corresponds to the wavenumber-0 component of $\mathbf{v}'$. The vector $\mathbf{v}_a'$ is the asymmetric portion of the perturbation wind; that is, wavenumbers 1 and higher.

Equation (3.2) can now be written as

$$\mathbf{v}_r(r,\ \lambda,\ z) = \underset{(i)}{\overline{\mathbf{v}_r}(z)} + \underset{(ii)}{\mathbf{v}_0'(r,\ z)} + \underset{(iii)}{\mathbf{v}_a'(r,\ \lambda,\ z)}, \qquad (3.6)$$

where (i) is the area-averaged storm-relative wind as a function of height, (ii) symmetric-vortex (wavenumber 0) circulation, and (iii) the asymmetric perturbation wind (wavenumbers 1 and higher).

2) STORM CENTER AND AREA-AVERAGED WIND PROFILE WITH HEIGHT

A key aspect of the wind partitioning is determining the center of the axisymmetric mean circulation at each altitude. The center is defined as the centroid of the inner core (radii <60 km) cyclonic vorticity maximum as defined by Marks et al. (1992). Their Fig. 10 showed that in Norbert the center varied by as much as 3 km with altitude, making an anticyclonic spiral with increasing altitude. Figure 3.3 shows the variation of the circulation center with altitude for Hurricanes Gilbert and Hugo. In the Gilbert case (Fig. 3.3a) the center varies 2–3 km with altitude tilting from southwest to northeast with increasing altitude. Below 7-km altitude the center made a cyclonic spiral within 1 km of the mean circulation center. From 7 to 12 km, the center was displaced up to 2 km from the mean center, making an anticyclonic spiral with increasing height. In the Hugo case (Fig. 3.3b) the magnitude of the deviation of the center with altitude was larger (~5 km), also tilting from southwest to northeast in an anticyclonic spiral with increasing altitude. The small magnitude of

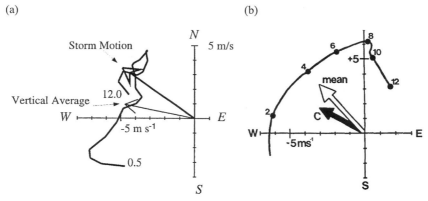

FIG. 3.4. Hodograph of mean winds from the three-dimensional Doppler analyses for Hurricanes (a) Gilbert on 14 Sep 1988 (Dodge et al. 1999) and (b) Hugo on 17 Sep 1989 (Roux and Mark 1996). Winds 12–25 km from the center were averaged for each level up to 12-km altitude. The altitude of points in the hodograph is labeled where the lowest level in 0.5 km and the highest level 12 km for both analyses. Arrows show the mean storm motion [(b) denoted as "C"] and the density-weighted mean wind (vertical average), respectively.

the deviation center with altitude for all three cases ($<$10 km), however, indicated that when viewed on the vortex scale ($\sim$100–300 km), the storm center would be nearly vertically aligned.

Figure 3.4 shows hodographs of $\overline{v}_r(z) + v_s$ computed using (3.3) for both storms. These hodographs provide an estimate of the shear of the flow in which the mean vortex is embedded. In each case the mean shear in the lower troposphere is southwest to northeast, with a steady veering of the wind with increasing altitude, from easterlies at altitudes below $\sim$5 km to southerlies or southwesterlies aloft. In both cases the vortex center tilted in the direction of the mean shear from 1- to 10-km altitude (cf. Figs. 3.3, 3.4). In the Gilbert case the magnitude of the shear from 1 to 10 km (5–6 m s^{-1}) is less than that for the Hugo case (9–10 m s^{-1}), suggesting that the tilt of the vortex with height is related to the magnitude, as well as the direction of the shear.

These two storms and the Norbert case had similar storm motions but distinctly different mean wind profiles. Marks et al. (1992; cf. Fig. 11) showed that Norbert had a mean wind profile that was aligned along the direction of motion of the vortex, with $\overline{v}_r(z) < v_s$ below 5-km altitude and $>v_s$ above that altitude. In contrast both the Hugo and Gilbert cases had mean wind profiles aligned across the track. In Hugo (Fig. 3.4b), the mean wind was closest to the storm motion at 3–4-km altitude, whereas, in Gilbert (Fig. 3.4a) the mean wind was closest to the storm motion at 7–8-km altitude, possibly indicating that Gilbert had a deeper vortex circulation. The magnitude of $\overline{v}_r(z) + v_s$ was never more than twice that of v_s. In Hugo, $\overline{v}_r(z) + v_s$ exceeded v_s at all altitudes $<$10-km altitude, whereas in Gilbert $\overline{v}_r(z) + v_s$ only exceeded v_s below 2- and above 9-km altitude.

All three cases illustrate the effect that vertical shear of the horizontal wind plays in vortex tilt. Each case indicates that the vortex tilts along the general direction of the shear vector with increasing height, with larger

tilts associated with larger shear magnitude. These cases also point out the lack of a clear relationship between storm motion and vertical wind shear. All three storms had storm motions within 40° of each other; however, they had drastically different shear patterns: from along-track shear in Norbert, to weak across-track shear in Gilbert, to relatively strong across-track shear in Hugo.

In many earlier studies, tropical cyclones were considered to move as a point vortex in a uniform, noninteracting fluid flow. That is, the storm moves with the speed and direction of a deep-layer mean environmental flow. Such a "steering" concept accounts for 30%–80% of the variability of the 24–72-h storm motion in the Atlantic (Neumann 1979). There is considerable uncertainty as to the atmospheric level or layer that determines the storm motion. Past studies by Neumann (1979) demonstrated the higher correlation of storm displacement with the flow at midtropospheric levels. However, the Doppler analyses showed the highest correlation with a mass-weighted, deep-layer mean flow.

Marks et al. (1992) showed that in Norbert the vertical, mass-weighted average of the mean horizontal wind throughout the depth of the analysis derived from the partitioning agreed with synoptic analyses, lying about 20° to the right of the storm motion. Figure 3.4 shows that this mass-weighted mean wind in Gilbert (Fig. 3.4a) was to the left of the storm motion, while in Hugo (Fig. 3.4b) the mean was to the right of the storm motion. Similar deviations of the mean storm motion were described in a number of studies (e.g., Neumann 1979; Dong and Neumann 1983; Dong 1986; Chan 1986). These studies indicate that the storms tracking toward the northwest (Norbert and Hugo) tend to move faster than, and 10°–20° to the left of, the mean flow, whereas storms moving more westerly (Gilbert) tend to move slower than, and to the right of, the mean flow.

(a)   (b)

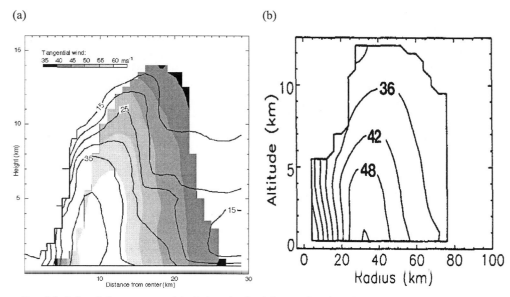

FIG. 3.5. Azimuthal mean tangential winds and reflectivity as a function of radius and height from the three-dimensional Doppler analyses for Hurricanes (a) Gilbert for the period 0902–1016 UTC 14 Sep 1988 (Dodge et al. 1999) and (b) Hugo for the period 2009–2055 UTC 17 Sep 1989 (Roux and Marks 1996). (a) Successively lighter shades of gray enclose higher wind speeds, starting at black for 35 m s^{-1}, in 5–m s^{-1} intervals. Solid lines are contours of reflectivity, in 5-dBZ steps starting at 15 dBZ. (b) The solid contours represent the tangential mean wind speed.

3) SYMMETRIC VORTEX

The tangential wind component of the symmetric vortex ($\mathbf{v}'_0(r, z)$), defined by (3.5), is illustrated in Fig. 3.5. In general, the tangential component of the symmetric vortex is a measure of storm intensity. Both profiles have characteristics common to those described in the literature (e.g., Jorgensen 1984a,b; Marks and Houze 1987; Marks et al. 1992). The tangential wind maximum is located close to the storm center in radius and at low altitude. It extends 8–12 km in radius and 5–10 km in altitude. The maximum slopes outward with increasing altitude.

The major differences between the tangential flow in the two storms are probably related to changes in the storm intensity. Gilbert is the more intense storm, with a tangential wind maximum exceeding 60 m s^{-1}, the 50–m s^{-1} contour reaching altitudes >11 km, and centered at 10-km radius (Fig. 3.5a). Hugo had mean tangential winds >50 m s^{-1} but a much shallower wind maximum, reaching just above 4-km altitude, centered at 30–35-km radius—almost twice that of Gilbert (Fig. 3.5b). As defined by the gradient wind equation and illustrated in these two profiles, a measure of storm intensity must include both the radius of maximum tangential wind (RMW) and the magnitude of the tangential wind maximum.

The radial wind component of the symmetric vortex ($\mathbf{v}'_0 (r, z)$), defined by (3.5), is illustrated in Fig. 3.6. As with the tangential wind, both profiles have characteristics common to those described in the literature (e.g., Jorgensen 1984a,b; Marks and Houze 1987; Marks et

al. 1992). They show weak symmetric radial flow inward or outward throughout most of the troposphere, except for a layer of outflow 4–8 m s^{-1} just outside the RMW above 8-km altitude. There is little evidence of a symmetric radial inflow in the lower troposphere primarily because of the lack of winds below 1-km altitude in both analyses. Hence, the vast majority of the inflow must be below that altitude (cf. discussion in section 2e). In contrast, the outflow layer near the RMW is rather deep, extending from 8- to 14-km altitude. In both cases the midtroposphere is characterized by weak inflow, where Gilbert (Fig. 3.6a) has weaker inflow than Hugo (Fig. 3.6b). The increased inflow in the Hugo case is likely related to the strengthening of the Hugo vortex at this time, whereas Gilbert was at peak intensity and slowly filling.

4) ASYMMETRIC STRUCTURE

Marks et al. (1992) showed that in Norbert the asymmetric perturbation field had a distinct signature at both lower and upper levels. At 1 km, the asymmetric perturbation had the form of a source–sink field. The source and sink were located at the RMW with the convergent sink on the leading, inflow side of the vortex and the divergent source on the trailing, outflow side of the vortex. At 3 km and higher, the asymmetric wind fields had the character of a vortex couplet with cyclonic and anticyclonic perturbations in the right-front and left-rear quadrants of the storm, respectively, and stagnation

(a)

(b)

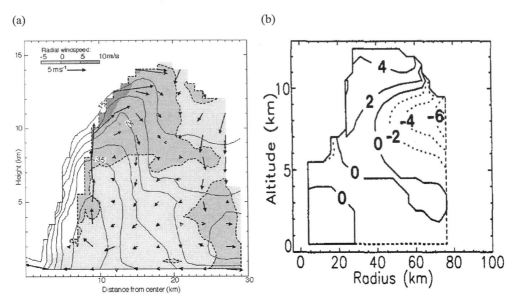

FIG. 3.6. Radius–height mean of the radial flow from the three-dimensional Doppler analyses for Hurricanes (a) Gilbert for the period 0902–1016 UTC 14 Sep 1988 (Dodge et al. 1999) and (b) Hugo for the period 2009–2055 UTC 17 Sep 1989 (Roux and Marks 1996). (a) Reflectivity (dBZ) is contoured as in Fig. 3.5, and the gray shades show the radial wind speed derived from the azimuthally averaged three-dimensional Doppler analysis. Black arrows map the secondary circulation; the vertical component was magnified two times to match the vertical scaling. The outflow regions are enclosed by a dashed line. (b) Contours represent the radial flow with solid contours corresponding to outflow and dashed contours inflow.

points at the RMW on the upwind-trailing and downwind-leading side of the storm.

Figure 3.7 shows the $\mathbf{v}'_a$ field at 3-km altitude for both storms. As in Norbert, the $\mathbf{v}'_a$ flow above 2 km is characterized by a wavenumber-1 vortex couplet aligned along the axis of the shear vector between 1- and 10-km altitude (Fig. 3.4). The vortex couplet is composed of a cyclonic perturbation in the downshear (right front) quadrant and an anticyclonic perturbation in the upshear (left rear) quadrant. The vortex couplet is centered on the radius of maximum wind. As in the Norbert analysis, this couplet extends vertically from 2 km to the top of the vortex, remaining centered on the radius of maximum wind as it slopes outward with increasing altitude.

(a)

(b)

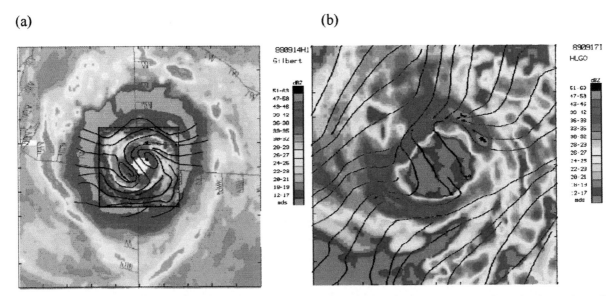

FIG. 3.7. The asymmetric perturbation wind (wavenumbers 1 and larger) at 3-km altitude from the three-dimensional Doppler analyses for Hurricanes (a) Gilbert for the period 0902–1016 UTC 14 Sep 1988 and (b) Hugo for the period 1746–1947 UTC 17 Sep 1989. Reflectivity (dBZ) is contoured as shown, and the streamlines map the perturbation circulation. From Marks (1991).

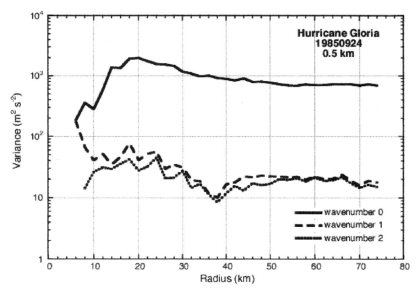

FIG. 3.8. Total variance at 0.5-km altitude for wavenumbers 0, 1, and 2 from the three-dimensional Doppler analysis from 0015 to 0032 UTC (east–west leg) in Hurricane Gloria, 24 Sep 1985. Adapted from Lee et al. (1994).

As in Norbert, the asymmetric perturbation below 2 km had the form of a source–sink field aligned along the direction of motion, similar to that described by Shapiro (1982).

As found by Marks et al. (1992), the perturbations observed in Fig. 3.7 are much smaller in horizontal scale than those discussed by Willoughby (1988) and Fiorino and Elsberry (1989). The small-scale perturbations are contained entirely within the inner core region of the storm, centered at the RMW, and likely result from a nonlinear interaction between the mean vortex and the storm motion. Although the horizontal flow varied with altitude, the perturbations were nearly aligned in the vertical above the boundary layer, suggesting that the interaction extended throughout the depth of the vortex.

5) AXISYMMETRY VERSUS ASYMMETRY

The relative importance of each wavenumber is illustrated in Fig. 3.8 where the explained variance by wavenumbers 0, 1, and 2 is plotted at 0.5-km altitude in Hurricane Gloria (1985) from Lee et al. (1994). A larger explained variance represents a larger amplitude in that wavenumber. It clearly shows that wavenumber-0 (axisymmetric mean) dominates the explained variance, representing >90% of the variance at all radii. It also shows that wavenumbers 1 and 2 dominate inside the radius of maximum wind ($\leq$16 km). Hence, while the large explained variance of wavenumber 0 at radii >10 km suggests that the storm is relatively axisymmetric outside the radius of maximum wind, the major asymmetries (wavenumbers 1 and 2) become important at and inside the radius of maximum wind. It should be noted that the explained variance could be a result of the tangential or radial components in this case.

The relative importance of the wavenumbers-0, -1, and -2 tangential flow, and the wavenumbers-0 and -1 radial flow in Hurricane Hugo is shown in Fig. 3.9 (Roux and Marks 1996). The amplitude of the wavenumber-0 tangential wind, VT-0 (Fig. 3.9a), is the largest over the whole domain, with maximum values >56 m s^{-1} at 30–40-km radii below 4-km altitude, showing that the storm is dominated by a strong symmetric circulation. The symmetric radial wind, VR-0 (Fig. 3.9b), is much smaller, especially below 5-km altitude, where it is <2 m s^{-1}. Wavenumber 1 for the tangential wind, VT-1 (Fig. 3.9c), displays features similar to VT-0 with a maximum at the radius of VT-0, but with values a factor of 7 smaller (<10 m s^{-1}), as in the Gloria case. In contrast, the features associated with wavenumber 1 for the radial wind, VR-1 (Fig. 3.9d), are very different from those associated with VR-0 and comparable in magnitude to VT-1. The maximum values (>6 m s^{-1}) are larger and found between 6- and 10-km altitude, where VR-0 is small, indicating that the secondary circulation is more complex than the primary one. The amplitude of wavenumber 2 for the tangential flow, VT-2 (Fig. 3.9e), is everywhere smaller, with maximum values >3 m s^{-1} at radii inside the maximum VT-0 (<25 km) below 5-km altitude, and at 60–80-km radii and 6–8-km altitude. A relative minimum of VT-2 is found at the location of the VT-0 and VT-1 maxima.

Recognition of the relative importance of the wavenumber 0 to the mean horizontal wind and higher wavenumber components of the vortex wind field led to changes in perspective for numerical and dynamical studies and also led to some new insights into vortex evolution (e.g., Franklin et al. 1993; Liu et al. 1997, 1999; Viltard and Roux 1998; Shapiro and Montgomery

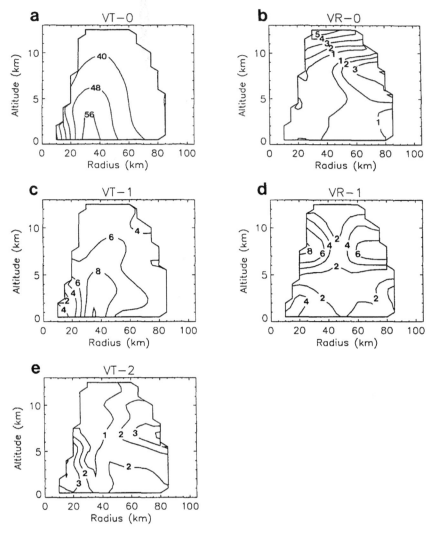

FIG. 3.9. Radius–height plot of the mean amplitude of the wind harmonics deduced from the three-dimensional Doppler analysis for Hurricane Hugo on 17 Sep 1989: (a) wavenumber 0 for the tangential wind (VT-0); (b) wavenumber 0 for the radial wind (VR-0); (c) wavenumber 1 for the tangential wind (VT-1); (d) wavenumber 1 for the radial wind (VR-1); and (e) wavenumber 2 for the tangential wind (VT-2). All velocity components are expressed in m s^{-1}. From Roux and Marks (1996).

1993). Smith and Montgomery (1995) used these concepts to introduce vortex "axisymmetrization" into discussions on tropical cyclone dynamics. They pointed out that the observed angular velocity (V/r) changes from the core to edge imply that the vortex timescale, defined as r/V, goes from 6 to 7 min (20 km/50 m s^{-1}) near the center to 5–6 h (500 km/25 m s^{-1}) in the periphery. This change in the rotational timescale results in increasing axisymmetrization as the flow reaches the core forcing horizontal variability into radial direction. This dynamic control on the various fields has broad impacts on such diverse activities as vortex initialization in numerical models to understanding the rain distribution in the storm.

These studies also pointed out that while the axisymmetric mean explains the majority of the horizontal wind variance, intensity and track changes are dependent on the scale of the asymmetries, which are much weaker and harder to measure. The evolution of the vortex asymmetries are now recognized as the key to understanding the vortex evolution and its interaction with its environment (e.g., Schubert et al. 1999; Moller and Montgomery 1999; Reasor et al. 2000). Before advances can be made in tropical cyclone intensity change, a better understanding of the temporal variability of the asymmetries is needed.

6) TEMPORAL VARIATIONS OF THE WIND COMPONENTS

The asymmetric dynamics of the tropical cyclone inner core region was examined through a novel analysis of high temporal resolution, three-dimensional wind fields in Hurricane Olivia (1994) that were derived from true dual-Doppler radar analyses of airborne dual-Doppler radar data (Reasor et al. 2000). Roux and Viltard (1995) had available a similar set of data for Hurricane Claudette (1991) spanning a period of about 7 h. In the Olivia case two NOAA WP-3D aircraft equipped with Doppler radar flew simultaneous, near-orthogonal flight tracks through the inner core of Hurricane Olivia on 25 September 1994. The use of two radar platforms allowed for true dual-Doppler sampling to be employed. The upper and lower aircraft flew orthogonal legs through the inner core (radii <100 km), providing near-simultaneous measurements of orthogonal components of the horizontal wind over a period of 10–15 min. Seven consecutive composites of Olivia's wind field were made during the period 2027–2355 UTC with 30-min time resolution to depict a weakening storm undergoing substantial structural changes. Dual-Doppler coverage was available out to a radius of 30 km from the storm center. In order to focus on the symmetric and asymmetric components of the vortex winds separately, and to obtain insight into the dynamics governing the evolution of these components of the total flow, an azimuthal Fourier decomposition of the wind field following Marks et al. (1992) was performed.

The observations provide a unique look at the role of vertical shear in producing structural changes to the inner core wind field. The seven consecutive wind analyses capture for the first time the asymmetric response of a tropical cyclone to dramatic changes in vertical shear over a relatively short time period. Figure 3.10 shows hodographs of area-averaged storm-relative wind ($\overline{v}_r(z) + v_s$) during the observation period. Initially the maximum local vertical shear over the 0.75–10.5-km depth is weak west-northwesterly, with values on the order of 3–5 m s^{-1}. Consistent with the weak shear, the vortex was nearly vertically aligned (cf. Fig. 9 in Reasor et al. 2000). Over the next 2.5 h the maximum shear increased from the west to 15 m s^{-1}. A west-to-east tilt of the vortex with height evolved over the same time, with a maximum displacement from low to middle levels of about 3 km. Inspection of the flow field from 6- to 10.5-km height, indicated that the inner core did not tilt more than 5 km.

Figure 3.11 from Black et al. (2002) shows that the eyewall reflectivity became increasingly asymmetric as the shear strengthened. The high reflectivity region was to the left of the shear, on the northern side of the eyewall (Fig. 3.11b). During the ~2.5 h between the first image (2021–2058 UTC) and the last 2334–2413 UTC), the outer rings of high reflectivity largely disappeared, so that convection in the northern eyewall was the dominant reflectivity feature of the storm. As the outer convection weakened, the outer wind maxima disappeared. By 2300 UTC, vigorous convective cells in the northern eyewall grew to large horizontal extent and exhibited reflectivities >50 dBZ. Black et al. (2002) showed that

at flight level and in vertical incidence Doppler radar data, an 8 m s^{-1} updraft occupied the downshear (eastern) side of the eyewall and a 5–m s^{-1} downdraft occupied the upshear (western) side. As the relative flow caused by increasing shear penetrated into the core, the earlier pattern of asymmetric radial flow became more pronounced. At 3-km altitude the radial inflow predominated on the downshear side and outflow on the upshear side, as the storm moved eastward relative to the surrounding low-level air.

Figure 3.12 shows the symmetric structure of Hurricane Olivia's primary circulation at the beginning and end of the observation period. At the RMW (~12–16 km) the tangential winds decrease by 5–10 m s^{-1} just above the boundary layer and by a more substantial 10–20 m s^{-1} above 6-km altitude. A decrease in tangential winds with time is observed near the RMW. The symmetric radial flow, also shown in Fig. 3.12, was variable over the observation period but was predominantly outward. The largest change in the symmetric radial flow between the two analyses was a decrease in altitude of the outflow near the RMW.

Reasor et al. (2000) examined the observed weakening trend in the primary circulation by using axisymmetric vortex spindown ideas. The departure of the flow from cyclostrophic balance in the vortex boundary layer due to the presence of friction drives a radial inflow. This radial inflow transports angular momentum into the inner core, compensating for the frictional losses of angular momentum. In the absence of diabatic forcing and the attendant radial inflow above the boundary layer, the free atmosphere radial flow will be outward at all levels (Willoughby 1979). By conservation of angular momentum the tangential winds above the boundary layer must decrease, and the vortex spins down. Although inflow is still observed up to 3.5-km height in Fig. 3.12b, the flow above this level is outward at all radii and the time average symmetric radial flow over the 3.5-h observation period was outflow. This prevalence of radial outflow was also observed by Marks et al. (1992) for weakening Hurricane Norbert, as opposed to the weak inflow seen in intensifying or steady-state storms such as Hurricanes Hugo and Gilbert (Fig. 3. 6).

The evolution of Olivia's vorticity derived from the dual-Doppler winds was also consistent with the vortex weakening. However, the vortex spindown may only be a partial explanation for the observed evolution. Such changes to the vorticity profile can also occur through asymmetric mechanisms. For example, the interaction of convectively forced vortex Rossby waves with the mean flow will lead to changes in the mean vorticity profile (e.g., Montgomery and Kallenbach 1997; Montgomery and Enagonio 1998; Moller and Montgomery 1999). Nonlinear mixing of vorticity through the barotropic instability mechanism will also erode sharp radial gradients of symmetric vorticity.

The role of asymmetric vorticity dynamics in explaining some of the physics of tropical cyclone inten-

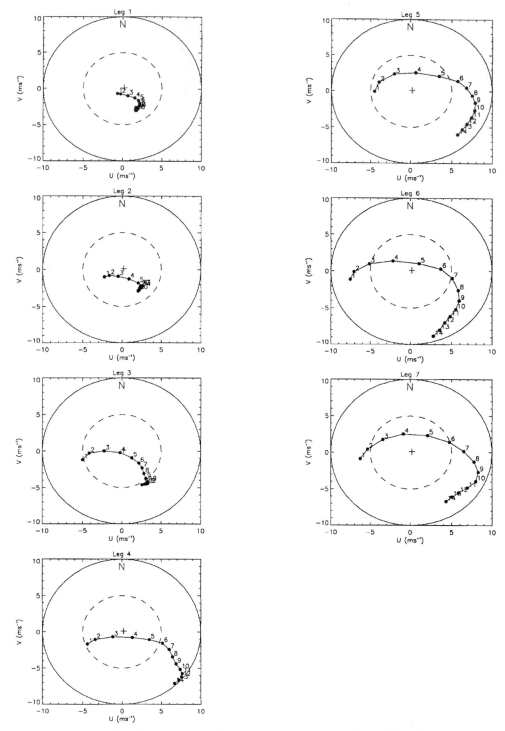

FIG. 3.10. Hodographs of the area-averaged storm-relative wind from 0.75- to 10.5-km height for each flight leg (seven legs, each roughly 30 min apart). From Reasor et al. (2000).

sity change motivated a special focus on Olivia's vorticity structure. Understanding the influence of vortex-scale dynamics on the distribution of convection and vorticity in the tropical cyclone inner core should aid in the prediction of the structure and evolution of a tropical cyclone as it enters or is embedded in different

environmental flows. Figure 3.13a shows that the vorticity takes the form of a ring of high values (12×10^{-3} s^{-1}) centered on the RMW (9.5-km radius) dropping off sharply at larger radii, with a hint that it drops below 8×10^{-3} s^{-1} at smaller radii. This vorticity structure resembles that presented in Schubert et al. (1999) for a

(a)   (b)

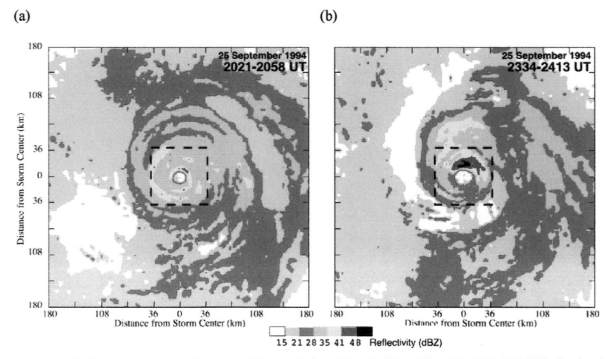

FIG. 3.11. Lower fuselage radar composite of Hurricane Olivia on 25 Sep 1994 at (a) 2021–2058 and (b) 2334–2413 UTC. The domain is 180 km × 180 km square centered on the storm. From Black et al. (2002).

hurricane-like vorticity ring. They predicted that some of the high vorticity of the ring is ultimately mixed into the center of the tropical cyclone vortex, forming a monotonic symmetric vorticity profile. High vorticity is also ejected outward in the form of vortex Rossby waves (e.g., Montgomery and Kallenbach 1997).

A clear relationship exists between the mesoscale asymmetry in convection, likely forced by the vertical shear of the environmental winds, and the asymmetric pattern of reflectivity (Fig. 3.10) and vorticity (Fig. 3.13). The largest values of reflectivity occur immediately downwind of the regions of enhanced convection, consistent with hydrometeors being carried up by the updrafts of the convective cells and simultaneously swept downwind by the much stronger primary circulation. Vertical shear increased dramatically during the observation period, leading to a strong projection of the convection onto an azimuthal wavenumber-1 pattern oriented along the maximum vertical shear vector. These observations are consistent with recent numerical simulations of hurricane-like vortices in vertical shear (Jones 1995; DeMaria 1996; Bender 1997; Frank and Ritchie 1999).

The perturbation vorticity at 6-km height in Fig. 3.13b that shows enhanced convection is also closely associated with wavenumber-1 vorticity features. The vertical velocity maximum is upwind of the wavenumber-1 vorticity perturbation. According to the mechanism first proposed by Jones (1995) for vertical velocity production in a vortex interacting with shear, the preferred location for enhanced convection due to vertical shear

effects should be downshear right in the east-to-southeast quadrant of the storm. Olivia was moving to the north-northeast at about 5 m s^{-1} during this time, so enhanced asymmetric boundary layer convergence is also expected in the north-northeast quadrant (Shapiro 1982). Figure 3.13b shows that the upward velocity is generally maximum downshear. As the vertical shear and vortex tilt increased, the maximum upward velocity increased from 1–2 to 12–16 m s^{-1} and the pattern of convection became more asymmetric. The storm motion during the analysis period is relatively steady and the direction of motion changes gradually by only 10°; hence, it is unlikely that the dramatic changes in vertical motion could be attributed to the asymmetric boundary layer convergence described by Shapiro (1982).

c. Rainbands, outer circulation, and severe weather

1) RAINBANDS

Early radar observations first pointed out that tropical cyclone rainbands have both stratiform and convective structure (e.g., Atlas et al. 1963). Rainbands have fewer vertically oriented cores of reflectivity and fewer organized updrafts than an eyewall. Radially outward from the eyewall, the rainbands are characterized by extensive horizontally homogeneous reflectivity patterns with bright bands just below the melting level (4.5–5.0-km altitude; cf. Fig. 3.2). Estimates showed that the stratiform precipitation in the rainbands covered areas 10 times larger than the convective precipitation.

(a)

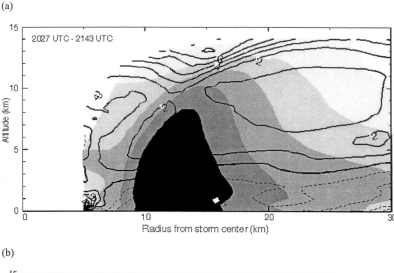

(b)

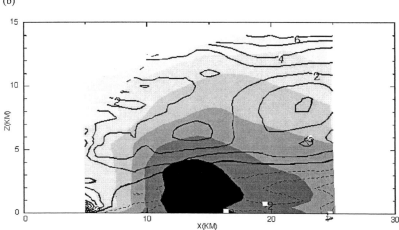

FIG. 3.12. Radius–height mean tangential and radial winds (m s^{-1}) winds from the three-dimensional Doppler analyses for Hurricanes Olivia on 25 Sep 1994 for the period (a) 2054–2117 UTC (leg 2) and (b) 2338–2359 UTC (leg 6). Tangential wind is depicted as successively darker shades of gray in 5-m s^{-1} intervals starting at (a) 35 m s^{-1} and (b) 30 m s^{-1}. Contours depict radial wind, in 1–m s^{-1} intervals, where solid contours represent positive values and dashed contours negative values.

Rainband structure appears to be more three-dimensional than that in the eyewall. Flight-level observations of rainbands show considerable variability between successive penetrations at the same level. In addition there is considerable variability in the structure of different rainbands. Thus, characteristic features of rainbands are much more difficult to specify than are those of the eyewall. Since 1982, aircraft studies of tropical cyclone rainbands have used airborne radar or airborne Doppler radar (e.g., Barnes et al. 1983; Marks and Houze 1984; Barnes and Stossmeister 1986; Barnes et al. 1991; Ryan et al. 1992; Barnes and Powell 1995; Powell 1990a,b). The Japanese are also making dual-Doppler observations of typhoon rainbands as they make landfall (e.g., Ishihara et al. 1986; Tabata et al. 1992; Tatehira et al. 1998; Suzuki et al. 2000). Single-Doppler radar and profiler observations of tropical cyclone rainbands are also being made in Australia (e.g., May et al. 1994; May

1996). A consensus on the evolving rainband structure is emerging from these studies.

Most of the rainband studies have examined the structure of features within the principal band. The observations generally show a cross-band circulation consisting of a mesoscale updraft sloping outward with height, with inflow at low levels and outflow at high levels (Fig. 3.14a). This basic cross-band flow is modified by the presence of convective features that move azimuthally along the rainband. These efforts pointed out that the upwind end of the principal band tends to be mainly convective and there is a transition to less convective and more stratiform precipitation toward the downwind end, in a manner similar to that first described by Atlas et al. (1963).

Airborne Doppler observations were used to show that the cross-band flow changed from the upwind to the downwind end of the principal band in Hurricane

(a)   (b)

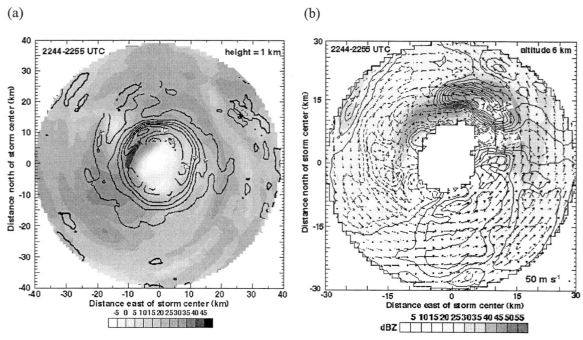

FIG. 3.13. Tail radar reflectivity composite (shading, dBZ) and (a) total vorticity (contours, 10^{-3} s^{-1}) at 1-km altitude and (b) wavenumber-1 vorticity perturbation (shading, 10^{-3} s^{-1}) and vertical velocity (contours, m s^{-1}) at 6-km altitude for Hurricane Olivia on 25 Sep 1994 at 2244–2255 UTC (leg 5). Solid contours represent positive w values and dashed contours negative w values.

Debby of 1982 (Powell 1990a). Upwind there was little cross-band flow at low levels, while the inflow from the outside of the band increased on the downwind end of the band. He also noted an increase in convergence into the band on the downwind end. However, the structure of a secondary band in Hurricane Irene of 1981 was different from that of the principal bands in other tropical cyclones (Barnes and Stossmeister 1986). There was no consistent cross-band circulation and the evolution of the band was strongly affected by the eyewall circulation. Their analysis indicated that downward motion from the eyewall was a likely cause for the rainband dissipation.

Samsury and Zipser (1995) put together a comprehensive dataset characterizing the composite wind and reflectivity for rainbands in 20 hurricanes. The database included 787 radial legs of storm-relative flight-level data in tropical cyclones of varying intensity. Their study was one of the first to analyze the kinematic structure associated with substantial secondary peaks of the horizontal wind. They identified 173 mesoscale secondary horizontal wind maxima and investigated their composite kinematic structure. The mesoscale fields of radial velocity, the convective-scale vertical velocity, and mass transport in these secondary wind maxima were similar to those found in the hurricane eyewall (e.g., Jorgensen 1984a,b; Shea and Gray 1973). As with the hurricane eyewall, there was radial convergence near the radius of the secondary wind maxima; the preferred location of updrafts was just inside this radius. These results used a large dataset to support the case study

findings of Barnes et al. (1983), Barnes and Powell (1995), May (1996), and Powell (1990a,b) related to rainband kinematic structure. For both the eyewalls and secondary wind maxima, these composite results show that the strong radial inflow is largely confined to the lowest 1 km.

In addition to the kinematic structure, they studied the relationship between secondary wind maxima and hurricane rainbands. In the radial legs with secondary wind maxima a mesoscale reflectivity feature (rainband) was identified within 20 km of the wind maximum. In contrast, over 70% of the rainbands that were found were not linked to any substantial maximum in the horizontal wind field. An important aspect of the relationship between secondary wind maxima and rainbands may be the convective or stratiform nature of the rainband. Results indicate that significant secondary wind maxima possess an inflow layer from both sides of wind maxima. It is hypothesized that this radial flow pattern would be more likely to occur in convective rainbands rather than in stratiform bands.

2) OUTER CIRCULATION

While the evolution of the inner core is dominated by interactions between the primary, secondary, and track-induced wavenumber-1 circulation, there is some indication that the local convective circulations in the rainbands may impact on intensity change (e.g., Barnes et al. 1983; Powell 1990a,b). Although precipitation in some bands is largely stratiform, condensation in most

a

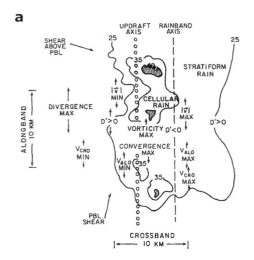

RAINBAND KINEMATIC STRUCTURE

b

RAINBAND KINEMATIC AND PRECIPITATION STRUCTURE

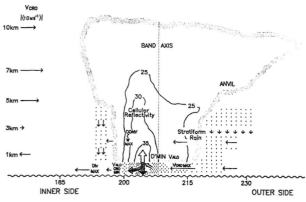

FIG. 3.14. Schematic model of rainband precipitation and kinematic structure. (a) Plan view, showing positions of relative maxima and minima of reflectivity and kinematic fields based on cross-band flight legs; inner side of rainband is to the left. (b) Rainband in cross-band height cross section. Outer solid line indicates the cloud boundary, contours represent radar reflectivity, horizontal arrows show cross-band component of the flow, and vertical arrows indicate convective updrafts and downdrafts. From Powell (1990a).

bands tends to be concentrated in convective cells rather than spread over wide mesoscale areas. As shown in Fig. 3.14, convective elements form, move through the bands, and dissipate as they move downwind. Doppler radar observations indicate that the roots of the updrafts lay in convergence between the low-level radial inflow and gust fronts that are produced by convective downdrafts (e.g., Barnes et al. 1991). This convergence may occur on either side of the band. A 20-K decrease in low-level θ_e was observed in a rainband downdraft and suggested that the draft acts as a barrier to inflow. Powell (1990b) pointed out that the reduction in boundary layer energy may be advected near the center, inhibit convection, and thereby alter storm intensity.

Moving bands, and other convective features, are frequently associated with cycloidal motion of the tropical cyclone center, and intense asymmetric outbursts of convection are observed to displace the tropical cyclone center by tens of kilometers (e.g., Muramatsu 1986). The bands observed by radar are often considered manifestations of internal gravity waves (e.g., Willoughby 1978; Gall et al. 1998), but these waves can exist only in a band of Doppler-shifted frequencies between the local inertia frequency (defined as the sum of the vertical component of the earth's inertial frequency f and the local angular velocity of the circulation, V/r) and the Brunt–Väisälä frequency.[5] Only two classes of trailing-spiral, gravity wave solutions lie within this frequency band: (i) waves with any tangential wavenumber that move faster than the swirling wind, and (ii) waves with tangential wavenumber ≥ 2 that move slower than the swirling wind.

Bands moving faster than the swirling wind with outward phase propagation are observed by radar. They are more like squall lines than linear gravity waves. Bands moving slower than the swirling wind should propagate wave energy and anticyclonic angular momentum inward, growing at the expense of the mean-flow kinetic energy, and reach appreciable amplitude if they are excited at the periphery of the tropical cyclone. Alternate explanations for these inward-propagating bands propagating slower than the wind involve filamentation of vorticity from the tropical cyclone environment, asymmetries in the radially shearing flow of the vortex, and high-order vortex Rossby waves (e.g., MacDonald 1968; Guinn and Schubert 1993; Montgomery and Kallenbach 1997). The vortex Rossby wave theory has so far provided a valuable tool to describe spiral rainbands mostly in idealized models and to investigate the interactions of these bands with the mean flow. Detailed observations of the vortex-scale rainband structure and wind field are necessary to determine which mechanisms play a role in rainband development and maintenance.

3) SEVERE WEATHER

There appear to be two regimes that are characterized by severe weather (i.e., tornadoes) and mesocyclones within a tropical cyclone. The first is just inside of the eyewall or incipient eyewall where the vorticity is a maximum, and the second regime is along the periphery of the storm within or outside the principal band where convective instability appears to be stronger. Marks and Houze (1984) first documented the development of mesocyclones associated with the wind maxima observed in rainbands of a developing tropical cyclone (Debby 1982). Stewart and Lyons (1996) investigated the radar structure of Typhoon Herb (1993) using the WSR-88D

---

[5] Natural gravity wave frequency, the square root of the static stability defined as $(g/\theta)\partial\theta/\partial z$.

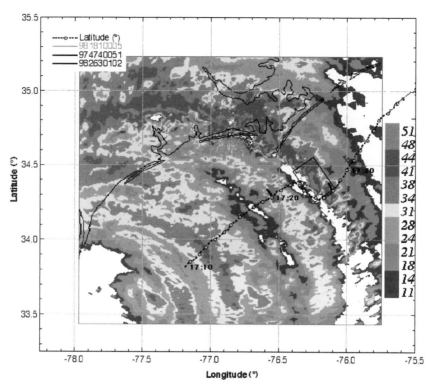

FIG. 3.15. Radar reflectivity from the Moorehead City, North Carolina, WSR-88D (KMHX) at 1720 UTC 26 Aug 1998. Color contours indicate reflectivity factor. The black dotted line indicates the aircraft flight track from 1710 to 1735 UTC. The thin solid line depicts the coastline of North Carolina and the Outer Banks. The black box denotes the region of the three-dimensional Doppler analysis in Fig. 3.17. The radar is located at 34.77°N, 76.87°W.

at Guam. They looked at a number of mesocyclones identified by the WSR-88D mesocyclone algorithm. They noted that three relatively long-lived (20 min) mesocyclones occurred near the rainband associated with the developing eyewall. They hypothesized, as did Marks and Houze (1984), that these mesocyclones were related to the intensification of the vortex.

Recent aircraft and radar observations have detected mesovortices along the inside edge of the eyewall of intensifying strong hurricanes (e.g., Marks and Black 1990; Black and Marks 1991; Willoughby and Black 1996). These inner regime mesovortices can be significant to the local warning meteorologist, producing intense localized damage streaks (e.g., Wakimoto and Black 1994). They are likely a cause of the local imbalances that occur during the vorticity mixing process described by Schubert et al. (1999). However, the majority of severe weather events (mesocyclones and subsequent tornadoes) in tropical cyclones are not confined to the region surrounding a developing eyewall but extend to the rainbands in the periphery of the storm.

Earlier studies of severe weather (mesocyclones and tornadoes) in tropical cyclones pointed out that the majority of these events typically occur in outer rainbands between 150 and 250 km from the center (e.g., Novlan and Gray 1974; Omoto 1982; McCaul 1987, 1991, 1993). More recently airborne and ground-based Dopp-

ler radar investigations of tropical cyclone rainband-embedded mesocyclones and tornadoes have advanced the understanding of severe weather life cycles (e.g., Spratt et al. 1997). An excellent example occurred while Hurricane Bonnie made landfall as a category-two hurricane along the North Carolina coast on 26 August 1998. Prior to landfall, two NOAA WP-3D aircraft conducted flights near the Morehead City, North Carolina, WSR-88D (radarstation KMHX). The aircraft were required to deviate around intense cells within a dominant outer rainband, 170–200 km northeast of the storm center (Fig. 3.15). The WSR-88D detected several mesocyclones in this rainband, one of which produced an F1 tornado around 1725 UTC as it crossed the coast near Beaufort, North Carolina.

WSR-88D data from KMHX revealed that many of the cells within the dominant band could be categorized at the lower bound of the tropical cyclone outer rainband mesocyclone spectrum identified by Sharp et al. (1997). The rain cells possessed reflectivity maxima of 50 dBZ extending to 3–4-km altitude. While the individual cells along the rainband were identifiable for well over 1 h, observed rotational couplets were much less persistent, lasting an average of only 15 min, and confined below 4 km.

Doppler radar data from the WSR-88D and airborne Doppler radar were used to derive the two-dimensional

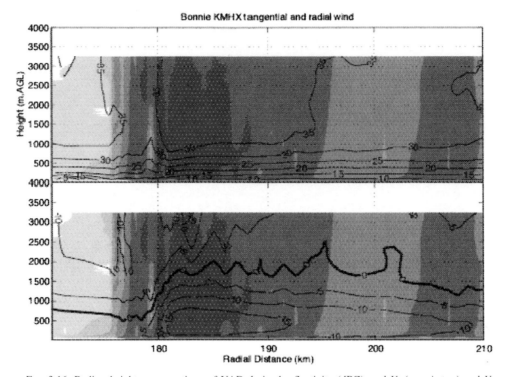

FIG. 3.16. Radius–height cross sections of VAD-derived reflectivity (dBZ), and V_θ (m s^{-1}, top) and V_r (m s^{-1}, bottom) from the KMHX WSR-88D when the radar was 170 km (1840 UTC) to 210 km (1500 UTC) northeast of Hurricane Bonnie on 26 Aug 1998. Reflectivity contours are color coded at 20, 25, 30, 35, 40, and 45 dBZ.

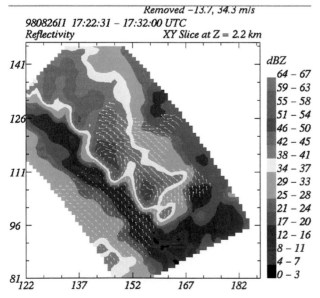

FIG. 3.17. Cell relative wind velocity (white arrows) at 1.5 km. Cell motion of 37 m s^{-1}, 330° ($u = -13.7$ m s^{-1}, $v = 34.3$ m s^{-1}), removed from total winds. Colors denote reflectivity factor from the NOAA WP-3D tail Doppler radar. Axes represent domain outlined in Figure 3.15. From Dodge et al. (2000).

structure of the rainband and the three-dimensional structure of the mesocyclones embedded within it. Figure 3.16 shows velocity-azimuth display–(VAD; Browning and Wexler 1968) derived tangential and radial wind components defined as by Marks et al. (1999) through the band as it crossed KMHX between 1500 and 1840 UTC (170–210 km radially from the storm center). VAD-derived horizontal winds were 35–40 m s^{-1} along the band, with sharp tangential gradients near 1×10^{-3} s^{-1} and strong convergence ($\sim 4 \times 10^{-3}$ s^{-1}) in the radial flow between 500 and 1000 m at the inside edge of the rainband.

Three analyses were done, for 1600, 1725, and 1805 UTC (Dodge et al. 2000). Figure 3.15 shows that the aircraft deviated around a strong cell in the rainband from 1720 to 1727 UTC before resuming the track northeast along the coast. Tracking reflectivity maxima at this time showed that the cell closest to the aircraft moved toward 330° at 37 m s^{-1}. When the cell motion is removed from the total wind velocities, mesoscale circulations are clearly revealed (Fig. 3.17) at the upwind ends of the reflectivity maxima. The vertical component of vorticity in these cells is $\sim 8 \times 10^{-3}$ s^{-1}, comparable to that seen in weaker mesocyclones in mid–latitude supercells.

Suzuki et al. (2000) documented the vertical structure of similar mesocyclones observed in a tropical cyclone 9019 making landfall in Japan on 19 September 1990.

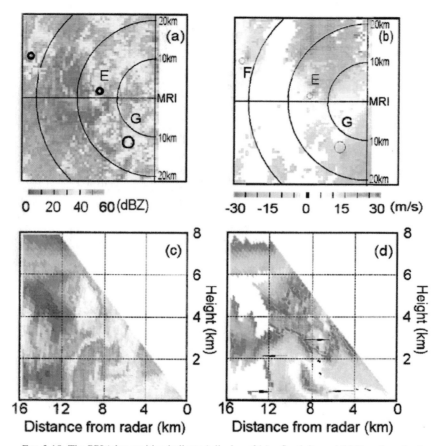

FIG. 3.18. The PPI (plan position indicator) display of (a) reflectivity and (b) Doppler velocity of storms E, F, and G with elevation angle of 2.2° at 2146 Japan Standard Time (JST). The RHI (range–height indicator) display of (c) reflectivity and (d) Doppler velocity of storm G with azimuth angle 250° at 2153 JST. The MCs are indicated by open circles in the PPI displays. (d) Arrows indicate wind directions relative the radar deduced from the Doppler velocity field. From Suzuki et al. (2000).

Three tornadoes formed within an outer rainband with characteristics similar to those shown in Hurricane Bonnie. Using single-Doppler radar observations they were able to capture the vertical structure of the mesocyclones that spawned the tornados (Fig. 3.18). Figure 3.18a shows that the reflectivity maxima associated with the mesocyclones were very similar to that observed in Hurricane Bonnie (cf. Fig. 3.17). Figure 3.18c shows that the tops of the intense reflectivity core were below 4-km altitude and a clear bounded weak echo region centered at 8 km from the radar and below 2-km altitude. Figure 3.18d shows strong radial velocity convergence within the weak echo region and strong radial velocity divergence just above the convergent region. This radial velocity structure appears to be a miniature version of that found in mid latitude supercells (e.g., Browning 1964). Similar shallow circulations were documented in previous studies of tropical cyclone rainbands (e.g., Spratt and Sharp 1999; Spratt et al. 1997; Saito 1992). Why these severe weather features are most prevalent in the core of developing storms and in the periphery of more intense storms remains a question.

d. Convective structure

Vertical velocities within a tropical cyclone are relatively weak compared to midlatitude convection. Jorgensen et al. (1985) found that only 2% of the total updrafts in a tropical cyclone are greater than 2 m s^{-1}. The strongest 10% of the updrafts have an average vertical velocity of 4 m s^{-1} with an average width of 3–4 km. Black et al. (1996) separated the vertical velocities derived from vertical incidence Doppler radar data (cf. Fig. 3.2b) in seven tropical cyclones along 120 radial legs into four regions: the eyewall, rainbands, stratiform, and other. Figure 3.19 shows the characteristics of the vertical velocity and radar structure as a function of altitude for the entire dataset and each of the four regions. In all of the regions, more than 70% of the vertical velocities range from −2 to 2 m s^{-1}. Averaged over the entire dataset, the mean vertical velocity is upward at all altitudes. Mean downward motion occurs only in the lower troposphere of the stratiform region. The eyewall showed the greatest variability, with updrafts reaching 20 m s^{-1}, yet even in the eyewall only 10% of the

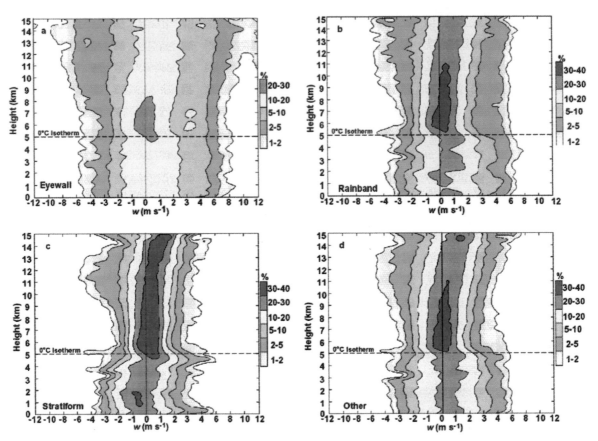

FIG. 3.19. Contoured frequency distributions of w (% per m s^{-1}) as a function of height for the (a) eyewall, (b) rainband, (c) stratiform, and (d) other regions. (a)–(d) The endpoints of the vertical velocity classes are labeled on the abscissa and the scale for the shaded contours is shown at the right. From Black et al. (1996).

drafts were >2 m s^{-1} through the depth of the storm. The remaining three regions show less variability and typically weaker vertical velocities.

In the lower and middle troposphere, the characteristics of the Doppler-derived vertical motions are similar to those described by Jorgensen et al. (1985), even though the horizontal resolution of the Doppler data is $\sim$750 m compared to $\sim$125 m from the in situ flight-level measurements. The Doppler data are available at higher altitudes than those reached by turboprop aircraft and provide information on vertical as well as horizontal variations. In a vertical plane along the radial flight tracks, Doppler updrafts and downdrafts were defined at each 300-m altitude interval as vertical velocities whose absolute values continuously exceed 1.5 m s^{-1}, with at least one speed having an absolute value greater than 3.0 m s^{-1}. Figure 3.20, from Black et al. (1996), shows that the properties of the Doppler drafts are log-normally distributed. In each of the regions, updrafts outnumber downdrafts by at least a factor of 2 and updrafts are wider and stronger than downdrafts. Updrafts in the eyewall slope radially outward with height and are significantly correlated over larger radial and vertical extends than in the other three regions. If the downwind (tangential) slope with height of updrafts varies little

among the regions, updrafts capable of transporting air with relatively large moist static energy from the boundary layer to the upper troposphere are primarily in the eyewall region. Downdrafts affect a smaller vertical and horizontal area than updraft and have no apparent radial slope.

Black et al. (1996) estimated the total upward or downward annular mass flux, defined as the flux produced by all of the upward or downward Doppler vertical velocities (drafts) within a 100-km radius (Fig. 3.21). The maximum upward mass flux in all but the "other" regions is near 1-km altitude, an indication that boundary layer convergence is efficient in producing upward motion. Above the sea surface, the downward mass flux decreases with altitude. At every altitude, the total net mass flux is upward, except for the lower troposphere in the stratiform region where it is downward. Figure 3.21 shows that the Doppler-derived updrafts and downdrafts, while a subset of the vertical velocity field that occupies small fractions of the total area, contribute a substantial fraction to the total mass flux. In the eyewall and rainband regions, for example, the Doppler updrafts cover less than 30% of the area but are responsible for $>$75% and $>$50% to the total upward mass flux, respectively. The Doppler downdrafts typically en-

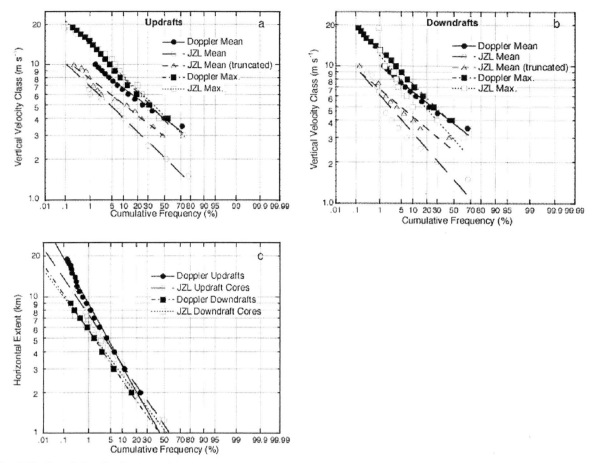

FIG. 3.20. Cumulative distributions of the mean and maximum strength of Doppler drafts and flight-level cores for (a) updrafts and (b) downdrafts. The truncated core distributions are calculated by removing the contribution of cores that are less than the threshold of the Doppler drafts ($|3|$ m s^{-1}). (c) Cumulative distribution of the horizontal extent of Doppler drafts and flight-level cores. The core data is from Fig. 1 of Jorgensen et al. (1985). The straight lines represent linear fits to the data. From Black et al. (1996).

compass less than 10% of the area yet provide ~50% of the total downward mass flux in the eyewall and ~20% of the total downward flux in the rainband, stratiform, and other regions.

While the Black et al. (1996) study focused on the vertical velocity characteristics over open water, a new vertically pointing X-band airborne Doppler radar system installed on the National Aeronautics and Space Administration (NASA) ER-2 flying at 20-km altitude can provide similar information over land (EDOP; Heymsfield et al. 1996). During August and September 1998 NASA joined efforts with NOAA's Hurricane Research Division in the third Convection and Moisture Experiment (CAMEX-3). During CAMEX-3 the ER-2 equipped with EDOP made a number of traverses of Hurricane Georges between 2000 and 2330 UTC on 22 September 1998 as the eye passed over the Cordillera Central on the island of Hispaniola, which was described by Geerts et al. (2000).

Figure 3.22 shows one of the three traverses as the ER-2 crossed the mountains (leg 3). Geerts et al. (2000) showed that a massive convective cell had mushroomed over the Cordillera Central, with a reflectivity of 35 dBZ as high as 12.6 km and updrafts above 19 m s^{-1}. As found by Black et al. (1994) in Hurricane Emily (1987), updrafts were surrounded by compensating downdrafts, suggesting that the rising bubbles were buoyancy driven. Orographic forcing was evident by the rising (descending) low-level motion over the mountain ridge as the terrain rises (sinks). On top of the massive convective cell are three towers, one of which reaches 16.2 km (close to the tropopause height), 2–3 km higher than the surrounding anvils. These EDOP observations complement those from the NOAA WP-3D Doppler radars, and they indicate that while the characteristics of the vertical velocity are similar in between storms over the open ocean and under orographic forcing, the probability of stronger updrafts and downdrafts occurring is much higher in the vicinity of the orographic features.

Black et al. (1996) also showed that precipitation is not well correlated with the strength and spatial distribution of the updrafts. High radar reflectivity cores cover only 10% of the total precipitation area of a tropical cyclone, with the heavy rain from each core occurring

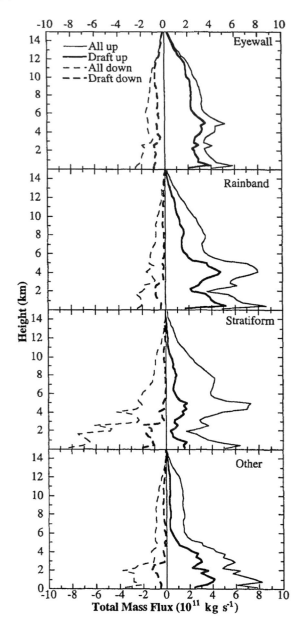

FIG. 3.21. Total upward (solid) and downward (dashed) mass flux normalized by the amount of missing data due to the range delay, as a function of altitude for each of the four regions. The thin lines are the total mass flux by all of the Doppler w estimates, and the solid lines represent the contribution of the drafts to the total mass flux. From Black et al. (1996).

on average over only a 50 km² area Parrish et al. (1982, 1984) showed that the average core was 3.7 km in radius and the lifetime of the cells was relatively short, with only 10% lasting longer than 8 min (roughly the time a 1-mm diameter raindrop takes to fall from the mean height of the 0°C isotherm at terminal velocity).

Figure 3.23 shows the autocorrelation of rainfall rate (R) and in situ 6-s mean vertical velocity ($\langle w \rangle$) from the microphysics probes in Hurricane Anita, described by Marks et al. (1993). The $\langle w \rangle$ and R values represent common 6-s intervals along the flight track (~900 m at 150 m s⁻¹ aircraft ground speed). The autocorrelations are computed for a given lag from each data point and the cross correlation of the two over the same lag. The auto correlation coefficients reduce to 0.5 within 12 s for $\langle w \rangle$ and 40 s for R, which indicates that the updrafts and rainshafts are only 3 and 9 km in width, respectively. The vertical velocities decorrelate approximately twice as fast as the rainfall rate, resulting in the rainshafts being approximately twice the size as the updrafts.

While there is in similarity in the timescales and space scales of R and $\langle w \rangle$, the cross correlation between the two indicate that they are not well correlated with each other likely due to substantial mixing of the precipitation around the vortex. Marks and Houze (1987) showed from azimuthal hydrometeor trajectories (Fig. 3.24) that hydrometeors lifted in the eyewall above the 0°C level can travel very far radially and azimuthally from the location of the updraft, even multiple times about the vortex before reaching the surface in the case of the ice particles. This type of behavior, characteristic of the stratiform rain process, leads to little correlation in time or space of the hydrometeors and the vertical motion. The only place where the spatial autocorrelation is significant is in the eyewall where larger precipitation particles, with greater fall speeds, move much shorter distances horizontally, arriving back on the ground in 10–15 min (Fig. 3.24b). Marks and Houze (1987) referred to this behavior as the "mix master," where the precipitation particles swirling around the upper levels of the vortex scavenge the cloud water and smaller particles as they slowly descend to the 0°C level. They suggested that the majority of the precipitation surrounding the eyewall, and likely the bulk of the clouds and ice in the central dense overcast, originated in the eyewall updrafts.

e. Atmospheric boundary layer structure

The advent of the operational WSR-88D Doppler radar network, and the construction of portable Doppler radars and profilers that can be moved to a location near tropical cyclone landfall, has also generated new and unique datasets that are enabling improved understanding of boundary layer wind structure as the storm moves from over the sea to the land. Wurman and Winslow (1998) analyzed portable Doppler radar data collected at the Wilmington, North Carolina, airport (~50 km upwind from KLTX WSR-88D radar). They showed that in the lowest 200 m the radial velocity data were characterized regions of maximum and minimum radial velocity aligned along the vertical shear vector that had dimensions <10 km parallel and <300 m perpendicular to the flow (Fig. 3.25). They stated that these linear features resembled atmospheric boundary layer (ABL) rolls and were characterized by large horizontal wind gradients on the order of 10^{-1} s⁻¹.

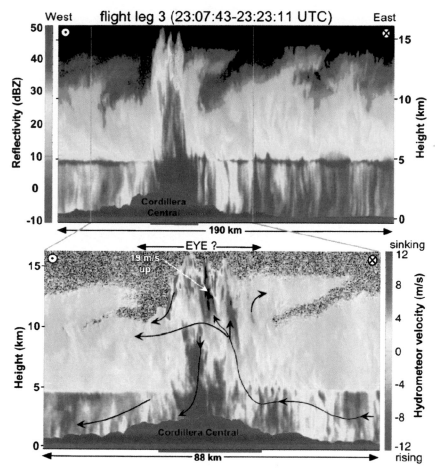

FIG. 3.22. EDOP nadir reflectivity (upper) and nadir Doppler velocity (lower) for an east–west section. The black lines with arrows are streamlines, estimated from EDOP's nadir and forward radial velocities. From Geerts et al. (2000).

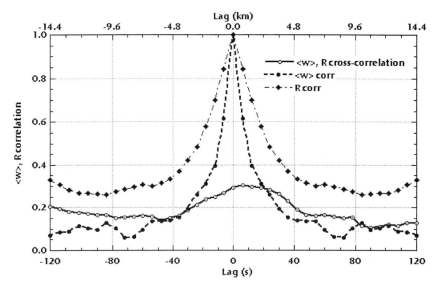

FIG. 3.23. Autocorrelation and cross-correlation of 6-s mean flight-level vertical velocity ($\langle w \rangle$), rainfall rate (R) at 3-km altitude in Hurricane Anita 1977.

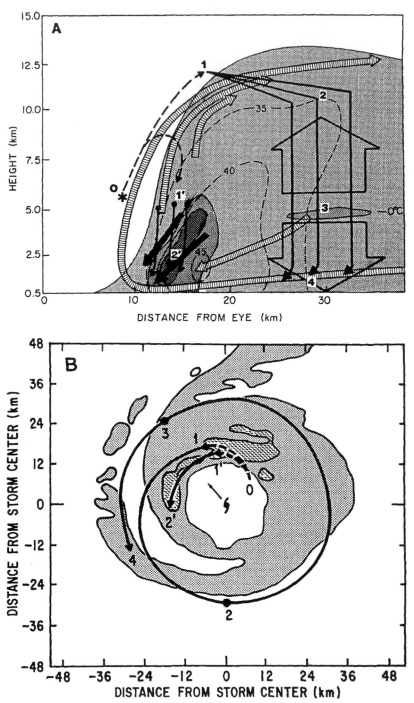

FIG. 3.24. (a) Schematic of the radius–height circulation of the inner core of Hurricane Alicia. Shading depicts the reflectivity field, with contours of 5, 30, and 35 dBZ. The primary circulation (azimuthal, m s^{-1}) is depicted by dashed lines and the secondary circulation by the wide hatched streamlines. The convective downdrafts are denoted by the thick solid arrows, while the mesoscale updrafts and downdrafts are shown by the broad arrows. The level of the 0°C isotherm is labeled. (b) A schematic plan view of the low-level reflectivity field in the inner core of Hurricane Alicia superimposed with the middle of the three hydrometeor trajectories in (a). (b) The reflectivity contours are 20 and 35 dBZ. Note that the storm center and direction are also shown. (a),(b) The hydro-meteor trajectories are denoted by dashed and solid lines labeled 0-1-2-3-4 and 0'-1'-2'. From Marks and Houze (1987).

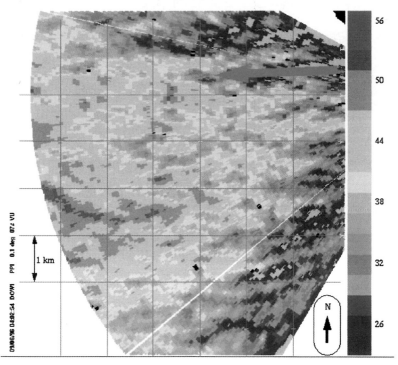

FIG. 3.25. High-resolution image of Doppler velocity at the 2° tilt angle to the west of Wilmington. Subkilometer-scale streaks modulate the easterly flow. Solid blue arrow represents the mean wind direction over the domain, while the blue grid has a resolution of 1 km. The radar origin is off the image to the right. All velocities are positive (receding from the radar) denoting the easterly wind component. Maximum and minimum wind speed values alternate from 33 to 50 m s^{-1} over distances less than 300 m near the center of the map. From Wurman and Winslow (1998).

WSR-88D radial velocity data from KLTX also indicate the presence of these linear ABL features below 500-m altitude near the coast and inland, suggesting that they are ubiquitous in the wind field of landfalling hurricanes. Morrison et al. (2002) developed a technique to locate the ABL features with operational WSR-88D radar and determine the characteristics of these features. The ABL features are identified in the WSR-88D data by subtracting a VAD-derived mean horizontal wind from the Doppler radial velocity display (Fig. 3.26a), yielding a "residual" radial velocity display (Fig. 3.26b). Using only elevation angles between 0.5° and 5.5°, they estimated the wavelength, length, depth, magnitude, and motion of these ABL features from the residual velocity display.

Estimates of the length, depth, and vertical extent of each feature are dependent on elevation angle, with higher elevation angles better at estimating the depth than the length of the ABL feature, and vice versa. The wavelength is given as a "half"-wavelength of each feature since the measurement is made from the highest positive to the lowest negative residual radial velocity. The intensity is the magnitude of the maximum or minimum residual radial velocity observed in each feature. Hence, the variation of the wind from a minimum to a maximum is twice the intensity estimate. The motion

of the features is determined by the change in location of the feature between each elevation angle, or a over a 30-s interval.

They examined these 172 ABL features in three storms: Fran (1996), Bonnie (1998), and Georges (1998) using the WSR-88D's from Wilmington, North Carolina (KLTX); Morehead City, North Carolina (KMHX); and Key West, Florida (KBYX), respectively. Their analysis focused on the period between the first identified ABL features in the residual velocity display and tropical cyclone landfall. All the features were identified between 25 and 120 km from the storm center.

Figure 3.27 shows the cumulative distribution of the characteristics for the 172 ABL features. The half-wavelength varied from 350 to 1100 m, with a median of 657 m, suggesting that the features were separated by 1.4 km. The length estimates ranged from 1 to 3 km, with a median of 1608 m. The depth estimates varied from 100 to 800 m, with a mean of 290 m. All of the features were observed to be centered below 1-km altitude. Hence, the depth estimates suggest that the ABL features are centered near 400–500 m above the surface.

Figure 3.27b shows estimates of the motion of the features varied from 23 to 50 m s^{-1}, with a median of 34.8 m s^{-1}. Comparison of the motion of the features to the VAD-derived mean wind profile showed that they

(a)

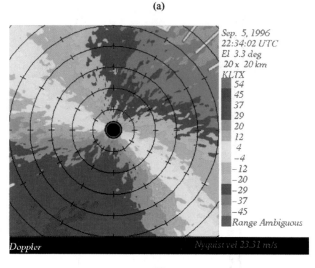

(b)

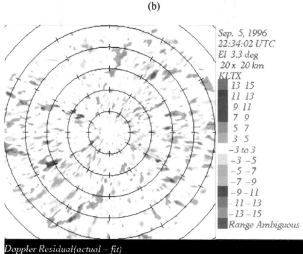

FIG. 3.26. (a) Plan view of the Doppler velocity and (b) perturbation Doppler velocity at the 3.3° tilt angle from the Wilmington, North Carolina, WSR-88D (KLTX) in Hurricane Fran at 2234 UTC 5 Sep 1996. Subkilometer-scale streaks modulate the flow. The radar origin is at the center of both images. Velocities are denoted by the color scales. Range rings are at 2-km intervals.

were moving close to the mean wind between 200- and 800-m altitude. The strength of the residual radial velocity or intensity of the feature ranged from 5 to 10 m s^{-1}, with a median of 7.3 m s^{-1}. A median feature separation of 1.4 km from maximum to minimum and a median velocity difference of twice the intensity (14 m s^{-1}) yields a median horizontal wind shear of 1×10^{-2} s^{-1}. Peak shear values were about half those observed by Wurman and Winslow (1998). These horizontal shear values are comparable to those observed in the mesocyclones and weak tornadoes observed in the rainbands and may explain some of the damage streaks found after landfall (e.g., Wakimoto and Black 1994). Katsaros et al. (2000) reported seeing similar linear streaks in tropical cyclones with synthetic aper-

ture radar (SAR) imagery from the Canadian Radar Satellite (RADARSAT). The streaks were spaced roughly 2 km apart and extended over a large radial extent around the storm.

These features may also explain recent GPS dropsonde observations of low-level jets in the eyewall ABL. Franklin et al. (1999), Powell et al. (1999), and Black and Franklin (2000) described jets in the vertical profile of the horizontal wind at altitudes <1 km in Hurricanes Guillermo (1997), Erika (1997), Georges (1998), and Mitch (1998) (cf. Fig. 3.28b). These eyewall soundings indicate numerous low-level wind maxima between 200 and 800 m. These dropsonde wind observations are consistent with Australian tower observations in the inner, high-wind core of tropical cyclones first reported by Wilson (1979) and recently discussed by Kepert and Holland (1997). The magnitude of these jets and the altitudes at which they are observed suggest that they may also be manifestations of the features in the WSR-88D and Doppler on Wheels (DOW) observations.

The presence of these ABL features near the KLTX WSR-88D may also partially explain a layer of nearly constant wind direction below 400-m altitude. Figure 3.28a shows the VAD-derived wind direction and speed profiles for KLTX at four times as Hurricane Fran passed over the radar on 5 September 1996. The vertical structure of the wind direction was very different from that of the wind speed. Rather than a log-linear profile below 1-km altitude exhibited by the wind speed, the wind direction was nearly constant to ~450-m altitude and characterized by strong veering of the wind to the altitude of the wind speed maximum (−50° over 1.5-km altitude). In time the veering was a maximum near 70° over 1.5-km altitude just outside the eyewall (2300 UTC). While the peak in the wind speed decreased in altitude with time, the veering of the wind direction with height was relatively steady, with only the 2100 UTC profile showing an increase in the depth of the constant direction layer. The mean vertical shear in the lowest 1 km is surprisingly constant near 1.7–2.0 × 10^{-2} s^{-1} and may be the cause of the ABL features.

3. Surface wind and wave observation

A major source of difficulty in past efforts to predict tropical cyclone wind fields and storm surge at landfall was the inability to estimate the surface wind field directly. Presently the surface wind field must be estimated from a synthesis of scattered surface ship or buoy observations and aircraft measurements at 1.5–3.0- km altitude (e.g., Powell et al. 1996; Powell and Houston 1996, 1998). A new suite of airborne remote sensing instruments is available on the WP-3D aircraft for the purpose of measuring surface winds in and around tropical cyclones. A C-band (5 GHz) scatterometer (C-SCAT; Donnelly et al. 1999) and a stepped frequency microwave radiometer (SFMR) were flown together since 1992. The C-SCAT measures the backscatter from

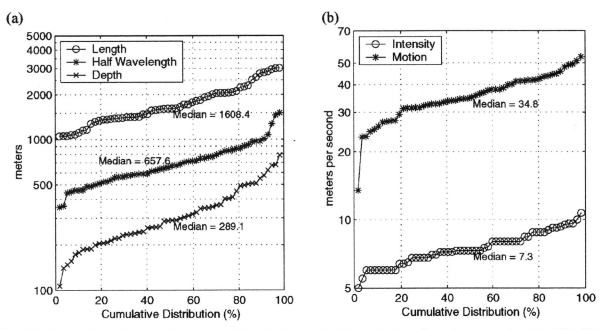

FIG. 3.27. Cumulative distributions of the characteristics of 172 atmospheric boundary layer linear features as viewed by WSR-88D radars for Hurricanes Fran (1996), Bonnie (1998), and Georges (1998). From Morrison et al. (2002).

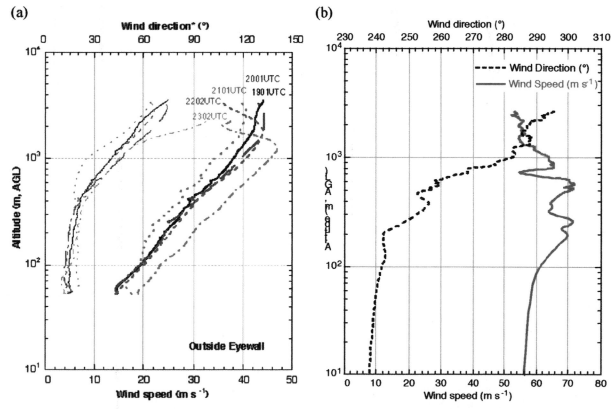

FIG. 3.28. (a) WSR-88D VAD-derived vertical profiles (Marks et al. 1999) and (b) GPS sonde derived of the total wind direction (°, thin lines) and wind speed (m s⁻¹, bold lines). The WSR-88D data is from Wilmington, North Carolina (KLTX) for the vicinity of the eyewall of Hurricane Fran at 1901 (solid), 2001 (long dash), 2101 (dot), 2202 (short dash), and 2302 UTC (dash–dot), 5 Sep 1996. The GPS sonde data are from the eyewall region of Hurricane Guillermo, 2000 UTC 3 Sep 1997.

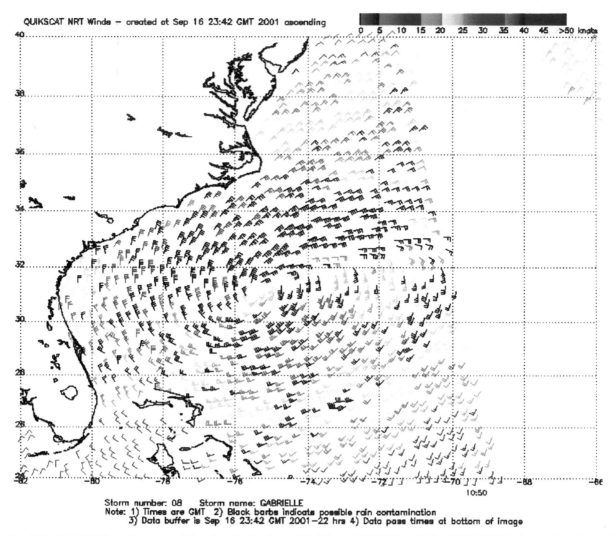

FIG. 3.29. QuickSCAT wind speeds and directions in Hurricane Gabrielle at 2342 UTC on 16 Sep 2001. Wind barbs are plotted every 25 km, using the standard convention: a pennant is 25 m s^{-1}, a barb represents 5 m s^{-1}, and a half-barb is 2.5 m s^{-1}. Colors denote increasing wind speed: blue, 5–7.5 m s^{-1}; green 7.5–10 m s^{-1}; yellow, 10–12.5 m s^{-1}; orange, 12.5–15 m s^{-1}; red, 15–17.5 m s^{-1}; and black denotes possible rain contamination.

the ocean surfaces, and SFMR measures brightness temperatures at six frequencies. The SFMR brightness measurements are used to estimate the emissivity of the ocean surface, derive estimates of the wind speed, wind stress, and rain rate, and to calculate two-way attenuation suffered by C-SCAT due to precipitation. Applying this correction to C-SCAT's backscatter measurements, the normalized radar cross section (NRCS) of the ocean surface is calculated (e.g., Donnelly et al. 1999). The SFMR surface wind estimates are being made operationally available to the hurricane specialists at the National Hurricane Center (NHC) for use in forecast guidance and in determining surface wind estimates at different locations surrounding the storm. These surface wind estimates will make a major impact on our understanding of high-wind, air–sea fluxes in conjunc-

tion with sea spray, surface wave, and oceanic mixing processes.

The NASA Seawinds scatterometer (QuickSCAT), carried on the Advanced Earth Observing Satellite (ADEOS), estimates surface wind speed and direction over a 1800-km-wide swath of ocean under the instrument's path with a 12.5-km resolution. QuickSCAT, launched in 1999, is a 13.4-GHz radar designed to measure winds over the oceans using a conically scanning antenna to cover. As shown in Fig. 3.29, QuickSCAT is extremely useful for mapping the overall surface wind field in tropical cyclones. The surface circulation in Gabrielle is clearly depicted, with peak winds of 20–25 m s^{-1}. However, with a frequency of 13.4 GHz, QuickSCAT wind estimates can be contaminated by heavy rain, which reduces the surface backscatter energy from the wind, increasing the uncertainty in both wind speed and direction. Figure 3.29 indicates that the possible rain-effected wind estimates cover a large portion of the storm's core, limiting the usefulness of the QuickSCAT data for defining the circulation. Research is underway to remove some of the ambiguity caused by the rain contamination.

The NASA/Goddard Space Flight Center (GSFC) has a long history of measuring the directional wave spectrum using the Surface Contour Radar (SCR; Walsh et al. 1985). The Ka-band (34 GHz) Scanning Radar Altimeter (SRA) replaced the SCR as the instrument of choice in the measurement of sea surface directional wave spectra. Both the SCR and SRA scan a narrow beam across the aircraft ground track, but the SRA has higher power and a wider swath (e.g., Wright et al. 2001).

Wright et al. (2001) described the sea surface directional wave spectrum with the SRA for the first time in all quadrants of a tropical cyclone's inner core over open water in Hurricane Bonnie on 24 August 1998. At that time, Bonnie was a large hurricane with 1-min sustained surface winds of nearly 50 m s^{-1}. The NOAA aircraft spent more than 5 h within 180 km of the storm center and made five eye penetrations at 1.5-km altitude. Wave topography maps showed individual waves as high as 19 m peak to trough. The dominant waves generally propagated at significant angles to the downwind direction. At some positions, three different wave fields of comparable energy crossed each other. Partitioning the SRA directional wave spectra enabled determination of the characteristics of the various components of the hurricane wave field and mapping of their spatial variation (cf. Fig. 3.30).

Figure 3.30 shows contours of the smoothed significant wave height (H_s) spatial variation measured by the SRA during the 5 h of measurement. The smallest waves are at the radius of maximum wind in the southwest quadrant, and the largest waves are in a similar position in the northeast quadrant. While accurate and valuable, Fig. 3.30 should be considered a simple characterization of the wave field. As simple as the characterization is, it demonstrates the tremendous variability of the wave structure around the storm. Future work will be directed to understanding the impact of this variability on processes at the air–sea interface.

4. Precipitation

A tropical cyclone drops copious amounts of precipitation daily. Over the oceans the rainfall is not a serious concern to society, but once a tropical cyclone makes landfall it can produce flooding, often well inland. Indeed, in the last 30 years the majority of deaths related to tropical cyclones were attributed to flooding (Rappaport 2000). Presently, accurate quantitative precipitation estimation (QPE) and forecasting (QPF) is poor worldwide, not only for the threat of flooding from land-

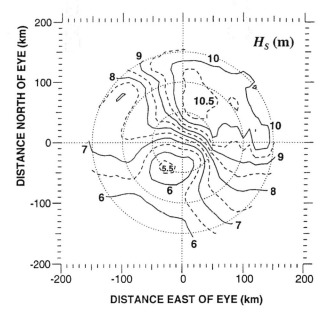

FIG. 3.30. Spatial variation in H_s measured by the SRA in Hurricane Bonnie on 24 Aug 1998. Contours for integer values of wave height (m) are solid and contours for integer values plus 0.5 m are dashed. From Wright et al. (2001).

falling tropical cyclones, but also on a daily basis as precipitation can disrupt commerce and recreational activities. The complexities of the precipitation process and the lack of microphysical data to support subgrid-scale parameterizations in numerical models have limited QPF. As pointed out in section 2c, much of the significant precipitation occurs in conjunction with convective clouds, but stratiform clouds also account for significant precipitation accumulations over extended intervals.

Recently developed observing systems provide opportunities for significant improvement in QPE. These include the WSR-88D radar network, various mesoscale networks of rain gauges and disdrometers, and a variety of satellite-based instruments. Included in the satellite-based instruments are passive microwave radiometers [e.g., special sensor microwave/imager (SSM/I) and Tropical Rainfall Measuring Mission (TRMM) microwave imager (TMI)] and an active precipitation radar (TRMM PR; Kummerow et al. 1998) (cf. Fig. 3.31a).

The standard precipitation products from the WSR-88D network are computed as 1-h, 3-h, and storm total precipitation on a 2 km × 1° polar grid. An example of the estimated storm total precipitation for Hurricane Danny (1997) is given in Fig. 3.31b. Measurements from rain gauges, disdrometers, or microphysics instrumentation may be used as validation for the radar estimates of rainfall, but comparisons among the different instruments are not straightforward. A fundamental issue is the probability distribution of rain at different temporal and spatial scales. Distributions for very small spatial scales can be computed from disdrometers (~1

(a)

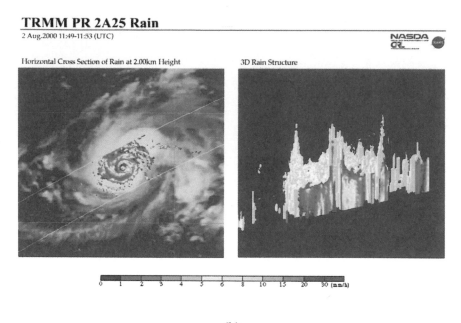

(b)

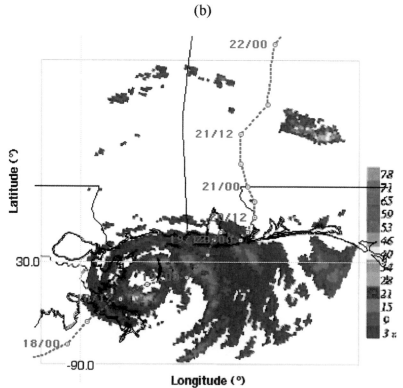

FIG. 3.31. (a) Plan view of 2-km and three-dimensional reflectivity factor from the TRMM precipitation radar in Typhoon Jelewat from 1149 to 1153 UTC 2 Aug 2000 (data available online at http://www.eorc.nasda.go.jp/TRMM/gallery/typh00_8/index.htm); (b) WSR-88D digital precipitation array 1-h rainfall amount (mm) from 2300 UTC 18 Jul to 0000 UTC 19 Jul 1997 for Mobile, Alabama, WSR-88D (KMOB) shown in the color contours (Marks et al. 2000). The dashed red line denotes the storm track.

m^2), or larger scales from radar estimates ($\sim$1 km^2), or even larger scales from satellite estimates ($>$10 km^2). Estimates can be accumulated over periods of seconds, hours, or even the entire duration of a rain event in the observational area. It is important to understand how the distributions at different scales relate to each other and how these distributions relate to whatever scales are of optimal interest. At present, relationships between these distributions at various scales are not well known. However, one thing known is that the mean rain amounts are more consistent between the different observation techniques when computed over large areas or long time periods, suggesting that while instantaneous rain estimates from any one technique may vary greatly, the 6-h, 12-h, 24-h, or storm total rain amounts are likely to be more consistent.

Marks et al. (2000) showed that the WSR-88D rain estimates over four days for Hurricane Danny were accurate when compared to the storm totals from the gauges. They showed that the distribution of radar-derived rain rates compares very well with the hourly gauge totals found in 10 Florida tropical cyclones by Miller (1958), and those from 19 gauges in Dade and Broward counties over two days from Hurricane Irene (1999), especially at the high rain rates. The major differences were between the Miller results and the more recent datasets when rain rates were $<$5 mm h^{-1}. The modern results had higher frequencies of rates $<$5 mm h^{-1}, likely due to an increased sensitivity to light rain for the radar and modern gauge estimates, which can detect rates $<$0.25–mm h^{-1} minimum threshold used in the Miller study. However, the agreement is extremely encouraging considering that these distributions should be expected to vary.

The TRMM PR is a major advancement, providing radar coverage at remote locations. An example of TRMM PR reflectivity is given in Fig. 3.31a for Typhoon Jelewat on 2 August 2000, which would otherwise not be sampled by ground-based radar. QPE using somewhat traditional radar methods can be accomplished with the TRMM PR and TMI (Lonfat et al. 2000). Unlike ground-based radar, satellites provide global coverage of the tropical cyclone basins, but because they are in low earth orbit ($<$800 km) they have relatively poor temporal coverage. These satellite-based approaches are useful in building a database of instantaneous rain-rate estimates in regions that would normally go unsampled. Ground-based approaches (e.g., radar and rain gauge networks) provide temporal continuity in the observations: the evolution of a rain field or rain distribution can be studied. The information from both approaches is necessary for understanding the rain distributions in tropical cyclones in a global context.

5. Transition of research to operations

a. Operational single-Doppler wind retrieval

The WP-3D airborne Doppler and WSR-88D data have also been instrumental in developing a suite of operational single-Doppler radar algorithms to objectively analyze a tropical cyclone's wind field. The algorithms were designed to determine the storm location and define the primary, secondary, and major asymmetric circulations. Some of these algorithms are used operationally on the WP-3D aircraft, as well as on the ground at NOAA's Tropical Prediction Center/NHC (e.g., Griffin et al. 1992; Burpee et al. 1994; Harasti et al. 2002) and at the Central Weather Bureau in the Republic of China (Taiwan).

In order to better depict the tropical cyclone wind field using single-Doppler radar data, several algorithms were investigated. These include the velocity track display (VTD; Lee et al. 1994), extended velocity track display (EVTD; Roux and Marks 1996), ground-based velocity track display (GBVTD; Lee et al. 1999), and the tracking echoes by correlation (TREC; Tuttle and Gall 1999) methods. A continuing cooperative effort among the National Center for Atmospheric Research, NOAA/Hurricane Research Division, and NHC led to the improvement and operational implementation of these and other methods.

One such application, the VTD in its various forms, is designed to extract the real-time three-dimensional primary circulation of a tropical cyclone in a form that can easily be communicated to forecast centers (e.g., Lee et al. 1994; Roux and Marks 1996; Lee et al. 1999). The goal is to transmit the real-time three-dimensional wind structure of the storm as a means of estimating intensity change, track, and environmental interaction, as well as for input into numerical models. The concept is best illustrated using the airborne Doppler radar, which scans in a track-orthogonal plane. As the aircraft moves, the radar beam sweeps out a three-dimensional domain 150-km wide centered on the aircraft track, and from the surface to 15–20 km in the vertical. The VTD originally referred to a display of the flight-level Doppler velocities as a function of time along the flight track (Lee et al. 1994).

A VTD display of Hurricane Gloria (1985), illustrated in Fig. 3.32, is characterized by a positive and negative Doppler velocity maxima, positive (negative) to the east (west) of the tropical cyclone center, and two zero Doppler velocity lines extending up and down through the tropical cyclone center.[6] The Doppler velocity maxima occur at points in the circulation where the radar looks tangentially along the storm circulation. Zero radial velocities are observed where the radar beam is oriented perpendicular to the storm circulation. Note in Fig. 3.32 that the centers of the velocity maxima are rotated counterclockwise from the flight track. Lee et al. (1994) proposed that such a shift results from a superposition of a mean radial flow on the tangential wind. One can

---

[6] Note that the Doppler velocities on the left side of the flight track were multiplied by -1 to remove the discontinuity across the flight track.

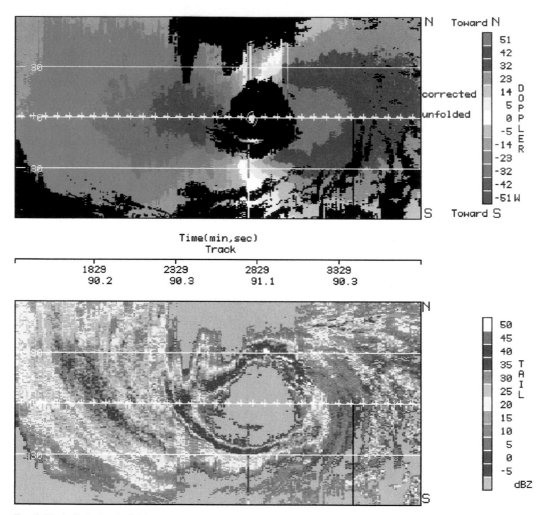

FIG. 3.32. A flight-level (6.5 km) VTD of Hurricane Gloria (1985). The aircraft moved from east (E) to west (W). From Lee et al. (1994).

estimate the inflow–outflow magnitude from the phase shift of the Doppler velocity maxima.

As pointed out previously, a real tropical cyclone contains not only the axisymmetric circulation (wavenumber 0), but also the asymmetric circulation (higher-order wavenumbers and mean motion). Therefore, a harmonic analysis can be used to decompose the horizontal projection of the Doppler velocity along a VTD ring into the magnitude and phase for each harmonic. Vertical profiles of the above kinematic properties can be estimated due to the three-dimensional coverage of a scanning airborne Doppler radar. The VTD technique processes an annulus of data, where the radial distance from the storm center determines the period necessary to satisfy the stationarity assumption. For the eyewall region, the time interval over which stationarity is assumed is normally <6 min, for an average aircraft speed of 130 m s^{-1}.

The coastal WSR-88D Doppler radar network in the United States enables continuous monitor of landfalling tropical cyclones within 200 km of the coastline. These coastal radars sample tropical cyclones approximately every 6 min. Since the average distance between radars is 250 km, deducing tropical cyclone circulation primarily relies on single-Doppler radar wind retrieval techniques. Lee et al. (1999, 2000) demonstrated that the primary circulation (tangential wind) can be retrieved reasonably well by a variation of the VTD technique they called GBVTD.

The importance of having an accurate tropical cyclone circulation center in decomposing tropical cyclone circulation into wavenumber domain was realized by Willoughby et al. (1982) and Marks et al. (1992). Lee and Marks (2000) illustrated that the GBVTD technique generates apparent asymmetric tropical cyclone structure if there is an error in the specified tropical cyclone center >1 km.

Current work is toward complementing the GBVTD algorithm with additional techniques, including TREC (Tuttle and Gall 1999). The TREC algorithm uses the reflectivity correlation method to derive a wind estimate, extending the wind analysis beyond the range of the

(a)   (b)

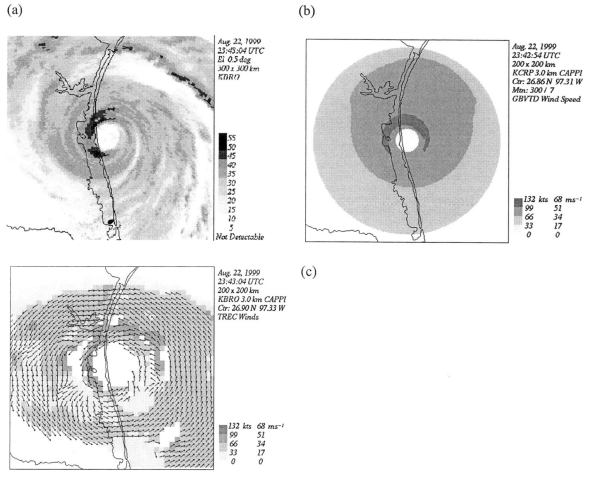

(c)

FIG. 3.33. (a) Radar reflectivity (dBZ); (b) GBVTD wind speed (m s⁻¹); and (c) TREC wind field (m s⁻¹) derived from the archive-IV data from WSR-88D radars at Corpus Christi (KCRP) and Brownsville (KBRO), Texas, in Hurricane Bret (1999). The reflectivity map is from the lowest tilt angle from KBRO. A CAPPI was constructed at 3 km from the KCRP data and used as input for the GBVTD analyses. A 3-km CAPPI was constructed at 3 km from the KBRO data and used as input for the TREC analyses. From Harasti et al. (2002).

Doppler velocities. It can be integrated into a combined product with GBVTD and other techniques. Harasti et al. (2002) applied the GBVTD and TREC techniques to the archive-IV WSR-88D data of Hurricane Bret (1999).

Figure 3.33a shows the reflectivity distribution and Fig. 3.33b shows the GBVTD results from the Corpus Christi, Texas, WSR-88D (KCRP). Figure 3.33c shows the TREC results from the Brownsville, Texas, WSR-88D (KBRO). All three were obtained from 3-km constant-altitude PPIs (CAPPIs). Overall, these results agreed reasonably well with each other, and with the results of a triple-Doppler radar wind synthesis using the KCRP, KBRO, and the airborne Doppler radar (Dodge et al. 2002). The GBVTD and TREC algorithms have also been applied with great success to the archive-IV WSR-88D datasets in Hurricane Debby (2000) and Tropical Storm Barry (2001). The fact that these algorithms work so well with the coarse archive-IV data bodes well for the future when full-resolution archive-II data become available in real time. However, the true benefit of these techniques will only be realized when we can assimilate the single-Doppler radial velocity data into the operational forecast models.

b. Operational precipitation estimation and forecasting

A major obstacle to improving QPF in tropical cyclones is a lack of a comprehensive climatology of tropical cyclone precipitation, that is, a description of the distribution of rain in space and time. Few precipitation climatologies exist for tropical cyclones in the United States, and other tropical cyclone basins have similarly limited climatologies. However, remote sensors such as those on the NASA TMI and PR and the WSR-88D DPA are providing a first cut at a credible tropical cyclone rain climatology (e.g., Marks et al. 2000; Lonfat et al. 2000). This climatology is precisely what is needed to develop a simple rainfall climatology and persistence (R-CLIPER) model, which can be used to validate numerical models and other QPF methods.

A tropical cyclone R-CLIPER was developed to use a global satellite-based tropical cyclone rainfall climatology based on rain estimates from the TMI and PR rain estimates developed by Lonfat et al. (2000). To date, this climatology includes global TMI rain estimates in 245 storms from December 1997 to December 2000, yielding 2121 events, where 64% of the events were tropical storms, 26% were category-1–2 hurricanes, and 10% were category 3 or higher.

The climatology provides a mean rain rate and the rain-rate probability distribution in a storm-centered coordinate system composed of 50 10-km-wide annuli. The results are stratified as a function of storm intensity. The results show that the mean rain rate increases by a factor of four (3 in. day^{-1} for tropical storms versus 12 in. day^{-1} for category 3 and higher tropical cyclones) within 50 km of the storm center with increasing intensity.

The climatology was combined with the operational track forecasts to compute an integrated rain distribution for each forecast interval to produce the R-CLIPER. An important use of the R-CLIPER is to provide a benchmark for the evaluation of other more-general QPF techniques and model rain forecasts. It is hoped that the R-CLIPER will provide a basis for estimating the true forecast skill of various QPF products.

6. Summary and outlook for the future

a. Summary

In the last 15 years, technological improvements such as the NOAA WP-3D tail airborne Doppler radar, the operational WSR-88D radar network, portable Doppler radars, and the first spaceborne radar system on the NASA TRMM satellite have produced a new generation of tropical cyclone data whose analysis and interpretation is rapidly expanding our understanding of these storms. The NOAA WP-3D airborne Doppler datasets led to improved understanding of the symmetric vortex and the major asymmetries. The ability to map the complete three-dimensional circulation of the storm in a short period enabled the partitioning of the wind to examine the roles of asymmetries in studies of storm motion and intensity change. The addition of a second airborne Doppler radar on other WP-3D enabled true dual-Doppler analyses, and the ability to study the evolution of the kinematic structure over 3–6 h.

The advent of the operational WSR-88D Doppler radar network, and the construction of portable Doppler radars that can be moved to the a location near tropical cyclone landfall, has also generated new and unique datasets. These new datasets are enabling improved understanding of 1) severe weather events associated with landfalling tropical cyclones, 2) boundary layer wind structure as the storm moves from over the sea to over land, and 3) spatial and temporal changes in the storm rain distribution. These datasets have revealed that there are two regions within the storm that are characterized by severe weather events: along the inside edge of the eyewall where vorticity mixing is present; and in the storm periphery, within or just outside the principal rainband, where mesocyclones and tornadoes occur as the band interacts with the land. These data have also revealed that the atmospheric boundary layer winds are characterized by long narrow streaks of high and low winds with lateral and vertical shears of 1×10^{-1} s^{-1}, on scales of 1–2 km wide and 3–10 km long, extending from the surface to 800-m altitude. These features in the wind are responsible for much of the local damage and suggest that the atmospheric boundary layer fluxes of heat and momentum are not omnidirectional but are maximized along the direction of the flow.

Scatterometers (C-SCAT and QuickSCAT) and other passive (SFMR) and active (SRA) remote sensors have enabled the mapping of the surface wind and wave fields in tropical cyclones. These observations are essential to improved understanding of the processes that transfer momentum and energy across the air–sea interface, essential for improved numerical models.

The WSR-88D rainfall data, together with new satellite microwave passive and active sensors on the NASA TRMM satellite, are proving useful in studies of the temporal and spatial variability of rain in tropical cyclones. The instantaneous satellite snapshots provide rain estimates to improve our understanding of tropical cyclone rain distributions globally, providing estimates from one instrument and common algorithms in each basin. The WSR-88D provides high–temporal resolution rain estimates (1 h), to improve our understanding of the temporal variability of the rain as the storm makes landfall. These datasets complement each other and the rain gauge information, serving as a key ingredient to our quantitative precipitation estimates and subsequent forecasts. The TRMM data have also been instrumental in developing a climatology and persistence rainfall forecast model, which will be used as a benchmark for comparison with model-generated rainfall forecasts.

The WP-3D airborne Doppler and WSR-88D data have also been instrumental in developing a suite of operational single-Doppler radar algorithms to objectively analyze a tropical cyclone's wind field. The algorithms were designed to determine the storm location and define the primary, secondary, and major asymmetric circulations. A potential benefit of the WSR-88D algorithms will be the opportunity to examine the kinematic structure at high temporal resolution (6 min). These algorithms are used operationally on the WP-3D aircraft, as well as on the ground at NOAA's Tropical Prediction Center/National Hurricane Center.

b. Where do we go from here?

1) Tropical cyclone structure and dynamics

The discussion in section 2 points to possible answers to important questions related to storm motion, intensity change, and the interaction of the vortex with its environment. Since the linear combination of the mean horizontal flow and the inner asymmetries determines the location of the updrafts and downdrafts in the eyewall, changes in either one can affect the location of the updrafts and downdrafts and thus alter the storm structure and intensity. If the recent studies are correct in that the mean-horizontal flow (large-scale outer gyres) changes with time over a storm's track, these changes could affect the storm intensity. If the inner asymmetries are instabilities of the mean vortex that also change over the storm's lifetime, these changes could also affect storm intensity. The instability causing the inner asymmetries may thus be dependent upon changes in the mean-horizontal flow or outer gyres. Future work in this area needs to examine the role shear plays in the vortex dynamics and subsequently in intensity change. Questions that need to be addressed are related to the role that shear plays in vortex tilt and the vorticity mixing mechanisms.

Determining the key factors that affect where severe weather will occur in tropical cyclones is another area of research that needs to be addressed. What are the mechanisms and processes that foster severe weather along the inside edge of the eyewall and along the periphery of the storm? What role do the vorticity mixing processes in the core play in the severe weather observed, and what are the characteristic timescales and space scales? What role does convective instability play in the outer region where severe weather occurs? Is it simply a superposition of strong atmospheric boundary layer vertical wind shear and increased convective instability, or is it more subtle?

Another major question is, what mechanisms produce the linear features in the atmospheric boundary layer, and how do their characteristics vary over water and over land? Are the mechanisms responsible related to the strong vertical wind shear in the boundary layer, or does static stability play a role in their generation and maintenance? Are these features present all of the time or just when the atmospheric boundary layer vertical wind shear exceeds a certain threshold? How do their characteristics change as the wind moves from over water to over land and vice versa? Does the scale and intensity of these features vary with the surface terrain and roughness? What role do these features play in enthalpy and momentum exchange in the boundary layer?

A topic that the radar meteorology community must address is the assimilation of Doppler radar data into mesoscale numerical models. As model resolution shrinks with advances in computing, the scale of information necessary to initialize the model approaches that of a Doppler radar. This community is the best equipped to tackle this task, and it seems that the tropical cyclone problem is the most tractable to attempt such an effort. In a tropical cyclone, the dynamic constraints are strongly controlled by the high vortex angular velocity. In a tropical cyclone the timescale of the divergent component of the wind, which typically dominates the variance in the Doppler winds, is one order of magnitude smaller than that of the mean vortex. A data assimilation technique needs to be designed that can take advantage of this difference; one that limits the aliasing of convective motions on the resultant initial condition. This assimilation system can be built upon the concepts outlined for dealing with the analysis of the tropical cyclone with single-Doppler radars (i.e., VTD, EVTD, and GBVTD). The problem seems tractable and should be a high priority over the next 5–10 years. If we cannot address this issue with tropical cyclones, it will be nearly impossible to do in situations where the divergent component of the wind dominates the variance in the Doppler winds.

2) TROPICAL CYCLONE PRECIPITATION

Opportunities to improve QPF in landfalling tropical cyclones seem particularly promising because of improved understanding of tropical cyclone precipitation mechanisms, quantitative precipitation estimation from remote sensors (both ground and space based), and improvements in horizontal resolution of operational and research models so that moist nonhydrostatic processes, which occur on small scales, may be represented better. However, a major obstacle to improved quantitative precipitation forecasting in tropical cyclones is a lack of a comprehensive climatology of tropical cyclone precipitation. Few precipitation climatologies exist for tropical cyclones in the United States, and other tropical cyclone basins have similarly limited climatologies. To date there is no attempt to compare these limited climatologies over a basin or between basins. New quantitative precipitation estimation opportunities (e.g., WSR-88D, TRMM) offer a unique opportunity to develop tropical cyclone rain climatologies for the United States and globally. These climatologies can be used to validate numerical models and thereby improve quantitative precipitation forecasting in tropical cyclones.

Due to the complexities of the precipitation process and the need to parameterize the microphysics within numerical models, quantitative precipitation forecasting poses a difficult challenge. However, probabilistic quantitative precipitation forecasting may be the direction to turn in the near future to provide useful forecasts to the public. The dynamic constraints present in a tropical cyclone, the mixing of hydrometeors around the vortex, and the recent advances in rainfall observing systems make the tropical cyclone a good candidate to study precipitation estimation and forecasting. We need to make better use of the TRMM PR, airborne radars, and microphysics datasets to help understand precipitation processes, particularly the processes that determine the vertical distribution of hydrometeors.

Another area of research that needs to be addressed is the role that shear plays on the rain distribution. Given the strong influence that the vertical shear of the horizontal wind plays in the kinematic structure of the vortex, how does it influence the rain distribution in space? Some recent work suggests that the vertical shear of the horizontal wind results in asymmetries in the rain distribution around the storm (e.g., Rogers et al. 2002). Does the shear cause the majority of the stratiform rain to fall preferably on one side of the storm or another, or does it produce a more favorable location for convection to occur on a certain side of the storm? Can this variability be modeled from the cases available?

As with the Doppler winds, the radar meteorology community needs to address the assimilation of radar reflectivity and rainfall data into mesoscale numerical models. A data assimilation technique needs to be defined that can take advantage of the unique characteristics of radar data. Novel ideas are needed, such as using the radar reflectivity data to delineate the raining from nonraining areas, or identifying the altitude of the 0°C level from the bright band. The vertical reflectivity structure provided globally by the TRMM PR is the best vehicle to use for attempting this effort. By virtue of being a single well-calibrated radar system, continually viewing different regions of the globe, it can be used without worrying about data format or calibration issues between different radar systems. By working with modelers and data assimilation experts, radar meteorologists can assist in the interpretation and use of radar data. The problem seems tractable and should be a high priority over the next 5–10 years.

Finally, new techniques need to be developed to present QPE information to forecasters. Probabilistic information that conveys an uncertainty in the rain estimates needs to developed and explained to the forecasters and the public. Rain is not a normally distributed variable. It is lognormally distributed. Hence, the mean rain rate is not a good measure of the rain distribution. A better measure of the natural rain distribution is the mean of the log of the rain amount. However, if one wants to know the peak storm-total rainfall it may be better to use the tail of the probability distribution of rain, such as the 95%, which varies little from storm to storm, as it is determined by the convective processes. Hence, just as much rain can fall from a tropical storm as a major hurricane. The above discussion brings up an interesting question: what does the forecaster need to predict? Is the forecaster interested in a good measure of the storm total rain distribution to ensure that the area and rough mean amounts are correct? On the other hand, is the forecaster interested in the likelihood that the storm total rainfall has some probability of exceeding some amount? The answer to these questions would determine what type of products should be produced.

Acknowledgments. I would like to thank my colleagues at NOAA's Hurricane Research Division and the pilots and crew who fly the NOAA research aircraft into hurricanes from the NOAA Aircraft Operations Center. Without their dedication and skill, much of the work reported on in this chapter would not be possible.

APPENDIX

Technological Developments since 1987

A review of radar meteorology's role in studies of tropical cyclones would be incomplete without a discussion of radar technology improvements. Donaldson and Atlas (1964) pointed out that the early advances in the study of tropical cyclones using radar observations came about through improvements (scientific and technological) in the ability of the radar to observe different aspects of the storms. These early radar studies provided the basis for understanding tropical cyclone structure and evolution. As pointed out by Marks (1989) technological advances in radar design and signal processing between 1974 and 1987, particularly the development of airborne Doppler radars, revolutionized our depiction of tropical cyclone structure and dynamics. Further technological advances in radar and antenna design over the last 15 years provided the tools necessary to greatly expand our knowledge of tropical cyclones.

Four technological advances were instrumental in improving observations of tropical cyclones in the last 15 years: 1) the development of new and improved airborne Doppler radar systems; 2) the development of operational ground-based Doppler radar and profiler networks in tropical cyclone–prone regions; 3) the development of mobile Doppler radar and profiler systems; and 4) the development of new active remote sensors for satellite applications. Table 3.A1 lists the years when these advances to studies of tropical cyclones took place.

a. Airborne Doppler radar systems

The two NOAA airborne Doppler radars have revolutionized the field of mesoscale meteorological research by providing a mobile platform for observing the three-dimensional structure of the precipitation and kinematic fields in precipitating weather systems. The airborne radar systems on the NOAA WP-3D aircraft were designed primarily for studies of tropical cyclones (Trotter 1978). In the last 25 years the airborne radars, both coherent and noncoherent, have allowed the first mapping of the three-dimensional kinematic and reflectivity structure of tropical cyclones away from the influence of land (e.g., Marks and Houze 1984, 1987; Marks et al. 1992).

The early analysis technique used single-platform sampling, which uses one Doppler radar on one aircraft that flies two consecutive, usually orthogonal, flight tracks. The antenna scans in the plane normal to the flight track. Two new developments in the NOAA airborne Doppler radar system reduced the Doppler error that results from evolution of weather systems. The first was that the antenna was programmed to scan ±25° fore or aft of the plane perpendicular to the long axis of the aircraft (dual-beam or fore-aft scanning technique, or FAST; e.g., Jorgensen et al. 1996). The second devel-

TABLE 3.A1. Technological developments in radar systems used to observe tropical cyclones since 1987.

Year	Technological development
1987	First 50-MHz MU profiler observations of the full wind field (three wind components) to within 100 km of Typhoon 8719 (Sato et al. 1991).
1988	First dual-aircraft 9-GHz Doppler radar dataset collected in Hurricane Gilbert (Dodge et al. 1999).
1990	TCM-90 deploys 50- and 915-MHz profilers at Saipan capturing data to within 115 km of the center of Typhoon Flo (May et al. 1994).
	First FAST observations in Hurricane Gustav (Gamache et al. 1995).
1991	First dual-beam airborne Doppler radar observations in Hurricane Claudette using the prototype ELDORA antenna developed by CETP in France (Roux and Viltard 1995).
	WSR-88D Doppler radar network deployed in the United States and at military bases in tropical cyclone–prone regions (Crum and Alberty 1993).
1992	First WSR-88D data in Hurricane Andrew from Melbourne, Florida (KMLB). Vortex core out of Doppler range (Willoughby and Black 1996).
	Level-II data recorders installed at a number of WSR-88D radar sites to collect research-quality Doppler radar data (Crum et al. 1993).
1993	First West Pacific WSR-88D Doppler radar data collected at Guam in Typhoon Ed (Stewart and Lyons 1996).
1995	First WSR-88D Level-II Doppler radar datasets collected in Hurricane Erin from Melbourne, Florida (KMLB), and Mobile, Alabama (KMOB).
1996	First DOW observations in Hurricane Fran (Wurman and Winslow 1998).
	First Taiwan WSR-88D Doppler radar dataset in Typhoon Herb (Tseng and Wang 1999; Kuo et al. 1999).
1997	NSCAT makes surface wind observations in tropical cyclones (Jones et al. 1999).
	TRMM PR is first satellite-based precipitation radar system capable of observing vertical reflectivity structure in tropical cyclones (Kummerow et al. 1998).
1998	First EDOP radar observations of tropical cyclone collected in Hurricane Bonne (Heymsfield et al. 2001).
	First ARMAR Doppler radar observations collected onboard the NASA DC-8 in Hurricane Bonnie (Heymsfield et al. 2001).
	First MIPS 915-MHz profiler observations of the eyewall of Hurricane Georges (Knupp 2000).
	First dual-mobile-Doppler radar dataset in Hurricane Georges made with DOW 2 and 3.
	First SCR measurements of the two-dimensional ocean wave structure collected onboard the NOAA WP-3D in Hurricane Bonnie (Wright et al. 2001).
	RADARSAT makes spaceborne SAR observations of tropical cyclones (Katsaros et al. 2000).

opment was that a Doppler radar was also placed on the second WP-3D aircraft in 1988 (dual-platform sampling; e.g., Gamache et al. 1995).

Another significant development was the installation of a X-band Doppler radar system on the NASA ER-2 to measure the vertical reflectivity structure and hydrometeor motions simultaneous with the radiometric measurements (EDOP; Heymsfield et al. 1996). EDOP, located in the nose of the ER-2, can look both in the nadir and along a 30° forward-directed beam with copolarized and cross-polarized receivers. The antennas have a 3° beamwidth and a spot size of about 1.2 km at the surface from 20-km aircraft altitude. The system is usually configured for 150-m gate spacing and a pulse repetition frequency of 2200 Hz. On tropical cyclone missions during the third and fourth NASA Convection and Moisture Experiments (CAMEX-3, −4) the ER-2 flew straight flight lines at 20-km altitude across the top of seven tropical cyclones (e.g., Geerts et al. 2000; Heymsfield et al. 2001). EDOP mapped out high-resolution time–height sections of nadir and forward beam reflectivity and Doppler velocity through the outer bands and eyewall, providing detailed vertical reflectivity and Doppler velocity information.

Another new airborne rain-mapping radar (ARMAR) was developed and operated on the NASA DC-8 aircraft (Durden et al. 1994) and was also used in tropical cyclone missions during the NASA CAMEX-3 and −4 experiments (e.g., Heymsfield et al. 2001). ARMAR was designed as a prototype for spaceborne rain radars, and its frequency and scanning geometry are identical to the NASA Tropical Rain Measurement Mission (TRMM) satellite Precipitation Radar (PR) (Kummerow et al. 1998). The radar operates at Ku-band (13.8-GHz), with a 3.8° beamwidth and 80-m range resolution. It scans ±20° about nadir across the aircraft track sweeping a swath roughly equal to the height of the aircraft above the ground (~9–10 km). ARMAR also has multiple polarization and Doppler velocity measurement capabilities, making it an excellent tool for observing the high-resolution vertical structure of precipitation in tropical cyclones.

A suite of airborne remote sensing instruments is available on the WP-3D aircraft for the purpose of measuring surface winds in and around tropical cyclones. A C-band (5 GHz) scatterometer (C-SCAT) and a stepped frequency microwave radiometer (SFMR) were flown together since 1992. C-SCAT conically scans the ocean surface obtaining backscatter measurements from 20° to 50° off nadir, and SFMR measures nadir brightness temperatures at six frequencies (4.6–7.2 GHz). The SFMR brightness measurements are used to estimate the emissivity of the ocean surface, derive estimates of the wind speed, wind stress, rain rate, and to calculate two-way attenuation suffered by C-SCAT due to precipitation. Applying this correction to C-SCAT's backscatter measurements, the normalized radar cross section (NRCS) of the ocean surface is calculated.

The NASA/Goddard Space Flight Center (GSFC) Ka-band (34 GHz) SRA measures the sea surface directional wave spectra. The SRA scans a narrow beam across the aircraft ground track (e.g., Wright et al. 2001). It measures the slant range to 64 points evenly spaced across the swath (at 8-m intervals for a 640-m altitude), converts them to surface elevations, and as the aircraft advances, displays the false-color coded topography on a monitor in real time. This grid of surface topography represents a snapshot of the wave field with along-track spacing of 12–13 m between points (with a 100–m s^{-1} aircraft groundspeed). These data over an along-track distance of 5–6 km and a cross-track swath of about 520 m are transformed into directional wave spectra by a two-dimensional fast Fourier transform.

b. Operational Doppler radar and profiler networks in tropical cyclone prone regions

NOAA deployed a state-of-the-art ground-based Doppler radar network over the entire United States (Crum and Alberty 1993). Nearly one-third of those radars are positioned along the portion of the coast threatened by tropical cyclones. At the same time operational ground-based Doppler radar networks have been developed in Taiwan, Japan, Hong Kong, and the French islands of La Réunion and Martinique. These Doppler data are extremely useful for locating and tracking tropical cyclones. The radar data also provide high temporal (e.g., 6 min for the WSR-88D) and spatial resolution (1° azimuthal × 1 km radial) wind fields useful in studies of the wind characteristics. These data can also be used to construct high–temporal resolution vertical profiles of the horizontal wind components over each radar site, providing details of the vertical structure of the wind as the storm moves over land.

Networks of wind profilers are being deployed in regions likely to be affected by tropical storms. Japanese studies have focused on the behavior of high-frequency wind variations associated with internal gravity waves (Sato et al. 1991). The Tropical Profiler Network (TPN) installed 50-MHz (VHF) and 915-MHz (UHF) profilers at a number of tropical island locations across the Pacific Ocean to study large-scale atmospheric teleconnections. Many of these islands are prone to tropical cyclones, and some of these profiler systems were used successfully to observe the vertical structure of tropical cyclones (May et al. 1994). The 50-MHz profiler provides hourly horizontal winds and the vertical wind every 4–5 min. The lowest observations are ~1.5 km and reach up to 16 km with 1–2-km resolution. The 915-MHz boundary layer profilers provide hourly observations from 100-m to as high as possible, with roughly 200-m resolution. By virtue of its operating frequency, the 915-MHz system is very sensitive to rain.

c. Mobile Doppler radars and profilers

A number of portable, pencil-beam, pulsed, Doppler, 5-GHz (5-cm wavelength) and 9-GHz (3-cm wavelength) radars were constructed in the last six years to study a wide variety of meteorological phenomena including tornadoes, severe storms, tropical cyclones, and boundary layer processes (e.g., Wurman et al. 1997). These new radar systems have full scanning capability,

real-time displays, and archiving capability, and are mounted on trucks for easy portability. This portability allows the radar to be brought to within a kilometer of rare meteorological phenomena. These radars have successfully collected data in several tropical cyclones and were used to detect boundary layer structures (e.g., Wurman and Winslow 1998).

A mobile integrated profiling system (MIPS) containing a 915-MHz boundary layer profiler mounted on a trailer was developed at the University of Alabama in Huntsville (Knupp 2000). The mobile system was developed with a single swithchable flatplate, microstrip phased-array antenna that is a 2-m square. Three independent beam orientations are produced by electronically changing the phasing. The wind profiler performance is greatly enhanced in the presence of precipitation, as in a tropical cyclone, but some errors may be introduced when the precipitation is inhomogeneous on the scales of the beam separation (~100 m). Clutter screens are not required in high signal-to-noise heavy precipitation environments common in tropical cyclones. The MIPS was deployed successfully in Hurricane Georges (1998) and Tropical Storm Gabrielle (2001).

d. Improved active microwave remote sensors for satellite application

The NASA TRMM satellite was launched 27 November 1997 into a near-circular orbit of approximately 350 km in altitude with an inclination of 35° to the equator and a period of 91.5 min. The satellite carries a unique set of precipitation-observing microwave sensors: a passive microwave imager (TMI) and an active PR (Kummerow et al. 1998). The TMI is a nine-channel, five-frequency, linearly polarized, passive microwave radiometric system. The instrument measures atmospheric and surface brightness temperatures at 10.7, 19.4, 21.3, 37.0, and 85.5 GHz. TMI is similar to the SSM/I instrument flown on the Defense Meteorological Satellite Program (DMSP) satellites with roughly twice as many pixels per scan.

The PR is a major advancement, providing radar coverage at remote locations, which would otherwise not be sampled by ground-based radar. Precipitation estimation using somewhat traditional radar methods can be accomplished with the PR, which is an active 13.8-GHz radar, recording energy reflected from atmospheric and surface targets. The PR electronically scans every 0.6 s with a swath width of 215 km.[A1] Each scan contains 49 rays sampled at 4.5-km resolution across track. For a given ray, the samples are recorded at 125-m intervals starting a fixed distance from the satellite to below the

---

[A1] All TRMM horizontal spatial scales at the earth's surface have increased by roughly 15% since August 2001 when the satellite's orbit was boosted from 350- to 405-km altitude to preserve fuel and lengthen the mission.

surface. The PR products are referenced to the level of the earth ellipsoid at 250-m resolution. These satellite-based approaches are useful in building a database of instantaneous rain estimates in regions that would normally go unsampled. The information from both approaches is necessary for understanding the rain distributions in tropical cyclones in a global context.

The NASA Scatterometer (NSCAT) and Seawinds scatterometer (QuickSCAT), carried on ADEOS, measures wind speeds and directions over at least 90% of the global oceans every 2 days. NSCAT, operational during 1997, was a 13.4-GHz radar designed to measure winds over the oceans using an array of antennas to scan two 600-km bands of ocean; one band on each side of the instrument's orbital path, separated by a gap of approximately 330 km (e.g., Jones et al. 1999). QuickSCAT, launched in 1999, is also a 13.4-GHz radar designed to measure winds over the oceans using an conically scanning antenna to cover a 1800-km-wide band of ocean under the instrument's orbital path. Both NSCAT and QuickSCAT collected backscatter and wind vector information nearly continuously. NSCAT generated wind estimates over the measurement swath at 25-km resolution and an accuracy of 2 m s^{-1}, while QuickSCAT reduced the resolution to 12.5 km.

Radar altimetry from NASA's Ocean Topography Experiment (TOPEX/Poseidon) Mission and the ERS-1 is used to precisely map changes of the sea surface height with accuracies of 2–5 cm in the vertical. These data are used to delineate the areal extent and time evolution of strong baroclinic features such as the Gulf Stream and warm core eddies. Of particular interest here is to infer variability in the oceanic mixed layer depth to properly initialize the ocean and coupled atmosphere–ocean models using data assimilation methods. Adaptive sample strategies are used to evaluate and validate the inferred mixed-layer depth variations in strong baroclinic current features for the purpose of estimated mixed-layer heat potential in the presence of advection.

REFERENCES

Atlas, D., K. Hardy, R. Wexler, and R. Boucher, 1963: The origin of hurricane spiral bands. *Geofis. Int.,* **3,** 123–132.

Barnes, G. M., and G. J. Stossmeister, 1986: The structure and decay of a rainband in Hurricane Irene (1981). *Mon. Wea. Rev.,* **114,** 2590–2601.

——, and M. D. Powell, 1995: Evolution of the inflow boundary layer of Hurricane Gilbert (1988). *Mon. Wea. Rev.,* **123,** 2348–2368.

——, E. J. Zipser, D. P. Jorgensen, and F. D. Marks, 1983: Mesoscale and convective structure of a hurricane rainband. *J. Atmos. Sci.,* **40,** 2125–2137.

——, J. F. Gamache, M. A. LeMone, and G. J. Stossmeister, 1991: A convective cell in a hurricane rainband. *Mon. Wea. Rev.,* **119,** 776–794.

Bender, M. A., 1997: The effect of relative flow on the asymmetric structure in the interior of hurricanes. *J. Atmos. Sci.,* **54,** 703–724.

Black, M. L., and J. F. Franklin, 2000: GPS dropsonde observations of the wind structure in convective and non-convective regions of the hurricane eyewall. Preprints, *23rd Conf. on Hurricanes and Tropical Meteorology,* Ft. Lauderdale, FL, Amer. Meteor. Soc., 448–449.

——, R. W. Burpee, and F. D. Marks, 1996: Vertical motion characteristics of tropical cyclones determined with airborne Doppler radial velocities. *J. Atmos. Sci.,* **53,** 1887–1909.

——, J. F. Gamache, F. D. Marks, C. E. Samsury, and H. E. Willoughby, 2002: Eastern Pacific Hurricanes Jimena of 1991 and Olivia of 1994: The effect of vertical shear on structure and intensity. *Mon. Wea. Rev.,* **130,** 2291–2312.

Black, P. G., and F. D. Marks, 1991: The structure of an eyewall meso-vortex in Hurricane Hugo (1989). Preprints, *19th Conf. on Hurricanes and Tropical Meteorology,* Miami, FL, Amer. Meteor. Soc., 579–582.

Black, R. A., and J. Hallett, 1986: Observations of the distribution of ice in hurricanes. *J. Atmos. Sci.,* **43,** 802–822.

——, H. B. Bluestein, and M. L. Black, 1994: Unusually strong vertical motions in a Caribbean hurricane. *Mon. Wea. Rev.,* **122,** 2722–2739.

Blackwell, K. G., 2000: The evolution of Hurricane Danny (1997) at landfall: Doppler-observed eyewall replacement, vortex contraction/intensification, and low-level wind maxima. *Mon. Wea. Rev.,* **128,** 4002–4016.

Browning, K. A., 1964: Airflow and precipitation trajectories within severe local storms which travel to the right of the winds. *J. Atmos. Sci.,* **21,** 634–668.

——, and R. Wexler, 1968: The determination of kinematic properties of a wind field using Doppler radar. *J. Appl. Meteor.,* **7,** 105–113.

Burpee, R. W., and Coauthors, 1994: Real-time guidance provided by NOAA's Hurricane Research Division to forecasters during Hurricane Emily of 1993. *Bull. Amer. Meteor. Soc.,* **75,** 1765–1783.

Chan, J. C.-L., 1986: Supertyphoon Abby—an example of present track forecast inadequacies. *Wea. Forecasting,* **1,** 113–126.

Crum, T. D., and R. L. Alberty, 1993: The WSR-88D and the WSR-88D Operational Support Facility. *Bull. Amer. Meteor. Soc.,* **74,** 1669–1687.

——, ——, and D. W. Burgess, 1993: Recording, archiving, and using WSR-88D data. *Bull. Amer. Meteor. Soc.,* **74,** 645–653.

DeMaria, M., 1996: The effect of vertical shear on tropical cyclone intensity change. *J. Atmos. Sci.,* **53,** 2076–2087.

Dodge, P. P., R. W. Burpee, and F. D. Marks, 1999: The kinematic structure of a hurricane with sea level pressure less than 900 mb. *Mon. Wea. Rev.,* **127,** 987–1004.

——, S. Spratt, F. D. Marks, D. Sharp, and J. Gamache, 2000: Dual-Doppler analyses of mesovortices in a hurricane rainband. Preprints, *24th Conf. on Hurricanes and Tropical Meteorology,* Ft. Lauderdale, FL, Amer. Meteor. Soc., 302–303.

——, M. L. Black, J. L. Franklin, J. F. Gamache, and F. D. Marks, 2002: High-resolution observations of the eyewall in an intense Hurricane Bret on 21–22 August 1999. Preprints, *25th Conf. on Hurricanes and Tropical Meteorology,* San Diego, CA, Amer. Meteor. Soc., 607–608.

Donaldson, R. J., Jr., and D. Atlas, 1964: Radar in tropical meteorology. *Proc. Symp. on Tropical Meteorology,* New Zealand Meteorological Service, 423–473.

Dong, K., 1986: The relationship between tropical cyclone motion and environmental geostrophic flows. *Mon. Wea. Rev.,* **114,** 115–122.

——, and C. J. Neumann, 1983: On the relative motion of binary tropical cyclones. *Mon. Wea. Rev.,* **111,** 945–953.

Donnelly, W. J., J. R. Carswell, R. E. McIntosh, P. S. Chang, J. Wilkerson, F. Marks, and P. G. Black, 1999: Revised ocean backscatter models at C and Ku band under high-wind conditions. *J. Geophys. Res.,* **104,** 11 485–11 497.

Durden, S. L., E. Im, F. K. Li, W. Ricketts, A. Tanner, and W. Wilson, 1994: ARMAR: An airborne rain-mapping radar. *J. Atmos. Oceanic Technol.,* **11,** 727–737.

Fiorino, M., and R. L. Elsberry, 1989: Some aspects of vortex structure related to cyclone motion. *J. Atmos. Sci.,* **46,** 975–990.

Frank, W. M., 1977: The structure and energetics of the tropical cyclone. Part I: Storm structure. *Mon. Wea. Rev.,* **105,** 1119–1135.

——, 1984: A composite analysis of the core of a mature hurricane. *Mon. Wea. Rev.,* **112,** 2401–2420.

——, and E. A. Ritchie, 1999: Effects of environmental flow upon tropical cyclone structure. *Mon. Wea. Rev.,* **127,** 2044–2061.

Franklin, J. F., M. L. Black, and S. E. Feuer, 1999: Wind profiles in hurricanes determined by GPS dropwindsondes. Preprints, *23rd Conf. on Hurricanes and Tropical Meteorology,* Dallas, TX, Amer. Meteor. Soc., 167–168.

Franklin, J. L., S. J. Lord, S. E. Feuer, and F. D. Marks, 1993: The kinematic structure of Hurricane Gloria (1985) determined from nested analyses of dropwindsonde and Doppler radar data. *Mon. Wea. Rev.,* **121,** 2433–2451.

Gall, R. L., J. D. Tuttle, and P. Hildebrand, 1998: Small-scale spiral bands observed in Hurricanes Andrew, Hugo, and Erin. *Mon. Wea. Rev.,* **126,** 1749–1766.

Gamache, J. F., R. A. Houze Jr., and F. D. Marks, 1993: Dual-aircraft investigation of the inner core of Hurricane Norbert. Part III: Water budget. *J. Atmos. Sci.,* **50,** 3221–3243.

——, F. Roux, and F. D. Marks, 1995: Comparison of three airborne Doppler sampling techniques with airborne in situ wind observations in Hurricane Gustav (1990). *J. Atmos. Oceanic Technol.,* **12,** 171–181.

Geerts, B., G. M. Heymsfield, L. Tian, J. B. Halverson, A. Guillory, and M. I. Mejia, 2000: Hurricane Georges's landfall in the Dominican Republic: Detailed airborne Doppler radar imagery. *Bull. Amer. Meteor. Soc.,* **81,** 999–1018.

Gray, W. M., and D. J. Shea, 1973: The hurricane's inner core region. Part II: Thermal stability and dynamic characteristics. *J. Atmos. Sci.,* **30,** 1565–1576.

Griffin, J. S., R. W. Burpee, F. D. Marks, and J. L. Franklin, 1992: Real-time airborne analysis of aircraft data supporting operational hurricane forecasting. *Wea. Forecasting,* **7,** 480–490.

Guinn, T. A., and W. H. Schubert, 1993: Hurricane spiral bands. *J. Atmos. Sci.,* **50,** 3380–3403.

Harasti, P. R., W.-C. Lee, J. Tuttle, C. J. McAdie, P. P. Dodge, S. T. Murrillo, and F. D. Marks, 2002: Operational implementation of single-Doppler radar algorithms for tropical cyclones. Preprints of *25th Conf. on Hurricanes and Tropical Meteorology,* San Diego, CA, Amer. Meteor. Soc., 487–488.

Hawkins, H. E., and D. T. Rubsam, 1968: Hurricane Hilda, 1964. II. Structure and budgets of the hurricane on October 1, 1964. *Mon. Wea. Rev.,* **96,** 617–636.

Hawkins, H. F., and S. M. Imbembo, 1976: The structure of a small, intense Hurricane—Inez 1966. *Mon. Wea. Rev.,* **104,** 418–442.

Heymsfield, G. M., and Coauthors, 1996: The EDOP radar system on the high-altitude NASA ER-2 aircraft. *J. Atmos. Oceanic Technol.,* **13,** 795–809.

——, J. B. Halverson, J. Simpson, L. Tian, and T. P. Bui, 2001: ER-2 Doppler radar investigations of the eyewall of Hurricane Bonnie during the Convection and Moisture Experiment-3. *J. Appl. Meteor.,* **40,** 1310–1330.

Houze, R. A., Jr., 1993: *Cloud Dynamics.* Academic Press, 573 pp.

——, F. D. Marks, and R. A. Black, 1992: Dual-aircraft investigation of the inner core of Hurricane Norbert. Part II: Mesoscale distribution of ice particles. *J. Atmos. Sci.,* **49,** 943–962.

Ishihara, M., Z. Yanagisawa, H. Sakakibara, K. Matsuura, and J. Aoyagi, 1986: Structure of typhoon rainband observed by two Doppler radars. *J. Meteor. Soc. Japan,* **64,** 923–939.

Jones, S. C., 1995: The evolution of vortices in vertical shear: Initially barotropic vortices. *Quart. J. Roy. Meteor. Soc.,* **121,** 821–851.

Jones, W., V. Cardone, W. Pierson, J. Zec, L. Rice, A. Cox, and W. Sylvester, 1999: NSCAT high-resolution surface wind measurements in Typhoon Violet. *J. Geophys. Res.,* **104,** 11 247–11 259.

Jordan, C., D. Hurt, and C. Lowery, 1960: On the structure of Hurricane Daisy on 27 August 1958. *J. Meteor.,* **17,** 337–348.

Jorgensen, D. P., 1984a: Mesoscale and convective-scale characteristics of mature hurricanes. Part I: General observations by research aircraft. *J. Atmos. Sci.,* **41,** 1268–1285.

——,1984b: Mesoscale and convective-scale characteristics of mature hurricanes. Part II: Inner core structure of Hurricane Allen (1980). *J. Atmos. Sci.,* **41,** 1287–1311.

——, and P. T. Willis, 1982: A Z–R relationship for hurricanes. *J. Appl. Meteor.,* **21,** 356–366.

——, E. J. Zipser, and M. A. LeMone, 1985: Vertical motions in intense hurricanes. *J. Atmos. Sci.,* **42,** 839–856.

——, T. Shepherd, and A. Goldstein, 1996: Multi-beam techniques for deriving wind fields from airborne Doppler radars. *Meteor. Atmos. Phys.,* **59,** 83–104.

Jou, B. J.-D., B.-L. Chang, and W.-C. Lee, 1994: Analysis of typhoon circulation using ground based Doppler radar. *Atmos. Sci.,* **22,** 163–187.

Katsaros, K., P. W. Vachon, P. G. Black, P. P. Dodge, and E. W. Uhlhorn, 2000: Wind fields from SAR: Could they improve our understanding of storm dynamics? *Johns Hopkins APL Tech. Dig.,* **21,** 86–93.

Kepert, J. D., and G. J. Holland, 1997: The Northwest Cape tropical cyclone boundary layer monitoring station. Preprints, *22nd Conf. on Hurricanes and Tropical Meteorology,* Ft. Collins, CO, Amer. Meteor. Soc., 82–83.

Kessler, E., 1957: Outer precipitation bands of Hurricanes Edna and Ione. *Bull. Amer. Meteor. Soc.,* **38,** 335–346.

——, 1958: Eye-region of Hurricane Edna, 1954. *J. Meteor.,* **15,** 264–270.

——, and D. Atlas, 1956: Radar-synoptic analysis of Hurricane Edna (1954). Geophysical Research Papers 50, AFGL, Bedford, MA, 113 pp.

Knupp, K. R., 2000: Doppler profiler observations of Hurricane Georges at landfall. *Geophys. Res. Lett.,* **27,** 3361–3364.

Kummerow, C., W. Barnes, T. Kozu, J. Shiue, and J. Simpson, 1998: The Tropical Rainfall Measuring Mission (TRMM) sensor package. *J. Atmos. Oceanic Technol.,* **15,** 809–817.

Kuo, H.-C., R. T. Williams, and J.-H. Chen, 1999: A possible mechanism for the eye rotation of Typhoon Herb. *J. Atmos. Sci.,* **56,** 1659–1673.

LaSeur, N. E., and H. F. Hawkins, 1963: An analysis of Hurricane Cleo (1958) based on data from research reconnaissance aircraft. *Mon. Wea. Rev.,* **91,** 694–709.

Lee, W.-C., and F. D. Marks, 2000: Tropical cyclone kinematic structure retrieved from single Doppler radar observations. Part II: The GBVTD-simplex center finding algorithm. *Mon. Wea. Rev.,* **128,** 1925–1936.

——, R. Carbone, and F. D. Marks, 1994: Velocity track display—A technique to extract real-time tropical cyclone circulations using a single airborne Doppler radar. *J. Atmos. Oceanic Technol.,* **11,** 337–356.

——, B. J.-D. Jou, P.-L. Chang, and S.-M. Deng, 1999: Tropical cyclone kinematic structure retrieved from single-Doppler radar observations. Part I: Interpretation of Doppler velocity patterns and the GBVTD technique. *Mon. Wea. Rev.,* **127,** 2419–2439.

——, ——, ——, and F. D. Marks, 2000: Tropical cyclone kinematic structure retrieved from single Doppler radar observations. Part III: Evolution and structures of Typhoon Alex (1987). *Mon. Wea. Rev.,* **128,** 3982–4001.

Liu, Y., D.-L. Zhang, and M. K. Yau, 1997: A multiscale numerical study of Hurricane Andrew (1992). Part I: Explicit simulation and verification. *Mon. Wea. Rev.,* **125,** 3073–3093.

——, ——, and ——, 1999: A multiscale numerical study of Hurricane Andrew (1992). Part II: Kinematics and inner-core structures. *Mon. Wea. Rev.,* **127,** 2597–2616.

Lonfat, M., F. D. Marks, and S. Chen, 2000: Study of the rain distribution in tropical cyclones using TRMM/TMI. Preprints, *24th Conf. on Hurricanes and Tropical Meteorology,* Ft. Lauderdale, FL, Amer. Meteor. Soc., 480–481.

MacDonald, N. J., 1968: The evidence for the existence of Rossby-like waves in the hurricane vortex. *Tellus,* **20,** 138–146.

Marks, F. D. 1985: Evolution of the structure of precipitation in Hurricane Allen (1980). *Mon. Wea. Rev.,* **113,** 909–930.

——, 1989: Meteorological radar observations of tropical weather systems. *Radar in Meteorology,* D. Atlas, Ed., Amer. Meteor. Soc., 401–425.

——, 1991: Kinematic structure of the hurricane inner core as revealed by airborne Doppler radar. Preprints, *Fifth Conf. on Mesoscale Processes,* Atlanta, GA, Amer. Meteor. Soc., 127–132.

——, and R. A. Houze, Jr., 1984: Airborne Doppler radar observations in Hurricane Debby. *Bull. Amer. Meteor. Soc.,* **65,** 569–582.

——, and ——, 1987: Inner core structure of Hurricane Alicia from airborne Doppler radar observations. *J. Atmos. Sci.,* **44,** 1296–1317.

——, and P. G. Black, 1990: Close encounter with an intense mesoscale vortex within Hurricane Hugo (September 15, 1989). Preprints, *Fourth Conf. on Mesoscale Meteorology,* Boulder, CO, Amer. Meteor. Soc., 114–115.

——, R. A. Houze, and J. Gamache, 1992: Dual-aircraft investigation of the inner core of Hurricane Norbert. Part I: Kinematic structure. *J. Atmos. Sci.,* **49,** 919–942.

——, D. Atlas, and P. T. Willis, 1993: Probability-matched reflectivity-rainfall relations for a hurricane from aircraft observations. *J. Appl. Meteor.,* **32,** 1134–1141.

——, P. Dodge, and C. Sandin, 1999: WSR-88D observations of hurricane atmospheric boundary layer structure at landfall. Preprints, *23rd Conf. on Hurricanes and Tropical Meteorology,* Dallas, TX, Amer. Meteor. Soc., 1051–1054.

——, L. Selevan, and J. Gamache, 2000: WSR-88D derived rainfall distributions in Hurricane Danny. Preprints, *24th Conf. on Hurricanes and Tropical Meteorology,* Ft. Lauderdale, FL, Amer. Meteor. Soc., 298–299.

May, P. T., 1996: The organization of convection in the rainbands of Tropical Cyclone Laurence. *Mon. Wea. Rev.,* **124,** 807–815.

——, G. J. Holland, and W. L. Ecklund, 1994: Wind profiler observations of Tropical Storm Flo at Saipan. *Wea. Forecasting,* **9,** 410–426.

Maynard, R. H., 1945: Radar and weather. *J. Meteor.,* **2,** 214–226.

McCaul, E. W., 1987: Observations of the hurricane "Danny" tornado outbreak of 16 August 1985. *Mon. Wea. Rev.,* **115,** 1206–1223.

——, 1991: Buoyancy and shear characteristics of hurricane–tornado environments. *Mon. Wea. Rev.,* **119,** 1954–1978.

——, 1993: Hurricane spawned tornadic storms. *The Tornado—Its Structure, Dynamics, Prediction, and Hazards, Geophys. Monogr.,* No. 79, Amer. Geophys. Union, 119–142.

Miller, B. I., 1958: Rainfall rates in Florida hurricanes. *Mon. Wea. Rev.,* **86,** 258–264.

Moller, J. D., and M. T. Montgomery, 1999: Vortex Rossby waves and hurricane intensification in a barotropic model. *J. Atmos. Sci.,* **56,** 1674–1687.

Montgomery, M. T., and R. Kallenbach, 1997: A theory for vortex Rossby-waves and its application to spiral bands and intensity changes in hurricanes. *Quart. J. Roy. Meteor. Soc.,* **123,** 435–465.

——, and J. Enagonio, 1998: Tropical cyclogenesis via convectively forced vortex Rossby waves in a three-dimensional quasigeostrophic model. *J. Atmos. Sci.,* **55,** 3176–3207.

Morrison, I., F. D. Marks, and S. Businger, 2002: WSR-88D observations of boundary layer rolls during hurricane landfall. Preprints, *25th Conf. on Hurricanes and Tropical Meteorology,* San Diego, CA, Amer. Meteor. Soc., 341–342.

Muramatsu, T., 1986: Trochoidal motion of the eye of Typhoon 8019. *J. Meteor. Soc. Japan,* **64,** 259–272.

Neumann, C. J., 1979: On the use of deep-layer-mean geopotential height fields in statistical prediction of tropical cyclone motion. Preprints, *Sixth Conf. on Probability and Statistics in Atmospheric Sciences,* Banff, AB, Canada, Amer. Meteor. Soc., 32–38.

Novlan, D. J., and W. M. Gray, 1974: Hurricane-spawned tornadoes. *Mon. Wea. Rev.,* **102,** 476–488.

Omoto, Y., 1982: On "tatsumaki" (tornadoes) associated with typhoons (in Japanese). *Tenki,* **29,** 967–980.

Parrish, J. R., R. W. Burpee, F. D. Marks, and R. Grebe, 1982: Rainfall patterns observed by digitized radar during the landfall of Hurricane Frederic (1979). *Mon. Wea. Rev.,* **110,** 1933–1944.

——, ——, and C. W. Landsea, 1984: Mesoscale and convective-scale characteristics of Hurricane Frederic during landfall. Preprints, *15th Conf. on Hurricanes and Tropical Meteorology,* Miami, FL, Amer. Meteor. Soc., 415–420.

Powell, M. D., 1990a: Boundary layer structure and dynamics in outer hurricane rainbands. Part I: Mesoscale rainfall and kinematic structure. *Mon. Wea. Rev.,* **118,** 891–917.

——, 1990b: Boundary layer structure and dynamics in outer hurricane rainbands. Part II: Downdraft modification and mixed layer recovery. *Mon. Wea. Rev.,* **118,** 918–938.

——, and S. H. Houston, 1996: Hurricane Andrew's landfall in south Florida. Part II: Surface wind fields and potential real-time applications. *Wea. Forecasting,* **11,** 329–349.

——, and ——, 1998: Surface wind fields of 1995 Hurricanes Erin, Opal, Luis, Marilyn and Roxanne at landfall. *Mon. Wea. Rev.,* **126,** 1259–1273.

——, ——, and T. A. Reinhold, 1996: Hurricane Andrew's landfall in south Florida. Part I: Standardizing measurements for documentation of surface wind fields. *Wea. Forecasting,* **11,** 304–328.

——, T. A. Reinhold, and R. D. Marshall, 1999: GPS sonde insights on boundary layer wind structures in hurricanes. *Wind Engineering into the 21st Century: Proceedings of the 10th International Conference on Wind Engineering,* A. Larsen, G. L. Larose, and F. M. Livesey, Eds., Balkema, 307–314.

Rappaport, E. N., 2000: Loss of life in the United States associated with recent Atlantic tropical cyclones. *Bull. Amer. Meteor. Soc.,* **81,** 2065–2074.

Reasor, P. D., M. T. Montgomery, F. D. Marks, and J. F. Gamache, 2000: Low-wavenumber structure and evolution of the hurricane inner core observed by airborne dual-Doppler radar. *Mon. Wea. Rev.,* **128,** 1653–1680.

Rogers, R., S. Chen, J. Tenerelli, and H. Willoughby, 2002: The role of vertical shear in determining the distribution of accumulated rainfall in high-resolution numerical simulations of tropical cyclones. Preprints, *25th Conf. on Hurricanes and Tropical Meteorology,* San Diego, CA, Amer. Meteor. Soc., 319–320.

Roux, F., and N. Viltard, 1995: Structure and evolution of Hurricane Claudette on 7 September 1991 from airborne Doppler radar observations. Part I: Kinematics. *Mon. Wea. Rev.,* **123,** 2611–2639.

——, and F. D. Marks, 1996: Extended velocity track display (EVTD): An improved processing method for Doppler radar observations of tropical cyclones. *J. Atmos. Oceanic Technol.,* **13,** 875–899.

Ryan, B. E., G. M. Barnes, and E. J. Zipser, 1992: A wide rainband in a developing tropical cyclone. *Mon. Wea. Rev.,* **120,** 431–437.

Saito, A., 1992: Mesoscale analysis of typhoon-associated tornado outbreaks in Kyushu Island on 13 October 1980. *J. Meteor. Soc. Japan,* **70,** 43–55.

Samsury, C. E., and E. J. Zipser, 1995: Secondary wind maxima in hurricanes: Airflow and relationship to rainbands. *Mon. Wea. Rev.,* **123,** 3502–3517.

Sato, T., N. Ao, M. Yamamoto, S. Fukao, T. Tsuda, and S. Kato, 1991: A typhoon observed with the MU radar. *Mon. Wea. Rev.,* **119,** 755–768.

Schubert, W. H., M. T. Montgomery, R. K. Taft, T. A. Guinn, S. R. Fulton, J. P. Kossin, and J. P. Edwards, 1999: Polygonal eyewalls, asymmetric eye contraction and potential vorticity mixing in hurricanes. *J. Atmos. Sci.,* **56,** 1197–1223.

Shapiro, L. J., 1982: The asymmetric boundary layer flow under a translating hurricane. *J. Atmos. Sci.,* **40,** 1984–1998.

——, and M. T. Montgomery, 1993: A three-dimensional balance

theory for rapidly rotating vortices. *J. Atmos. Sci.,* **50,** 3322–3335.

Sharp, D. W., J. Medlin, S. M. Spratt, and S. J. Hodanish, 1997: A spectrum of outer spiral rainband mesocyclones associated with tropical cyclones. Preprints, *22nd Conf. on Hurricanes and Tropical Meteorology,* Ft. Collins, CO, Amer. Meteor. Soc., 117–118.

Shea, D. J., and W. M. Gray, 1973: The hurricane's inner core region. I: Symmetric and asymmetric structure. *J. Atmos. Sci.,* **30,** 1544–1564.

Smith, G. B., and M. T. Montgomery, 1995: Vortex axisymmetrization: Dependence on azimuthal wavenumber or asymmetric radial structure changes. *Quart. J. Roy. Meteor. Soc.,* **121,** 1615–1650.

Spratt, S. M., and D. W. Sharp, 1999: Dominant tropical cyclone rainbands related to tornadic and non-tornadic mesoscale circulation families. Preprints, *23rd Conf. on Hurricanes and Tropical Meteorology,* Dallas, TX, Amer. Meteor. Soc., 455–458.

——, ——, P. Welsh, A. Sandrik, F. Alsheimer, and C. Paxton, 1997: A WSR-88D assessment of tropical cyclone outer rainband tornadoes. *Wea. Forecasting,* **12,** 479–501.

Stewart, S. R., and S. W. Lyons, 1996: A WSR-88D radar view of Tropical Cyclone Ed. *Wea. Forecasting,* **11,** 115–132.

Suzuki, O., H. Niino, H. Ohno, and H. Nirasawa, 2000: Tornado-producing mini supercells associated with Typhoon 9019. *Mon. Wea. Rev.,* **128,** 1868–1882.

Tabata, A., H. Sakakibara, M. Ishihara, K. Matsuura, and Z. Yanagisawa, 1992: A general view of the structure of Typhoon 8514 observed by dual-Doppler radar: From outer rainbands to eyewall clouds. *J. Meteor. Soc. Japan,* **70,** 897–917.

Tatehira, R., H. Seko, and T. Suzuki, 1998: Estimation of upper wind in typhoon region from single-Doppler radar. *Tenki,* **45,** 633–642.

Trotter, R. L., 1978: Design considerations for the NOAA airborne meteorological radar and data system. Preprints, *18th Conf. on Radar Meteorology,* Atlanta, GA, Amer. Meteor. Soc., 405–408.

Tseng, C.-H., and C. T. Wang, 1999: The structure of Typhoon Herb (1996) during its landfall by the dual Doppler analysis. *Atmos. Sci.,* **27,** 295–318.

Tuttle, J., and R. Gall, 1999: A single-radar technique for estimating the winds in tropical cyclones. *Bull. Amer. Meteor. Soc.,* **80,** 653–668.

Viltard, N., and F. Roux, 1998: Structure and evolution of Hurricane Claudette on 7 September 1991 from airborne Doppler radar observations. Part II: Thermodynamics. *Mon. Wea. Rev.,* **126,** 281–302.

Wakimoto, R. M., and P. G. Black, 1994: Damage survey of Hurricane Andrew and its relationship to the eyewall. *Bull. Amer. Meteor. Soc.,* **75,** 189–202.

Walsh, E. J., D. W. Hancock, D. E. Hines, R. N. Swift, and J. F. Scott, 1985: Directional wave spectra measured with the surface contour radar. *J. Phys. Oceanogr.,* **15,** 566–592.

Wexler, H., 1947: Structure of hurricanes as determined by radar. *Ann. N. Y. Acad. Sci.,* **48,** 821–844.

Willoughby, H. E., 1978: A possible mechanism for the formation of hurricane rainbands. *J. Atmos. Sci.,* **35,** 838–848.

——, 1979: Forced secondary circulations in hurricanes. *J. Geophys. Res.,* **84,** 3173–3183.

——, 1988: The dynamics of the tropical cyclone core. *Aust. Meteor. Mag.,* **36,** 193–191.

——, 1990: Temporal changes of the primary circulation in tropical cyclones. *J. Atmos. Sci.,* **47,** 242–264.

——, and M. B. Chelmow, 1982: Objective determination of hurricane tracks from aircraft observations. *Mon. Wea. Rev.,* **110,** 1298–1305.

——, and P. G. Black, 1996: Hurricane Andrew in Florida: Dynamics of a disaster. *Bull. Amer. Meteor. Soc.,* **77,** 543–549.

——, J. Clos, and M. Shoribah, 1982: Concentric eyewalls, secondary wind maxima, and the evolution of the hurricane vortex. *J. Atmos. Sci.,* **39,** 395–411.

——, F. D. Marks, and R. J. Feinberg, 1984: Stationary and propagating convective bands in asymmetric hurricanes. *J. Atmos. Sci.,* **41,** 3189–3211.

Wilson, K. J., 1979: Characteristics of the subcloud layer wind structure in tropical cyclones. Preprints, *Int. Conf. on Tropical Cyclones,* Perth, Australia, *Roy. Meteor. Soc. (Australia),* 15 pp.

Wright, C. W., and Coauthors, 2001: Hurricane directional wave spectrum spatial variation in the open ocean. *J. Phys. Oceanogr.,* **31,** 2472–2488.

Wurman, J., and J. Winslow, 1998: Intense sub-kilometer boundary layer rolls in Hurricane Fran. *Science,* **280,** 555–557.

——, J. Straka, E. Rasmussen, M. Randall, and A. Zahrai, 1997: Design and deployment of a portable, pencil-beam, pulsed, 3-cm Doppler radar. *J. Atmos. Oceanic Technol.,* **14,** 1502–1512.

Chapter 4

Forcing and Organization of Convective Systems

DAVID P. JORGENSEN

NOAA/National Severe Storms Laboratory, Boulder, Colorado

TAMMY M. WECKWERTH

National Center for Atmospheric Research, Boulder, Colorado

Jorgensen   **Weckwerth**

1. Introduction

Perhaps as in no other area of meteorology, radar has proven to be *the* key tool in modern detection and forecasting, as well as in identifying and understanding the physics, of convective storms and convective systems. From its initial deployment as a research tool following the second World War [see the excellent review of the history of radar meteorology in chapter 1 of the AMS monograph *Radar in Meteorology,* edited by Dave Atlas (Fletcher 1990)], radar has played a fundamental role in increasing our understanding of the forces that initiate and organize severe storms and larger convective systems that are composed of a conglomeration of convective storm cells. Early radar observations were primarily descriptive and showed the tremendous variety of types and sizes of precipitating moist convection (see reviews by Browning 1990; Parsons et al. 1990; Ray 1990; Carbone et al. 1990b; Smull 1995). Examples of types include single convective storms, longer-lived multicellular storms, fast-moving squall lines, slower-moving linear and nonlinear convective systems, and long-lived supercell storms. Although this review emphasizes the role played by radar in observational studies of convective storms and systems, the ever-increasing use of numerical models in explaining the dynamics

of precipitating convective systems has resulted in the inclusion of many of those studies. Moreover, we believe numerical models will play an increasing role in extending our understanding of these phenomena by providing information on variables and in areas that are unobserved by radar and other instruments. Likewise, the accuracy of simulations that involve unavoidable parameterization of unresolved or poorly understood physics must be verified against data from the same observing systems.

For the purposes of this review we adopt a definition of convective system that is fairly broad; that is, we follow that of Emanuel (1994) in which convection encompasses a thermally direct circulation that results from the action of gravity upon an unstable vertical distribution of moist air. This definition includes convective storms (multicellular as well as supercellular) of diameters of $\sim$1–10 km and larger mesoscale convective systems with diameters of $\sim$100–200 km.

Groups of individual convective storms (and their associated stratiform rain components) are often organized on the mesoscale, exhibit behaviors that are long lasting (3–12-h lifetimes), and propagate as discrete entities. Early radar observations by Ligda (1956) first noted the differing structure of convective and stratiform rain regions within convective systems. These organized precipitation systems are usually termed mesoscale convective systems (MCSs; Zipser 1982) and are observed in the Tropics as well as in higher latitudes, and over both oceanic and continental locations. MCSs are global phenomena, as pointed out by Laing and Fritsch (2000), who noted MCS genesis environments in Africa, Australia, China, and South America that are similar to those documented in the United States. In most regions of the globe, MCSs are the dominant contributor to the annual precipitation. MCSs are both forced by, and influence, their large-scale environments (Stensrud 1996a). Consequently, their parameterization in terms of heat, moisture, and momentum vertical redistribution in numerical forecast models (Nicholls et al. 1991; LeMone and Moncrieff 1994) is important. Moreover, as pointed out by Doswell et al. (1996), the precipitation organization and propagation of MCSs is a critical forecast issue for flash flooding situations. Clearly, improved understanding of the factors that lead to MCS genesis would benefit forecasts and warnings.

Soon following radar observation of the various convective organization modes, it was recognized that the type of convective system was strongly dependent on the environment in which it was embedded. Research based on theory (e.g., Rotunno et al. 1988; Emanuel 1986), radar and sounding observations (e.g., Houze et al. 1990; Parker and Johnson 2000; Schiesser et al. 1995; Hane and Jorgensen 1995; LeMone et al. 1998), and numerical simulations (e.g., see reviews by Schlesinger 1982; Brooks and Wilhelmson 1993; McCaul and Weisman 2001; and the seminal work by Weisman and Klemp 1982) determined that two environmental factors are particularly important in determining convective system organization and evolution. These factors are the vertical profile of the horizontal wind, particularly at low levels (i.e., vertical wind shear) and the potential instability of the air feeding the system [i.e., the convective available potential energy (CAPE)].

This review of convective systems and forcing will focus on recent research work since about 1990. For a comprehensive review of earlier results see the *Radar in Meteorology* monograph that resulted from the Battan Memorial Radar Conference held 9–13 November 1987.

2. Types of convective systems and their relationship to shear and CAPE

In their panel discussion and review of "convective dynamics" for the *Radar in Meteorology* monograph, Carbone et al. (1990b) provide an excellent synthesis of the organizational modes of various convective systems in terms of vertical wind shear and CAPE. They define the concept of "*R* space," where *R* is the bulk Richardson number expressed as a ratio of shear to CAPE. The Carbone schematic is adapted and extended in Fig. 4.1. Because the domain of individual convective storms cells typically overlaps that of mesoscale con-

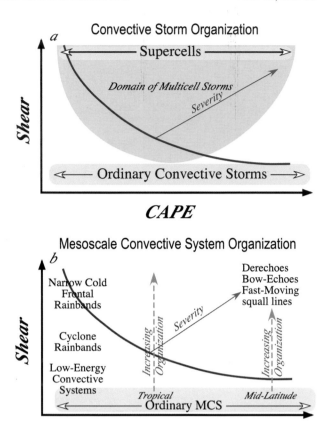

FIG. 4.1. (a) Schematic depiction of convective storm types usually associated with typical vertical wind shear and CAPE environments: the approximate range of multicellular storms is shown by the shaded region; (b) mesoscale convective system organization usually associated with similar CAPE–shear regimes. The curved line running from upper left to lower right of each plot indicates the approximate dividing line between severe and nonsevere storms. Although the axis values are meant to be schematic, a typical range of CAPE is probably 0–5000 J kg^{-1}, and the range of shear is probably 0–20 m s^{-1} over the lowest 2.5 km of height. After Carbone et al. (1990a).

vective systems, the two types have been separated for clarity. Figure 4.1a shows the shear–CAPE domain of convective storms while Fig. 4.1b shows the domain of mesoscale convective systems. The blue curved line from upper left to lower right in each figure indicates a rough demarcation of "severe" systems from "ordinary" ones. Severity in this context implies the presence of significant hail, strong damaging surface wind, and/or the collocation of tornadoes with the convective system. The location of such a demarcation on a diagram like Fig. 4.1 is somewhat subjective; however, some statistical evidence for such a curved line is presented by Johns et al. (1993) in their study of 242 tornadic storms in the warm and cool seasons. Likewise, the severity of MCSs increases toward the upper right of the diagram where some of the strongest systems (e.g., bow echoes and convective systems/storms with damaging straight-line surface winds called derechoes if a size and longevity criteria is reached) reside. The degree of organization of MCSs tends to increase with increasing shear (Houze et al. 1990).

The caveats to such a relatively simple classification scheme detailed by Carbone et al. (1990b) apply here as well. Namely, the placing of a specific convective system type in a particular location on the shear–CAPE coordinate system relates to its most commonly observed mode or behavior; however, a particular system type may span a range of shear–CAPE values. Moreover, other factors, such as the vertical profile of water vapor, the convection initiation mechanism, the mesoscale forcing brought on by upper-level lifting due to short waves and gravity waves, neighboring-storm interactions (e.g., Bluestein and Weisman 2000), and interactions between microphysics and updrafts/downdrafts may be important in determining the specific convective structure in a particular shear–CAPE environment. The role of such factors in influencing convective system organization, evolution, and severity and their influence on our ability to forecast MCSs is a subject of considerable ongoing research.

a. Low CAPE regimes

At the low end of the shear–CAPE spectrum in Fig. 4.1b, lie the weak convective systems that usually do not produce severe weather (tornadoes, strong straight-line winds, or large hail) but might produce locally heavy rainfall or snowfall. As pointed out by Carbone et al. (1990b), examples of this system type include fair weather cumulus clouds (Joseph and Cahalan 1990), rainbands near tropical islands (e.g., Hawaii; Austin et al. 1996; Szumowski et al. 1997; Carbone et al. 1998), high-based thunderstorms (Colman 1990), rainbands in the warm sector of midlatitude cyclones (Hertzman and Hobbs 1988; Matejka et al. 1980), and lake-effect snowstorms (Steenburgh et al. 2000; Hjelmfelt 1990). Owing to their ubiquity, however, these systems may have a profound impact on heat, moisture, and momentum redistribution on a global scale, as well as a significant role in the local rainfall climatology. An example of the behavior of these low-energy systems is shown in Fig. 4.2 from the Hawaiian Rainband Project in 1990 (Austin et al. 1996). These slow-moving storms propagated toward the island and were typically observed to develop well upstream of the offshore low-level wind reversal zone (Smolarkiewicz et al. 1988) that is nocturnally generated by the offshore land breeze (Carbone et al. 1995). As the bands intersected the flow reversal zone they frequently organized into an arc shape that mimicked the coastline geometry and sometimes strengthened to produce radar reflectivities in excess of 60 dBZ, but with storm depths that rarely exceeded 3 km. Cold pools produced by downdrafts from these bands are rarely evident, due to the very moist nature of the Hawaiian trade wind environment. So "cold pool" dynamics, which are hypothesized to contribute to the lon-

gevity and severity in other squall-line systems, are not a factor in these storms.

Slightly stronger in shear strength than the low-energy systems are extratropical cyclone rainbands that have been observed by radar to exhibit complex banded structures (Shields et al. 1991). For additional information on the precipitation structure of extratropical cyclones, see the chapter by Browning (2003) in this volume. Earlier studies by Houze and Hobbs (1982) and Browning (1985) identified six general types of rainbands within cyclones based on their location and orientation with respect to synoptic-scale fronts: 1) warm frontal, 2) warm sector, 3) wide cold frontal, 4) narrow cold frontal, 5) prefrontal cold surge, and 6) postfrontal. Many theories have been advanced to explain the presence and motion of these bands. Some of the suggested dynamical forcing mechanisms for these rainbands include gravity waves (Bosart et al. 1998), conditional symmetric instability (Lindzen and Tung 1976; Sanders and Bosart 1985), quasigeostrophic circulations and transverse ageostrophic circulations due to fronts (Eliassen 1962), cold fronts aloft (Locatelli et al. 1998), boundary layer convergence associated with frontogenesis, vertical shear instability, and instability arising from melting hydrometeors. The complexity of rainband structure in extratropical cyclones is shown in Fig. 4.3, a series of plan position indicator (PPI) or horizontal radar displays, adapted from Shields et al. (1991). They documented using radar the existence of three mesoscale precipitation band types within this cyclone during its life cycle. Early in the cyclone's life cycle, narrow bands were oriented along a frontal boundary and associated with surface confluence (bands A and B). Later, bands along a region of surface convergence associated with frontogenetical forcing were evident (band C). Later still, multiple parallel bands (bands D–F) arose in a region of near-neutral conditional symmetric stability. Shields et al. (1991) argued that continued frontogenetic forcing along the developing cold front would lead to the eventual destabilization of the region and periodic release of conditional symmetric instability, which would lead to the development of the precipitation bands.

Convective bands within cyclones (but usually not close to the cyclone center), which typically occur in strong low-level wind shear with low instability environments, occur in conjunction with rapidly advancing cold fronts. These bands have been termed narrow cold frontal rainbands (NCFRs; Hobbs and Persson 1982; Carbone 1982; Roux et al. 1993; Parsons et al. 1987; Lemaitre et al. 1989; Wakimoto and Bosart 2000). The advancing pool of low-level cold air that forces ascent and precipitation along its leading edge sustains these bands. Thermodynamic instability above 2 km is usually marginal, so the vertical extent of the radar echo is usually much less than that of convective lines in the midwest United States that produce comparable surface maximum reflectivity. Although the basic structure of

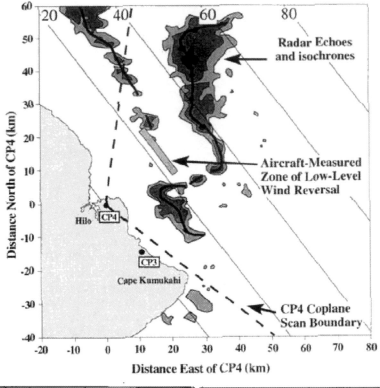

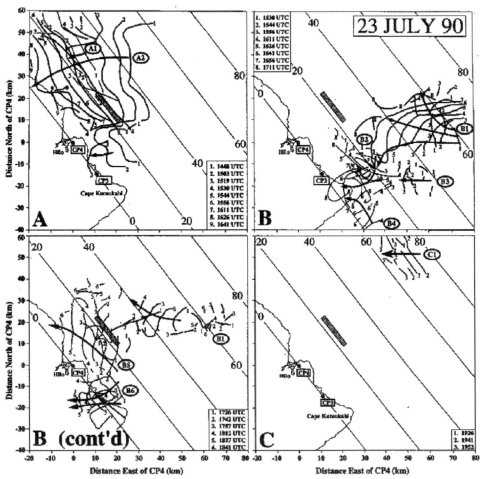

NCFRs has been known for about 20 years, recent observations by airborne Doppler radar has provided some new insights. Jorgensen et al. (2003) examines a fast-moving winter NCFR off the coast of California using the National Oceanic and Atmospheric Administration (NOAA) P-3 and observes low-to-middle level circulations near gap regions in the NCFR that resembled "bookend" vortices seen in midlatitude bow-echo systems (Fig. 4.4). Wakimoto and Bosart (2000) have provided some refinement in NCFR structure based on airborne Doppler radar data collected in the eastern North Atlantic during the Fronts and Atlantic Storm Track Experiment (FASTEX; Joly et al. 1997). Although the FASTEX case had structures producing much weaker maximum surface reflectivity than the Carbone (1982), Hobbs and Persson (1982), or Jorgensen et al. (2003) cases, the structure of the NCFRs was similar in the cases. Horizontal shearing instability was hypothesized to be responsible for the discontinuities, or gaps, in the NCFR horizontal reflectivity pattern. These gap regions are shown in the radar reflectivity and updraft schematic of Fig. 4.5, produced by Wakimoto and Bosart (2000), who contrast an earlier model of precipitation core structure proposed by Hobbs and Persson (1982) with the new airborne Doppler radar observations. The refinement in the conceptual model is primarily in the precipitation core–wind shift discontinuity relationship and in the preferred location of the updraft zone. A more abrupt discontinuity is evident on the southern region of the core. The modification of the NCFR gap region model implies that a shearing instability mechanism alone would not account for the observed structure. Other more recent research on NCFRs has also called into question the shearing instability mechanism (Locatelli et al. 1995; Browning and Roberts 1996; Brown et al. 1998, 1999). The numerical simulations of Brown et al. (1999) indicated that a trapped gravity wave was responsible for the precipitation core–gap regions, although no such indication of gravity waves was found in pressure perturbation retrievals calculated by Wakimoto and Bosart (2000) in the FASTEX NCFR. Braun et al. (1997) suggested that the spacing of gaps and cores of a NCFR approaching the coast were influenced by the orography of the coast as the overall shape of the NCFR mimicked the shape of the coast.

Another point of contention raised by the FASTEX case is the method of NCFR propagation. Although earlier studies indicated a density current mechanism for NCFR motion, Wakimoto and Bosart (2000) noted that the frontal boundary structure is not consistent with a hydraulic head structure characteristic of density currents in mean vertical cross sections, although reflec-

tivity structure in individual radar cross sections did resemble hydraulic heads. In addition, calculations of density current motion using aircraft in situ and Doppler data input into the classic theory equation did not match the observed motion of the front as a whole, although local regions of the front in the vicinity of the strong reflectivity cores did appear to propagate as a density current in the direction perpendicular to the major axis of the core. Clearly, more research (and improved datasets) is needed to understand more fully the dynamics of convective rainbands (including NCFRs) within extratropical cyclones.

b. Moderate CAPE MCS systems

Within the region of Fig. 4.1b marked by moderate to high CAPE and moderate to large shear lies the domain of tropical and midlatitude MCSs. Both tropical and subtropical systems usually exist in comparable shear and CAPE environments. Zipser and LeMone (1980), Jorgensen and LeMone (1989), and Lucas et al. (1994) have argued that the weaker vertical velocity strengths of oceanic convective systems versus continental thunderstorms possibly result from slightly lower tropospheric instability (and more efficient entrainment processes) over ocean regions than any inherent differences in tropical versus subtropical MCS structure. Indeed, the bulk radar structure of tropical and midlatitude squall-line systems and the character of the environments in which they are embedded seem identical (e.g., Houze 1981; Smull and Houze 1985). This similarity in structure and environment implies that similar dynamic processes, shaped by shear and CAPE, are operating regardless of latitude or surface character (i.e., ocean versus land).

The character of the initial convective forcing mechanism clearly plays a strong role in defining the character of the convective system. Indeed, though convective lines tend to be longer lived than individual storms (Bluestein and Jain 1985), they apparently coexist in the same environment (Parker and Johnson 2000). For example, Bluestein and Parker (1993), in their survey of 16 years of conventional radar data, found a variety of storm modes that formed along drylines, from isolated cells to cell clusters and convective lines, yet noted no significant differences in the local environments supporting the events. Hence, our understanding of the precise environmental mechanisms that lead to different modes of convective systems is still elusive and under active research. The climatology of Bluestein and Parker (1993) generally shows CAPE values of 1400–2260 J kg^{-1} for ordinary and severe squall lines, and CAPE

←

FIG. 4.2. (top) Radar domain used for the analysis of Hawaiian rainband radar data from 1418 UTC 23 Jul 1990. Shading represents the 0-, 10-, and 30-dBZ radar-echo contours from the CP-4 radar. Diagonal lines represent range (km) from the CP-3 and CP-4 radar baseline. (bottom) Isochrones of 0-dBZ echo (leading edge) at selected times. From Austin et al. (1996).

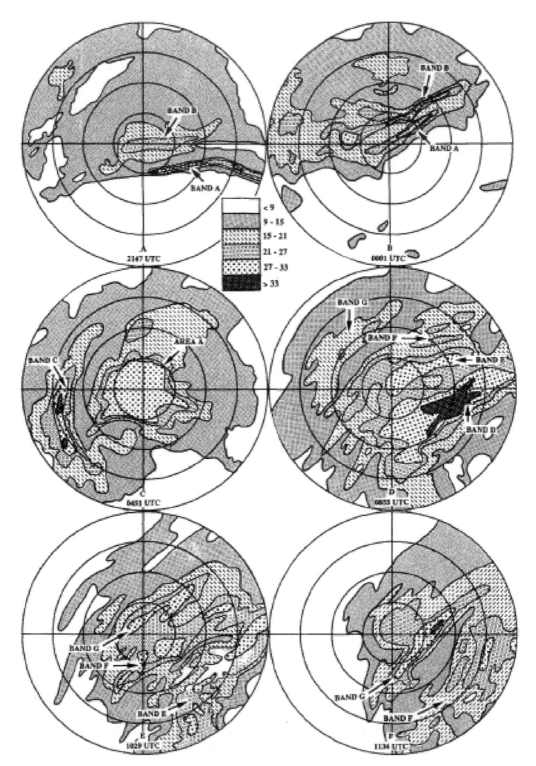

Fig. 4.3. Series of six 1.3° elevation angle PPI reflectivity images at various times during the 10–11 Feb 1988 cyclone over east-central Illinois. Various rainbands are labeled. The reflectivity factor is given in dBZ with the shading scale in the top-middle of the figure. The times are at the bottom of each PPI display. From Shields et al. (1991).

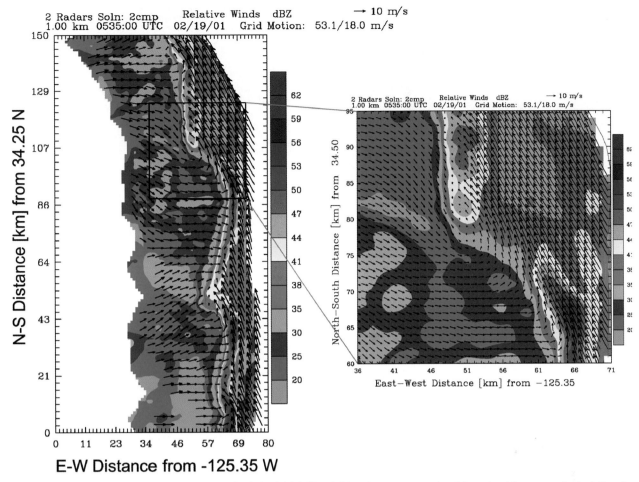

FIG. 4.4. Horizontal storm-relative winds and reflectivity field (left) at 1.0 km above mean sea level from the airborne pseudo-dual-Doppler analysis of the NCFR observed by the P-3 aircraft on 19 Feb 2001. The P-3 flight track is shown as the solid red line running approximately south to north along the right-hand side of the plot. A zoomed section of the plot, centered on the gap region of the NCFR, is shown in the right panel.

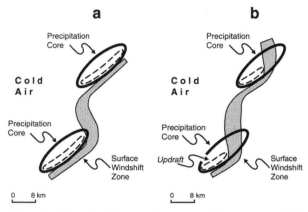

FIG. 4.5. Schematic diagram of two precipitation cores and their updraft areas within a narrow cold frontal rainband along with the associated surface wind shift zone along the band (a) adapted from Hobbs and Persson (1982) from a Pacific Northwest case; (b) revised schematic based on airborne Doppler radar data collected within an eastern North Atlantic cyclone by Wakimoto and Bosart (2000). The shaded region indicates the wind shift zone and the strong updrafts enclosed by dashed lines (from Wakimoto and Bosart 2000).

values of 1000–3200 J kg^{-1} for supercells and low precipitation (LP) thunderstorm cells. For that reason the domain of multicellular convective storms is fairly broad in Fig. 4.1a, with supercells alone shown at the highest level of shear.

1) TROPICAL MCSs

It is well known that the precise organization and evolution of convective lines depends critically on the interaction and orientation of the line with the vertical wind shear. Barnes and Sieckman (1984) studied the behavior of eastern Atlantic tropical convective lines during the Global Atmospheric Research Programme Atlantic Tropical Experiment (GATE) using conventional radar and sounding data. They found that for fast-moving lines (translation speed > 7 m s^{-1}) the vertical shear is approximately normal to the leading edge of the line. Slow-moving lines (translation speed < 3 m s^{-1}), on the other hand, are oriented more parallel to the shear vector. Although boundary layer thermody-

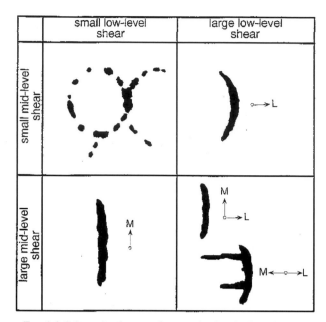

FIG. 4.6. Schematic of expected convective band orientation on the mesoscale as related to the vertical shear in a lower layer (1000–800 hPa) and midlevels (800–400 hPa) based on aircraft observations in TOGA COARE. Length of schematic radar echo is ∼100–300 km; (upper-left) line segments are up to 50 km in length. Cutoff between "strong" and "weak" shear for lower layer is 4 m s^{-1} from 1000 to 800 hPa (2 m s^{-1} per 100 hPa); for upper layer, 5 m s^{-1} from 800 to 400 hPa (1.25 m s^{-1} per 100 hPa). Arrows marked **L** are shear vectors for the lower layer; arrows marked **M** are shear vectors for the upper layer. From LeMone et al. (1998).

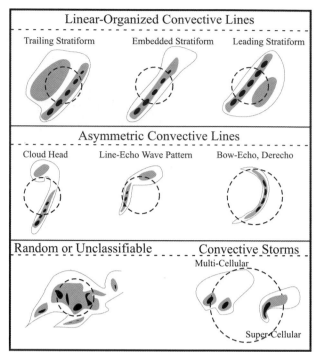

FIG. 4.7. Schematic base scan radar reflectivity for common archetypes of convective organization within warm sector mesoscale convective systems during the mature phase of each system's life cycle. Transformation of one type of system to another type is typical, usually in response to changes in environmental wind. Dashed circle represents the 50-km range ring. Levels of shading roughly indicate 20, 40, and 50 dBZ. Based on results presented by Parker and Johnson (2000), Fujita (1978), and Houze et al. (1989, 1990).

namic values and CAPE were similar for both fast- and slow-moving lines, the fast movers had a more pronounced minimum in equivalent potential temperature (θ_e) at midlevels, implying there was a greater potential to generate a more robust cold pool through evaporatively driven convective and mesoscale downdrafts (Zipser 1977). These results are verified and extended by LeMone et al. (1998), in a study of about 20 western Pacific MCSs that were investigated by Doppler-equipped aircraft during the Tropical Ocean Global Atmosphere Coupled Ocean–Atmosphere Response Experiment (TOGA COARE). They documented the relationship between convective line orientation and vertical shear of the horizontal wind and schematically summarized their results in Fig. 4.6. They noted that nearly all the convective bands occurring in environments with appreciable shear below the low-level jet maxima (1000–800 mb) are oriented nearly normal to the shear. Those bands tended to propagate in the direction of the low-level shear at a speed close to that of the wind maximum (e.g., Jorgensen et al. 1997). Conversely, if there was appreciable shear at midlevels (800–400 mb), convective bands form parallel to the shear and propagate more slowly than shear-perpendicular lines (Alexander and Young 1992; Lewis et al. 1998). When large shear occurs in both low-level and midlevel layers (lower right panel) the band orientation

depends on the alignment of the shear vectors. For midlevel shear normal to the low-level shear (upper figure in the lower right panel) the primary band remains two-dimensional with the band orientation parallel to the midlevel shear (and orthogonal to the low-level shear). If the two shear vectors oppose each other (lower figure in the lower right panel), the primary band is oriented normal to the low-level shear, but secondary bands extending rearward develop 3–4 h into the system's life cycle. For weak low-level and midlevel shear (upper left panel) convection often appeared "random" or to consist of short-lived band segments that formed along cold pools of previously existing systems.

2) MIDLATITUDE MCSs

With the advent of a network of operational radars covering most of the midwest United States, it became possible to more easily examine the evolution of many storm systems over several seasons, to devise a classification of common organization, and to relate the organizational modes to their environments. Figure 4.7 schematically illustrates an example of common organizational types observed in the spring and summer seasons in the midwest United States. Within the warm sector of midlatitude cyclones, usually observed in their mature stage late at night, and often associated with low-level jets (Augustine and Caracena 1994; Stensrud 1996b), the most frequently observed mesoscale organizational mode is that of a linear organized convective line (sometimes referred to as a "squall line"). If the line is propagating relatively rapidly perpendicular to the low-level wind shear vector, it often leaves behind a region of relatively horizontally homogeneous precipitation, termed "stratiform" rain, consisting of hydrometeors initially consisting of water or ice detrained from upper levels of the line convection. With time, this horizontally more uniform precipitation develops a vertical circulation of its own, which promotes further precipitation development (Rutledge 1986; Biggerstaff and Houze 1991; Gallus and Johnson 1995). Depending on the magnitude of the upper-level cross-line wind flow or vertical wind shear relative to the moving line, the stratiform region encompasses the convective line or even extends ahead of the convective line.

An individual MCS can evolve during its lifetime from one type to another, usually in response to its moving into a changed environment or having its circulation acted on by Coriolis forces (Houze et al. 1990; Scott and Rutledge 1995; Loehrer and Johnson 1995). Typically, an MCS will move from a symmetric to an asymmetric structural type as it ages, reflecting the action of Coriolis accelerations in amplifying its northern bookend cyclonic vortex over its southern anticyclonic one.

In their study of 88 *linear* MCSs that occurred in the spring of 1996 and 1997 in the central United States, Parker and Johnson (2000) found that virtually all of the warm sector systems (64 of the 88) occurred near a linear synoptic boundary, usually a warm or stationary front (27%), surface pressure trough (21%), cold front (19%), outflow boundary (9%), or dryline (7%). These results confirm earlier findings of Bluestein and Jain (1985) that the shape of lower-troposphere convergence is a critical ingredient for triggering linear MCSs. The Parker and Johnson (2000) 2-yr climatology documented 33 systems with trailing stratiform rain, 12 leading stratiform systems, and 12 embedded or parallel stratiform systems. Nearby wind profiler data were composited to produce mean wind fields to illustrate the differences in environmental flow for the three squall-line system types, as illustrated in Fig. 4.8. Not unexpectedly, the trailing stratiform MCS cases exhibited negative line-perpendicular storm-relative winds at all levels, consistent with case studies of these systems (e.g., Biggerstaff and Houze 1991). The parallel stratiform cases showed much stronger line-parallel winds than either of the other cases. It was also noted that the upper-tropospheric shear in leading stratiform MCSs was not significantly stronger than that in the other two classes, suggesting that upper-level shear in itself is not sufficient cause for leading stratiform rain. Indeed, Weisman et al. (1988) showed that low-level shear also significantly impacts the upper-level outflow. Other cases that have been investigated, however, have disclosed

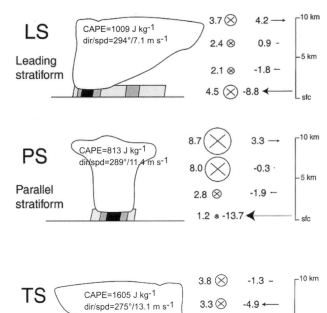

FIG. 4.8. Vertical profiles of layer-mean storm-relative pre-MCS winds for linear MCS classes. Wind vectors depicted as line-parallel (X) and line-perpendicular (→) components in m s^{-1}. Layers depicted are 0–1, 2–4, 5–8, and 9–10 km. Typical base scan radar reflectivity patterns (shading) and hypothetical cloud outlines are drawn schematically for reference. Average CAPE and propagation speed and direction are indicated inside the cloud schematic. MCSs's leading edges are to the right. From Parker and Johnson 2000.

stronger upper-level shear in leading stratiform squall-line cases (Grady and Verlinde 1997). Forward advection of hydrometeors is certainly aided by the stronger rear-to-front flow near storm top. The trailing stratiform systems tended on average to have the highest instability and most rapid west-to-east propagation. Consistent with the higher CAPE values (and stronger cold pools) of trailing stratiform cases was Grady and Verlinde's observation that those systems lasted nearly twice as long as either leading or parallel stratiform MCSs.

Vortices associated with MCS have been seen in recent Doppler radar studies (e.g., Jorgensen and Smull 1993; Chong and Bousquet 1999; Verlinde and Cotton 1990; Weisman 2001; Yu at al. 1999; Scott and Rutledge 1995). These vortices [termed mesoscale convective vortices (MCVs)] are sometimes long lived (i.e., they outlive their convective line origins) and have been known to be the seed regions of subsequent convective storms. An example of such a vortex associated with an oceanic squall line, observed in the Southern Hemisphere during the TOGA COARE experiment, is shown in Fig. 4.9. Various physical processes have been hypothesized for MCV formation. Most hypotheses have stressed the importance of vertical wind shear, originating from the undisturbed environment and/or con-

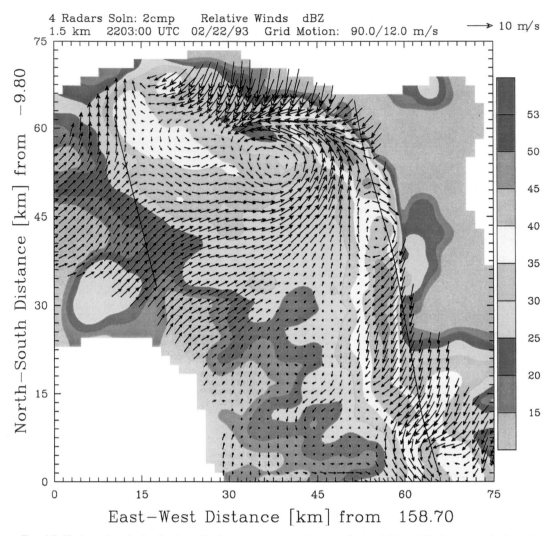

FIG. 4.9. Horizontal analysis of radar reflectivity and system-relative airflow at 1.5 km MSL from a tropical squall-line system observed by the two NOAA P-3 Doppler radar–equipped aircraft during TOGA COARE. The radar reflectivity color scale and the 10–m s⁻¹ scaling vector for winds are shown to the right and above the plot, respectively. Wind field is calculated using the "quad-Doppler methodology" (Jorgensen et al. 1996) from the fore and aft beams of the two NOAA P-3 aircraft radars, the tracks of which are shown as the two black lines running approximately from north-northwest to south-southeast on each side of the main reflectivity band. Adapted from Jorgensen et al. (1997).

vectively induced mesoscale flows, in the initiation of MCVs. The growth of these vortices, originally closely associated with the ends of convective lines or the locally "bowing" segments of convective lines, into larger vortices probably is related to interactions of the descending rear inflow current, which in turn is strengthened by sublimation, melting, and evaporative cooling in the stratiform region. Finally, in the Northern Hemisphere, Coriolis accelerations would tend to favor, over time, the strengthening of cyclonic vortices, typically seen at the northern end of bow-shaped line segments, a fact first noted in idealized simulations (Skamarock et al. 1994; Weisman and Davis 1998). Figure 4.10 illustrates the low-level circulations within idealized squall-line simulations. The northern bookend vortex is

clearly seen. The northern vortex shown in Fig. 4.9, however, was a Southern Hemisphere squall line, so the large northern vortex was not amplified by Coriolis accelerations.

Theoretical arguments as well as numerical simulations of idealized two- and three-dimensional squall-line systems have provided considerable insights into the dynamics of linear convective systems and their associated vortices. Ray (1990), starting with the early work of Thorpe et al. (1982) and Moncrieff and Miller (1976), reviews the significant work that has led to the recognition that shear is important in the maintenance of long-lived convective systems. Weisman et al. (1988) performed the most comprehensive and systematic numerical model investigation of the effect of shear orienta-

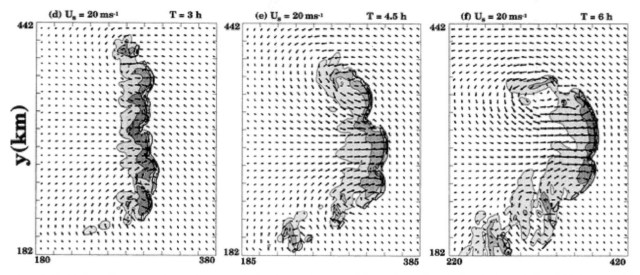

FIG. 4.10. Horizontal cross sections of system-relative flow, rainwater mixing ratio, and vertical velocity at 2 km AGL from model simulations at (d) 3, (e) 4.5, and (f) 6 h. The low-level wind shear is 20 m s^{-1} over the lowest 2.5 km. Coriolis accelerations are included in the model. Vectors are presented every four grid points (8 km), with a vector length of 8 km equal to a wind magnitude of 20 m s^{-1}. The rainwater is contoured for magnitudes greater than 1 g kg^{-1} (lightly shaded) and magnitudes greater than 3 g kg^{-1} (darkly shaded). The vertical velocity is contoured at 5-m s^{-1} intervals, with the zero contours omitted. Tick marks are 20 km apart. A domain speed of u = 18.5 m s^{-1}; v = −2.5 m s^{-1} has been subtracted from the flow field. From Weisman and Davis (1998).

tion, strength, and geometry on squall-line dynamics. Their results extended earlier findings that convective-scale downdrafts, originating near the leading edge thunderstorms owing to drag from raindrops and evaporative cooling, produce or augment a pool of cold air that spreads along the ground. Convergence and lifting produced by collision of the advancing cold pool with the warm and moist environmental air triggers new storms cells. These new storm cells are limited in location to the leading edge by the interaction of the low-level convergence with the vertical wind shear, particularly the shear in the lowest 2–3 km above the ground (Fig. 4.11). A theoretical explanation for this behavior is given by Rotunno et al. (1988) who hypothesized that the "optimal" conditions for triggering new storm cells along a spreading cold pool occur when there is a balance between horizontal vorticity produced by the cold pool with the vorticity inherent in the ambient shear. When that balance occurs, a vertically erect updraft is created at the leading edge that produces the deepest vertical lifting of the boundary layer air. Figure 4.11 also represents and helps explain the life cycle of convective systems in numerical simulations from a down-shear-tilted convective line to upshear-tilted convective elements as the MCS ages and its cold pool becomes stronger (Weisman 1993).

Asymmetrically organized systems (termed a "cloud head" from their resemblance to cyclonic storms) often are associated with slightly stronger 0–6-km vertical wind shear, principally in the along-line direction, than symmetric systems (Houze et al. 1990). This stronger along-line component of shear would favor the development of supercell storms, particularly on the south-

western end of the line giving the line an asymmetric appearance on radar. Additionally, modeled convective lines often generate counterrotating midlevel vortices at their ends in response to the development of a strong rear inflow jet at the midpoint of the line (Weisman and Davis 1998). With time, Coriolis acceleration acting on the inflowing air would decay the southern vortex, leading to the dominance of the northern one. This vortex would promote asymmetric precipitation structure by advecting dry midlevel air into the southern half of the system while advecting hydrometeors from the convective line rearward within the northern half of the system.

Line-echo wave pattern echoes (Fig. 4.7) are also a commonly observed convective storm type that are seen along prefrontal squall lines and dryline environments and along stationary or warm frontal surface boundaries, where the precipitation ahead of the advancing north–south line is formed as a result of the low-level jet overrunning the warm front (Smull and Augustine 1993; Hane and Jorgensen 1995). Often, when the shear is very weak, a random or "unclassifiable" pattern of precipitation results. Houze et al. (1990) noted in their climatology of springtime rainstorms in Oklahoma that these nonlinear chaotic systems produced as much severe weather as the linearly organized cases. When compared to linear cases, they also noted a higher frequency of hail occurrence and less likelihood of tornadoes with these random convective patterns.

Often, radar observations of MCSs show a characteristic "bow-echo" shape (Fujita 1978). Frequently these systems produce long swaths of straight-line wind damage and some small tornadoes (Przybylinski 1995). Weisman (2001) has provided an excellent review of

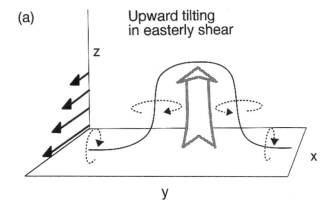

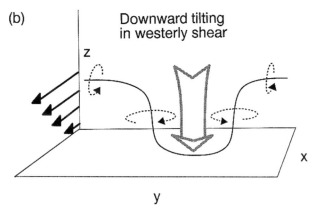

FIG. 4.11. Schematic illustration of vertical vorticity generation through vortex tilting within a finite length convective line. For an updraft in easterly shear (a) the ascending motion in the convective leading line pushes the vortex lines upward resulting in cyclonic rotation on the north end and anticyclonic rotation on the south end of the system. Downward motion in westerly shear can contribute to the same vorticity pattern (b), analogous to mechanisms proposed for rotation generation in downdrafts of splitting supercells. From Weisman and Davis (1998).

bow-echo structure and dynamics starting from the early work of T. T. Fujita. The life cycle of bow-echo systems was documented through the pioneering work of Fujita (1978) to include the kinematics of the MCS. Fujita's conceptual evolution included the generation of stong straight-line surface winds (sometimes termed a "derecho;" Johns and Hirt 1987; Bernardet and Cotton 1998) at the apex of the bow at the mature stage, and a rotating cloud-head echo at its northern region during its dissipating stage. Evans and Doswell (2001) documented the environments of derecho and nonsevere bow-echo systems over an 11-yr period. Their results show that derechoes develop in a wide range of shear and instability conditions, which suggests that ambient wind shear and instability conditions alone are not sufficient to distinguish derecho environments from nonsevere MCS conditions. Moreover, the strength of the

mean flow seems a better discriminator between derechoes and nonsevere events given similar thermodynamic environments for both. Also, when synoptic-scale forcing is strong, derechoes can sometimes develop and persist within environments with little CAPE.

More recent simulations of bow-echo systems by Weisman (1993) documented that the strongest, most organized bow echoes occur primarily in high CAPE (i.e., >2000 J kg^{-1}) and moderate to strong vertical wind shear (>15 m s^{-1} over the lowest 2.5–5.0 km). These systems also had elevated rear inflows and well-defined bookend or counterrotating midlevel vortices at the ends of the convective line. It is also interesting to note that the bow echoes were most favored in simulations when the shear was confined to the lowest 2.5 km. Simulations with a deeper shear layer produced more isolated supercellular storms rather than a bow-shaped MCS.

Many observational and numerical modeling studies have shown that long-lived vortices can arise from MCSs, even from initially linear convective lines (e.g., Menard and Fritsch 1989; Bartels and Maddox 1991; Fritsch et al. 1994; Bartels et al. 1997; Davis and Weisman 1994; Weisman and Davis 1998; Zhang and Fritsch 1988; Trier et al. 2000). Furthermore, these vortices are often the initiation mechanism for subsequent regeneration of convection during the next diurnal cycle. The small scale of these convectively generated vortices makes them difficult to resolve in numerical forecast models; thus, subsequent precipitation forecasts can be adversely affected. Weisman and Davis (1998) and Skamarock et al. (1994) clarified, in a series of numerical simulations, the role that Coriolis forces play in the development of a dominant cyclonic vortex arising from an initially linear convective line (Fig. 4.10). The model was initialized with five convective cells along a 150-km-long line in a moderate CAPE environment (220 J kg^{-1}) with strong low-level vertical wind shear (20 m s^{-1} over the lowest 2.5. km). Little difference in the 2-km wind and precipitation structure between simulations with and without Coriolis terms resulted after 3 h of integration, but by 6 h in the non-Coriolis simulation a strong bow echo was produced by the model with cyclonic and anticyclonic vortices behind the northern and southern ends, respectively, of the convective line. The addition of Coriolis terms caused the northern cyclonic line-end vortex to strengthen with time, and the southern anticyclonic vortex to weaken, resulting in a highly asymmetric system after 6 h. One of the outstanding issues still to be resolved involves the details of how these original northern-end vortices grow upscale to become a large MCV that is often seen in early morning satellite pictures of MCSs after the convection has dissipated (Bartels and Maddox 1991).

These bow-echo simulations, as well as other squall-line simulations (e.g., Weisman 1993; Chin and Wilhelmson 1998) and radar observations of squall lines and bow echoes (Przybylinski 1995), show the development of a "rear-inflow" jet as the systems mature. This flow is often indicated on radar imagery by a notch of weaker echo in the trailing stratiform rain (Braun and Houze 1997; Smull and Houze 1985). The flow can sometimes be so strong that it affects the mesoscale surface pressure pattern (Loehrer and Johnson 1995) and creates a "wake low" or "heat burst" that is far to the rear of the convective line. Strong, elevated rear inflow is characteristic of strongly sheared bow-echo systems. Sometimes this strong flow can reach the surface and produce damaging surface winds. In a two-dimensional framework, Weisman (1993) proposed a hypothesis for the development of rear inflow based on the Rotunno et al. (1988) and Lafore and Moncrieff (1989) ideas for the "relative" balance of opposing circulations produced by cold pools and low-level vertical wind shear. According to this hypothesis (Fig. 4.11) the tilt and rear inflow of convective systems evolve as the cold pool strengthens. Initially, before the development of a strong cold pool (Fig. 4.11a), the convection is downshear tilted but evolves to an upshear-tilted system (Fig. 4.11c), depending on the relative strengths of the cold pool–induced vorticity and the vorticity in the ambient low-level wind shear. As the system tilts upshear, a rear-inflow jet is generated in response to the mesoscale low pressure that is hydrostatically induced just to the rear of the convective line, owing to the buoyant front-to-rear ascending motion flowing back over the cold pool (LeMone et al. 1984; Jorgensen et al. 1997; Trier et al. 1997). For most convective systems, this rear inflow descends as it approaches the leading convective line and enhances the spreading cold pool. However, as pointed out by Weisman (1993), for bow-echo systems, with their stronger sheared, larger CAPE environments, the rear inflow can remain elevated and helps to promote stronger leading edge convergence and a more vigorous and longer-lived system. Moreover, the development of line-end vortices also contributes to system severity by enhancing local regions of midlevel rear inflow between the vortices, and by enhancing surface outflow and leading edge convergence. Weisman (1993), Weisman and Davis (1998), and Trier et al. (1997) studied the origin of vertical vorticity within simulated convective systems. Although the initial source of vorticity for the line-end vortices in strongly shear systems was downward tilting of ambient westerly shear within the cells at the end of the line, as the system matured and tilted upshear, the primary source for vorticity generation was the upward tilting of system-generated easterly shear. These two mechanisms are schematically illustrated in Fig. 4.12.

The sensitivity of physical processes to modeled arc-shaped convective line elements and storm cells that were observed on 4–5 September during the GATE experiment was documented by Chin and Wilhelmson (1998). Their results are shown in Fig. 4.13. Although large-scale lifting in the numerical simulation is critically important in determining the initial structure of

EVOLUTIONARY STAGES

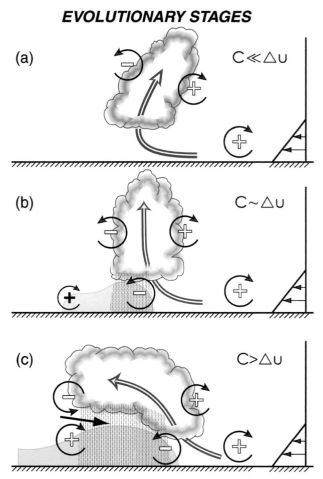

FIG. 4.12. Three stages in the evolution of an idealized two-dimensional convective line developing in a strongly sheared, large-CAPE environment. Here C is the circulation (i.e., vorticity) associated with the cold pool, Δu the circulation associated with the ambient environmental low-level wind shear. In the early stages (a) C is weak compared to the Δu; (b) the two circulations are comparable; and (c) C is strong compared to Δu. The updraft is denoted by the thick, double-lined flow vector, with (c) the rear-inflow current denoted by the thick solid vector. The shading denotes the surface cold pool. The thin, circular arrows depict the most significant sources of horizontal vorticity, which are either associated with the ambient shear or with the cold pool. Regions of lighter or heavier rainfall are indicated by the more sparsely or densely packed vertical lines, respectively. The scalloped line denotes the outline of the cloud. Adapted from Weisman (1993).

the convective band, once the band is formed the subsequent evolution is nearly identical to that in the control simulation when the forcing is turned off after 2 h (Fig. 4.13b). Much more important is the role of evaporative cooling in band evolution (Fig. 4.13c). With evaporative cooling turned off, no cold pool developed and the early cells died without forming the arc-shaped band. On the other hand, when a drier midlevel sounding (Fig. 4.13d) was used, the band was qualitatively similar to the control run, but precipitation was more intense and the system was longer lived. The organization of convective elements was dramatically influenced by the low-level

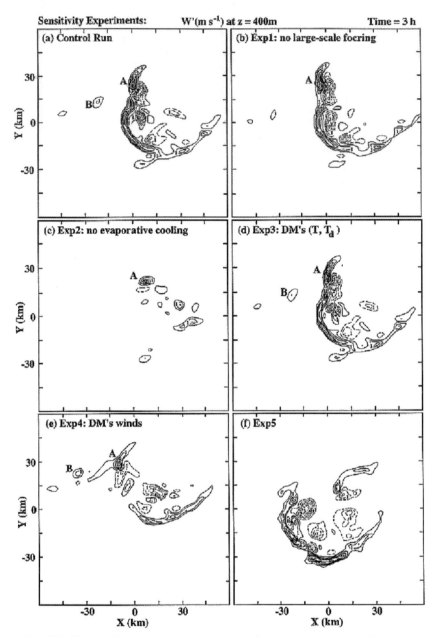

Fig. 4.13. Horizontal cross sections of vertical velocity (m s^{-1}) at 3 h and at 400-m height for a series of numerical experiments investigating squall-line sensitivity to moisture and environmental wind: (a) the control run with full physics, (b) for large-scale forcing turned off after 2 h of simulation, (c) without evaporation of rainwater, (d) using a modified sounding with a drier midlevel layer, (e) using a modified wind profile with weaker low-level veering and stronger linear shear above, and (f) using a modified wind profile with no 1000–850-hPa wind veering. Labels "A" and "B" refer to Supercell-like storm cells. From Chin and Wilhelmson (1998).

wind profile (Figs. 4.13e,f). Weakening the wind speed of the veering profile (while maintaining the hodograph curvature) strongly influenced the band evolution, but the supercell-like cells remained. Eliminating the strong curvature in the 1000–850-hPa wind profile eliminated the strong cells while still allowing the arc-shaped band to evolve. Clearly, these results reinforce the concept

that the dynamical evolution of convective lines is strongly dependent on the environment and the character of the initial forcing.

The examination of small and shallow supercell-like storms in rainbands associated with landfalling hurricane environments has been documented observationally by McCaul (1991) and simulated by McCaul and

Weisman (1996). These environments have been known to produce weak tornadoes. In spite of limited CAPE, the strong low-level vertical wind shear associated with landfalling tropical systems is sufficient to lead to perturbation pressure minima comparable to those resulting from simulations of Great Plains storms. However, the strong low-level updrafts found in simulated storms need to be verified by direct observation.

c. High CAPE regimes

As pointed out by Carbone et al. (1990b), the most organized forms of deep convective storms occur in moderate to high shear environments. This is the shaded regime labeled "domain of multicell storms" on Fig. 4.1a. Ray (1990) provides a comprehensive review of the state knowledge of moderate to high shear storms. Church et al. (1993) provide a comprehensive review of the structure and dynamics of tornadoes. Within this volume, Bluestein and Wakimoto (2003) review the results of recent field projects that have examined high shear/CAPE convective storms, particularly the results from the Verification of the Origins of Rotation in Tornadoes Experiment (VORTEX).

Thunderstorms evolving in weakly sheared environments (labeled "ordinary convective storms" in Fig. 4.1a) have been studied since the early days of radar meteorology (Byers and Braham 1949). These storms, historically referred to as "airmass thunderstorms," are noted for their single-cellular nature and their ability to produce microbursts from intense downdraft air in (sometimes) deep moisture environments. Occasionally, weak to moderate strength tornadoes have been observed with these storm types. A study by Kingsmill and Wakimoto (1991), using Doppler radar and photogrammetry data collected during the Microburst and Severe Thunderstorm (MIST) project during the summer of 1987 in Alabama, documented the life cycle and structure of such a thunderstorm. They confirmed the life cycle features originally documented by the Thunderstorm Project and added several new features to that conceptual model. One new feature was a midlevel downdraft associated with a weak echo trench driven by negative thermal buoyancy. Their conceptual model of the three Byers and Braham stages (developing or cumulus, mature, and dissipating) of the thunderstorm life cycle is show in Fig. 4.14. In the early stages, the storm is dominated by an updraft with a vorticity couplet at cloud base in spite of the weak vertical wind shear. A simple vortex tilting effect such as that shown in Fig. 4.14a may be acting to create the cloud-base vorticity couplet shown in Fig. 4.14. The mature stage is still dominated by the updraft (Fig. 4.14b), but the descending precipitation core splits the updraft, while the midlevel downdraft exists higher in the storm. Byers and Braham (1949) in their classic study may have mistakenly identified this downdraft as a deep, precipitation-laden downdraft that reaches the surface. Also, in the

mature stage of the storm (Fig. 4.14b), midlevel inflow creates a visible cloud constriction. In the dissipation stage (Fig. 4.14c) the precipitation-associated microburst downdraft (largely irrotational) accelerates downward, previously driven by water loading, but at this stage driven by negative thermal buoyancy as it becomes displaced from the precipitation core.

3. Boundary layer processes that initiate and organize convective systems

It has long been known that boundary layer convergence zones are precursors to convective development and organization. These low-level convergence zones often act to locally deepen the moist layer and create conditions favorable for deep convection. Such boundaries may provide the initial forcing to any of the convective systems shown in Fig. 4.1. The subsequent organization is then dictated by the CAPE–shear values, as described in section 2.

An example of a boundary layer convergence zone initiating thunderstorms is shown in Fig. 4.15. The top panel shows a well-defined radar reflectivity thin line while the bottom panel shows deep convective development at the same location less than 2 h later. Byers and Braham (1949) were the first to note a systematic relationship between low-level convergence zones and convection initiation. Using surface data from the Thunderstorm Project in Florida, they observed these zones 20–30 min prior to the appearance of precipitation echoes. Subsequent studies also showed that surface mesonet anemometers detected convergence 15–90 min prior to the onset of convective rainfall (Ulanski and Garstang 1978; Garstang and Cooper 1981; Achtemeier 1983; Watson and Blanchard 1984). Purdom (1982) used satellite imagery to conclusively show that boundaries, as indicated by shallow linear cloud bands, occur prior to deep convective development.

The utility of weather radars in detecting clear air motions has increased since the development of more sensitive research radars and the Weather Surveillance Radar (WSR-88D) network. Wilson and Carbone (1984) proposed the use of sensitive Doppler radars to monitor convergence zones in both cloudy and clear air. Using Doppler radar data in the Denver area, Wilson and Schreiber (1986) observed fine lines of enhanced reflectivity, which are known to be indicative of boundary layer convergence zones. Figure 4.16a, from Wilson and Schreiber (1986), shows the apparently random geographic location of first echoes in the region. Figure 4.16b, however, shows that when the cell generation locations are plotted relative to convergence lines, there is strong evidence that the location of convection initiation is deterministic. In fact, they found that 80% of thunderstorms in the area were initiated close to boundary layer convergence zones. This was one of the groundbreaking studies clearly illustrating the utility of

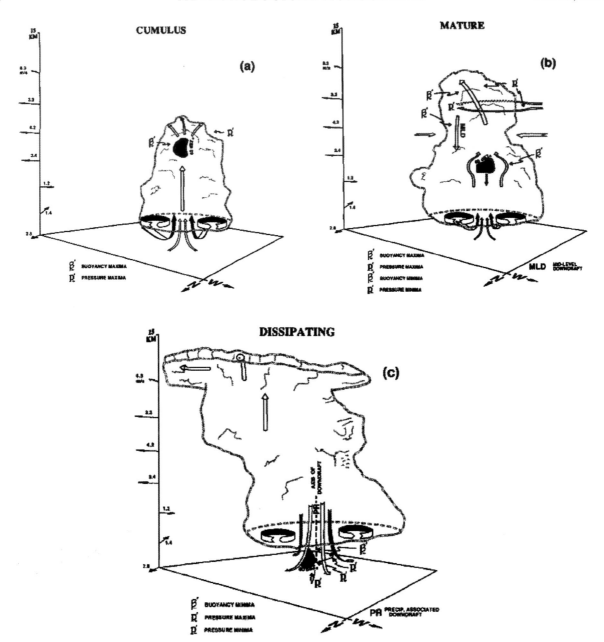

FIG. 4.14. Conceptual model for the various life cycle stages of a convective storm in a weakly sheared environment: (a) developing or cumulus; (b) mature; and (c) dissipating. Size of arrows is indicative of the magnitude of the airflow in a particular region. Regions of maximum or minimum perturbation pressure and thermal buoyancy are also indicated. From Kingsmill and Wakimoto (1991).

radar data in detecting boundaries and identifying likely regions for convective development.

Early studies suggested that the cause of the clear air weather radar return was index of refraction gradients (e.g., Atlas 1960), but subsequent research using dual-wavelength and polarimetric radars clearly showed that scatterers within the mixed layer are insects (e.g., Wilson et al. 1994). This is a fortunate characteristic for meteorologists so they may use the insects as "markers" of the boundary layer clear air convergence zones. Above the mixed layer, the clear air return is likely due to refractive index gradients (Wilson et al. 1994). If one assumes that the insects are carried by the wind, then continuity dictates that the insects would be carried into the low-level convergence zones, lifted by the mesoscale updrafts, and spread by divergence at the top of the convective boundary layer. Thus, there would not be a concentration of insects to create an enhanced low-level reflectivity zone. Achtemeier (1991) showed, however, that the insects are indeed carried into the low-level convergence zones but then fold their wings to strive to remain at low levels where the temperatures are

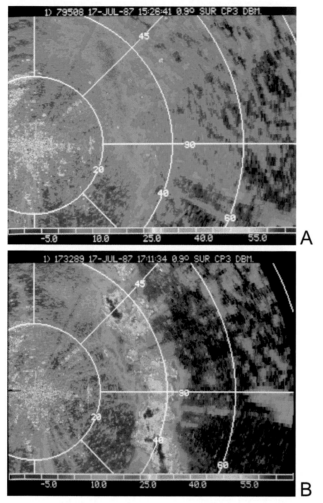

FIG. 4.15. Radar example of convection initiation along a boundary in the Colorado Front Range: (a) low-level reflectivity fine line at 1526 UTC 17 Jul 1987; (b) convection generated along boundary shown in (a) at 1711 UTC on the same day. Figures courtesy of Jim Wilson.

warmer. Thus, there is a net convergence of insects in the column of the mesoscale updraft and a corresponding increase in radar return in that region.

Although the radar aptly identifies locations of boundaries, it is also necessary to have detailed thermodynamic measurements to obtain an accurate assessment of convective potential. Crook (1996) performed model sensitivity studies of the convection initiation processes associated with convergence zones. He found that convection initiation is sensitive to the magnitude of the low-level vertical gradients of temperature and moisture. In particular he noted that changing the magnitude of the vertical temperature gradient by only 1°C made the difference between no simulated storm and intense convection. The strength of the simulated storm varied significantly when the magnitude of the low-level moisture gradient changed by only 1 g kg^{-1}. These critical temperature and moisture variations are within the range

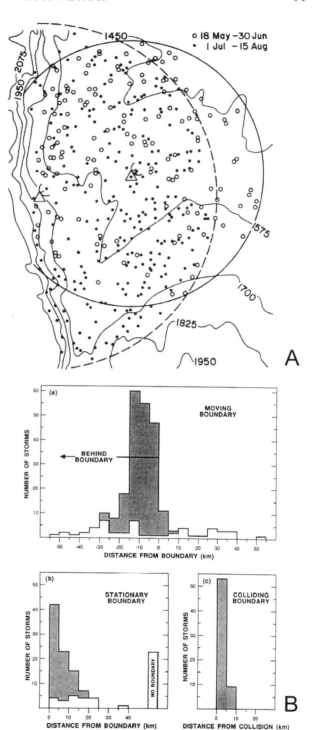

FIG. 4.16. (a) Storm initiation locations in the Colorado Front Range during the summer of 1984. The contours are ground elevation (m). (b) Number of storms initiated relative to moving, stationary, and colliding boundaries. The shading represents the cases classified as being initiated by boundaries. From Wilson and Schreiber (1986).

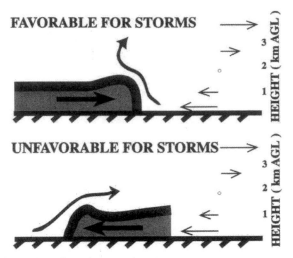

FIG. 4.17. Schematic illustration of conditions favorable (top) and unfavorable (bottom) for storm development. The wind vectors on the right represent the environmental wind profile, which is the same for both cases. The dark shading represents a density current. The arrow in the density current represents its motion. The curved arrow ahead of the density current represents the updraft tilt. From Wilson et al. (1998).

of typical observational boundary layer variability (e.g., Weckwerth et al. 1996).

In addition to knowing the locations of boundaries and the detailed thermodynamics to understand convection initiation processes, it is also necessary to assess the ambient wind shear profile. Numerical simulations (e.g., Moncrieff and Miller 1976; Droegemeier and Wilhelmson 1985; Rotunno et al. 1988) and observational studies (Wilson and Megenhardt 1997) have shown that the lifetime of storms and squall lines is dependent on the relationship between the environmental wind profile, the storm motion, and the motion of the boundary that forced the convection. Figure 4.17 schematically illustrates that if the boundary moves with the storms, then long-lived convective systems are likely, but if the boundary moves away from the storms, the storms tend to dissipate.

a. Boundary intersections

Purdom and Marcus (1982) used satellite imagery to determine that 73% of afternoon thunderstorms in the southeastern United States are triggered by the interactions of outflow boundaries. Wilson and Schreiber's (1986) radar study found that 71% of boundary collisions in the Denver area initiated thunderstorms. The resulting storms were relatively intense and often evolved into squall lines. Several detailed observational studies have clearly shown that thunderstorms developed near the collision regions of two boundaries (e.g., Droegemeier and Wilhelmson 1985; Mahoney 1988; Carbone et al. 1990a; Intrieri et al. 1990).

Rotunno et al. (1988) showed that boundary collisions are more likely to initiate storms if the horizontal vor-

ticity associated with the boundaries are in approximate balance. This creates vertical updrafts and therefore more optimal conditions for convective development. In particular, Carbone et al. (1990a) observed such a balance after the collision of two boundaries. This vorticity balance, however, is not a sufficient condition for deep convective development. Numerical simulations examining the intersections between boundary layer convergence lines by Lee et al. (1991) showed that storm initiation is sensitive to the amount of moisture, strength of convergence, and shear values. They found that increased low-level moisture creates stronger and taller modeled storms, and that variations in boundary layer convergence affected the timing and character of the modeled storms while a threshold of shear above the boundary layer inhibited the convective development of the modeled storm.

Boundary collisions, however, do not always produce convection (Wilson and Schreiber 1986; Stensrud and Maddox 1988). Kingsmill (1995) clearly illustrated this in a case study of a sea-breeze front-gust front intersection. He noted that convection along the separate boundaries was more likely as the boundaries approached existing cumulus clouds. After the boundaries collided, however, the depth of the convergence zone decreased due to the effects of the relatively shallow sea-breeze front. Convective initiation did not occur along the boundary without pre-existing clouds to intercept. The existence of pre-existing small cumulus clouds was also previously shown to be a key ingredient for deep convective development with moving boundaries (e.g., Wilson and Mueller 1993; Kingsmill 1995; Hane et al. 1987; May 1999).

Convection initiation associated with some specific mesoscale boundary layer forcing mechanisms will be discussed. These mechanisms include gust fronts, sea-breeze fronts, drylines, horizontal convective rolls, and topographically induced boundaries.

b. Gust fronts

The rain-cooled downdraft air of thunderstorms creates a low-level cold pool characterized by density differences between it and the ambient air. Gust fronts are the leading edges of such cold air outflows. These density currents have been studied using tower data (e.g., Charba 1974; Goff 1976), Doppler radar data (e.g., Wakimoto 1982; Mueller and Carbone 1987), wind profilers with Radio Acoustic Sounding System (RASS; May 1999), and numerical modeling (e.g., Seitter 1986; Droegemeier and Wilhelmson 1985). Of all the boundaries examined in the High Plains by Wilson and Schreiber (1986), gust fronts were most often associated with convection initiation.

Waves are often produced along gust fronts and the interactions between the waves and the outflow boundary have been shown to influence convective development (e.g., May 1999). Carbone et al. (1990a) showed that vertices in a line-echo wave pattern (a series of 80–150-km long arcs in radar reflectivity) along a gust front were the preferred locations for surface convergence and for convective development. Deep convection occurred along the gust front only after interaction with the dryline and/or low-level jet. This outflow boundary first propagated as a density current and evolved toward an internal undular bore. Dual-Doppler radar analyses also showed waves generated atop the cold air outflow (e.g., Mueller and Carbone 1987; Weckwerth and Wakimoto 1992). Weckwerth and Wakimoto (1992) found that internal gravity waves and Kelvin–Helmholtz waves influenced the initiation and organization of small convective cells generated at the gust front.

Proper representation of such cold pools in numerical models has been shown to be essential for an accurate prediction of deep convective events, especially when the synoptic-scale forcing is not strong (e.g., Stensrud and Fritsch 1994; Stensrud et al. 1999). Thus, simulations must either explicitly resolve cold pool circulations, or parameterizations in models with lesser resolution must accurately represent the convergence, and circulations within, and the variability of the gust fronts.

c. Sea-breeze circulations

The sea-breeze circulation occurs nearly daily in many coastal regions. This is a locally induced onshore (sea breeze) or offshore (land breeze) flow driven by differential heating between land and sea. In the sea-breeze circulation, the transition region between the warm, dry continental air and the cool, moist marine air often exhibits changes in wind direction and is called the sea-breeze front. The propagation speed, inland penetration, and strength of the frontal circulation are related to the synoptic flow (Arritt 1993; Atkins and Wakimoto 1997). Simpson (1994) provides a detailed description of the physics and characteristics of this circulation.

It is well-known that the sea-breeze front can produce sufficient lifting for convection initiation (e.g., Pielke 1974; Pielke and Mahrer 1978). Convective development along the sea-breeze front is influenced by the shape of the coast (e.g., Neumann 1951; Purdom 1976), other small-scale low-level circulations, and interactions with other boundaries and waves. Local updraft maxima were simulated at points where the convex curvature of the coastline accentuated the sea-breeze convergence zone (Pielke 1974). Idealized numerical simulations of Florida convection associated with a sea-breeze front showed that the distribution of initial soil moisture, the curvature of the coastline, and the circulations associated with Lake Okeechobee all influenced the timing and location of subsequent precipitation (Baker et al. 2001). Laird et al. (1995) performed a detailed Doppler radar, aircraft, sounding, and surface mesonet analysis of the Cape Canaveral sea-breeze circulation and observed that variations in the direction of the sea breeze

in the vicinity of irregular coastlines can lead to persistent convergence zones behind the sea-breeze front. These zones can locally increase the depth of the onshore low-level sea-breeze air and create vertical motions atop the sea breeze, which can support deep convective development. Carbone et al. (2000) found that all MCSs generated over the Tiwi Islands off the coast of Australia can be traced backward in time to convection initiated by the sea-breeze circulations. The evolution of the MCSs is then dependent upon the location and movement of interactions among sea breezes, gust fronts, cumulus clouds, and existing storms (Wilson et al. 2001).

With high-resolution numerical models, Kelvin–Helmholtz billows have been simulated along sea-breeze fronts (e.g., Sha et al. 1991; Dailey and Fovell 1999). Similar to the Kelvin–Helmholtz instability billows along gust fronts, these waves create enhanced updraft regions atop the cold air behind the front and may enhance convective development (Rao et al. 1999).

d. Drylines

The dryline is a zone of moisture gradient(s) separating the warm, moist air flowing northward from the Gulf of Mexico and the hot, dry air flowing eastward from the elevated terrain of the southwest United States and northern Mexico. Another way to view the dryline is that it represents the intersection of the top of the moist layer to the east with the terrain that slopes upward to the west. The dryline typically ranges from 500 to 1000 km in length. Strong cross-dryline gradients in moisture occur on a scale of 2–20 km that cannot be adequately sampled by the traditional sounding network (e.g., Parsons et al. 1991). Multiple dryline gradients frequently occur with "secondary" drylines located to the east of the stronger moisture gradient (e.g., Hane et al. 1993; Crawford and Bluestein 1997; Hane et al. 2001). Dryline evolution is closely linked to the diurnal cycle as daytime heating produces strong vertical mixing in the lower levels, which preferentially brings down drier air on the west side of the dryline and leads to a tightening of the low-level moisture gradient and enhances low-level convergence (Schaefer 1986).

The dryline has been the object of numerous investigations, likely due to the frequency of convective development along it. Nearly 40 years ago, Rhea (1966) called attention to the frequency of convection initiation along the dryline during the spring and early summer months. Bluestein and Parker (1993) used 16 years of radar data to identify modes of isolated severe storm development along the dryline. Critical conditions necessary for severe convective development include small values of convective inhibition (CIN), high CAPE values, and deep tropospheric wind shear. A 3D modeling study showed that storms form along the dryline when moisture convergence, due to the thermally direct secondary dryline circulations, is sufficient to destabilize

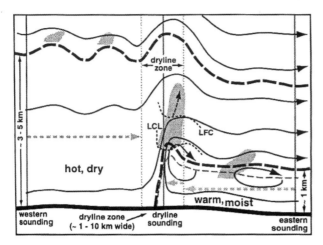

Fig. 4.18. Conceptual model of dryline environment during the afternoon and early evening. Dryline position is shown relative to cumulus clouds (gray shading) and streamlines (solid arrows). The black, dashed lines represent boundaries of the region, which includes the dry convective boundary layer (west of dryline) and elevated residual layer (east of dryline and above moist layer). Gray, dashed arrows indicate the surface of zero westerly wind. The three vertical gray lines denote the locations of soundings used to deduce the conceptual model. From Ziegler and Rasmussen (1998).

the local sounding to a state supportive of deep convection (Ziegler et al. 1997). Ziegler and Rasmussen (1998) summarized their case study in a schematic (Fig. 4.18), which shows the dryline relative to developing clouds and airflow streamlines. For development of deep convection it is essential that the moist boundary layer air parcels reach their lifted condensation level (LCL) and level of free convection (LFC) prior to leaving the mesoscale updraft zone.

Similar to that observed along other boundary layer convergence zones, convection is seldom initiated uniformly along the entire length of the dryline. Preferred locations for convective development along the line include (i) larger-scale features creating capped zones and sometimes subsequent internal gravity waves, (ii) mesoscale low pressure areas along the dryline, and (iii) boundary layer circulations intersecting the dryline. McCarthy and Koch (1982) and Koch and McCarthy (1982) attributed along-line variability in convection to perturbations associated with gravity waves. The dryline and gravity wave interactions created enhanced moisture convergence. Sanders and Blanchard (1993) also found a relationship between dryline convection and gravity waves. They found that localized lifting of a capping inversion created waves that influenced convective development along a small sector of a western Kansas dryline. Bluestein et al. (1988) observed localized differences in low-level diabatic heating, which led to pressure decreases in the dry air that locally enhanced low-level convergence and led to convection initiation. Along-line variations have been the focus of some recent radar studies. Hane et al. (1997) observed the formation of a cloud line in the dry air west of a dryline. The

cloud line apparently formed owing to landscape and land use differences. The intersection of the cloud line with the dryline produced enhanced convergence that led to development of tornadic storms in northwest Oklahoma. Hane et al. (2002) in another case observed convection initiation at the intersection of a cloud line with a rapidly advancing dryline. The cloud line developed over a region that had received the heaviest rainfall during the previous night. Along another section of this dryline, no storms formed owing to cooler boundary layer air resulting from reduced daytime heating over an area that had received significant rainfall during the previous night. Along a third section of the dryline, deep convection developed owing to enhanced convergence from backed winds in locally moist air in response to decreased pressure in the warm air to the northwest. Atkins et al. (1998) found that variations in cloud development along the dryline were due to intersections with horizontal convective rolls. They observed reflectivity maxima and sometimes cloud formation at the intersection points.

Observations of intense convection have been made with fronts merging and/or approaching the dryline (e.g., Schaefer 1986; Parsons et al. 1991). Bluestein et al. (1990) observed locally enhanced convergence and local deepening of the moist layer at the intersection between a dryline and outflow boundary. Cumulus congestus and a small cumulonimbus formed near the intersection. Convection initiation often occurs at the "triple point," which is the intersection region between a baroclinic boundary and dryline. In a case study of a dryline–gust front intersection, Weiss and Bluestein (2002) found that the dryline circulation being lifted upward by the approaching gust front possibly caused convection initiation. Alternatively, they theorized that a cause of the locally deep updrafts might have been that the gust front acted as a primer for deep convection by lifting air parcels to their LCL prior to the interaction of the two boundaries. Convective development does not always occur at intersection regions, however. In a severe squall-line case study, initiation did not occur at the intersection between a cold front and dryline (Koch and Clark 1999) but rather they observed the development of a gravity current at the leading edge of the cold front. This produced a bore, which propagated ahead on the surface-based stable layer. The gravity current structure along the cold front played an active and crucial role in the development of a severe squall line along the part of the front that produced the prefrontal bore.

e. Horizontal convective rolls

Horizontal convective rolls are a common form of boundary layer convection manifested as counterrotating vortices about the horizontally oriented axis. Clouds often occur atop the updraft branches of rolls (e.g., Kuettner 1959, 1971; Christian and Wakimoto 1989).

Rolls and cloud streets can extend hundreds of kilometers and last several hours. The conditions necessary for roll development and maintenance are surface-layer heat flux, some minimal low-level wind shear, and somewhat uniform surface characteristics (e.g., LeMone 1973; Grossman 1982; Weckwerth et al. 1997).

Intersection regions between convergence zones and horizontal convective roll updraft zones are also preferred locations for enhanced updrafts and cloud formation (e.g., Kessinger and Mueller 1991; Wilson et al. 1992). Intersections between the Denver Convergence Zone and rolls were observed (Wilson et al. 1992) and simulated (Crook et al. 1991) to be optimal locations for convective development. In a case study using Doppler radars, aircraft, mesonets, soundings, and a 2D numerical model, Fankhauser et al. (1995) found that sea-breeze front–horizontal convective roll intersections were insufficient to initiate deep convection but the interactions between the same rolls with a thunderstorm outflow did trigger storms.

In a study of the sea-breeze front intersecting horizontal convective rolls, enhanced updrafts and clouds were observed due to increased lifting at the intersection regions (Wakimoto and Atkins 1994; Atkins et al. 1995). Two detailed case studies compared and contrasted normal (offshore flow case) and parallel (onshore flow case) orientations of horizontal convective rolls relative to the sea-breeze front (Atkins et al. 1995). In the offshore flow case, Fig. 4.19a shows that the rolls are tilted upward by the frontal updraft, thereby creating stronger, deeper updrafts and deeper convection at the intersection zones. Furthermore, clouds that occurred at periodic intervals along the rolls intensified as the sea-breeze front intercepted them. An onshore flow case with rolls and sea-breeze front nearly parallel is schematically shown in Fig. 4.19b. Extended sections of the front merge with the rolls, thereby strengthening the front when like-sign vortices interact. Clouds form along the intensified portions of the front and at the locations of periodic enhancements of the rolls that were present prior to the merger. Simulations with a 3D cloud model similarly showed perpendicular (Dailey and Fovell 1999) and parallel (Fovell and Dailey 2000) interactions between rolls and the sea-breeze front and their positive effect on convection initiation. It has also been suggested that the interactions of internal gravity waves propagating along the inversion layer at the top of the boundary layer with roll circulations within the boundary layer can initiate deep convection (Balaji and Clark 1988).

In a study of 13 days in east-central Florida during the Convection and Precipitation/Electrification (CaPE; e.g., Wakimoto and Atkins 1994) project, it was shown that rolls are capable of initiating deep convection in the absence of intersections with another boundary (Weckwerth 2000). This study noted that roll updraft measurements (in contrast to general boundary layer measurements) are necessary to ascertain whether the environment is suitable for deep convective development. Previous studies have shown the degree of moisture variability associated with horizontal roll circulations (e.g., LeMone and Pennell 1976; Reinking et al. 1981). The roll updraft branches force moist, surface air upward into the boundary layer while the roll downdraft branches force dry, inversion-level air downward (e.g., Weckwerth et al. 1996). Since clouds and thunderstorms form atop roll updraft branches, it is essential that this updraft air be sampled to obtain an accurate assessment of the potential for deep moist convection (Fig. 4.20).

f. Orographic and topographic effects

Banta and Schaaf (1987) provided a schematic diagram showing four mechanisms that may lead to moist convection in the mountains (Fig. 4.21). These mechanisms include (i) orographic lifting, (ii) leeside convergence, (iii) channeling into a valley, and (iv) wake effects. Wake effects might be manifested in such phenomena as turbulence, gravity waves, and obstacle flow. A number of studies using radar and lightning data showed that mountain thunderstorms tend to occur in preferred topographic and orographic regions (e.g., Kuo and Orville 1973; Schaaf et al. 1988; Lopez and Holle 1986). Satellite data showed that the upper-level wind direction was a good predictor of mountain thunderstorm initiation, where thermally forced upslope flow interacted with the prevailing wind to produce leeside convergence zones (e.g., Banta and Schaaf 1987; Schaaf et al. 1988). Redelsperger and Clark (1990) used 2D and 3D numerical simulations to show that, in the Rocky Mountain area, the preferred orographic region for both dry and moist convection was the eastern slopes of the Rocky Mountains, where terrain effects result in enhancement of the low-level shear. Raymond and Wilkening (1982) observed convection initiation along mountain-induced ridge-top convergence zones.

Segal and Arritt (1992) showed that significant spatial heterogeneities in daytime sensible heat flux are common over land. These heterogeneities can lead to thermally induced circulations due to land use variations, contrasts in soil moisture owing to antecedent precipitation, natural landform variations, contrasts in cloudiness, or contrasts in snow cover. Modeling studies have shown circulations with intensities comparable to the sea breeze, while observations suggest reduced intensity compared to the model results.

Convergence boundaries arising from the thermally induced circulations frequently occur in numerous locations around the world. Many of these boundaries provide environmental conditions suitable for deep convective development. In particular the quasi-stationary Piedmont Front has been shown to be influential in initiating thunderstorms (Businger et al. 1991; Koch and Ray 1997). Numerical modeling (Crook et al. 1991) and observational studies (e.g., Szoke et al. 1984; Wilson

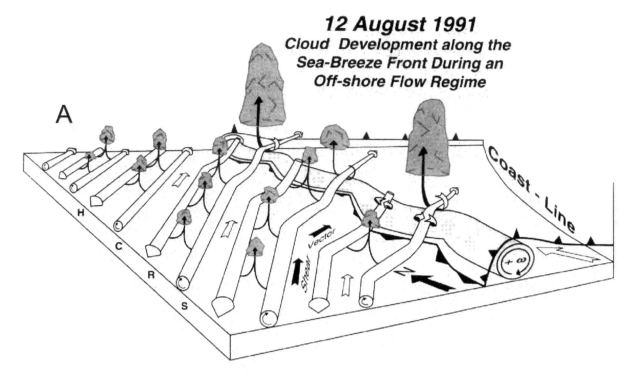

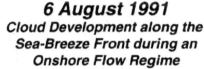

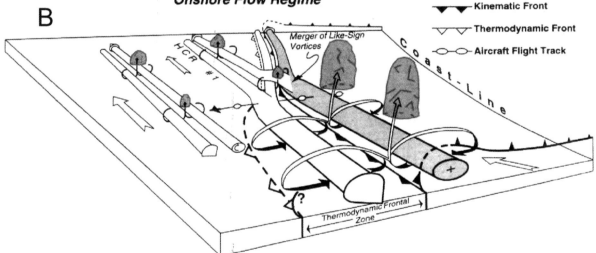

FIG. 4.19. Schematic diagrams showing the interaction between the sea-breeze front and horizontal convective rolls during (a) an offshore flow regime resulting in nearly perpendicular intersections and (b) an onshore flow regime resulting in parallel, periodically encountered interactions. The sea-breeze front is shown by the heavy, barbed line, and its head circulation is lightly shaded. The horizontal vorticity vectors associated with the counterrotating rolls are indicated by cylindrical white arrows. The clouds along the rolls and at the intersections are shaded gray. The shear vectors (solid, horizontal arrows) and low-level winds (white arrows) are also shown. (b) The kinematic (heavy solid) and thermodynamic fronts (heavy white) are separate features. From Atkins et al. (1995).

et al. 1988, 1992) have shown that the Denver Convergence Zone along the Colorado Front Range provides an environment that is conducive to convective development.

Knupp et al. (1998) observed the initiation of an MCS within a synoptically benign environment. They found that convection initiation resulted from pre-existing cloud streets and variability in surface heat flux, which

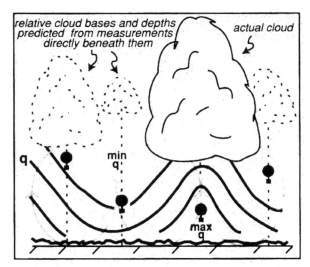

FIG. 4.20. Schematic diagram showing boundary layer moisture contours (solid black lines) within and outside the horizontal convective roll updraft (roll circulations depicted in gray). Actual cloud bases and depths atop roll updraft branches shown by the solidly bounded cloud. Clouds, bounded by dashes, represent the relative cloud bases and depths expected if stability parameters were estimated from boundary layer moisture values directly beneath those clouds. The variability shown here illustrates the difficulty in using radiosonde measurements to predict convective potential. From Weckwerth et al. (1996).

produced solenoidally driven mesoscale circulations. The heat flux variations were due to cloud shading and antecedent rainfall. They determined that it is essential for mesoscale models to include accurate representation of the atmospheric boundary layer and associated surface flux variables if they are to produce an accurate prediction of convective development.

The importance of monitoring boundaries and improving observations of stability parameters was illustrated in a status report dealing with nowcasting of thunderstorms (Wilson et al. 1998). There remains a basic lack of knowledge on the details of storm initiation and evolution that results in a lack of precision in predicting the time and location of convective development (e.g., Wilson and Mueller 1993; Ziegler et al. 1997). This lack of precision occurs even when boundary layer convergence zones are present. Such boundaries depict preferred, but not definite, locations for deep convective development. It is still largely unknown why some boundaries initiate deep moist convection while others do not. Additionally there is a need for improved measurements of boundary layer variables (particularly moisture fields) and more detailed observations of cumulus cloud location and evolution to augment the radar observations of convergence zones.

4. Summary and future work

Substantial progress has been made over the last 10–15 years toward better understanding of the dynamics and kinematics of convective systems, based on the

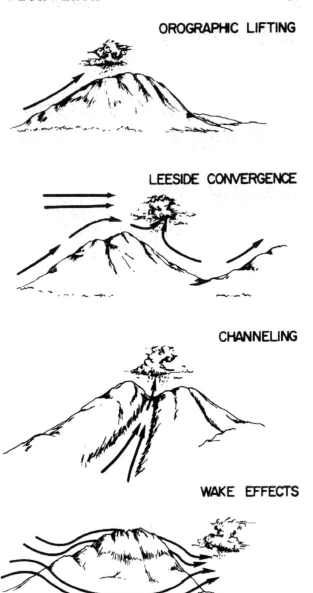

FIG. 4.21. Mechanisms that lead to moist convection in the mountains. Included are upslope advection of moist air or orographic lifting, leeside convergence zone, channeling into a convergent valley, and updrafts induced by wake phenomena, that is, turbulence, gravity waves, and obstacle flow. From Banta and Schaaf (1987).

combined use of Doppler radar, (and other in situ and remote sensors) and high-resolution two- and three-dimensional numerical simulations. It has become apparent that convective system organization and behavior depend critically on CAPE and the wind shear profile (particularly the shear in the lowest 3 km) representative of the environment in which the convective system is embedded. The character of the initial boundary layer and upper-troposphere environments that initiate convective systems is becoming better seen with improved Doppler radars and satellite sensors.

A remaining daunting challenge is to capitalize on

recent research results for the purposes of substantially improving near term (6–12 h) forecasts of hazardous weather associated with convective systems. For example, although there has clearly been much research to date on the morphology, dynamics, and severe weather potential of bow-echo convective systems, especially using numerical mesoscale convective models, it is still an open question how more isolated convection evolves upscale into the larger, more coherent bow-shaped segments. The ability to anticipate this upscale growth represents a critical gap in our ability to forecast and warn for such events. Forecasting of severe surface winds within bow echoes is limited by lack of knowledge of the relationships between the development of a strong, deep cold pool and associated mesohigh, which may act to accelerate the surface flow. Numerical simulations indicate that the strength of the cold pool also depends on the structure and strength of the rear-inflow jet, which can entrain drier midlevel air into the mesoscale downdraft, enhancing evaporative effects, and additionally transport higher momentum air down to the surface. The microphysical makeup of the stratiform precipitation region has also been hypothesized to be important in this regard. The relative magnitudes and importance of these various contributions to surface outflow strength have never been clearly established, making it quite difficult to anticipate the development of severe winds in any given event. The conditions that lead to the development of descending, rather than elevated, rear-inflow jets especially need to be clarified. In order to address these issues, documentation of the evolution of the buoyancy fields throughout the convective system, both within the cold pool at the surface and within the stratiform regions aloft, needs to be observationally determined. There is also a need to clarify the relative roles of convective versus mesoscale effects in the production of the severest winds. One of the more intriguing mysteries is how severe bow echoes can occur at night in the presence of a stable nocturnal boundary layer that does not as readily support the generation of the strong surface cold pool.

Substantial capital investments have recently been made by operational weather agencies in the deployment of national Doppler weather radar networks, improved satellite imaging and sounding systems, wind profilers, surface observations on 50–100-km spatial scales, and next-generation operational mesoscale prediction models. To realize the potential gains from this major observational and modeling infrastructure investment requires research directed toward forecast improvements that will substantially mitigate the effects of weather disasters.

As pointed out in Bluestein and Wakimoto (2003), there are many new and exciting ground-based and airborne radar systems that have been recently developed or are under development that offer researchers many new innovative approaches to observing convective systems. These systems include polarization-diversity and multifrequency radars, networks of ground-based mobile Doppler platforms, rapid-scan radars (e.g., SPY-1 radar technology originally developed for defense purposes), and eventually radars mounted on satellites. Field projects are being planned to capitalize on these new tools and use them in a coordinated approach toward the problems of understanding and forecasting convection. For example, the International H$_2$O Project (IHOP 2002), conducted in spring 2002, brought to bear new water vapor sensing technology with existing mobile and fixed-based platforms in the southern Great Plains to improve the characterization of the four-dimensional distribution of water vapor and its application to improving the understanding and prediction of convection, including convection initiation and quantitative precipitation forecasting. Another project, the Bow-Echo and Mesoscale Convective Vortex Experiment (BAMEX), planned for the spring 2003 severe weather season, would focus airborne Doppler radar and dropsonde platforms, coupled to several ground-based mobile radar and sounding systems, on severe convective systems that often produce damaging straight-line winds from bow-echo convective systems in the Midwest region of the United States. BAMEX will address many scientific issues concerning bow echoes that will lead to improvement in forecast skill. Specific questions include the following. 1) What are the physical processes acting to produce strong surface winds? 2) What are the dominant physical processes that control the size of severe wind producing systems? 3) Are the dynamics of nocturnal systems significantly different from systems occurring over heated terrain? 4) Why are tornadoes often associated with the line segment northward of the apex of the bow?

Future studies of convective system will need to rely more heavily on multisensor approaches and not just the use of Doppler radar. The combined use of airborne radar, dropsondes, and other remote sensing instruments (e.g., airborne lidars and radiometers for thermodynamic profiling) with ground-based mobile and fixed-base radars, lidars, and sounding platforms will result in datasets of unprecedented detail. These datasets will allow the kinematics, microphysics, and thermodynamics of convective systems to be studied in much more detail, and the interplay between microphysical and dynamical processes to be understood in more depth than has previously been possible. As has been done in the past, these new observations should spur the continued development of numerical models to guide the models toward greater realism and improved forecast accuracy.

Acknowledgments. We wish to acknowledge the assistance provided by Morris Weisman in the preparation of Fig. 4.1, as well as his many constructive comments as a reviewer. The paper was also greatly improved by a review provided by Carl Hane and another anonymous reviewer.

REFERENCES

Achtemeier, G. L., 1983: The relationship between the surface wind field and convective precipitation over the St. Louis area. *J. Appl. Meteor.*, **22**, 982–999.

——, 1991: The use of insects as tracers for "clear-air" boundary-layer studies by Doppler radar. *J. Atmos. Oceanic Technol.*, **8**, 746–765.

Alexander, G. D., and G. S. Young, 1992: The relationship between EMEX mesoscale precipitation feature properties and their environmental characteristics. *Mon. Wea. Rev.*, **120**, 554–564.

Arritt, R. W., 1993: Effects of the large-scale flow on characteristic features of the sea breeze. *J. Appl. Meteor.*, **32**, 116–125.

Atkins, N. T., and R. M. Wakimoto, 1997: Influence of the synoptic-scale flow on sea breezes observed during CaPE. *Mon. Wea. Rev.*, **125**, 2112–2130.

——, ——, and T. M. Weckwerth, 1995: Observations of the sea-breeze front during CaPE. Part II: Dual-Doppler and aircraft analysis. *Mon. Wea. Rev.*, **123**, 944–969.

——, ——, and C. L. Ziegler, 1998: Observations of the finescale structure of a dryline during VORTEX 95. *Mon. Wea. Rev.*, **126**, 525–550.

Atlas, D., 1960: Radar detection of the sea breeze. *J. Meteor.*, **17**, 244–258.

Augustine, J. A., and F. Caracena, 1994: Lower-tropospheric precursors to nocturnal MCS development over the central United States. *Wea. Forecasting.*, **9**, 116–135.

Austin, G. R., R. M. Rauber, H. T. Ochs III, and L. J. Miller, 1996: Trade-wind clouds and Hawaiian rainbands. *Mon. Wea. Rev.*, **124**, 2126–2151.

Baker, R. D., B. H. Lynn, A. Boone, W.-K. Tao, and J. Simpson, 2001: The influence of soil moisture, coastline curvature, and land-breeze circulations on sea-breeze-initiated precipitation. *J. Hydrometeor.*, **2**, 193–211.

Balaji, V., and T. L. Clark, 1988: Scale selection in locally forced convective fields and the initiation of deep cumulus. *J. Atmos. Sci.*, **45**, 3188–3211.

Banta, R. M., and C. B. Schaaf, 1987: Thunderstorm genesis zones in the Colorado Rocky Mountains as determined by traceback of geosynchronous satellite images. *Mon. Wea. Rev.*, **115**, 463–476.

Barnes, G. M., and K. Sieckman, 1984: The environment of fast- and slow-moving tropical mesoscale convective cloud lines. *Mon. Wea. Rev.*, **112**, 1782–1794.

Bartels, D. L., and R. A. Maddox, 1991: Midlevel cyclonic vortices generated by mesoscale convective systems. *Mon. Wea. Rev.*, **119**, 104–118.

——, J. M. Brown, and E. J. Tollerud, 1997: Structure of a midtropospheric vortex induced by a mesoscale convective system. *Mon. Wea. Rev.*, **125**, 193–211.

Bernardet, L. R., and W. R. Cotton, 1998: Multiscale evolution of a derecho-producing mesoscale convective system. *Mon. Wea. Rev.*, **126**, 2991–3015.

Biggerstaff, M. I., and R. A. Houze Jr., 1991: Kinematic and precipitation structure of the 10–11 June 1985 squall line. *Mon. Wea. Rev.*, **119**, 3034–3065.

Bluestein, H. B., and M. H. Jain, 1985: Formation of mesoscale lines of precipitation: Severe squall lines in Oklahoma during the spring. *J. Atmos. Sci.*, **42**, 1711–1732.

——, and S. S. Parker, 1993: Modes of isolated, severe convective storm formation along the dryline. *Mon. Wea. Rev.*, **121**, 1354–1372.

——, and M. L. Weisman, 2000: The interaction of numerically simulated supercells initiated along lines. *Mon. Wea. Rev.*, **128**, 3128–3149.

——, and R. M. Wakimoto, 2003: Mobile radar observations of severe convective systems. *Radar and Atmospheric Science: A Collection of Essays in Honor of David Atlas, Meteor. Monogr.*, No. 52, Amer. Meteor. Soc., 105–136.

——, E. W. McCaul Jr., G. P. Byrd, and G. R. Woodall, 1988: Mobile sounding observations of a tornadic storm near the dryline: The Canadian, Texas, storm of 7 May 1986. *Mon. Wea. Rev.*, **116**, 1790–1804.

——, ——, ——, R. L. Walko, and R. P. Davies-Jones, 1990: An observational study of splitting convective clouds. *Mon. Wea. Rev.*, **118**, 1359–1370.

Bosart, L. F., W. E. Bracken, and A. Seimon, 1998: A study of cyclone mesoscale structure with emphasis on a large-amplitude inertia-gravity wave. *Mon. Wea. Rev.*, **126**, 1497–1527.

Braun, S. A., and R. A. Houze Jr., 1997: The evolution of the 10–11 June 1985 PRE-STORM squall line: Initiation, development of rear inflow, and dissipation. *Mon. Wea. Rev.*, **125**, 478–504.

——, and B. F. Smull, 1997: Airborne dual-Doppler observations of an intense frontal system approaching the Pacific northwest coast. *Mon. Wea. Rev.*, **125**, 3131–3156.

Brooks, H. E., and R. B. Wilhelmson, 1993: Hodograph curvature and updraft intensity in numerically modeled supercells. *J. Atmos. Sci.*, **50**, 1824–1833.

Brown, S. A., J. D. Locatelli, and P. V. Hobbs, 1998: Organization of precipitation along cold fronts. *Quart. J. Roy. Meteor. Soc.*, **124**, 649–652.

——, M. T. Stoelinga, and P. V. Hobbs, 1999: Numerical modeling of precipitation cores on cold fronts. *J. Atmos. Sci.*, **56**, 1175–1196.

Browning, K. A., 1985: Conceptual models of precipitating systems. *Meteor. Mag.*, **114**, 293–319.

——, 1990: Organization and internal structure of synoptic and mesoscale precipitation systems in midlatitudes. *Radar in Meteorology*, D. Atlas, Ed., Amer. Meteor. Soc., 433–460.

——, 2003: Mesoscale substructure of extratropical cyclones observed by radar. *Radar and Atmospheric Science: A Collection of Essays in Honor of David Atlas, Meteor. Monogr.*, No. 52, Amer. Meteor. Soc., 7–32.

——, and N. M. Roberts, 1996: Variation of frontal and precipitation structure along a cold front. *Quart. J. Roy. Meteor. Soc.*, **122**, 1845–1872.

Businger, S., W. H. Bauman, and G. F. Watson, 1991: The development of the Piedmont front and associated outbreak of severe weather on 13 March 1986. *Mon. Wea. Rev.*, **119**, 2224–2251.

Byers, H. R., and R. R. Braham Jr., 1949: *The Thunderstorm.* U.S. Government Printing Office, 287 pp.

Carbone, R. E., 1982: Severe frontal rainband. Part 1: Stormwide dynamic structure. *J. Atmos. Sci.*, **39**, 258–279.

——, J. W. Conway, N. A. Crook, and M. W. Moncrieff, 1990a: The generation and propagation of a nocturnal squall line. Part I: Observations and implications for mesoscale predictability. *Mon. Wea. Rev.*, **118**, 26–49.

——, and Coauthors, 1990b: Convective dynamics: Panel report. *Radar in Meteorology*, D. Atlas, Ed., Amer. Meteor. Soc., 391–400.

——, W. A. Cooper, and W. C. Lee, 1995: On the forcing of flow reversal along the windward slopes of Hawaii. *Mon. Wea. Rev.*, **123**, 3416–3480.

——, J. D. Tuttle, W. A. Cooper, V. Grubisic, and W. C. Lee, 1998: Trade wind rainfall near the windward coast of Hawaii. *Mon. Wea. Rev.*, **126**, 2847–2863.

——, J. W. Wilson, T. D. Keenan, and J. M. Hacker, 2000: Tropical island convection in the absence of significant topography. Part I: Life cycle of diurnally forced convection. *Mon. Wea. Rev.*, **128**, 3459–3480.

Charba, J., 1974: Application of gravity current model to analysis of squall-line gust front. *Mon. Wea. Rev.*, **102**, 140–156.

Chin, H.-N. S., and R. B. Wilhelmson, 1998: Evolution and structure of tropical squall line elements within a moderate CAPE and strong low-level jet environment. *J. Atmos. Sci.*, **55**, 3089–3133.

Chong, M., and O. Bousquet, 1999: A mesovortex within a near-equatorial mesoscale convective system during TOGA COARE. *Mon. Wea. Rev.*, **127**, 1145–1156.

Christian, T. W., and R. M. Wakimoto, 1989: The relationship between radar reflectivities and clouds associated with horizontal roll convection on 8 August 1982. *Mon. Wea. Rev.,* **117,** 1530–1544.

Church, C., D. Burgess, C. Doswell, and R. Davies-Jones, Eds., 1993: *The Tornado: Its Structure, Dynamics, Prediction, and Hazards, Geophys. Monogr.,* No. 79, Amer. Geophys. Union, 637 pp.

Colman, B. R., 1990: Thunderstorms above frontal surfaces in environments without positive CAPE. Part II: Organization and instability mechanisms. *Mon. Wea. Rev.,* **118,** 1123–1144.

Crawford, T. M., and H. B. Bluestein, 1997: Characteristics of dryline passage during COPS-91. *Mon. Wea. Rev.,* **125,** 463–477.

Crook, N. A., 1996: Sensitivity of moist convection forced by boundary layer processes to low-level thermodynamic fields. *Mon. Wea. Rev.,* **124,** 1768–1785.

——, T. L. Clark, and M. W. Moncrieff, 1991: The Denver Cyclone. Part II: Interaction with the convective boundary layer. *J. Atmos. Sci.,* **48,** 2109–2126.

Dailey, P. S., and R. G. Fovell, 1999: Numerical simulation of the interaction between the sea-breeze front and horizontal convective rolls. Part I: Offshore ambient flow. *Mon. Wea. Rev.,* **127,** 858–878.

Davis, C. A., and M. L. Weisman, 1994: Balanced dynamics of mesoscale vortices produced in simulated convective systems. *J. Atmos. Sci.,* **51,** 2005–2030.

Doswell, C. A., III, H. E. Brooks, and R. A. Maddox, 1996: Flash flood forecasting: An ingredients-based methodology. *Wea. Forecasting,* **11,** 560–581.

Droegemeier, K. K., and R. B. Wilhelmson, 1985: Three-dimensional numerical modeling of convection produced by interacting thunderstorm outflows. Part I: Control simulation and low-level moisture variation. *J. Atmos. Sci.,* **42,** 2381–2403.

Eliassen, A., 1962: On the vertical circulation in frontal zones. *Geophys. Publ.,* **27,** 1–15.

Emanuel, K. A., 1986: Some dynamical aspects of precipitating convection. *J. Atmos. Sci.,* **43,** 2183–2198.

——, 1994: *Atmospheric Convection.* Oxford University Press, 580 pp.

Evans, J. S., and C. A. Doswell III, 2001: Examination of derecho environments using proximity soundings. *Wea. Forecasting,* **16,** 329–342.

Fankhauser, J. C., N. A. Crook, J. Tuttle, L. J. Miller, and C. G. Wade, 1995: Initiation of deep convection along boundary layer convergence lines in a semitropical environment. *Mon. Wea. Rev.,* **123,** 291–313.

Fletcher, J. O., 1990: Early developments of weather radar during World War II. *Radar in Meteorology,* D. Atlas, Ed., Amer. Meteor. Soc., 3–6.

Fovell, R. G., and P. S. Dailey, 2001: Numerical simulation of the interaction between the sea-breeze front and horizontal convective rolls. Part II: Alongshore ambient flow. *Mon. Wea. Rev.,* **129,** 2057–2072.

Fritsch, J. M., J. D. Murphy, and J. S. Kain, 1994: Warm core vortex amplification over land. *J. Atmos. Sci.,* **51,** 1780–1807.

Fujita, T. T., 1978: Manual of downburst identification for project Nimrod. Satellite and Mesometeorology Res. Pap. 156, Dept. of Geophysical Sciences, University of Chicago, 104 pp. [NTIS PB-286048.]

Gallus, W. A., Jr., and R. H. Johnson, 1995: The dynamics of circulations within the trailing stratiform regions of squall lines. Part II: Influence of the convective line and ambient environment. *J. Atmos. Sci.,* **52,** 2188–2211.

Garstang, M., and H. J. Cooper, 1981: The role of near surface outflow in maintaining convective activity. *Proc. Nowcasting-I Symp.,* Copenhagen, Denmark, European Space Agency, 161–168.

Goff, R. C., 1976: Vertical structure of thunderstorm outflows. *Mon. Wea. Rev.,* **104,** 1429–1440.

Grady, R. L., and J. Verlinde, 1997: Triple Doppler analysis of a discretely propagating, long-lived, high-plains squall line. *J. Atmos. Sci.,* **54,** 2729–2748.

Grossman, R. L., 1982: An analysis of vertical velocity spectra obtained in the BOMEX fair-weather, trade-wind boundary layer. *Bound.-Layer Meteor.,* **23,** 323–357.

Hane, C. E., and D. P. Jorgensen, 1995: Dynamic aspects of a distinctly three-dimensional mesoscale convective system. *Mon. Wea. Rev.,* **123,** 3194–3214.

——, C. J. Kessinger, and P. S. Ray, 1987: The Oklahoma squall line of 19 May 1977. Part II: Mechanisms for maintenance of the region of strong convection. *J. Atmos. Sci.,* **44,** 2866–2886.

——, C. L. Ziegler, and H. B. Bluestein, 1993: Investigation of the dryline and convective storms initiated along the dryline: Field experiments during COPS-91. *Bull. Amer. Meteor. Soc.,* **74,** 2133–2145.

——, H. B. Bluesein, T. M. Crawford, M. E. Baldwin, and R. M. Rabin, 1997: Severe thunderstorm development in relation to along-dryline variability: A case study. *Mon. Wea. Rev.,* **125,** 231–251.

——, M. E. Baldwin, H. B. Bluestein, T. M. Crawford, and R. M. Rabin, 2001: A case study of severe storm development along a dryline within a synoptically active environment. Part I: Dryline motion and an Eta Model forecast. *Mon. Wea. Rev.,* **129,** 2183–2204.

——, R. M. Rabin, T. M. Crawford, H. B. Bluestein, and M. E. Baldwin, 2002: A case study of severe storm development along a dryline within a synoptically active environment. Part II: Multiple boundaries and convection initiation. *Mon. Wea. Rev.,* **130,** 900–920.

Hertzman, O., and P. V. Hobbs, 1988: The mesoscale and microscale structure and organization of clouds and precipitation in midlatitude cyclones. Part XIV: Three-dimensional airflow and vorticity budget of rainbands in a warm occlusion. *J. Atmos. Sci.,* **45,** 893–914.

Hjelmfelt, M. R., 1990: Numerical study of the influence of environmental conditions on lake-effect snowstorms over Lake Michigan. *Mon. Wea. Rev.,* **118,** 138–150.

Hobbs, P. V., and P. O. G. Persson, 1982: The mesoscale and microscale structure and organization of clouds and precipitation in mid-latitude cyclones. Part V: The substructure of narrow cold frontal rainbands. *J. Atmos. Sci.,* **39,** 280–295.

Houze, R. A., Jr., 1981: Structure of atmospheric precipitation systems—A global survey. *Radio Sci.,* **16,** 671–689.

——, and P. V. Hobbs, 1982: Organization and structure of precipitating cloud systems. *Advances in Geophysics,* Vol. 24, Academic Press, 225–315.

——, S. A. Rutledge, M. I. Biggerstaff, and B. F. Smull, 1989: Interpretation of Doppler weather radar displays of midlatitude mesoscale convective system. *Bull. Amer. Meteor. Soc.,* **70,** 608–619.

——, B. F. Smull, and P. Dodge, 1990: Mesoscale organization of springtime rainstorms in Oklahoma. *Mon. Wea. Rev.,* **118,** 613–654.

Intrieri, J. M., A. J. Bedard, and R. M. Hardesty, 1990: Details of colliding thunderstorm outflows as observed by Doppler lidar. *J. Atmos. Sci.,* **47,** 1081–1098.

Johns, R. H., and W. D. Hirt, 1987: Derechos: Widespread convectively induced windstorms. *Wea. Forecasting,* **2,** 32–49.

——, J. M. Davies, and P. W. Leftwich, 1993: Some wind and instability parameters associated with strong and violent tornadoes. 2: Variations in the combinations of wind and instability parameters. *The Tornado: Its Structure, Dynamics, Prediction, and Hazards, Geophys. Monogr.,* No. 79, Amer. Geophys. Union, 583–590.

Joly, A., and Coauthors, 1997: The Fronts and Atlantic Storm Track Experiment (FASTEX): Scientific objectives and experimental design. *Bull. Amer. Meteor. Soc.,* **78,** 1917–1940.

Jorgensen, D. P., and B. F. Smull, 1993: Mesovortex circulations seen by airborne Doppler radar within a bow-echo mesoscale convective system. *Bull. Amer. Meteor. Soc.,* **74,** 2146–2157.

——, and M. A. LeMone, 1989: Vertical velocity characteristics of oceanic convection. *J. Atmos. Sci.,* **46,** 621–644.

——, T. Matejka, and J. D. DuGranrut, 1996: Multi-beam techniques for deriving wind fields from airborne Doppler radars. *J. Meteor. Atmos. Phys.,* **59,** 83–104.

——, M. A. LeMone, and S. B. Trier, 1997: Structure and evolution of the 22 February 1993 TOGA COARE squall line: Aircraft observations of precipitation, circulation, and surface energy fluxes. *J. Atmos. Sci.,* **54,** 1961–1985.

Jorgensen, D. P., Z. Pu, P. O. G. Persson, and W.-K. Tao, 2003: Variations associated with cores and gaps of a pacific narrow cold frontal rainband. *Mon. Wea. Rev.,* in press.

Joseph, J. H., and R. F. Cahalan, 1990: Nearest neighbor spacing of fair weather cumulus clouds. *J. Appl. Meteor.,* **29,** 793–805.

Kessinger, C. J., and C. K. Mueller, 1991: Background studies and nowcasting Florida thunderstorm activity in preparation for the CaPOW forecast experiment. Preprints, *25th Conf. on Radar Meteorology,* Paris, France, Amer. Meteor. Soc., 67–70.

Kingsmill, D. E., 1995: Convection initiation associated with a sea-breeze front, a gust front, and their collision. *Mon. Wea. Rev.,* **123,** 2913–2933.

——, and R. M. Wakimoto, 1991: Kinematic, dynamic, and thermodynamic analysis of a weakly sheared severe thunderstorm over northern Alabama. *Mon. Wea. Rev.,* **119,** 262–297.

Knupp, K. R., B. Geerts, and S. J. Goodman, 1998: Analysis of a small, vigorous mesoscale convective system in a low-shear environment. Part I: Formation, radar echo structure, and lightning behavior. *Mon. Wea. Rev.,* **126,** 1812–1836.

Koch, S. E., and J. McCarthy, 1982: The evolution of an Oklahoma dryline. Part II: Boundary layer forcing of mesoconvective systems. *J. Atmos. Sci.,* **39,** 237–257.

——, and C. A. Ray, 1997: Mesoanalysis of summertime convergence zones in central and eastern North Carolina. *Wea. Forecasting,* **12,** 56–77.

——, and W. L. Clark, 1999: A nonclassical cold front observed during COPS-91: Frontal structure and the process of severe storm initiation. *J. Atmos. Sci.,* **56,** 2862–2890.

Kuettner, J. P., 1959: The band structure of the atmosphere. *Tellus,* **11,** 267–294.

Kuo, J.-T., and H. D. Orville, 1973: A radar climatology of summertime convective clouds in the Black Hills. *J. Appl. Meteor.,* **12,** 357–368.

Lafore, J., and M. W. Moncrieff, 1989: A numerical investigation of the organization and interaction of the convective and stratiform regions of tropical squall lines. *J. Atmos. Sci.,* **46,** 521–544.

Laing, A. G., and J. M. Fritsch, 2000: The large-scale environments of the global populations of mesoscale convective complexes. *Mon. Wea. Rev.,* **128,** 2756–2776.

Laird, N. F., D. A. R. Kristovich, R. M. Rauber, H. T. Ochs, and L. J. Miller, 1995: The Cape Canaveral sea and river breezes: Kinematic structure and convective initiation. *Mon. Wea. Rev.,* **123,** 2942–2956.

Lee, B. D., R. D. Farley, and M. R. Hjelmfelt, 1991: A numerical case study of convection initiation along colliding convergence boundaries in northeast Colorado. *J. Atmos. Sci.,* **48,** 2350–2366.

Lemaitre, Y., G. Scialom, and P. Amayenc, 1989: A cold frontal rainband observed during the LANDES-FRONTS 84 experiment: Mesoscale and small-scale structure inferred from dual-Doppler radar analysis. *J. Atmos. Sci.,* **46,** 2215–2235.

LeMone, M. A., 1973: The structure and dynamics of horizontal roll vortices in the planetary boundary layer. *J. Atmos. Sci.,* **30,** 1077–1091.

LeMone, M. A., and W. T. Pennell, 1976: The relationship of trade wind cumulus distribution to subcloud layer fluxes and structure. *Mon. Wea. Rev.,* **104,** 524–539.

——, and M. W. Moncrieff, 1994: Momentum and mass transport by convective bands: Comparisons of highly idealized dynamical models to observations. *J. Atmos. Sci.,* **51,** 281–305.

——, G. M. Barnes, E. J. Szoke, and E. J. Zipser, 1984: The tilt of the leading edge of mesoscale tropical convective lines. *Mon. Wea. Rev.,* **112,** 510–519.

——, E. J. Zipser, and S. B. Trier, 1998: The role of environmental shear and thermodynamic conditions in determining the structure and evolution of mesoscale convective systems during TOGA COARE. *J. Atmos. Sci.,* **55,** 3493–3518.

Lewis, S. A., M. A. LeMone, and D. P. Jorgensen, 1998: Evolution and dynamics of a late-stage squall line that occurred on 20 February 1993 during TOGA COARE. *Mon. Wea. Rev.,* **126,** 3189–3212.

Ligda, M. G. H., 1956: The radar observation of mature prefrontal squall lines in the Midwestern United States. *Swiss Aero Rev.,* **11/12,** 153–155.

Lindzen, R. S., and K. K. Tung, 1976: Banded convective activity and ducted gravity waves. *Mon. Wea. Rev.,* **104,** 1602–1617.

Locatelli, J. D., J. E. Martin, and P. V. Hobbs, 1995: Development and propagation of precipitation cores on cold fronts. *Atmos. Res.,* **38,** 177–206.

——, M. T. Stoelinga, and P. V. Hobbs, 1998: Structure and evolution of winter cyclones in the central United States and their effect of the distribution of precipitation. Part V: Thermodynamic and dual-Doppler radar analysis of a squall line associated with a cold front aloft. *Mon. Wea. Rev.,* **126,** 860–875.

Loehrer, S. M., and R. H. Johnson, 1995: Surface pressure and precipitation life cycle characteristics of PRE-STORM mesoscale convective system. *Mon. Wea. Rev.,* **123,** 600–621.

Lopez, R. E., and R. L. Holle, 1986: Diurnal and spatial variability of lightning activity in northeastern Colorado and central Florida during the summer. *Mon. Wea. Rev.,* **114,** 1288–1312.

Lucas, C., E. J. Zipser, and M. A. LeMone, 1994: Vertical velocity in oceanic convection off tropical Australia. *J. Atmos. Sci.,* **51,** 3183–3193.

Mahoney, W. P., 1988: Gust front characteristics and the kinematics associated with interacting thunderstorm outflows. *Mon. Wea. Rev.,* **116,** 1474–1491.

Matejka, T. J., R. A. Houze Jr., and P. V. Hobbs, 1980: Microphysics and dynamics of clouds associated with mesoscale rainbands in extratropical cyclones. *Quart. J. Roy. Meteor. Soc.,* **106,** 29–56.

May, P. T., 1999: Thermodynamic and vertical velocity structure of two gust fronts observed with a wind profiler/RASS during MCTEX. *Mon. Wea. Rev.,* **127,** 1796–1807.

McCarthy, J., and S. E. Koch, 1982: The evolution of an Oklahoma dryline. Part I: A mesoscale and subsynoptic-scale analysis. *J. Atmos. Sci.,* **39,** 225–236.

McCaul, E. W., Jr., 1991: Buoyancy and shear characteristics of hurricane tornado environments. *Mon. Wea. Rev.,* **119,** 1954–1978.

——, and M. L. Weisman, 1996: Simulations of shallow supercell storms in landfalling hurricane environments. *Mon. Wea. Rev.,* **124,** 408–429.

——, and M. L. Weisman, 2001: The sensitivity of simulated supercell structure and intensity to variations in the shapes of environmental buoyancy and shear profiles. *Mon. Wea. Rev.,* **129,** 664–687.

Menard, R. D., and J. M. Fritsch, 1989: A mesoscale convective complex-generated inertially stable warm core vortex. *Mon. Wea. Rev.,* **117,** 1237–1261.

Moncrieff, M. W., and M. J. Miller, 1976: The dynamics and simulation of tropical cumulonimbus and squall lines. *Quart. J. Roy. Meteor. Soc.,* **102,** 373–394.

Mueller, C. K., and R. E. Carbone, 1987: Dynamics of a thunderstorm outflow. *J. Atmos. Sci.,* **44,** 1879–1898.

Neumann, J., 1951: Land breezes and nocturnal thunderstorms. *J. Meteor.,* **8,** 60–67.

Nicholls, M. E., R. A. Pielke, and W. R. Cotton, 1991: Thermally forced gravity waves in an atmosphere at rest. *J. Atmos. Sci.,* **48,** 1869–1884.

Parker, M. D., and R. H. Johnson, 2000: Organizational modes of midlatitude mesoscale convective systems. *Mon. Wea. Rev.,* **128,** 3413–3436.

Parsons, D. B., C. G. Mohr, and T. Gal-Chen, 1987: A severe frontal rainband. Part III: Derived thermodynamic structure. *J. Atmos. Sci.,* **44,** 1615–1631.

——, B. F. Smull, and D. K. Lilly, 1990: Mesoscale organization and processes: Panel report. *Radar in Meteorology,* D. Atlas, Ed., Amer. Meteor. Soc., 461–476.

——, M. A. Shapiro, R. M. Hardesty, R. J. Zamora, and J. M. Intrieri, 1991: The finescale structure of a west Texas dryline. *Mon. Wea. Rev.,* **119,** 1242–1258.

Pielke, R. A., 1974: A three-dimensional numerical model of the sea-breezes over south Florida. *Mon. Wea. Rev.,* **102,** 115–139.

——, and Y. Mahrer, 1978: Verification analysis of the University of Virginia three-dimensional mesoscale model prediction over south Florida for 1 July 1973. *Mon. Wea. Rev.,* **106,** 1568–1589.

Przybylinski, R. W., 1995: The bow echo: Observations, numerical simulations, and severe weather detection methods. *Wea. Forecasting,* **10,** 203–218.

Purdom, J. F. W., 1976: Some uses of high-resolution GOES imagery in the mesoscale forecasting of convection and its behavior. *Mon. Wea. Rev.,* **104,** 1474–1483.

——, 1982: Subjective interpretations of geostationary satellite data for nowcasting. *Nowcasting,* K. Browning, Ed., Academic Press, 149–166.

——, and K. Marcus, 1982: Thunderstorm trigger mechanisms over the southeast U.S. Preprints, *12th Conf. on Severe Local Storms,* San Antonio, TX, Amer. Meteor. Soc., 487–488.

Rao, P. A., H. E. Fuelberg, and K. K. Droegemeier, 1999: High-resolution modeling of the Cape Canaveral area land-water circulations and associated features. *Mon. Wea. Rev.,* **127,** 1808–1821.

Ray, P. S., 1990: Convective dynamics. *Radar in Meteorology,* D. Atlas, Ed., Amer. Meteor. Soc., 348–390.

Raymond, D. J., and M. H. Wilkening, 1982: Flow and mixing in New Mexico mountain cumuli. *J. Atmos. Sci.,* **39,** 2211–2228.

Redelsperger, J.-L., and T. L. Clark, 1990: The initiation and horizontal scale selection of convection over gently sloping terrain. *J. Atmos. Sci.,* **47,** 516–541.

Reinking, R. F., R. J. Doviak, and R. O. Gilmer, 1981: Clear-air roll vortices and turbulent motions as detected with an airborne gust probe and dual-Doppler radar. *J. Appl. Meteor.,* **20,** 678–685.

Rhea, J. O., 1966: A study of thunderstorm formation along drylines. *J. Appl. Meteor.,* **5,** 58–63.

Rotunno, R., J. B. Klemp, and M. L. Weisman, 1988: A theory for strong, long-lived squall lines. *J. Atmos. Sci.,* **45,** 463–485.

Roux, F., V. Marécal, and D. Hauser, 1993: The 12/13 January 1988 narrow cold-frontal rainband observed during MFDP/FRONTS 87. Part I: Kinematics and thermodynamics. *J. Atmos. Sci.,* **50,** 951–974.

Rutledge, S. A., 1986: A diagnostic modeling study of the stratiform region associated with a tropical squall line. *J. Atmos. Sci.,* **43,** 1356–1377.

Sanders, F., and L. F. Bosart, 1985: Mesoscale structure in the megalopolitan snowstorm of 11–12 February 1983. Part I: Frontogenetical forcing and symmetric instability. *J. Atmos. Sci.,* **42,** 1050–1061.

——, and D. O. Blanchard, 1993: The origin of a severe thunderstorm in Kansas on 10 May 1985. *Mon. Wea. Rev.,* **121,** 133–149.

Schaaf, C. B., J. Wurman, and R. M. Banta, 1988: Thunderstorm-producing terrain features. *Bull. Amer. Meteor. Soc.,* **69,** 272–277.

Schaefer, J. T., 1986: The dryline. *Mesoscale Meteorology and Forecasting,* P. S. Ray, Ed., Amer. Meteor. Soc., 549–572.

Schiesser, H. H., R. A. Houze Jr., and H. Huntrieser, 1995: The mesoscale structure of severe precipitation systems in Switzerland. *Mon. Wea. Rev.,* **123,** 2070–2097.

Schlesinger, R. E., 1982: Three-dimensional numerical modeling of convective storms: A review of milestones and challenges. Preprints, *12th Conf. on Severe Local Storms,* San Antonio, TX, Amer. Meteor. Soc., 506–515.

Scott, J. D., and S. A. Rutledge, 1995: Doppler radar observations of an asymmetric mesoscale convective system and associated vortex couplet. *Mon. Wea. Rev.,* **123,** 3437–3457.

Segal, M., and R. W. Arritt, 1992: Nonclassical mesoscale circulations caused by surface sensible heat-flux gradients. *Bull. Amer. Meteor. Soc.,* **73,** 1593–1604.

Seitter, K. L., 1986: A numerical study of atmospheric density current motion including the effect of condensation. *J. Atmos. Sci.,* **43,** 3068–3076.

Sha, W., T. Kawamura, and H. Ueda, 1991: A numerical study on sea/land breezes as a gravity current: Kelvin–Helmholtz billows and inland penetration of the sea-breeze front. *J. Atmos. Sci.,* **48,** 1649–1665.

Shields, M. T., R. M. Rauber, and M. K. Ramamurthy, 1991: Dynamical forcing and mesoscale organization of precipitation bands in a Midwest winter cyclonic storm. *Mon. Wea. Rev.,* **119,** 936–964.

Simpson, J. E., 1994: *Sea Breeze and Local Wind.* Cambridge University Press, 234 pp.

Skamarock, W. C., M. L. Weisman, and J. B. Klemp, 1994: Three-dimensional evolution of simulated long-lived squall lines. *J. Atmos. Sci.,* **51,** 2563–2584.

Smolarkiewicz, P. K., R. M. Rasmussen, and T. L. Clark, 1988: On the dynamics of Hawaiian cloud bands: Island forcing. *J. Atmos. Sci.,* **45,** 1872–1905.

Smull, B. F., 1995: Convectively-induced mesoscale weather systems in the tropical and warm-season midlatitude atmosphere. *Rev. Geophys.,* **33** (Suppl.), 897–906.

——, and R. A. Houze Jr., 1985: A midlatitude squall line with a trailing region of stratiform rain: Radar and satellite observations. *Mon. Wea. Rev.,* **113,** 117–133.

——, and J. A. Augustine, 1993: Multiscale analysis of a mature mesoscale convective complex. *Mon. Wea. Rev.,* **121,** 103–132.

Steenburgh, W. J., S. F. Halvorson, and D. J. Onton, 2000: Climatology of lake-effect snowstorms of the Great Salt Lake. *Mon. Wea. Rev.,* **128,** 709–727.

Stensrud, D. J., 1996a: Effects of persistent, midlatitude mesoscale regions of convection on the large-scale environment during the warm season. *J. Atmos. Sci.,* **53,** 3503–3527.

——, 1996b: Importance of low-level jets to climate: A review. *J. Climate,* **9,** 1698–1711.

——, and R. A. Maddox, 1988: Opposing mesoscale circulations: A case study. *Wea. Forecasting,* **3,** 189–204.

——, and J. M. Fritsch, 1994: Mesoscale convective systems in weakly forced large-scale environments. Part II: Generation of mesoscale initiation condition. *Mon. Wea. Rev.,* **122,** 2068–2083.

——, G. S. Manikin, E. Rogers, and K. E. Mitchell, 1999: Importance of cold pools to NCEP mesoscale Eta Model forecasts. *Wea. Forecasting,* **14,** 650–670.

Szoke, E. J., M. L. Weisman, J. M. Brown, F. Caracena, and T. W. Schlatter, 1984: A subsynoptic analysis of the Denver tornadoes of 3 June 1981. *Mon. Wea. Rev.,* **112,** 790–808.

Szumowski, M. J., R. M. Rauber, H. T. Ochs III, and L. J. Miller, 1997: The microphysical structure and evolution of Hawaiian rainband clouds. Part I: Radar observations of rainbands containing high reflectivity cores. *J. Atmos. Sci.,* **54,** 369–385.

Thorpe, A. J., M. J. Miller, and M. W. Moncrieff, 1982: Two-dimensional convection in non-constant shear: A model of midlatitude squall lines. *Quart. J. Meteor. Soc.,* **108,** 739–762.

Trier, S. B., W. C. Skamarock, and M. A. LeMone, 1997: Structure and evolution of the 22 February 1993 TOGA COARE squall line: Organizational mechanisms inferred from numerical simulations. *J. Atmos. Sci.,* **54,** 386–407.

——, C. A. Davis, and W. C. Skamarock, 2000: Long-lived mesoconvective vortices and their environment. Part II: Induced thermodynamic destabilization in idealized simulations. *Mon. Wea. Rev.,* **128,** 3396–3412.

Ulanski, S. L., and M. Garstang, 1978: The role of surface divergence and vorticity in the life cycle of convective rainfall. Part I: Observations and analysis. *J. Atmos. Sci.,* **35,** 1047–1062.

Verlinde, J., and W. R. Cotton, 1990: A mesoscale vortex couplet observed in the trailing anvil of a multicellular convective complex. *Mon. Wea. Rev.,* **118,** 993–1010.

Wakimoto, R. M., 1982: The life cycle of thunderstorm gust fronts as viewed with Doppler radar and rawinsonde data. *Mon. Wea. Rev.*, **110**, 1060–1082.

——, and N. T. Atkins, 1994: Observations of the sea-breeze front during CaPE. Part I: Single-Doppler, satellite, and cloud photogrammetric analysis. *Mon. Wea. Rev.*, **122**, 1092–1114.

——, and B. L. Bosart, 2000: Airborne radar observations of a cold front during FASTEX. *Mon. Wea. Rev.*, **128**, 2447–2470.

Watson, I. W., and D. O. Blanchard, 1984: The relationship between total area divergence and convective precipitation in south Florida. *Mon. Wea. Rev.*, **112**, 673–685.

Weckwerth, T. M., 2000: The effect of small-scale moisture variability on thunderstorm initiation. *Mon. Wea. Rev.*, **128**, 4017–4030.

——, and R. M. Wakimoto, 1992: The initiation and organization of convective cells atop a cold-air outflow boundary. *Mon. Wea. Rev.*, **120**, 2169–2187.

——, J. W. Wilson, and R. M. Wakimoto, 1996: Thermodynamic variability within the convective boundary layer due to horizontal convective rolls. *Mon. Wea. Rev.*, **124**, 769–784.

——, ——, ——, and N. A. Crook, 1997: Horizontal convective rolls: Determining the environmental conditions supporting their existence and characteristics. *Mon. Wea. Rev.*, **125**, 505–526.

Weisman, M. L., 1993: The genesis of severe, long-lived bow echoes. *J. Atmos. Sci.*, **50**, 645–670.

——, 2001: Bow echoes: A tribute to T. T. Fujita. *Bull. Amer. Meteor. Soc.*, **82**, 97–116.

——, and J. B. Klemp, 1982: The dependence of numerically simulated convective storms on vertical wind shear and buoyancy. *Mon. Wea. Rev.*, **110**, 504–520.

——, and C. A. Davis, 1998: Mechanisms for the generation of mesoscale vortices within quasi-linear convective systems. *J. Atmos. Sci.*, **55**, 2603–2622.

——, J. B. Klemp, and R. Rotunno, 1988: Structure and evolution of numerically simulated squall lines. *J. Atmos. Sci.*, **45**, 1990–2013.

Weiss, C. C., and H. B. Bluestein, 2002: Airborne pseudo-dual-Doppler analysis of a dryline-outflow boundary intersection. *Mon. Wea. Rev.*, **130**, 1207–1226.

Wilson, J. W., and R. Carbone, 1984: Nowcasting with Doppler radar: The forecaster–computer relationship. *Nowcasting II*, K. Browning, Ed., European Space Agency, 177–186.

——, and W. E. Schreiber, 1986: Initiation of convective storms at radar-observed boundary-layer convergence lines. *Mon. Wea. Rev.*, **114**, 2516–2536.

——, and C. K. Mueller, 1993: Nowcasts of thunderstorm initiation and evolution. *Wea. Forecasting*, **8**, 113–131.

——, and D. L. Megenhardt, 1997: Thunderstorm initiation, organization, and lifetime associated with Florida boundary layer convergence lines. *Mon. Wea. Rev.*, **125**, 1507–1525.

——, J. A. Moore, G. B. Foote, B. Martner, A. R. Rodi, T. Uttal, and J. M. Wilczak, 1988: Convection initiation and downburst experiment (CINDE). *Bull. Amer. Meteor. Soc.*, **69**, 1328–1348.

——, G. B. Foote, N. A. Crook, J. C. Fankhauser, C. G. Wade, J. D. Tuttle, C. K. Mueller, and S. K. Krueger, 1992: The role of boundary layer convergence zones and horizontal rolls in the initiation of thunderstorms: A case study. *Mon. Wea. Rev.*, **120**, 1785–1815.

——, T. M. Weckwerth, J. Vivekanandan, R. M. Wakimoto, and R. W. Russell, 1994: Boundary layer clear-air radar echoes: Origin of echoes and accuracy of derived winds. *J. Atmos. Oceanic Technol.*, **11**, 1184–1206.

——, N. A. Crook, C. K. Mueller, J. Sun, and M. Dixon, 1998: Nowcasting thunderstorms: A status report. *Bull. Amer. Meteor. Soc.*, **79**, 2079–2099.

——, R. E. Carbone, J. D. Tuttle, and T. D. Keenan, 2001: Tropical island convection in the absence of significant topography. Part II: Nowcasting storm evolution. *Mon. Wea. Rev.*, **129**, 1637–1655.

Yu, C.-K., B. J.-D. Jou, and B. F. Smull, 1999: Formative stage of a long-lived mesoscale vortex observed by airborne Doppler radar. *Mon. Wea. Rev.*, **127**, 838–857.

Zhang, D. L., and J. M. Fritsch, 1988: A numerical investigation of a convectively generated, inertially stable, extratropical warm-core mesovortex over land. Part I: Structure and evolution. *Mon. Wea. Rev.*, **116**, 2660–2687.

Ziegler, C. L., and E. N. Rasmussen, 1998: The initiation of moist convection at the dryline: Forecasting issues from a case study perspective. *Wea. Forecasting*, **13**, 1106–1131.

——, T. J. Lee, and R. A. Pielke Sr., 1997: Convection initiation at the dryline: A modeling study. *Mon. Wea. Rev.*, **125**, 1001–1026.

Zipser, E. J., 1977: Mesoscale and convective-scale downdrafts as distinct components of squall-line circulation. *Mon. Wea. Rev.*, **105**, 1568–1589.

——, 1982: Use of a conceptual model of the life cycle of mesoscale convective systems to improve very-short-range forecasts. *Nowcasting*, K. Browning, Ed., Academic Press, 191–204.

——, and M. A. LeMone, 1980: Cumulonimbus vertical velocity events in GATE. Part II: Synthesis and model core structure. *J. Atmos. Sci.*, **37**, 2458–2469.

Chapter 5

Mobile Radar Observations of Severe Convective Storms

HOWARD B. BLUESTEIN

School of Meteorology, University of Oklahoma, Norman, Oklahoma

ROGER M. WAKIMOTO

Department of Atmospheric Sciences, University of California, Los Angeles, Los Angeles, California

Bluestein   **Wakimoto**

1. Introduction

Fixed-site ground-based radars seldom are able to collect data with sufficient resolution for research purposes owing to the small-scale nature, rapid translation, and infrequency of severe convective storms and the small-scale phenomena they spawn. Accordingly, in recent years, radars mounted on mobile[1] platforms have played an important role in advancing our understanding of the behavior of these phenomena. Ground-based mobile platforms such as vans and trucks, and airborne mobile platforms such as aircraft, have been used to transport various types of radars close to storms as they evolve. In addition to increasing substantially the likelihood of collecting datasets in which convective phenomena are sampled adequately, it is possible to document simultaneously the evolution of severe storm features photographically. Satellites have also been used to carry radars that can view a wide region of the globe.

Radars mounted within aircraft have the ability to seek out target storms easily since they are able to cover a large domain in a single mission. For example, the National Oceanic and Atmospheric Administration (NOAA) P-3 aircraft has a range of approximately 4300 km and an endurance of about 8–10 h. It is now possible to collect high–spatial resolution data following a severe storm phenomenon that, heretofore, was not possible. This increased resolution, however, is not achieved without a price. The airborne platforms can track a feature for an extended period of time, but with a slightly degraded temporal resolution compared to that of a fixed ground-based platform (e.g., see Wakimoto et al. 1998). In addition, airborne radars typically have larger beamwidths than ground-based radars and, therefore, need to fly closer to the target of interest to achieve comparable spatial resolution.

Ground-based mobile platforms often have excellent temporal resolution while they are parked in one location but lose continuity if it is necessary to redeploy to another location. The spaceborne platforms offer the greatest areal coverage over the earth but nominally view a single region only a few times per day. Attenuation is a problem in heavy rain situations if the wavelengths are within the X band (e.g., Hildebrand et al. 1981) or shorter. These relatively short wavelengths are often required in order to keep the antenna diameter small enough to be transportable in a truck or van, or in an aircraft.

---

[1] Bluestein et al. (2001) attribute "mobile" to those instruments that are deployable either while moving or within a few minutes after stopping.

These limitations, however, have not prevented mobile radars from recording unprecedented data in severe storms in recent years (e.g., Wakimoto et al. 1998; Bluestein and Pazmany 2000; Wurman and Gill 2000). The instruments, their platforms, and some of the major scientific findings advanced as a result of field experiments are detailed. This article should be considered an update of the review by Hildebrand and Moore (1990), which provides an excellent historical perspective on mobile radars. Airborne radars are highlighted in section 2, spaceborne radars are discussed in section 3, and mobile ground-based radars are detailed in section 4. In section 5 a summary of evolving radar systems is presented. Section 6 contains a brief summary and a look to the future.

2. Airborne platforms

The concept of an airborne meteorological radar was originally proposed by Lhermitte (1971). Subsequently, a major workshop on multiple Doppler radar was held in 1979 with the general objective to examine the data acquisition, processing, and analysis techniques in use at various laboratories. Particular emphasis was placed upon identification of problems associated with the estimation of vertical air motion (Harris and Carbone 1980). One of the strongest recommendations that emerged from the meeting was the need for an airborne Doppler radar that was vertically pointing in order to provide independent observations of vertical velocities derived from the continuity equation using dual-Doppler-derived horizontal winds (Hildebrand and Carbone 1980). With this strong community support, only a few years elapsed before the first airborne platforms onboard a NOAA P-3 were collecting Doppler radar data. A number of meteorological research problems that were almost intractable became attainable. The capabilities of airborne radars were enhanced as scientists worked with engineers and technicians to implement a number of improvements. The following section provides enough background information to allow the reader to appreciate the airborne radars' characteristics resulting in the collection of a series of unique datasets on severe local storms. The details of the characteristics of these radars can be found in a number of publications that are cited. It is ironic that the workshop did not envision the multiple-view side scanning airborne radars in use today. These airborne platforms utilize essentially the same methodology that was questioned back then.

a. NOAA P-3 and the ELDORA platforms

The first Doppler airborne system to be introduced into the community was the X-band (3.2-cm wavelength) radar capable of scanning vertically onboard the NOAA P-3 (Jorgensen et al. 1983). The basic methodology of deriving horizontal and vertical winds using a vertically scanning radar is given by Jorgensen et al.

(1996). This airborne platform initially became an important tool for studying the internal structure of tropical cyclones (reported in another chapter in this monograph), but soon its use was expanded to study mesoscale convective systems (e.g., Jorgensen et al. 1991), frontal systems within extratropical cyclones (e.g., Wakimoto et al. 1992), and severe local storms (e.g., Dowell et al. 1997). Details of the P-3 Doppler radar characteristics can be found in Jorgensen et al. (1983) and Jorgensen (1984). Recognizing the success of the P-3 platform, the National Center for Atmospheric Research (NCAR) and the Centre de Recherche en Physique de l'Environment Terrestre et Planetaire developed a unique airborne radar (~3.15-cm wavelength) called the Electra Doppler Radar (ELDORA) that significantly improved the collection of data around precipitating systems, especially for severe local storms.

The initial design of the P-3 tail Doppler radar required it to scan in a plane perpendicular to the aircraft track by slewing the antenna's pointing directions with changes in the aircraft drift angle. The motion of the aircraft results in a corkscrew pattern being swept out by the beam as the aircraft flies by a storm. This scanning strategy requires some judicious selection of flight tracks in order to collect Doppler velocities that could be used in a dual-Doppler wind synthesis. There are a number of possible flight tracks that the aircraft can execute but one of the basic types is illustrated in Fig. 5.1a. Sometimes referred to as an L pattern, the radar beams from the two legs intersect at right angles as illustrated in the figure. This geometric angle is optimal for recovering the wind field. The disadvantage of the pattern shown in Fig. 5.1a is that time stationarity must be assumed for a long period of time (except for data collected near the turn) and the track may not be executable if there is intervening convection. Indeed, an aircraft would have a difficult time flying this pattern at the leading edge of a long, linear squall line because of the hazards to the aircraft if it were to fly through the line.

The problems mentioned above were mitigated with the introduction of the fore-aft scanning technique on the P-3 (FAST) (Jorgensen et al. 1996) as illustrated in Figs. 5.1b and 5.2. [This type of scan was first implemented by the National Aeronautics and Space Administration (NASA) in the early 1980s for an airborne Doppler lidar system (see Bilbro et al. 1984) and used to produce an analysis of the horizontal winds in clear air in a gust front produced in a convective storm (McCaul et al. 1987).] This technique requires the antenna onboard the aircraft to be mechanically steered between ±25° in alternate scans. Accordingly, the angle of intersection of the radar scans is nominally 50°.[2] The

---

[2] In 1991, the French provided a prototype ELDORA antenna to NOAA that consisted of two flat plate antennae mounted back to back. The chief advantage of this system is the switching between fore and aft scans is accomplished nearly instantaneously rather than mechanical slewing of a single antenna. ELDORA uses this approach but has two radars that transmit fore and aft scans simultaneously.

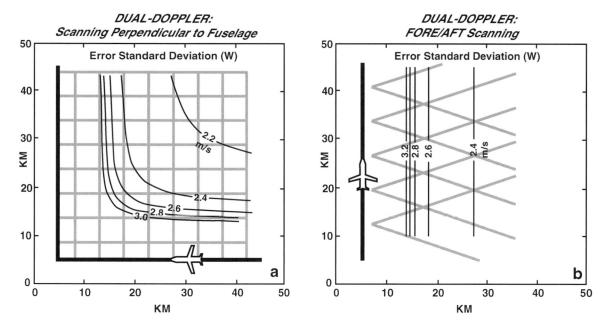

FIG. 5.1. Errors in vertical velocity based on airborne dual-Doppler wind syntheses expressed as standard deviations. Aircraft flight altitude is assumed to be 5 km and storms are considered to have tops at 10 km. (a) The "L"-shaped flight pattern with the radar scanning in a vertical plane perpendicular to the ground track and (b) the straight flight leg with two antennas scanning, one scanning ~18.5° fore of the aircraft axis and the other scanning ~18.5° aft of the axis. Black lines represent the flight track of the aircraft and the gray lines represent beam scanning directions. Based on figures from Ray and Jorgensen (1988).

geometry of the two conical scans is illustrated in Fig. 5.2a. This angle was chosen to minimize the time between the points where the fore and aft scans intersect while still ensuring that the angle of intersection between the beams is large enough to lower the standard deviations of the estimates of the three-dimensional wind field to an acceptable level. The initial design plans for ELDORA included the capability of using the FAST method with one antenna ~18.5° fore of the aircraft axis and the other scanning ~18.5° aft of the axis. The results presented in Fig. 5.1 show that the standard deviations of the estimates of the vertical velocity are comparable for the two scan types.

The major advantage of the FAST scan is that the aircraft flies in a straight line while collecting data. The major disadvantage of FAST is that it requires more postprocessing of the velocity data before it can be inputted into a wind synthesis. The Doppler velocities recorded during FAST are contaminated by the platform component of motion as has been discussed by Lee et al. (1994), Testud et al. (1995), and Bosart et al. (2002). Removal of this velocity component is nontrivial yet critical for obtaining a successful wind synthesis. For example, an error in the tilt angle of only a few tenths of a degree can lead to significant errors in wind direction and vertical velocity calculations, even in stratiform precipitation (Bosart et al. 2002). Examples of how errors in antenna pointing angle appear on a real-time display of Doppler velocity are presented by Bosart et al. (2002).

Early missions to collect data on severe storms en-

countered aliased velocities (e.g., Wakimoto and Atkins 1996; Ziegler et al. 2001) due to the small Nyquist velocity. For a typical airborne radar, the wavelength would be 3.2 cm and transmit at a pulse repetition frequency (PRF) of 2 KHz. This choice of PRF and wavelength would result in an unambiguous velocity of ±16 m s^{-1}, a number that is too small for collecting data from severe storms. Accurate dealiasing of these velocities can be onerous, extremely time consuming, and lead to confusion whether the true "unfolded" velocities have been recovered. The P-3 (Jorgensen et al. 2000) and ELDORA (Frush 1991; Loew and Walther 1995; Wakimoto et al. 1996) have circumvented this problem by introducing dual PRFs that can increase the Nyquist velocity from a single PRF by a factor of 4–5 times. The implementation of the dual-PRF schemes differs slightly for the P-3 and ELDORA and has been explained in detail by Jorgensen et al. (2000). A problem that is occasionally encountered when using dual PRF is the dealiasing errors that the processor makes in regions characterized by broad spectra. Fortunately, these regions are easily identified and removed and nominally account for no more than 2%–3% of the total range bins in a typical sweep.

A challenging problem for a radar platform flying at high speeds is averaging a sufficient number of independent samples in order to calculate a good estimate of the Doppler velocity. In the past, this restriction would have required a relatively slow antenna scanning rate (e.g., 75° s^{-1}), which seriously degrades the along-track resolution. ELDORA represented a significant im-

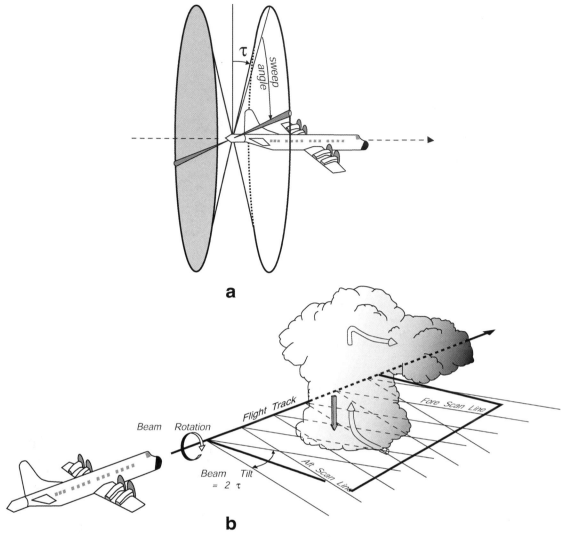

Fig. 5.2. (a) The FAST showing the dual radar beams tilted fore and aft of a plane normal to the fuselage; here τ is the tilt angle of the antenna from the vertical direction; the dashed line represents the flight track. (b) Sampling of a hypothetical storm; the intersection of the fore and aft beams in a horizontal plane is illustrated. Based on figures from Hildebrand et al. (1996).

provement in airborne radar performance by increasing the accuracy of the radar data by averaging more independent samples in the radar pulse volume. This increased accuracy was achieved while increasing the antenna scanning rate (e.g., up to $133° \ s^{-1}$). This apparent contradiction was resolved by the simultaneous transmission of pulses at multiple frequencies (wavelengths). A full scan with ELDORA can be completed in less than 3 s, which at a typical aircraft speed corresponds to a beam separation of about 300 m in the horizontal. Increased sensitivity was achieved by the selection of a longer pulse length (i.e., transmits more power) and allows ELDORA to be the first (and still the only) airborne radar to detect echoes and velocities within the clear air (Wakimoto et al. 1996). Since ELDORA also transmits the fore and aft beams simultaneously rather than switching between fore and aft scans, there is a

factor of 2 increase in data density over the single radar approach used by the NOAA P-3. For more information on this technique and other details of the radar design, the reader is referred to Hildebrand et al. (1994, 1996) and Wakimoto et al. (1996).

1) WEAK ECHO REGIONS WITHIN SEVERE STORMS

The radar vault or weak echo region (WER) of a severe storm has been extensively discussed in the literature for nearly 40 years. Browning and Ludlam (1962) and Browning and Donaldson (1963) correctly interpreted the WER as being due to an updraft so strong that precipitation particles large enough to be detected by the radar did not have enough time to form. The maiden flight of ELDORA when it was operating up to its full capabilities during the Verification of the Origins

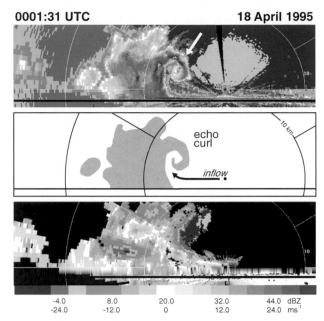

0001:31 UTC   **18 April 1995**

echo curl

inflow

-4.0	8.0	20.0	32.0	44.0 dBZ
-24.0	-12.0	0	12.0	24.0 ms⁻¹

FIG. 5.3. Range height indicator (RHI) cross section through a severe thunderstorm at 0001:31 UTC 18 Apr 1995: (top) radar reflectivity and (bottom) Doppler velocity. A prominent curl in radar reflectivity is shown by the white arrow. Range rings and azimuths are every 10 km and 30°, respectively. A schematic of the flow pattern is shown in the middle. Figure from Wakimoto et al. (1996).

of Rotation in Tornadoes Experiment (VORTEX) provided impressive views of the WERs. An example of a range height indicator (RHI) cross section through a cell contained within a squall line on the evening of 17 April 1995 over southern Oklahoma is shown in Fig. 5.3. Striking in the figure is the identification of a pronounced curl of echo located within the overhang echo. This curl not only protrudes into the WER but wraps completely around it. This type of echo pattern was frequently recorded during VORTEX even though comparable observations, surprisingly, had not been reported earlier in the refereed literature. It is thought that this echo curl may be related to hail trajectories, as first hypothesized by Browning and Foote (1976).

2) MICROBURST

The downburst has been defined as an area of strong winds produced by a downdraft over an area from <1 to 10 km in horizontal dimension. Downbursts can be further subdivided into macrobursts and microbursts with the following definitions (Fujita 1985):

Microburst. Small downburst, less than 4 km in outflow diameter at the ground, with peak winds lasting only 2–5 min. It may induce dangerous tailwind and downflow wind shears that can reduce aircraft performance.

Macroburst. Large downburst with 4 km or larger outflow diameter at the ground, with damaging winds lasting 5–20 min. Intense macrobursts cause

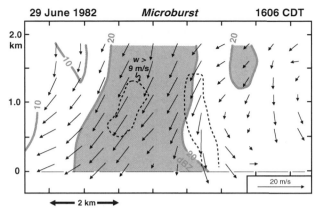

29 June 1982   *Microburst*   **1606 CDT**

FIG. 5.4. Vertical cross section of triple-Doppler winds, through a microburst, on 29 Jun 1982 over Colorado. The analysis includes the P-3 radar data combined with two ground-based Doppler radars. Gray lines are radar reflectivity from the tail radar with values >20 dBZ shaded gray. The dashed lines denote areas where the downdrafts are >9 m s⁻¹. Based on figure from Mueller and Hildebrand (1985).

tornado-force damage up to F3 (Fujita 1981) intensity.

The microburst definition has been modified by radar meteorologists so that peak-to-peak differential Doppler velocity across the divergent center must be greater than 10 m s⁻¹ (Wilson et al. 1984). While many of these events do not produce damaging winds at the surface, the small temporal and spatial scales of the microburst are particularly hazardous to aircraft operations during takeoff and landings (Fujita and Caracena 1977).

Dual-Doppler syntheses using fixed ground-based radars have been used extensively to define the kinematic structure of microbursts (e.g., Wilson et al. 1984; Hjelmfelt 1988; Kingsmill and Wakimoto 1991). Mueller and Hildebrand (1985) were able to illustrate the advantage of combining ground-based and P-3 radar data to recover the detailed vertical structure within a strong microburst (Fig. 5.4). The aircraft flew directly over the microburst at a height of 4.5 km above ground level (AGL; hereafter, all heights are in AGL) so it directly sampled the vertical component of Doppler velocities. As a result, the microburst downdrafts shown in Fig. 5.4 were much stronger near the ground when compared with winds synthesized using ground-based radar alone (not shown). Historically, this study is important since it is believed to have been the first published airborne radar study of a severe convective storm. Moreover, it is still one of the only studies to utilize fully the triple Doppler solution methodology using a ground-based radar coupled to the fore-aft scans of an airborne radar.

A well documented case of an intense microburst occurring over Alabama during the Microburst and Severe Thunderstorm (MIST) Project was presented by Fujita (1992). The P-3 executed a triangular-shaped track around the parent storm (Fig. 5.5) while three ground-based Doppler radars scanned the storm. A series of illuminating photographs were taken of the parent storm

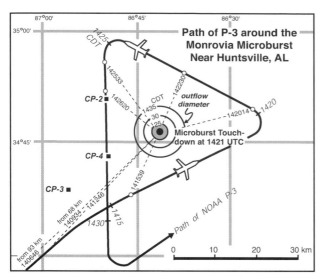

FIG. 5.5. Triangular-shaped path of the NOAA P-3 flying at ~5500 m AGL around a microburst-producing storm on 20 Jul 1986. Black line represents the flight track of the P-3. Black circles represent the leading edge of the outflow at select times. Dashed lines represent the viewing angles of a series of photographs taken from the aircraft. The locations of the CP-2, -3, and -4 ground-based Doppler radars are indicated on the figure. Based on a figure from Fujita (1992).

from the aircraft during the data collection period. Of particular interest, was the visual identification of a pronounced constriction of the cloud at midlevels that appeared to signal the descent of the precipitation core. The descending core would lead to the formation of a microburst when it impacted the ground (Fujita 1992). The CP-2 radar was a dual-polarization radar; that is, it transmitted both horizontally and vertically polarized signals. Comparison of the radar reflectivity factor from both polarizations provides information of hydrometeor type (e.g., Vivekanandan et al. 1999). The importance of high water contents within the storm in order to produce strong downdrafts in environments that are more statically stable was verified. The role of ice (Srivastava 1987; Proctor 1989) in the production of strong surface winds was also shown. The combination of visual, ground, and airborne Doppler radar data, and dual-polarization information makes this case, arguably, one of the best-documented examples of a single-cell thunderstorm since the Thunderstorm Project (Byers and Braham 1949).

High-resolution data were collected using ELDORA at the leading edge of a microburst during VORTEX as reported by Wakimoto et al. (1996). The tail radar resolved the presence of a thin line (Wilson and Megenhardt 1997) of radar reflectivity and Doppler velocities within the boundary layer (not shown; see Fig. 5 in Wakimoto et al. 1996). A pronounced horizontal rotor was present (Fujita 1985; Droegemeier and Wilhelmson 1987; Kessinger et al. 1988) just to the rear of the cold pool. The aircraft was located near the descending branch of the rotor and experienced strong downdrafts

that were of microburst intensity. This case demonstrated the ability of this airborne radar to detect boundary layer convergence zones in the optically clear air.

3) SUPERCELLS, WALL CLOUDS, AND TORNADOES

Although there have been several pioneering studies on supercell storms using Doppler radars (e.g., Brandes 1978, 1981; Ray et al. 1981), there is still a fundamental lack of understanding of the processes that lead to tornado development within this type of convective storm. A contributing factor to this lack of understanding is the relatively small number of case studies on the supercell with dual-Doppler radar observations collected at high resolution. Multi-Doppler radar studies of supercells began in Oklahoma in the 1970s; however, nearly 20 years of research produced only approximately 10 cases (Brandes 1993). These ground-based studies were also limited by the relatively low temporal and spatial resolutions of the wind fields. Typical radar volume scans were separated by 9–10 min and the horizontal grid resolution was approximately 1 km. These problems are not surprising in view of the low probability that a supercell storm will pass through a dual-Doppler radar lobe.

A major opportunity to overcome these aforementioned obstacles was possible with the introduction of airborne Doppler radars. In recent years, as the radar designs improved and skill in flying near these destructive storms increased, these radars on these platforms have collected unique datasets that have resolved supercell structure in greater detail than was previously possible and have also raised a number of new questions concerning tornadogenesis. The first known study to prove the utility of using airborne Doppler radars around supercell storms was by Ray and Stephenson (1990). Although the increased spatial resolution was noted in their study, they also highlighted several disadvantages discussed earlier in this paper. The Nyquist velocity was too small, the flight legs required the stationary assumption to be long, and there were severe attenuation problems due to the short transmitted wavelength. The first two problems have been alleviated with the introduction of dual PRFs and the FAST scanning strategy, respectively. The attenuation problem has not been resolved, although there have been studies that have tried to correct for attenuation using a "stereo radar" approach (Testud et al. 1995; Kabèche and Testud 1995). Attenuation results in a decrease in the magnitude of the reflectivity field and a distortion in the shape of the reflectivity field around the highest values. Fortunately, attenuation does not affect the radial velocity field unless the signal strength drops below the minimum detectable value.

The analysis by Ray and Stephenson (1990) was followed by a series of studies using the airborne Doppler radar onboard either the P-3 or ELDORA to examine the storm-scale structure of the supercell (Bluestein et

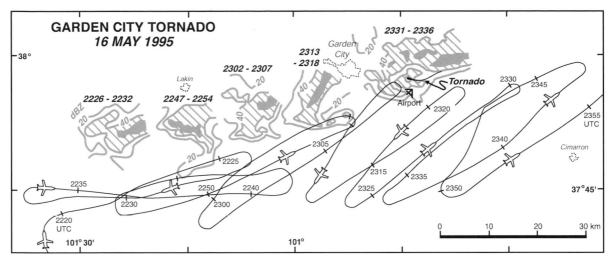

Fig. 5.6. Evolution of the Garden City echo at 600 m AGL based on radar reflectivity data collected by ELDORA. The time interval for each echo composite is labeled on the figure. The flight track of the Electra is plotted. Radar reflectivity is drawn as gray lines, with values greater than 40 and 50 dBZ hatched and shaded gray, respectively. The tornado track is indicated by the black line located north of the Garden City airport. Based on a figure from Wakimoto et al. (1998).

al. 1997a,b; Dowell et al. 1997; Wakimoto et al. 1998; Wakimoto and Cai 2000; Bluestein and Gaddy 2001; Ziegler et al. 2001; Dowell and Bluestein 2002a,b). There are two fundamental flight strategies near supercells. The first is to maintain a large distance away from the storm so that the wind field can be synthesized over the entire storm but the spatial resolution will be coarse. Another flight pattern, which was the primary track during VORTEX, was one in which the aircraft is flown close to the storm and at low altitudes. The latter pattern resolves the hook and mesocyclone in detail; however, a wind synthesis can only be performed on the lower half of the storm because the wind data collected at high elevation angles are highly influenced by the vertical velocities and terminal fall speeds of the hydrometeors.

An example of the typical flight pattern near a supercell during VORTEX is shown in Fig. 5.6. The best location for the aircraft is within the inflow region of the storm, which is nominally situated southeast of the main echo. The flight legs are 5–6 min long and are 8–10 km from the location of the hook echo/mesocyclone. The altitude of the aircraft shown in Fig. 5.6 was 300 m. This region is characterized by less turbulence compared to other regions of the storm (Marwitz et al. 1972) as long as the aircraft does not penetrate one of the nearby gust fronts.

(a) Overview of the Garden City supercell

Figure 5.6 shows the evolution of the Garden City, Kansas, supercell as it moved across the state of Kansas.

Bluestein (1999) has stated that this may be the best-documented case of the incipient stages of supercell formation leading to low-level mesocyclogenesis and the development of a tornado. A series of horizontal cross sections of radar reflectivity and single-Doppler velocities illustrating the low-level evolution of this supercell are presented in Fig. 5.7. The analysis for the first flyby past the storm is shown in Fig. 5.7a. Wakimoto et al. (1998) showed that a well-defined mesocyclone and mesoanticyclone were already present at midlevels at this time. A weak downdraft (not shown) had developed as suggested by the dark green region in single-Doppler velocities in Fig. 5.7a. The storm-relative winds, however, do not indicate any outstanding flow features of note at this time. There is a rapid evolution in the kinematic fields at low levels in only 15 min (two aircraft passes later) as shown in Fig. 5.7b. An echo appendage appears to form on the southeast flank of the storm and the single-Doppler velocities are beginning to outline the region of the developing rear-flank gust front (note the region of velocities toward the radar between 5 and 10 m s^{-1} located south of the main storm echo). The echo appendage is still apparent in the radar reflectivity pattern in Fig. 5.7c while the velocities denoting the location of the rear-flank gust front have become well defined. The aircraft flew parallel to and along the leading edge of this gust front.

The dynamic evolution of the appendage–hook echo is shown in the syntheses in Figs. 5.7d,e and the first indication of a storm-relative circulation is evident in

→

Fig. 5.7. Horizontal cross sections of storm-relative winds (white arrows) superimposed on (left) radar reflectivity and (right) single-Doppler velocities at (a) 2220:00–2223:32, (b) 2234:31–2238:26, (c) 2242:02–2246:15, (d) 2247:31–2252:28, (e) 2301:42–2307:02, (f) 2308:20–2312:42, (g) 2319:01–2323:54, (h) 2324:30–2329:31, and (i) 2331:02–2335:55 UTC 16 May 1995. The dashed black line represents the flight track and the black arrow is the radar viewing angle. Color scales for the reflectivity and Doppler velocities are shown.

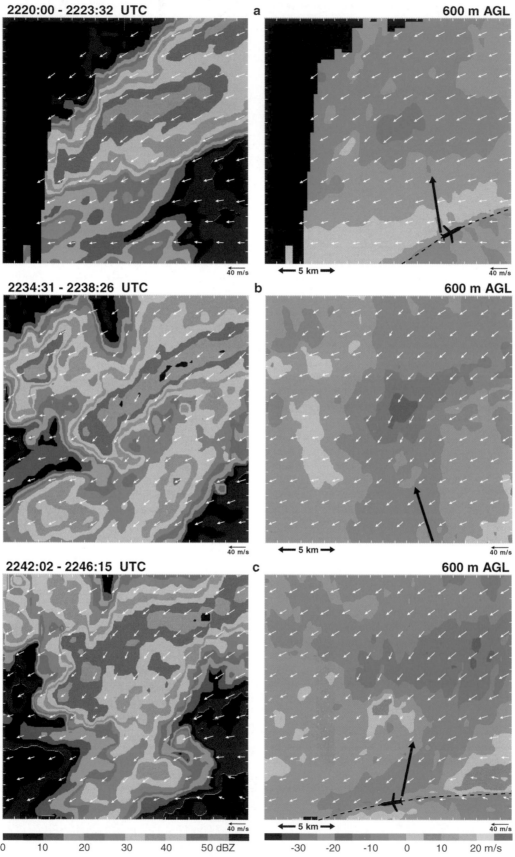

FIG. 5.7. (*Continued*)

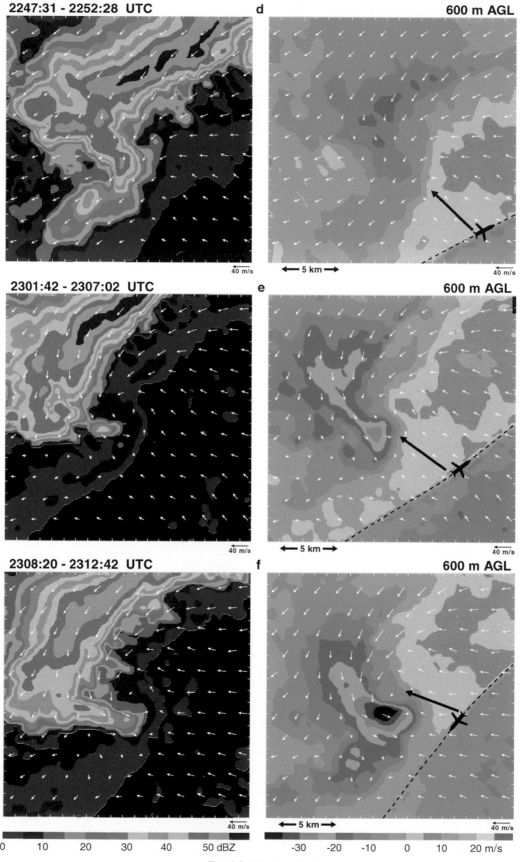

FIG. 5.7. (*Continued*)

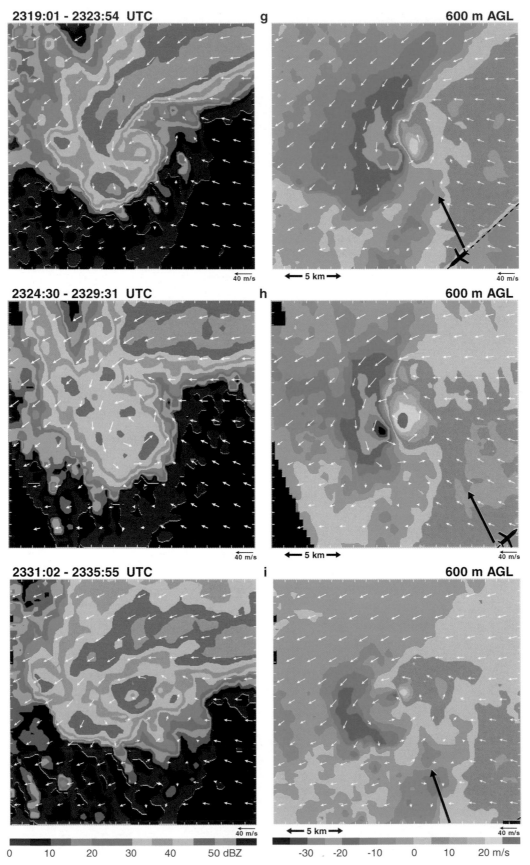

Fig. 5.7. (*Continued*)

the wind field shown at 2301–2307. The key feature within the velocity field is the development of a jet of strong flow toward the radar within the rear-flank gust front just to the southwest of the mesocyclone. The weak echo areas, with reflectivity between 5 and 10 dBZ, outlining the location of the gust fronts in Fig. 5.7e have assumed an approximate "S-shaped" pattern in response to the strong outflow jet and the persistent inflow into the storm east of the mesocyclone, respectively.

The jet of strong Doppler velocities toward the radar (>35 m s^{-1}) collocated with the echo appendage is seen in Fig. 5.7f. This pass corresponds to the time when photographs of a well-defined wall cloud were taken from the airplane. The rotational couplet in Doppler velocities indicative of the mesocyclone circulation (Brown et al. 1978) can be seen in Fig. 5.7g and is accompanied by a "doughnut-shaped" hole in radar reflectivity. The low-level mesocyclone intensifies and reaches its peak resolvable value of vertical vorticity at 2324–2329 and is characterized by a classic rotational couplet embedded within the echo appendage.

Wakimoto and Liu (1998) noted a striking evolution in the echo and velocity plots during the next pass of the aircraft shown in Fig. 5.7i. The hook echo is apparent as a ring of high reflectivity surrounding a weak echo hole. The larger-scale mesocyclone circulation appears to have collapsed with only a small and weak rotational couplet evident. The transition shown in Figs. 5.7h,i suggest that the mesocyclone had weakened. Instead, it was shown that a tornado had formed and was located within the weak echo eye of the hook echo. The false reduction in vorticity was related to the narrowness of the weak, F1 tornado, the flow structure of which was not fully resolved by the airborne radar. The possible mechanism that triggered tornadogenesis will be discussed later.

(b) Wall cloud and mesocyclogenesis

The wall cloud is a lowered cloud base located at the base of the updraft (Davies-Jones 1986) and is recognized as a characteristic feature of most supercell storms (Golden and Purcell 1978; Bluestein 1980). It is such a common attendant phenomenon with the supercell that it is surprising that there have been relatively few studies that have quantitatively examined its visual features. Fujita, in a seminal study of the Fargo, North Dakota, tornadoes [reproduced by Fujita (1992)] and his examination of the Plainfield, Illinois, supercell (Fujita 1993), produced the only detailed studies of wall clouds prior to VORTEX.

A pronounced wall cloud was visible during the 2308–2312 pass by the Garden City storm described earlier (Fig. 5.7f). A horizontal analysis of this pass at low levels is shown in Fig. 5.8. The aircraft passed to the southeast of the echo and within a few kilometers of the echo appendage. The location of the wall cloud on the figure is based on a sequence of photographs

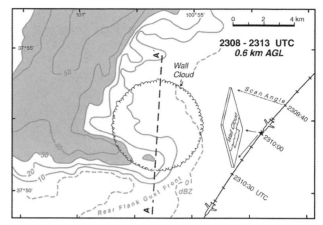

FIG. 5.8. Horizontal cross section at 600 m AGL at 2308–2313 UTC 16 May 1995 of radar reflectivity and the position of the wall cloud. Radar reflectivities are drawn as gray lines with values greater than 30 dBZ shaded gray. The flight track is shown by the black line. A schematic diagram of the viewing angle of the photograph shown in Fig. 5.9 is indicated. The black star indicates the position of the camera site. The location of the vertical cross section shown in Fig. 5.9 is indicated by the dashed black line labeled AA'. Based on a figure from Wakimoto and Liu (1998).

taken from the Electra. One of the photographs is shown in Fig. 5.9 and the position along the track when the photograph was shot is noted schematically on Fig. 5.8.

The fore antenna scans and radar footprint have been photogrammetrically superimposed onto the wall cloud picture in Fig. 5.9a to provide an indication of the resolution of the radar data in the present case. The wall cloud was shown to be approximately 5.5 km in diameter and 1.4 km lower than the overall cloud base. The results presented in Fig. 5.8 illustrate that a substantial fraction of the wall cloud is located outside of the echo appendage. This finding suggests that only a portion of the wall cloud is composed of hydrometeors that are detectable by the radar. This can also be seen in Fig. 5.9c, which presents a vertical cross section through the center of the wall cloud and echo appendage. The vertical structure of the ground-relative single-Doppler velocities is shown in Fig. 5.9b. The geometry of this presentation is such that negative and positive velocities represent flow out of and into the photograph, respectively.

The strong asymmetry in the ground-relative flow across the wall cloud is apparent. This was also independently confirmed by the visual observations of dust that was raised by the strong winds from the middle of the wall cloud and extending to the south (visible in the photograph shown in Fig. 5.9). The component of flow in the plane of the cross section is shown in Fig. 5.9c. Horizontally convergent flow at low levels and the strong updrafts in excess of 30 m s^{-1} within the wall cloud are evident.

Klemp and Rotunno (1983), Rotunno and Klemp (1985), and Davies-Jones and Brooks (1993) have shown, based on numerical simulations, that two sep-

ELDORA - Fore Antenna Scans

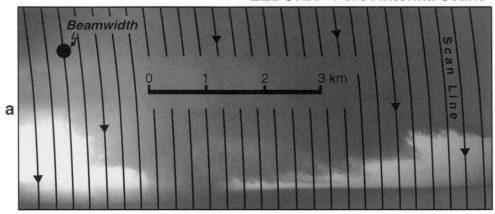

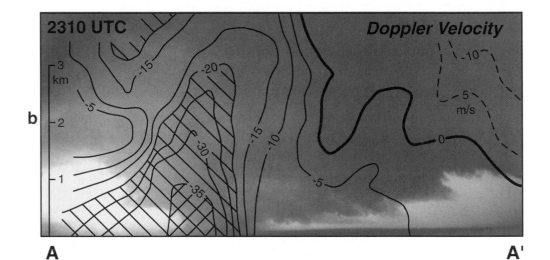

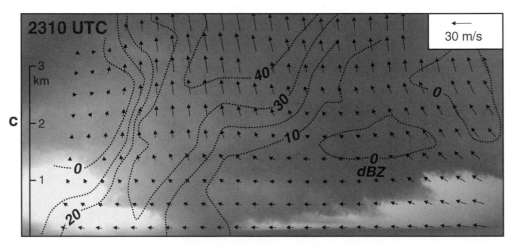

FIG. 5.9. Photograph looking west of the wall cloud at 2310:00 UTC 16 May 1995. (a) Superposition of the individual fore antenna scans from ELDORA based on a photogrammetric analysis of the picture; the beamwidth of 1.8° is indicated along with a length scale that is valid at the distance of the wall cloud. (b) Single-Doppler velocities along the cross section indicated in Fig. 5.8 superimposed on the picture of the wall cloud; positive and negative Doppler velocities are indicated by dashed and solid black lines, respectively. Velocity values less than −20 and −35 m s^{-1} are hatched and cross-hatched, respectively. (c) Radar reflectivity and storm-relative wind field in the plane of the cross section shown in Fig. 5.8; reflectivity values are drawn as dashed black lines. Based on a figure from Wakimoto and Liu (1998).

arate mechanisms are required to generate the low-level and midlevel mesocyclones. This is contrary to earlier hypotheses that suggested that mesocyclone rotation first developed at midlevels and was advected down to low levels by the rear-flank downdraft (e.g., Lemon and Doswell 1979). Observational studies to confirm the ideas advanced by these simulations have been difficult to achieve owing to poor spatial and temporal resolutions until the introduction of airborne radar platforms. Wakimoto et al. (1998) and Ziegler et al. (2001) have both presented vertical cross sections through two different supercells (not shown) that suggest that separate low-level and midlevel mesocirculations strengthen and merge to produce a single column of strong vorticity during the mature stage of the storm.

(c) Tornadogenesis

As previously mentioned, the along-track resolution of the ELDORA data can be as small as 300–400 m depending on the rotation rate of the antenna. Unfortunately, even this high-density data is insufficient to resolve the internal structure of many tornadoes. Accordingly, airborne radar data have proven effective in generating hypotheses concerning the possible mechanisms that trigger tornadogenesis (e.g., Wakimoto and Atkins 1996; Wakimoto and Liu 1998; Trapp 1999; Wakimoto and Cai 2000; Ziegler et al. 2001; Dowell and Bluestein 2002a,b). While dual-Doppler analyses of tornadoes have been difficult, several examples of single radar scans through tornadoes have been documented (see Wakimoto and Atkins 1996; Wakimoto et al. 1996).

Wakimoto and Liu (1998) and Trapp (1999) examined the Garden City mesocyclone using ELDORA and P-3 data, respectively. Both studies arrived at the same conclusion that intense rotation within the mesocyclone at low levels prior to tornadogenesis results in a downward-directed perturbation pressure gradient that induces downflow near the central axis of the mesocyclone. This downflow terminates in low-level radial outflow that turns vertical in an annular updraft at an outer radius. This downflow has been referred to as the occlusion downdraft by Klemp and Rotunno (1983). The development of the occlusion downdraft is illustrated in Fig. 5.10.

Initially, strong updrafts are collocated with the intensifying midlevel mesocyclone (Fig. 5.10a). Note that separate low-level and midlevel mesocyclones are evident as discussed in the previous section. As the rotation at low levels strengthens, the updraft develops a bimodal structure as shown in the vertical velocity plot in Fig. 5.10b. Figure 5.10c presents an analysis for the aircraft pass that immediately preceded tornadogenesis. A downdraft is apparent in the figure and has replaced the upward motion within the mesocyclone. This flow has created a two-celled vortex that Rotunno (1984) has shown can be unstable to three-dimensional perturbations. The instability is manifest greatest in azimuthal

wavenumber 2. Subsidiary vortices form in an annular ring that surrounds the central downdraft. Wakimoto and Liu (1998) and Trapp (1999) suggest that this mechanism could have triggered tornadogenesis in the larger-scale mesocyclone. Trapp (1999) also states that this hypothesis is supported by visual observations of tornadogenesis near the periphery of the mesocyclone rather than in its center.

Wakimoto and Atkins (1996), using single-Doppler velocities, proposed that the Newcastle, Texas, tornado that developed on 29 May 1994 initially formed from a low-level shear feature along the flanking line of a supercell. Vortex stretching of this feature to tornadic intensity occurred under the influence of an intense updraft from a rapidly growing storm along the flanking line. This scenario is similar to one for nonsupercell tornadogenesis proposed by Wakimoto and Wilson (1989). In contrast, Ziegler et al. (2001) used dual-Doppler syntheses to suggest that the Newcastle storm midlevel circulation could be classified as a mesocyclone based on accepted criteria and, therefore, the parent storm was a supercell.

Perhaps one of the most important findings from VORTEX was new insight into the nontornadic supercell, which has been given little attention over the years although it is the most common type of supercell (see Burgess et al. 1993). These storms contribute to high false alarm rates for tornado warnings so they are important to understand. One plausible hypothesis before the field phase of VORTEX was that supercells that did not spawn tornadoes were ones that simply did not form low-level mesocyclones. This hypothesis was quickly refuted when it was shown on several occasions that tornadoes did not develop even in the presence of a low-level mesocyclone (Trapp 1999; Wakimoto and Cai 2000; Bluestein and Gaddy 2001).

Trapp (1999) examined three tornadic and three nontornadic supercells using airborne radar data. His results suggest that at the time of tornado formation or failure, tornadic mesocyclones have smaller core radii and stronger low-level vertical vorticity, and are associated with stronger vortex stretching. Wakimoto and Cai (2001) examined the kinematic structure of a low-level mesocyclone from a nontornadic supercell and compared it with that of a tornadic storm (Fig. 5.11). There are strong similarities between the plots shown in Fig. 5.11. The horizontal dimensions and intensity (as measured by the vertical vorticity) were nearly identical. It was concluded that the presence of a low-level mesocyclone, occlusion downdraft, and updraft–downdraft structure that spirals cyclonically around the circulation are not sufficient conditions for tornadogenesis. In light of these results, Markowski et al. (2002) have suggested that the thermodynamic characteristics of the rear-flank downdraft may be the key to distinguishing tornadic versus nontornadic supercells.

While some supercells produce one tornado that goes through a well-defined life cycle, in rarer instances, su-

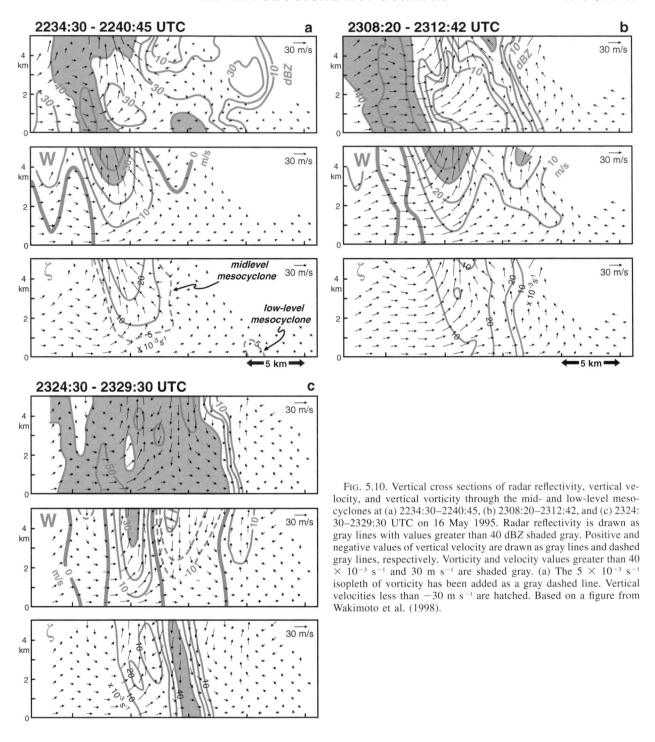

FIG. 5.10. Vertical cross sections of radar reflectivity, vertical velocity, and vertical vorticity through the mid- and low-level mesocyclones at (a) 2234:30–2240:45, (b) 2308:20–2312:42, and (c) 2324:30–2329:30 UTC on 16 May 1995. Radar reflectivity is drawn as gray lines with values greater than 40 dBZ shaded gray. Positive and negative values of vertical velocity are drawn as gray lines and dashed gray lines, respectively. Vorticity and velocity values greater than 40 × 10⁻³ s⁻¹ and 30 m s⁻¹ are shaded gray. (a) The 5 × 10⁻³ s⁻¹ isopleth of vorticity has been added as a gray dashed line. Vertical velocities less than −30 m s⁻¹ are hatched. Based on a figure from Wakimoto et al. (1998).

percells produce a series of tornadoes in a process referred to as "cyclic tornadogenesis" (Burgess et al. 1982). Cyclic tornadogenesis is thought to account for the types of damage tracks found in some tornado families (e.g., Fujita et al. 1970). Adlerman et al. (1999) have numerically simulated cyclic mesocyclogenesis and related it to cyclic tornadogenesis. During VORTEX a dataset documenting cyclic tornadogenesis was col-

lected by ELDORA. Pseudo-dual-Doppler analyses captured, for the first time, the formation and mature stages of three successive tornadoes (Fig. 5.12), one of which was rated as F4 intensity by the National Weather Service (Dowell and Bluestein 2002a). The second author, however, on the basis of an exhaustive ground and aerial damage survey, rated the damage as F5. When a tornado formed, a localized region of outflow about 3–5 km wide

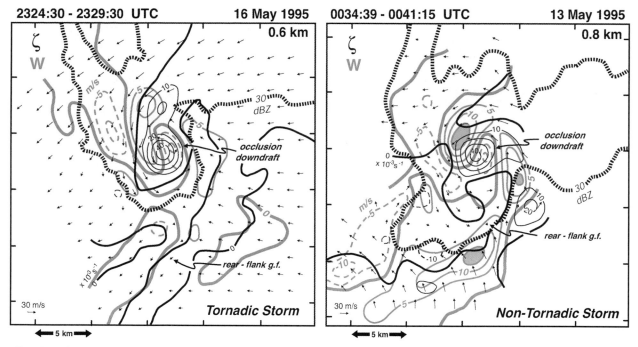

FIG. 5.11. Comparison of the radar reflectivity, wind synthesis, and the vertical vorticity and velocity fields at low levels for the Garden City (tornadic) and Hays (nontornadic) supercells. The thick dashed line represents the 30-dBZ isopleth. The thin black and dashed lines represent positive and negative vertical vorticity, respectively. The gray and dashed gray lines represent positive and negative vertical velocities, respectively, with values greater than 15 m s^{-1} shaded gray. The locations of the occlusion downdraft and the rear-flank gust front are indicated on the figure. Based on a figure from Wakimoto and Cai (2000).

surged ahead of the tornado and produced a bulge in the nearby gust front. A new vorticity maximum formed near the leading edge of the outflow, which then developed into a new tornado. It was found (Dowell and Bluestein 2002b) that air parcels at low elevation entered the rear portion of the main storm updraft from its left side and underwent tornadogenesis as a result of the tilting of horizontal vorticity onto the vertical, followed by the stretching of vertical vorticity. The nature of the horizontal vorticity that was tilted onto the vertical varied with time as the location of the vortex with respect to the updraft changed. Early on, the interaction of the updraft with the environmental low-level horizontal vorticity produced a vorticity column whose intensity increased with height. Later, as the vortex matured, vorticity increased greatly at low levels and exceeded that aloft. The tornado vortex was located near the rear side of the updraft, where the surrounding low-level horizontal vorticity had been modified by weak baroclinity within the storm. The most important conclusion regarding the ELDORA observations of cyclic tornadogenesis is that it appears to be associated with a mismatch between the horizontal motion of successive tornadoes and the horizontal velocity of the main storm-scale updraft and downdraft. Low-level relative flow seems to be the most important factor in determining tornado motion.

Another intense vortex that has not received as much attention as the tornado is the waterspout. Quantitative observations of waterspouts and their parent storms are rare due to the difficulty of collected measurements over large bodies of water (e.g., Golden 1974; Wakimoto and Lew 1993; Golden and Bluestein 1994). These sparse observations have suggested that some waterspouts undergo a life cycle that is similar to nonsupercell tornadoes as documented by Wakimoto and Wilson (1989). Verlinde (1997) also suggested a low-level vorticity source for a waterspout observed by the P-3 and EL-DORA during the Tropical Oceans Global Atmosphere Coupled Ocean–Atmosphere Response Experiment (TOGA COARE). The waterspout circulation was difficult to resolve in the wind synthesis as in the VORTEX analyses; however, a single radar scan did cut through the vortex.

(d) Supercell propagation

The main defining feature of the supercell is storm-scale rotation; in addition, a supercell may split in its early stages of development into a rightward-propagating cyclonic and a leftward-propagating anticyclonic thunderstorm (e.g., Hitschfield 1960; Fujita and Grandoso 1968; Charba and Sasaki 1971; Klemp and Wilhelmson 1978). The terms "right" and "left" are relative to the mean direction of the environmental winds. Newton and Newton (1959) were among the first to suggest that the perturbation pressure forces were playing an important role in explaining the latter behavior.

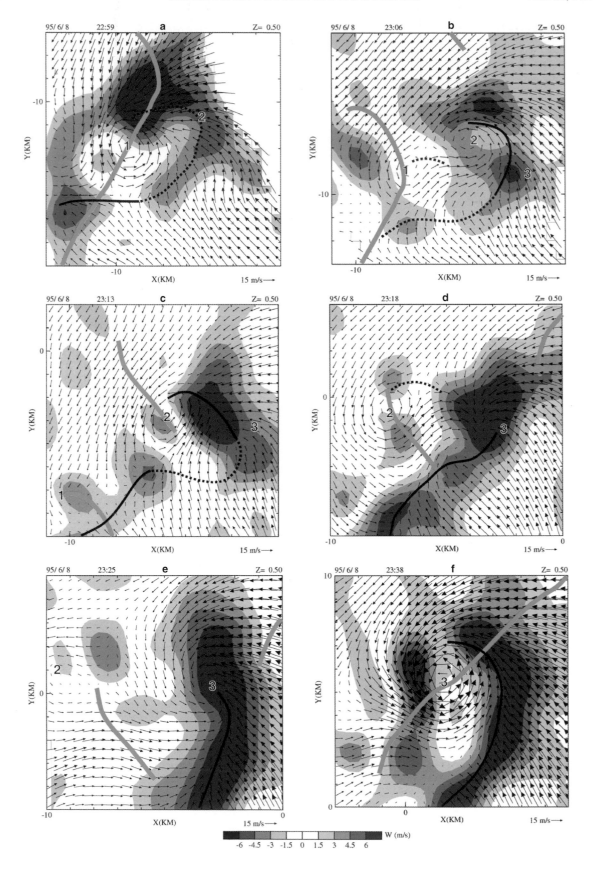

Schlesinger (1980) and Rotunno and Klemp (1982, 1985) have provided critical insight into the process using numerical simulations. When the environmental shear vector is straight, low perturbation pressure centers are produced along the flanks of the storm by nonlinear effects within the mesocyclone and mesoanticyclone. In addition, the perturbation pressure deficits are reduced further by the effects of horizontal vorticity being generated by buoyancy gradients developing along the updraft flanks. Rotunno and Klemp (1982, 1985) show that these two nonlinear effects combine to force positive vertical motion along the updraft flanks and lead to storm splitting. Furthermore, they show that when the wind shear vector veers with height, the updraft splitting occurs as before but the right-moving storm is stronger. It is only the linear perturbation pressure effect that can account for this rightward bias. The linear effect from the perturbation pressure equation is proportional to the interaction of the convective updraft with the vertical shear of the horizontal wind within the environment.

Observational confirmation of these numerical results has been rare due to the difficulty of collecting comprehensive pressure data within and along the periphery of intense convective storms. Aircraft have provided the most reliable pressure measurements (Marwitz 1973; Raymond 1978; LeMone et al. 1988); however, the data are collected only along the flight path of the aircraft, time stationarity assumptions must be invoked to reconstruct a two- or three-dimensional picture, and it is difficult to collect data within regions of heavy precipitation.

Multi-Doppler radar syntheses have proven to be useful in retrieving the perturbation pressure and density patterns using a least squares method proposed by Gal-Chen (1978). The retrieval technique treats the pressure and temperature as unknown variables in the anelastic momentum equation and solves the derived Poisson equation using the Doppler wind field. Additional details of the retrieval scheme can be found in Roux (1985, 1988). Cai and Wakimoto (2001) applied this scheme to the Garden City wind synthesis to assess the role of the linear and nonlinear perturbation pressure terms in determining the motion of the supercell.

Figure 5.13 is a horizontal plot at 3.0 km for a time during the storm's mature stage. The updraft is greater than 15 m s^{-1} and is near the location of the mesocyclone and a developing echo appendage (Figs. 5.13a,b). A prominent mesoanticyclone is also apparent. The lo-

cation of the two mesocirculations is consistent with the orientation of the wind shear vector as first shown by Klemp and Wilhelmson (1978). The vertical gradient of perturbation pressure based on the retrieved fields is shown in Fig. 5.13c and reveals a bias to the right of the shear vector as would be expected. The partitioning of the perturbation pressure gradient into linear and nonlinear components is shown in Fig. 5.13d. It is the nonlinear component that contributes to the rightward bias as predicted by Rotunno and Klemp (1982, 1985). The linear pressure gradient is downward and upward directed in the upshear and downshear regions of the storm, respectively. Rotunno and Klemp have shown that this pattern is to be expected at heights below the maximum vertical velocity.

b. ER-2 Doppler radar

Previous airborne Doppler radars were restricted to mid- and low altitudes. Most of these aircraft radars scan about the aircraft's longitudinal axis as discussed in the previous section and base their wind syntheses on side-looking views of the convection. Recently, an X-band radar referred to as the ER-2 Doppler radar (EDOP; Heymsfield et al. 1996) has been installed in the ER-2 high-altitude aircraft (20 km). It is a dual-beam (nonscanning) Doppler radar with one beam pointed at nadir and the other pointed at approximately 33.5° forward of nadir. A schematic diagram illustrating the scanning methodology is shown in Fig. 5.14. Nadir beams can be separated by 100 m providing high-resolution along-track measurements. Dual-Doppler winds and the echo patterns in a vertical plane along the aircraft motion vector can be created. The wind component normal to the aircraft track cannot be obtained.

One of the most difficult measurements with Doppler radars has been the vertical air velocity. This is a derived quantity for ground-based and airborne radars flying at low altitudes and is estimated by integrating the mass continuity equation using the horizontal winds directly measured by the radars. EDOP samples the vertical air motion while in the nadir mode in precipitation regions if the effects of platform motion and hydrometeor fall speeds can be successfully removed (Heymsfield 1989). The high altitude of ER-2 flight track is advantageous since it can fly over virtually all cloud tops and suffers from fewer aircraft traffic control restrictions. It should be noted that Lhermitte (1971) originally proposed that

---

←

FIG. 5.12. Pseudo-dual-Doppler analyses of dominant-vortex-relative horizontal wind field (vectors; scale indicated at bottom right-hand side of each panel) and vertical velocity (shaded regions; color scale indicated at the bottom of the figure) at 0.5 km AGL, based on ELDORA data, for 8 Jun 1995 for a several-minute time interval centered at (a) 2259, (b) 2306, (c) 2313, (d) 2318, (e) 2325, and (f) 2338 UTC. The green swaths indicate the damage paths of each tornado; the solid black line denotes gust fronts, wind-shift lines with or without baroclinic zones; the dotted lines indicate the approximate location of weak or retreating gust fronts. The numbers label each major cyclonic vortex. The distance scale is given in km; the spacing between tick marks represents 1 km; the x axis (y axis) points toward the east (north). Courtesy of D. Dowell; adapted from Dowell and Bluestein (2002a).

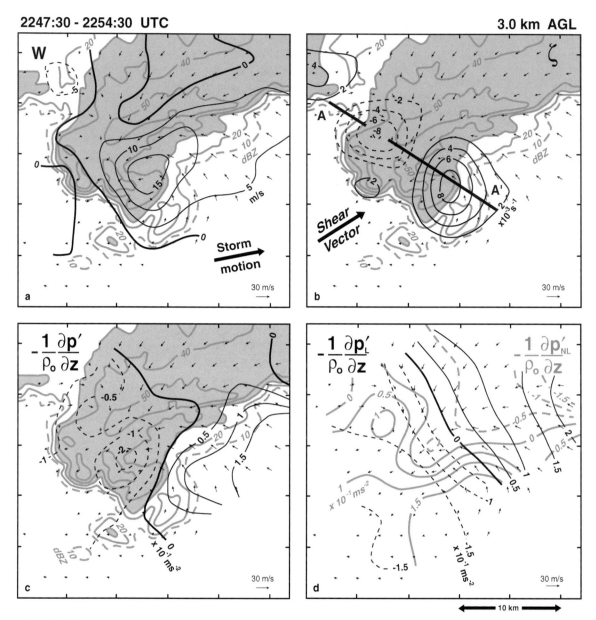

FIG. 5.13. Horizontal cross sections at 3.0 km AGL for 2247:30–2254:30 UTC of (a) radar reflectivity and vertical velocity (positive and negative values are drawn as black and dashed gray lines, respectively), (b) radar reflectivity and vertical vorticity (positive and negative values are drawn as black and dashed lines, respectively), (c) radar reflectivity and the vertical perturbation pressure gradient (positive and negative values are drawn as black and dashed lines, respectively), and (d) vertical gradients of the linear (positive and negative values are drawn as black and dashed lines, respectively) and nonlinear (positive and negative values are drawn as gray and dashed lines, respectively) components of the perturbation pressure. The storm motion and shear vector are indicated on the figure. Based on a figure from Cai and Wakimoto (2001).

a downward-pointing radar would be the optimum scanning mode for an airborne Doppler radar.

An example of EDOP images through a severe winter squall line from Heymsfield et al. (1999) is shown in Fig. 5.15. Also superimposed on the figure are relative humidity, zonal and longitudinal winds, and vertical-velocity (ω) contours interpolated from a run by the nested grid model (NGM). The radar reflectivity and vertical velocities clearly depict the convective and stratiform regions of the squall line (e.g., see Smull and

Houze 1987a). The hydrometeor velocities in the stratiform region increase from 1–2 m s^{-1} in the snow layer to greater than 6 m s^{-1} in the rain layer below the melting layer (shown as a "bright band" of reflectivity). The panel displaying the vertical air motions reveals more detail than has been previously possible with other radar platforms. In addition, the largest vertical velocities within the stratiform region are nearly twice those reported in earlier studies except for a case reported by Jorgensen et al. (1997). Owing to instrument problems,

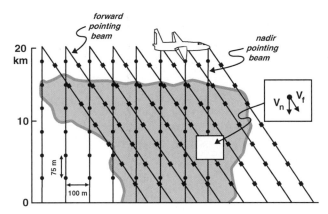

FIG. 5.14. A schematic of the EDOP scanning strategy. The nadir- and forward-pointing beams are shown on the figure. The nadir- and forward-pointing components of the Doppler velocity are indicated, respectively, as v_n and v_f on the figure. The vertical and horizontal resolutions are 75 and 100 m, respectively, in the example shown in the figure. Based on a figure from Heymsfield et al. (1996).

the squall-line-relative zonal winds were restricted to the higher reflectivity regions. Even with this limitation, the rear-inflow jet from west to east is seen descending toward the leading convective line (Smull and Houze 1987b).

3. Spaceborne platforms

Placing radars on spaceborne platforms to sample convective events has been gaining more interest in recent years. The high altitude and speed of the platform would appear to preclude these platforms for studying severe convective storms. Serious limitations of a spaceborne platform are associated with the short wavelength used (a long wavelength would require a large antenna, which would be prohibitively expensive) and the narrow swath with respect to the earth. These radars are also limited to a couple of views over the same geographic regions during a 24-h period. However, there have been a couple of instances when these radars have provided unique views of severe weather phenomenon that would not have been possible with airborne or mobile ground-based radars.

a. Synthetic aperture radar (SAR)

A synthetic-aperture radar (SAR) achieves high resolution in the along-track direction by taking advantage of the motion of the spaceborne platform to synthesize the effect of a large antenna aperture. SAR permits the attainment of high resolution by using the motion of the platform to generate the antenna aperture sequentially rather than simultaneously as with a conventional array antenna (e.g., see Skolnik 1980). The imaging of the earth's surface by SAR can be used to measure the sea state and ocean wave conditions.

The SEASAT SAR was the first imaging radar flown

in space and was designed to study wave spectra on the ocean. It was soon apparent that these radars could see variations in sea surface scatter produced by strong surface winds from nearby convection. These regions often appear as semicircular features on the sea surface. The radar is not able to detect the precipitation itself, but the rain and outflow winds associated with the storm affect the sea state. Atlas (1994), Atlas and Black (1994), and Atlas et al. (1995) have suggested that rain damps the surface waves and causes echo-free holes in the SAR data. Strong downdrafts (microbursts) accompanying the heavy rain subsequently diverge outward at the surface and generate waves and associated echoes surrounding the holes.

An example of an analysis of SAR features is shown in Fig. 5.16. The gray regions are the locations of the echo-free hole and the gray lines represent the outline of the storm-induced disturbances that are believed to be caused by microburst outflows. Atlas and Black (1994) used the WSR-57 radar data from Miami (S-band) to superimpose precipitation echoes on the SAR features shown in Fig. 5.16. Unfortunately, the beamwidth of the radar was 2° and the distance to the echoes was 200–230 km. In spite of these limitations, there appears to be a strong connection between the convective cells and the echo-free holes lending credence to the ability of a SAR to detect microburst activity over the ocean which is free of the complexity of the underlying land.

b. Tropical Rainfall Measuring Mission (TRMM) radar

The TRMM rain mapping radar launched in 1997 made it possible for quantitative measurement of tropical rainfall to be obtained on a continuing basis over the entire global Tropics and subtropics (Kummerow et al. 1998). The inclination of the radar is 35° and it orbits at a height of 350 km. The radar is a phased-array system operating at 13.8 GHz (subject to attenuation in heavy precipitation areas) and scans in the cross-track direction over ±17° (215-km swath).

The TRMM radar in concert with a lightning imaging sensor (LIS) and the TRMM microwave imager (TMI) have provided, for the first time, a global dataset of severe convection. This is especially true in regions that are remote and inaccessible. Yorty et al. (2001) have presented preliminary analyses of the distribution of the most intense storms over the 2-yr period of March 1998–February 2000 (not shown). Vertical profiles of reflectivity were used as a proxy for intensity. Their findings include the following. 1) Oceans are devoid of intense convection. Here "intense" was arbitrarily determined to be storms where the 20-dBZ height was ≥18.5 km and the 40-dBZ height was ≥15.5 km MSL. 2) Tropical Africa has many more extreme events than tropical South America. 3) Subtropical North and South America have more extreme events than subtropical Africa, Asia,

EDOP Images & NGM Model Contours

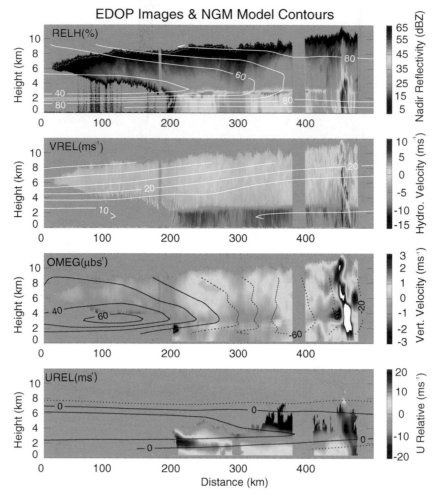

FIG. 5.15. EDOP vertical cross section through a squall line on 13 Jan 1995 from 1730 to 1807 UTC with superimposed 1800 forecast contours from the National Centers for Environmental Prediction (NCEP). Shown from top to bottom are nadir reflectivity, vertical hydrometeor velocity, vertical velocity, and horizontal (along track) winds. Contoured NCEP forecast from top to bottom are relative humidity, relative velocity parallel to the squall line, omega, and relative velocity perpendicular to the squall line. White area in vertical velocity panel denotes values greater than 3 m s⁻¹. Based on a figure from Heymsfield et al. (1999).

or Australia. 4) Desert fringe areas often have extreme events (e.g., Sahel, Texas–Mexico, Pakistan, and the shore of the Red Sea).

4. Ground-based platforms

The development of ground-based, mobile radars for the study of tornadoes was originally suggested by Zrnic et al. (1985). The advantages of a ground-based mobile radar over that of an airborne radar are that it can probe scatterers much closer to the ground, it can be brought safely much closer to severe phenomena such as tornadoes, and that much higher temporal resolution can be attained. Ground-based mobile radars are thus much more suitable for the detailed study of tornadoes, since the most dynamically important region of a tornado is near the ground (e.g., Rotunno 1984; Davies-Jones

1986; Davies-Jones et al. 2001). Furthermore, hypotheses for tornadogenesis have involved aspects of the wind and thermodynamic fields near the ground (e.g., Wakimoto and Wilson 1989; Roberts and Wilson 1995; Davies-Jones et al. 2001). In addition, when there is little if any precipitation in or adjacent to a tornado, it is necessary that the radar is as sensitive as possible to clear air return; thus, it is advantageous to be at as close range as possible.

A number of X-band radar systems and a W-band radar system have been developed. These systems are now detailed.

a. The LANL portable CW/FM-CW X-band radar

The first portable Doppler radar was developed by staff members of the Mechanical and Electronic Engi-

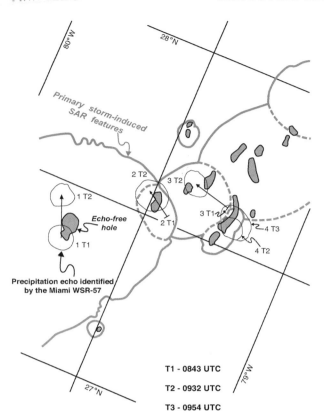

FIG. 5.16. The gray lines are outlines of the primary storm-induced SAR features. The echo-free holes are shaded. The thin black lines represent the precipitation echoes as seen by the Weather Surveillance Radar WSR-57 at Miami. Each such echo is marked by a number and time (e.g., 1T2 represents echo 1 and time T2); times are indicated at the lower right. Where the initial precipitation echo would confuse the drawing, the center is located by a tick mark. The arrows show the movement of the precipitation echoes between the two times. Based on a figure from Atlas and Black (1994).

FIG. 5.17. Photograph of the LANL portable CW/FM-CW Doppler radar system and OU graduate students on 13 May 1989, probing a tornado near Hodges, Texas. Copyright H. Bluestein; from Bluestein and Golden (1993).

neering Division of the Los Alamos National Laboratory (LANL). Originally designed for nonmeteorological targets, it was modified for meteorological targets at LANL by W. Unruh and his group and first used in collaboration with the School of Meteorology at the University of Oklahoma (OU) and NSSL (National Severe Storm Laboratory) in 1987 (Bluestein and Unruh 1989). The X-band, CW (continuous wave), solid-state radar had separate 5° half-power beamwidth parabolic antennas for receiving and for transmitting 1 W of continuous power. Operated while mounted on a tripod (Fig. 5.17), the radar system could be set up outside the van in which it was transported and simultaneous video was recorded while scanning its meteorological target from a range of up to 10–15 km. The radar was scanned manually. Although there was no real-time display, the signals from each channel could be monitored as audio signals; the pitch of the audio signal was proportional to the wind speed, while the receding and approaching spectra were separately monitored. Since the beamwidth of the antennas was relatively broad, radar volumes on the order of several hundred meters across, containing

much of the circulations of tornadoes, could be resolved at ranges of 2–10 km. While ranging information was not available during the early years of the radar's use, Doppler wind spectra were collected in many tornadoes, from which the maximum wind speed could be estimated (Bluestein et al. 1993). The number of tornado wind speed spectrum datasets collected in a few seasons far exceeded the number collected by NSSL's fixed-site radar over many seasons.

The most important result of the initial field experiments conducted in 1987–94 was that the thermodynamic speed limit in tornadoes (Lilly 1969; Snow and Pauley 1984) appeared to be exceeded in almost every case, most likely owing to nonhydrostatic pressure-gradient forces resulting from the interaction between the tornado vortex and the ground (Fiedler and Rotunno 1986). In addition, wind speeds of 120–125 m s^{-1} were measured under cloud base in one tornado, thus confirming that F5 wind speeds can indeed occur in tornadoes (Fig. 5.18). Prior to these radar measurements, estimates of F5 wind speeds came from uncalibrated

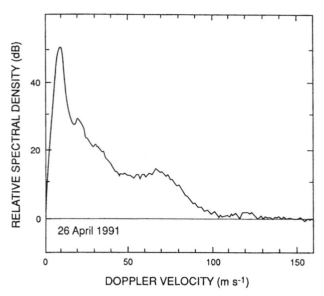

FIG. 5.18. Evidence of F5 wind speeds in a tornado, from data collected by the LANL radar. Receding portion of the Doppler spectrum averaged over a 1.2-s period, for the right-hand side of a tornado, north-northeast of Ceres, Oklahoma, on 26 Apr 1991: subjectively determined noise level (solid line). The height of the center of the radar beam was approximately 150–190 m; the approximate width of the half-power beamwidth at the range of the tornado was 140 m. Power from ground clutter at very low wind speeds has been suppressed. From Bluestein et al. (1993).

damage surveys and photogrammetric estimates made from movies of debris circulating in suction vortices well above the ground (Forbes 1978; Bluestein and Golden 1993).

There were several other significant findings. It was found that strong winds may occur in tornadoes, even during the "shrinking stage" (Davies-Jones et al. 2001), as inferred previously from a damage survey and photographic analysis of the Union City, Oklahoma, storm of 24 May 1973 (Davies-Jones et al. 1978). Furthermore, it was noted that wind speed estimates based upon damage estimates alone must be viewed with caution as had been suggested by Doswell and Burgess (1988). In addition it was found, based on comparisons of actual Doppler wind spectra with idealized spectra, that the maximum radar reflectivity lies well outside the radius of maximum wind, as had been predicted by Snow (1984). The latter finding was hypothesized to be due to the effects of centrifuging radially outward of the most reflective scatterers. Finally, it was found that the damage swath of a tornado is wider than the width of the tornado condensation funnel at cloud base.

The LANL portable radar was modified to operate also in the frequency modulated (FM)-CW mode in 1988 (Bluestein and Unruh 1993) and datasets were first successfully collected in a tornado during VORTEX-94 (Bluestein et al. 1997b). Doppler wind spectra were computed in range gates separated by 78 m; the spectra (Fig. 5.19) represent measurements on both the scale of the tornado and its parent circulation. Maximum wind

speeds in the tornado were around 65 m s^{-1}; some smaller areas had wind speeds possibly as high as 75 m s^{-1}.

The spatial resolution of the LANL portable X-band radar was limited in the cross-beam direction mainly by the size of the antenna, which in turn was limited by the ability of the users to carry an antenna from the van to the tripod. The along-the-beam resolution was limited by the sweep width of 1.92 MHz. The along-the-beam resolution could have been improved by increasing the sweep width. In the FM-CW mode, however, there was no real-time display, save for a video display that allowed the user to monitor signal strength and quality. Wind spectra had to be postprocessed. The performance of the radar was also limited by the sweep repetition frequency of 15.75 kHz, which corresponds to a maximum unambiguous range of 5 km (and, incidentally, a maximum unambiguous velocity of ±115 m s^{-1}). The FM-CW parameters were fixed by the use of a video cassette recorder (VCR) and could have been improved, but at the expense of the ability to monitor the signal quality. It was therefore decided that the next-generation mobile radars should have improved spatial resolution either by using larger antennas, which would have to be mounted on the mobile platform itself, and/or by using higher-frequency radars, which would not require much larger antennas. In addition, higher-power radars could be used in the more traditional pulsed mode, which would permit easier real-time signal processing and real-time display. A group at the University of Massachusetts—Amherst chose the latter approach, of developing a higher-frequency, pulsed radar; a group from OU, NCAR, and NSSL chose the former approach, of developing a pulsed, X-band radar having a much larger antenna and much more powerful transmitter.

b. The University of Massachusetts W-band mobile Doppler radar

The first mobile, pulsed, W-band Doppler radar was designed and constructed at the Microwave Remote Sensing Laboratory at the University of Massachusetts—Amherst by A. Pazmany and his group. This radar system was first used in 1993 in the southern plains by Bluestein et al. (1995) and is often referred to as the "U. Mass. radar." The first version was mounted in an OU van and used a 30.5-cm lens antenna having a half-power beamwidth of 0.7°. The antenna poked through a hole in the roof of the van in which it was housed (Fig. 5.20). In addition to using a smaller antenna but still achieving higher spatial resolution, a W-band system is more sensitive to Rayleigh scattering by small scatterers because the scattering efficiency is inversely proportional to λ^{-4}, where λ is the wavelength.

On the other hand, for scatterers such as raindrops whose diameters are on the order of and wider than 3 mm (the wavelength of the radar), radar return is due to Mie scattering. It therefore is not easy to relate rainfall rate to radar reflectivity by a simple Z–R relationship.

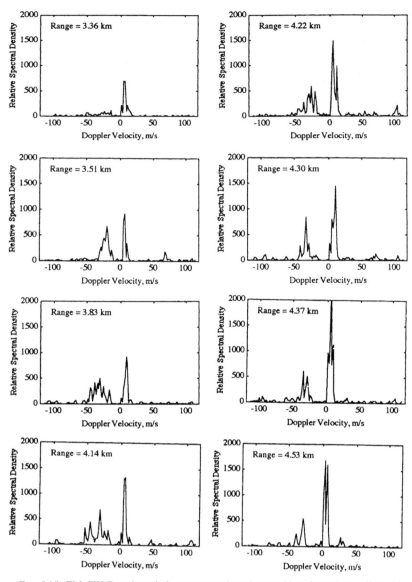

FIG. 5.19. FM-CW Doppler wind spectra at selected range bins, from data collected by the LANL radar in a tornado near Northfield, Texas, on 25 May 1994. From Bluestein et al. (1997b).

In addition, the dynamic range of the radar reflectivity is compressed relative to what it is at 3-cm wavelengths (Lhermitte 1990). Furthermore, attenuation at 3-mm wavelengths is more severe than it is at 3-cm wavelengths; the attenuation coefficient for a rainfall rate of 2.5 cm h^{-1} is approximately 10 dB km^{-1} (Ulaby et al. 1981). Finally, the tradeoff between maximum unambiguous velocity and range at 3-mm wavelengths is much more difficult than it is at 3-cm wavelengths. For example, if the maximum unambiguous range is 10 km, the maximum unambiguous velocity is only 12 m s^{-1}. Such a low unambiguous velocity range is unacceptable for applications to tornadoes, in which there are velocity gradients of 10 m s^{-1} per 10–100 m.

To alleviate the latter problem, the polarization di-versity pulse-pair (PDPP) technique (Doviak and Sirmans 1973), in which pairs of identically polarized and orthogonally polarized pulses are interleaved, was implemented (Pazmany et al. 1999); the former are separated by 66.6 μs and the latter are separated by 10 μs. In the PDPP mode a maximum unambiguous velocity of approximately ±80 m s^{-1} is attained. The drawbacks of the PDPP technique are that it is noisier than the conventional pulse-pair technique and that it does not work for scatterers within about 1.5-km range, owing to the finite polarization isolation of the switch network and the antenna. However, operating the radar within 1.5 km of a tornado is considered dangerous and is thus to be avoided. It is possible to make use of the strengths of both the PDPP technique and conventional pulse-pair

FIG. 5.20. The first version of the U. Mass. mobile W-band radar, mounted in a van, probing the edge of a precipitation shaft in a convective storm near Abilene, Kansas, on 7 Jun 1993. Photograph copyright H. Bluestein; from Bluestein et al. (1997a).

processing by using the PDPP data to unfold the conventional pulse-pair data when possible, for example, outside the core of a tornado.

The first version of the radar had a range resolution of 30 m, so that at a range of 2–3 km the radar volumes were 30 m on a side. In 1999, the radar was reinstalled in a small pickup truck with a 1.2-m dish antenna (Fig. 5.21), which has a half-power beamwidth of 0.18°. The pulse length was decreased so that at 2–3-km range the radar volumes were only about 8 m × 8 m × 15 m. A spatial resolution on the order of 10 m is required to sample adequately the Doppler velocity field in tornadoes whose diameters are on the order of 200–1000 m across. Since the resolution volumes are so small, it was necessary to install load levelers in the van and pickup truck to provide stability.

In spring field programs conducted in 1993 and 1994 data were collected with the first version of the radar in a mesocyclone located near the intersection of two squall lines, in a low-precipitation supercell, and in the hook echo of a supercell. It was found that attenuation, while substantial in face of heavy precipitation, was not damaging enough to limit the range of the radar to closer than a few kilometers. During VORTEX-95 the W-band radar resolved counterrotating vortices 500 m in diameter along the edge of a rear-flank downdraft gust front in a supercell (Bluestein et al. 1997a); each vortex was accompanied by a mirror-image hook echo separated by about 1 km. It was speculated that the vorticity associated with each of the vortices could be the source

of vorticity in tornadoes, even though no tornadoes occurred in this case.

High-resolution data were collected at an elevation angle as close to the ground as possible, at close range (in some cases within a few kilometers), in a number of tornadoes during the spring of 1999 (Bluestein and Pazmany 2000). The W-band radar was able to penetrate the tornadoes with no limiting effects from attenuation. The tornadoes were scanned at one elevation angle to minimize the time between scans. Even during the scanning period of 10–20 s, features changed significantly. Eyes and spiral bands in the radar reflectivity field were ubiquitous (Fig. 5.22); many of the reflectivity analyses looked like scaled down hurricanes [also noted by Fujita (1981)]. The existence of eyes is consistent with the findings of Bluestein et al. (1993) based upon data collected with the LANL radar. It was hypothesized that the eyes are caused by centrifuging of the largest scatterers outward from the center of the tornado. Since there was no strong convergence signature in the Doppler velocity data, either the radar beam did not sample the surface layer of the tornado or centrifuging contaminated the along-line-of-sight velocities. Evidence of multiple vortices along the edge of the eyewall were found in a tornado on 3 May (Fig. 5.23) whose parent circulation redeveloped into the F5 tornado that about an hour later moved through parts of the Oklahoma City area. The tangential wind speed as a function of radius was approximately in solid body rotation within the core and dropped off slowly outside the core (Fig. 5.24),

FIG. 5.21. Photograph of the latest version of the U. Mass. mobile W-band truck-mounted radar, probing a tornado near Stockton, Kansas, on 15 May 1999. Photograph copyright H. Bluestein; from Bluestein et al. (2001).

similar to the wind speed profile of a combined-Rankine vortex (Davies-Jones 1986).

A dataset collected on 5 June 1999 captured the entire life cycle of a tornado as it formed about 5–6 km away and moved to within a few kilometers away before dissipating. It was found that the incipient tornado vortex was of very small scale. The tornado Doppler velocity data are being subjected to the ground-based velocity track display (GBVTD) of Lee et al. (1999) to estimate the mean properties of the tornado vortices. Detailed results from an analysis of this case and from that of another on 15 May 1999 are forthcoming.

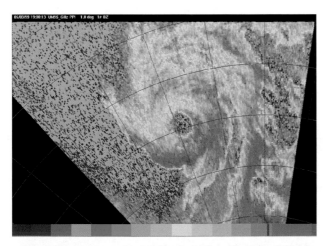

FIG. 5.22. An eye and spiral bands in the reflectivity field of a tornado, on 3 May 1999, near Verden, Oklahoma. Data collected by the U. Mass. mobile W-band radar. Range markers are shown at 500-m intervals. Reflectivity scale color coded below, but not to be interpreted quantitatively because much of the scattering may be in the Mie range. Courtesy of H. Bluestein and A. Pazmany.

Another use of the W-band radar is collecting high-resolution boundary layer velocity-azimuth displays (VADs). A storm-environment VAD collected just upstream from a supercell about to produce a tornado showed very strong vertical shear in the boundary layer associated with horizontal vorticity intense enough to be associated with a significant mesocyclone had it been tilted onto the vertical (Bluestein and Pazmany 2000).

c. The Doppler on Wheels

The first mobile Doppler radar to map out the Doppler wind field in a tornado was an X-band truck-mounted system, named the Doppler on Wheels (DOW; Wurman et al. 1997), designed and built by J. Wurman and his colleagues at OU, NCAR, and NSSL; it was first used in the spring of 1995 during VORTEX. The first DOW was built from a surplus transmitter from the NCAR CP-2 radar. The antenna was a 1.8-m diameter parabolic dish having a half-power beamwidth of 1.2°. Using a pulse-length equivalent to 75-m resolution in the radial direction, radar volumes at a range of 2–3 km are approximately 50 m × 50 m × 75 m. The radar signals were processed by the NCAR PC Integrated Acquisition processor (PIRAQ). The DOW (Fig. 5.25), like the U. Mass. W-band radar, also had load levelers to provide stability. Care was taken to ensure that the antenna pedestal and antenna were mounted stably on the back of the flatbed truck.

The first tornado datasets were collected during the second half of VORTEX-95. A detailed analysis of radar reflectivity data from the Dimmitt, Texas, tornado on 2 June 1995 showed an eye and spiral bands. Unlike the U. Mass. W-band radar, which scanned tornadoes only

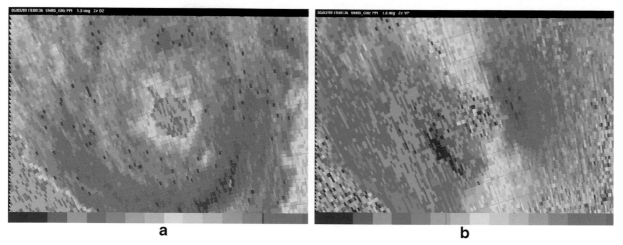

FIG. 5.23. Evidence of waves/multiple vortices along the inner edge of the eye in the same tornado depicted in Fig. 5.22. Data are from the U. Mass. mobile W-band radar: (a) reflectivity and (b) Doppler velocity. Range markings are shown at 100-m intervals. (a) The notches seen in (a) may be associated with vorticity maxima. Compare the largest notch, to the upper left, with the vortex signature (small region of brown pixels next to green pixels) evident in (b). The maximum receding (red) and approaching (purple) Doppler velocities in the main tornado vortex are seen; the color scale varies from -45 m s^{-1} at the left (purple) to $+67.5$ m s^{-1} at the right (red). The main vortex signature is divergent, which could be an artifact that results from the centrifuging outward of scatterers. Courtesy of H. Bluestein and A. Pazmany.

at one elevation angle to minimize the time between consecutive scans, the DOW collected volume scans through the tornado to reveal the vertical structure of the tornado (Fig. 5.26) every 100 s. It is seen in Fig. 5.26 how the reflectivity maximum surrounding the eye, which is thought to represent airborne debris, leans outward with height. Strongest wind speeds were found about 100 m AGL. The DOW data also did not indicate any strong convergence signature at low elevation angle, presumably owing to centrifuging contamination. Some

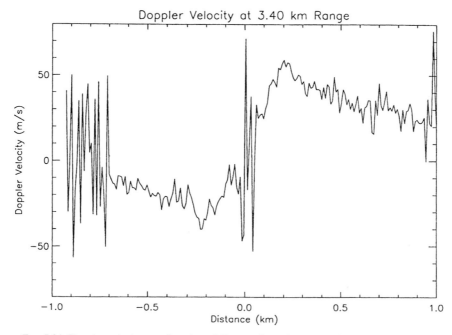

FIG. 5.24. Doppler velocity as a function of distance from the center of a tornado vortex, at 3–4-km range, on 3 May 1999. Data are from the U. Mass. mobile W-band radar. Positive (negative) distances from the center of the tornado are to the right (left) with respect to the view of the radar. The spikes at distances greater than 0.7 km (1 km) to the left (right) of the center of the tornado and near the center of the vortex are probably erroneous, owing to a relatively low signal level. From Bluestein and Pazmany (2000).

FIG. 5.25. Photograph of one of the DOW systems in Italy in 1999. Photograph by S. Richardson.

evidence was found of a central downdraft when the tornado was most intense.

In years subsequent to VORTEX, many tornado datasets have been collected by the DOW. Improved versions of the DOW prototype, the DOW2 and DOW3, were constructed in later years. Perhaps the most significant datasets collected with these radars have been that of the Spencer, South Dakota, tornado in 1998 and the Oklahoma City (Burgess et al. 2002) and Mulhall, Oklahoma (Wurman 2002) tornadoes on 3 May 1999. The former dataset is noteworthy owing to the damage that was well documented while the radars were collecting data, so that a comparison could be made between F-scale ratings and radar-derived wind speeds. The latter two datasets are noteworthy because extremely high wind speeds were indicated southwest of

Oklahoma City[3] and because multiple vortices were detected in the very wide Mulhall tornado (Fig. 5.27). In addition, analysis of the Oklahoma City tornado using the GBVTD technique of Lee et al. (1999) has been used to estimate the swirl ratio in the tornado vortex.

Bluestein et al. (1995) suggested that two storm-intercept vehicles, each equipped with a mobile Doppler radar, could be positioned so as to collect data from two different viewing angles and thereby make a dual-Doppler analysis of the horizontal find field in a tornado or in a developing tornado circulation possible. J. Wurman and colleagues have done so for several tornadoes, beginning in 1997; the results of the analyses are in progress. It is especially difficult to collect mobile dual-Doppler datasets of tornadoes, owing to the difficulty in positioning both radars at suitable sites on short notice. Extreme care must be taken to ensure that the locations of each radar, their viewing angles, and times of data collection are known with very high precision and that nearby objects do not provide too much ground clutter. However, it may be the best method currently

---

[3] Although mean Doppler wind speeds of 133 m s^{-1} were indicated by the DOW in one low-level scan, it is not possible to ascertain definitively whether or not actual wind speeds near the ground were in the upper F5 region or not because (a) the mean wind speed was valid for a time interval of only about 1/100 s, not several seconds as demanded by National Weather Service (NWS) criteria; (b) the radar volume was about 50–100 m AGL, not 3 m AGL as demanded by NWS criteria; (c) the spectral width of the data was 15–25 m s^{-1}, so that the mean wind speed measurement could have been overly biased by a small, highly reflective piece of rapidly moving or spinning debris; and (d) the motion of the debris targeted by the radar could have been different from the air motion. Scans just prior to and just after the high–wind speed scan were not available; it was therefore not possible to establish temporal continuity in the measurements. These measurements have been widely erroneously interpreted and reported by the media to indicate wind speeds as high as the F6 range (J. Wurman 2001, personal communication).

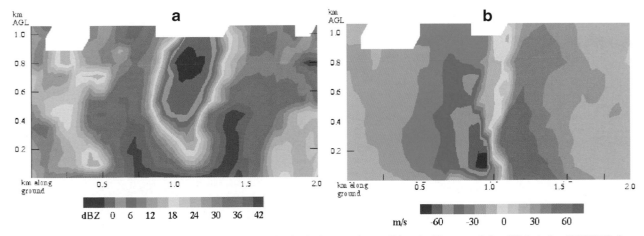

FIG. 5.26. Vertical cross section of (a) the radar reflectivity field of a tornado near Dimmitt, Texas, on 3 Jun 1995, during VORTEX; from data collected by the first DOW, at 3-km range. Note the eye and low-level debris-related shield at low levels. (b) Corresponding Doppler velocity field. Maximum Doppler velocities are found near 100 m AGL. Courtesy of J. Wurman; similar images seen in Fig. 19 of Wurman and Gill (2000).

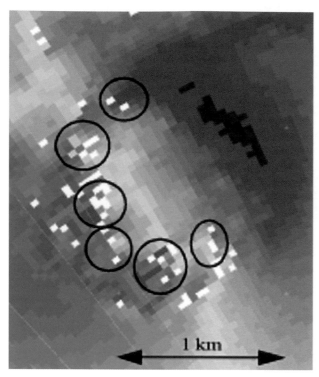

FIG. 5.27. Doppler radar signatures of multiple vortices (encircled) in a tornado near Mulhall, Oklahoma, on 3 May 1999; from a DOW mobile radar. From Wurman (2002, Fig. 19).

available to determine the wind field near the ground in tornadoes.

5. Evolving radar systems

A summary of mobile meteorological radar systems in existence as of February 2000 is found in Bluestein et al. (2001). Major new mobile radar systems under development or already developed, but not yet having been used to probe severe convective storms, include C-band systems, polarization diversity radars, and rapid-scan radars. The development of these systems has sparked a great deal of excitement in the research community. In addition, these radars will provide a unique and invaluable opportunity for "hands-on" experience for undergraduate and graduate students in future university course offerings. The educational aspects of radar meteorology are discussed in detail in another chapter in this monograph.

a. C-band systems

W-band radar systems are appropriate for probing nearby features only, because attenuation may severely limit the maximum range to a few kilometers in heavy precipitation. X-band systems are suitable for probing portions of severe convective storms, but not usually for the entire storm, owing to attenuation. C-band systems are even less vulnerable than X-band systems to

attenuation, but a very large, unwieldy antenna is needed to achieve high spatial resolution (i.e., small beamwidth). A compromise is to develop a mobile, ground-based C-band system with less spatial resolution than that of the DOWs and the U. Mass. W-band radar, and use it for the purpose of better analyzing storm-scale motions in the presence of heavy precipitation.

Two truck-mounted C-band systems have recently been built by M. Biggerstaff and colleagues at Texas A&M University, NSSL, OU, and Texas Tech University (Biggerstaff and Guynes 2000). These radars are known as the Shared Mobile Research and Teaching Radars (SMART-Rs); they will also have polarization diversity capability for precipitation physics studies. Their use in field programs focused on severe convective storms is anticipated. The design of a platform for the SMART-Rs is different from that of a platform for the DOWs and the U. Mass. W-band radar because the radar system is heavier. However, range and velocity ambiguity problems at 5-cm wavelengths are less serious.

Another truck-mounted, C-band system, the Seminole Hurricane Hunter, has been developed by P. Ray and colleagues at Florida State University (P. Ray 2001, personal communication); this radar system has dual-polarization capability and is suitable for use in severe storm studies requiring polarization diversity measurements.

b. Polarization diversity radars

No consensus has emerged yet on what wavelength mobile radars should operate to make the best use of polarization diversity capability for precipitation studies in convective storms (Bluestein et al. 2001). Although the U. Mass. W-band radar is already equipped for polarization diversity, it is used for signal-processing purposes (Pazmany et al. 1999) only and is not appropriate for precipitation studies. The SMART-Rs and Seminole Hurricane Hunter will provide a C-band polarization diversity capability. J. Wurman has built a polarization diversity X-band mobile, the X-POL, for the government of Greece. A. Pazmany and colleagues at the University of Massachusetts—Amherst are developing another truck-mounted X-band Doppler radar with polarization capability to be called the Polarized Doppler on Wheels (POW), which is expected to be operational during the spring of 2002 (A. Pazmany 2001, personal communication). In the next 5–10 years we will likely see many studies of precipitation processes in severe convective storms using these polarization diversity radars.

c. Rapid-scan mobile radars

A shortcoming of current mobile ground-based Doppler radar systems is that it takes 100 s or longer to scan a volume containing a convective storm. Since storm evolution takes place on 100-s timescales and tornado evolution takes place on 10-s timescales, it is necessary to scan more rapidly. Airborne radar systems at best have been able to scan a volume every 5 min, owing to the time it takes to fly by the storm and turn around and execute the next pass. If a volume scan contains both Doppler velocity and radar reflectivity, one can use adjoint methods to retrieve the two-dimensional wind field at low elevation angles from a sequence of scans (Xu and Qiu 1995). A. Shapiro (2001, personal communication) suggests that it may be possible to use the adjoint method to retrieve the full three-dimensional wind field from a sequence of rapid-scan Doppler velocity and reflectivity data.

J. Wurman is developing the rapid DOW, a rapid-scanning mobile X-band radar (Bluestein et al. 2001). This radar system is expected to scan, in about 10 s or less, volumes that the current DOW takes several minutes to scan. It will transmit at five slightly different frequencies dispersed in time so that the signal at each frequency samples an independent volume. The technique is an extension of the multiple-frequency techniques used in ELDORA (Hildebrand et al. 1996). However, the system will not synchronize the different frequencies as in ELDORA. It is anticipated that the rapid DOW will be field tested in 2003 (J. Wurman 2002, personal communication).

A. Pazmany and colleagues are currently converting an AN/MPQ-64 X-band phased-array radar developed by the U.S. Army and owned by the Naval Postgraduate School, for use with meteorological (distributed) targets (Bluestein et al. 2001; A. Pazmany 2001, personal communication). The radar had been developed for tracking multiple hard targets confined to within a 22° vertical wedge around a full 360° circle. The radar system is mounted on a truck and adjoining small trailer. Although a full scan can be accomplished in only a few seconds, it is likely that the signals will be averaged in time for a longer period (e.g., 10–100 s) so as to increase the radar's sensitivity. The half-power beamwidth of the antenna is 1.8° in azimuth and 2° in elevation; the range resolution set by the pulse length is 150 m. At ranges of 10–50 km, the radar is appropriate for rapidly scanning convective storms and providing storm-scale wind and reflectivity fields. It is hoped that the new radar system will be available for use during the spring of 2003.

d. Dual-frequency mobile radar systems and other radars

In the spring of 2001, H. Bluestein (OU) probed a small tornado and a number of mesocyclones simultaneously with both the U. Mass. W-band mobile Doppler radar and a new U. Mass. truck-mounted, X-band radar. The latter mobile radar, whose antenna has a half-power beamwidth of 1.2°, was designed and constructed by A. Pazmany from an inexpensive, commercial, marine radar. It is called the Massachusetts–Oklahoma Observing

System (MOOSE) but will be renamed the POW in 2002 after it has been given polarization diversity (and Doppler) capability. The data collected are being processed with an artificial neural network algorithm to estimate liquid water content (LWC) and the drop sizes of precipitation particles. This information may be of use in trying to subtract out quantitatively the radially outward-directed speeds of scatterers centrifuged in strong vortices and thus determine the actual radially directed windspeeds. It is also hoped to see if there is indeed any size sorting as predicted by Snow (1984).

6. Summary and outlook

In the last decade and a half, substantial progress has been made in our knowledge of the dynamics and kinematics of severe convective storms and their sub-storm-scale features based on the analysis of data collected by radars mounted on airborne and ground-based vehicle platforms. Precipitation curls within the weak echo region were seen for the first time and were suggestive of possible hail trajectories. Data were collected within microburst downdrafts and the evolution of a microburst storm was documented. Perhaps the biggest impact of mobile radar platforms has been the collection of high-resolution data on supercell storms. New information on supercell propagation, mesocyclones, wall clouds, and cyclic tornadogenesis has been obtained. These platforms have also been able to probe the windfields within tornadoes and resolve the elusive multiple vortex phenomenon.

High-altitude aircraft have provided direct sampling of the vertical velocity using a nadir-pointing beam. This has been missing in mid- and low-level aircraft and ground-based studies since this component of air motion was always derived by integrating the equation of continuity. At even higher altitudes, spaceborne platforms have begun to enhance our understanding of the global characteristics of severe convection.

Future studies will involve the use of polarization diversity mobile radars, networks of ground-based mobile Doppler radars, rapid-scan radars, mobile radars operating at C band and other bands, and perhaps, in the more distant future, radars mounted on more exotic mobile platforms such as unmanned aircraft and helicopters.

In light of the major accomplishments by mobile radar systems during the past few years, it is surprising to note that we have yet to coordinate airborne and ground-based platforms in a comprehensive and simultaneous collection of data on a supercell even though both platforms were deployed during VORTEX. This combination will result in a dataset that will resolve both the overall storm-scale structure and the finescale features of the tornado and mesocyclone. Accordingly, this should be a future high priority goal of "mobile radar observations of severe convective systems."

Acknowledgments. This paper and some of the research results were supported by the National Science Foundation under Grants ATM-9422499, 9801720, and 0121048 (through RMW) and Grants ATM-8902594, 9019821, 9302379, 9612674, and 9912097 (through HBB).

REFERENCES

Adlerman, E. J., K. K. Droegemeier, and R. P. Davies-Jones, 1999: A numerical simulation of cyclic mesocyclogenesis. *J. Atmos. Sci.,* **56,** 2045–2069.

Atlas, D., 1994: Footprints of storms on the sea: A view from spaceborne synthetic aperture radar. *J. Geophys. Res.,* **99,** 7961–7969.

——, and P. G. Black, 1994: The evolution of convective storms from their footprints on the sea as viewed by synthetic aperture radar from space. *Bull. Amer. Meteor. Soc.,* **75,** 1183–1190.

——, T. Iguchi, and H. F. Pierce, 1995: Storm-induced wind patterns on the sea from spaceborne synthetic aperture radar. *Bull. Amer. Meteor. Soc.,* **76,** 1585–1592.

Biggerstaff, M. I., and J. Guynes, 2000: A new tool for atmospheric research. Preprints, *20th Conf. on Severe Local Storms,* Orlando, FL, Amer. Meteor. Soc., 277–280.

Bilbro, J., G. Ficht, D. Fitzjarrald, M. Krause, and R. Lee, 1984: Airborne Doppler lidar wind field measurements. *Bull. Amer. Meteor. Soc.,* **65,** 348–359.

Bluestein, H. B., 1980: The University of Oklahoma Severe Storms Intercept Project—1979. *Bull. Amer. Meteor. Soc.,* **61,** 560–567.

——, 1999: A history of severe-storm-intercept field programs. *Wea. Forecasting,* **14,** 558–577.

——, and W. P. Unruh, 1989: Observations of the wind field in tornadoes, funnel clouds, and wall clouds with a portable Doppler radar. *Bull. Amer. Meteor. Soc.,* **70,** 1514–1525.

——, and J. H. Golden, 1993: A review of tornado observations. *The Tornado: Its Structure, Dynamics, Prediction, and Hazards, Geophys. Monogr.,* No. 79, Amer. Geophys. Union, 319–352.

——, and W. P. Unruh, 1993: On the use of a portable FM-CW Doppler radar for tornado research. *The Tornado: Its Structure, Dynamics, Prediction, and Hazards, Geophys. Monogr.,* No. 79, Amer. Geophys. Union, 367–376.

——, and A. L. Pazmany, 2000: Observations of tornadoes and other convective phenomena with a mobile, 3-mm wavelength, Doppler radar: The spring 1999 field experiment. *Bull. Amer. Meteor. Soc.,* **81,** 2939–2952.

——, and S. G. Gaddy, 2001: Airborne pseudo-dual-Doppler analysis of a rear-inflow jet and deep convergence zone within a supercell. *Mon. Wea. Rev.,* **129,** 2270–2289.

——, J. LaDue, H. Stein, D. Speheger, and W. P. Unruh, 1993: Doppler radar wind spectra of supercell tornadoes. *Mon. Wea. Rev.,* **121,** 2200–2221.

——, A. L. Pazmany, J. C. Galloway, and R. E. McIntosh, 1995: Studies of the substructure of severe convective storms using a mobile 3-mm wavelength Doppler radar. *Bull. Amer. Meteor. Soc.,* **76,** 2155–2169.

——, S. G. Gaddy, D. C. Dowell, A. L. Pazmany, J. C. Galloway, R. E. McIntosh, and H. Stein, 1997a: Mobile, 3-mm wavelength, pulsed Doppler radar observations of substorm-scale vortices in a supercell. *Mon. Wea. Rev.,* **125,** 1046–1059.

——, W. P. Unruh, D. C. Dowell, T. A. Hutchinson, T. M. Crawford, A. C. Wood, and H. Stein, 1997b: Doppler radar analysis of the Northfield, Texas, tornado of 25 May 1994. *Mon. Wea. Rev.,* **125,** 212–230.

——, B. A. Albrecht, R. M. Hardesty, W. D. Rust, D. Parsons, R. Wakimoto, and R. M. Rauber, 2001: Ground-based mobile instrument workshop summary, 23–24 February 2000, Boulder, Colorado. *Bull. Amer. Meteor. Soc.,* **82,** 681–694.

Bosart, B. L., W.-C. Lee, and R. M. Wakimoto, 2002: Procedures to improve the accuracy of airborne Doppler radar data. *J. Atmos. Oceanic Technol.,* **19,** 322–339.

Brandes, E. A., 1978: Mesocyclone evolution and tornadogenesis: Some observations. *Mon. Wea. Rev.,* **106,** 995–1011.

——, 1981: Finestructure of the Del City–Edmond tornadic mesocirculation. *Mon. Wea. Rev.,* **109,** 635–647.

——, 1993: Tornadic thunderstorm characteristics determined with Doppler radar. *The Tornado: Its Structure, Dynamics, Prediction, and Hazards, Geophys. Monogr.,* No. 79, Amer. Geophys. Union, 97–105.

Brown, R. A., L. R. Lemon, and D. W. Burgess, 1978: Tornado detection by pulsed Doppler radar. *Mon. Wea. Rev.,* **106,** 29–38.

Browning, K. A., and F. H. Ludlam, 1962: Airflow in convective storms. *Quart. J. Roy. Meteor. Soc.,* **88,** 117–135.

——, and R. J. Donaldson Jr., 1963: Airflow and structure of a tornadic storm. *J. Atmos. Sci.,* **20,** 533–545.

——, and G. B. Foote, 1976: Airflow and hail growth in supercell storms and some implications for hail suppression. *Quart. J. Roy. Meteor. Soc.,* **102,** 499–533.

Burgess, D. W., V. T. Wood, and R. A. Brown, 1982: Mesocyclone evolution statistics. Preprints, *12th Conf. on Severe Local Storms,* San Antonio, TX, Amer. Meteor. Soc., 422–424.

——, R. J. Donaldson Jr., and P. R. Desrochers, 1993: Tornado detection and warning by radar. *The Tornado: Its Structure, Dynamics, Prediction, and Hazards, Geophys. Monogr.,* No. 79, Amer. Geophys. Union, 203–231.

——, M. Magsig, J. Wurman, D. Dowell, and Y. Richardson, 2002: Radar observations of the 3 May 1999 Oklahoma City tornado. *Wea. Forecasting,* **17,** 456–471.

Byers, H. R., and R. R. Braham Jr., 1949: *The Thunderstorm.* U.S. Government Printing Office, 287 pp.

Cai, H., and R. M. Wakimoto, 2001: Retrieved pressure field and its influence on the propagation of a supercell thunderstorm. *Mon. Wea. Rev.,* **129,** 2695–2713.

Charba, J., and Y. Sasaki, 1971: Structure and movement of the severe thunderstorms of 3 April 1964 as revealed from radar and surface mesonetwork analysis. *J. Meteor. Soc. Japan,* **49,** 191–213.

Davies-Jones, R. P., 1986: Tornado dynamics. *Thunderstorm Morphology and Dynamics,* E. Kessler, Ed., *Thunderstorms: A Social, Scientific, and Technological Documentary,* 2d ed., Vol. 2, University of Oklahoma Press, 197–236.

——, and H. E. Brooks, 1993: Mesocyclogenesis from a theoretical perspective. *The Tornado: Its Structure, Dynamics, Prediction, and Hazards, Geophys. Monogr.,* No. 79, Amer. Geophys. Union, 557–571.

——, D. W. Burgess, L. R. Lemon, and D. Purcell, 1978: Interpretation of surface debris patterns from the 24 May 1973 Union City, Oklahoma, tornado. *Mon. Wea. Rev.,* **106,** 12–21.

——, R. J. Trapp, and H. B. Bluestein, 2001: Tornadoes and tornadic storms. *Severe Convective Storms, Meteor. Monogr.,* No. 50, Amer. Meteor. Soc., 167–221.

Doswell, C. A., III, and D. W. Burgess, 1988: On some issues of United States tornado climatology. *Mon. Wea. Rev.,* **116,** 495–501.

Doviak, R. J., and D. Sirmans, 1973: Doppler radar with polarization diversity. *J. Atmos. Sci.,* **30,** 737–738.

Dowell, D. C., and H. B. Bluestein, 2002a: The 8 June 1995 McLean, Texas, storm. Part I: Observations of cyclic tornadogenesis. *Mon. Wea. Rev.,* **130,** 2626–2648.

——, and ——, 2002b: The 8 June 1995 McLean, Texas, storm. Part II: Cyclic, tornado formation, maintenance, and dissipation. *Mon. Wea. Rev.,* **130,** 2649–2670.

——, ——, and D. P. Jorgensen, 1997: Airborne radar analysis of supercells during COPS-91. *Mon. Wea. Rev.,* **125,** 365–383.

Droegemeier, K., and R. Wilhelmson, 1987: Numerical simulation of thunderstorm outflow dynamics. Part I: Outflow sensitivity experiments and turbulence dynamics. *J. Atmos. Sci.,* **44,** 1180–1210.

Fiedler, B. H., and R. Rotunno, 1986: A theory for the maximum wind speeds in tornado-like vortices. *J. Atmos. Sci.,* **43,** 2328–2340.

Forbes, G. S., 1978: Three scales of motion associated with tornadoes. Final Rep. NUREG/CR-0363, U.S. Nuclear Regulatory Commission, 359 pp. [NTIS PB-288 291.]

Frush, C. L., 1991: A graphical representation of the radar velocities dealiasing problems. Preprints, *25th Int. Conf. on Radar Meteorology,* Paris, France, Amer. Meteor. Soc., 885–888.

Fujita, T. T., 1981: Tornadoes and downbursts in the context of generalized planetary scales. *J. Atmos. Sci.,* **38,** 1511–1534.

——, 1985: The downburst. SMRP Res. Pap. 210, University of Chicago, 122 pp. [NTIS PB-148880.]

——, 1992: Mystery of severe storms. SMRP Res. Pap. 239, University of Chicago, 298 pp. [NTIS PB92-64168.]

——, 1993: The Plainfield tornado of August 28, 1990. *The Tornado: Its Structure, Dynamics, Prediction, and Hazards, Geophys. Monogr.,* No. 79, Amer. Geophys. Union, 1–17.

——, and H. Grandoso, 1968: Split of a thunderstorm into anticyclonic and cyclonic storms and their motion as determined from numerical model experiments. *J. Atmos. Sci.,* **25,** 416–439.

——, and F. Caracena, 1977: An analysis of three weather-related aircraft accidents. *Bull. Amer. Meteor. Soc.,* **58,** 1164–1181.

——, D. L. Bradbury, and C. F. van Thullenar, 1970: Palm Sunday tornadoes of April 11, 1965. *Mon. Wea. Rev.,* **98,** 29–69.

Gal-Chen, T., 1978: A method for the initialization of the anelastic equations: Implications for matching models with observations. *Mon. Wea. Rev.,* **106,** 587–606.

Golden, J. H., 1974: The life cycle of Florida Keys' waterspouts. I. *J. Appl. Meteor.,* **13,** 676–692.

——, and D. Purcell, 1978: Life cycle of the Union City, Oklahoma tornado and comparison with waterspouts. *Mon. Wea. Rev.,* **106,** 3–11.

——, and H. B. Bluestein, 1994: The NOAA–National Geographic Society waterspout expedition (1993). *Bull. Amer. Meteor. Soc.,* **75,** 2281–2288.

Harris, F. I., and R. E. Carbone, 1980: The multiple Doppler radar workshop, November 1979. Part I: Workshop impetus and objectives. *Bull. Amer. Meteor. Soc.,* **61,** 1170–1173.

Heymsfield, G. M., 1989: Accuracy of vertical air motions from nadir-viewing Doppler airborne radars. *J. Atmos. Oceanic Technol.,* **6,** 1079–1082.

——, and Coauthors, 1996: The EDOP radar system on the high-altitude NASA ER-2 aircraft. *J. Atmos. Oceanic Technol.,* **13,** 795–809.

——, J. B. Haverson, and I. J. Caylor, 1999: A wintertime Gulf Coast squall line observed by EDOP airborne Doppler radar. *Mon. Wea. Rev.,* **127,** 2928–2949.

Hildebrand, P. H., and R. E. Carbone, 1980: The multiple Doppler radar workshop, November 1979. Part V: Verification of results. *Bull. Amer. Meteor. Soc.,* **61,** 1189–1194.

——, and R. K. Moore, 1990: Meteorological radar observations from mobile platforms. *Radar in Meteorology,* D. Atlas, Ed., Amer. Meteor. Soc., 287–314.

——, R. A. Oye, and R. E. Carbone, 1981: X-band vs C-band aircraft radar: The relative effects of beamwidth and attenuation in severe storm situations. *J. Appl. Meteor.,* **20,** 1353–1361.

——, C. A. Walther, C. L. Frush, J. Testud, and F. Baudin, 1994: The ELDORA/ASTRAIA airborne Doppler weather radar: Goals, design, and first field tests. *Proc. IEEE,* **82,** 1873–1890.

——, and Coauthors, 1996: The ELDORA/ASTRAIA airborne Doppler weather radar: High resolution observations from TOGA COARE. *Bull. Amer. Meteor. Soc.,* **77,** 213–232.

Hitschfield, W., 1960: The motion and erosion of convective storms in severe vertical wind shear. *J. Meteor.,* **17,** 270–282.

Hjelmfelt, M. R., 1988: Structure and life cycle of microburst outflows observed in Colorado. *J. Climate Appl. Meteor.,* **27,** 900–927.

Jorgensen, D. P., 1984: Mesoscale and convective-scale characteristics of mature hurricanes. Part I: General observations by research aircraft. *J. Atmos. Sci.,* **41,** 1268–1285.

——, P. H. Hildebrand, and C. L. Frush, 1983: Feasibility test of an airborne pulse-Doppler meteorological radar. *J. Appl. Meteor.,* **22,** 744–757.

——, M. A. LeMone, and B. J.-D. Jou, 1991: Precipitation and kinematic structure of an oceanic mesoscale convective system. Part I: Convective line structure. *Mon. Wea. Rev.,* **119,** 2608–2637.

——, T. Matejka, and J. D. DuGranrut, 1996: Multi-beam techniques for deriving wind fields from airborne Doppler radars. *J. Meteor. Atmos. Phys.,* **59,** 83–104.

——, M. A. LeMone, and S. B. Trier, 1997: Structure and evolution of the 22 February 1993 TOGA COARE squall line: Aircraft observations of precipitation, circulation, and surface energy fluxes. *J. Atmos. Sci.,* **54,** 1961–1985.

——, T. R. Sheperd, and A. S. Goldstein, 2000: A dual-pulse repetition frequency scheme for mitigating velocity ambiguities of the NOAA P-3 radar. *J. Atmos. Oceanic Technol.,* **17,** 585–594.

Kabèche, A., and J. Testud, 1995: Stereoradar meteorology: A new unified approach to process data from airborne or ground-based meteorological radars. *J. Atmos. Oceanic Technol.,* **12,** 783–799.

Kessinger, C. J., D. B. Parsons, and J. W. Wilson, 1988: Observations of a storm containing misocyclones, downbursts, and horizontal vortex circulations. *Mon. Wea. Rev.,* **116,** 1959–1982.

Kingsmill, D. A., and R. M. Wakimoto, 1991: Kinematic, dynamic, and thermodynamic analysis of a weakly sheared severe thunderstorm over northern Alabama. *Mon. Wea. Rev.,* **119,** 262–297.

Klemp, J. B., and R. B. Wilhelmson, 1978: Simulations of right and left-moving storms produced through storm splitting. *J. Atmos. Sci.,* **35,** 1097–1110.

——, and R. Rotunno, 1983: A study of the tornadic region within a supercell thunderstorm. *J. Atmos. Sci.,* **40,** 359–377.

Kummerow, C., W. Barnes, T. Kozu, J. Shiue, and J. Simpson, 1998: The tropical rainfall measuring mission (TRMM) sensor package. *J. Atmos. Oceanic Technol.,* **15,** 809–817.

Lee, W.-C., P. Dodge, F. D. Marks, and P. H. Hildebrand, 1994: Mapping of airborne Doppler radar data. *J. Atmos. Oceanic Technol.,* **11,** 572–578.

——, B. J.-D. Jou, P.-L. Chang, and S.-M. Deng, 1999: Tropical cyclone kinematic structure retrieved from single-Doppler radar observations. Part I: Interpretation of Doppler velocity patterns and the GBVTD technique. *Mon. Wea. Rev.,* **127,** 2419–2439.

Lemon, L. R., and C. A. Doswell III, 1979: Severe thunderstorm evolution and mesocyclone structure as related to tornadogenesis. *Mon. Wea. Rev.,* **107,** 1184–1197.

LeMone, M. A., G. M. Barnes, J. C. Fankhauser, and L. F. Tarleton, 1988: Perturbation pressure fields measured by aircraft around the cloud-base updraft of deep convective clouds. *Mon. Wea. Rev.,* **116,** 313–327.

Lhermitte, R. M., 1971: Probing of atmospheric motion by airborne pulse-Doppler radar techniques. *J. Appl. Meteor.,* **10,** 234–246.

——, 1990: Attenuation and scattering of millimeter wavelength radiation by clouds and precipitation. *J. Atmos. Oceanic Technol.,* **7,** 464–479.

Lilly, D. K., 1969: Tornado dynamics. NCAR Manuscript 69-117, 52 pp. [Available from NCAR, P.O. Box 3000, Boulder, CO 80307.]

Loew, E., and C. A. Walther, 1995: Engineering analysis of dual pulse interval radar data obtained by the ELDORA radar. Preprints, *27th Conf. on Radar Meteorology,* Vail, CO, Amer. Meteor. Soc., 710–712.

Markowski, P. M., J. M. Straka, and E. N. Rasmussen, 2002: Direct surface thermodynamic obervations within the rear-flank downdrafts of nontornadic and tornadic supercells. *Mon. Wea. Rev.,* **130,** 1692–1721.

Marwitz, J. D., 1973: Trajectories within the weak echo regions of hailstorms. *J. Appl. Meteor.,* **12,** 1174–1182.

——, A. H. Auer, and D. L. Veal, 1972: Locating the organized updraft on severe thunderstorms. *J. Appl. Meteor.,* **11,** 236–238.

McCaul, E. W., Jr., H. B. Bluestein, and R. J. Doviak, 1987: Airborne Doppler lidar observations of convective phenomena in Oklahoma. *J. Atmos. Oceanic Technol.,* **4,** 479–497.

Mueller, C. K., and P. H. Hildebrand, 1985: Evaluation of meteorological airborne Doppler radar. Part II: Triple-Doppler analyses of air motion. *J. Atmos. Oceanic Technol., 2,* 381–392.

Newton, C. W., and H. R. Newton, 1959: Dynamical interactions between large convective clouds and environment with vertical shear. *J. Meteor., 16,* 483–496.

Pazmany, A. L., J. C. Galloway, J. B. Mead, I. Popstefanija, R. E. McIntosh, and H. B. Bluestein, 1999: Polarization diversity pulse pair technique for millimeter-wave Doppler radar measurements of severe-storm features. *J. Atmos. Oceanic Technol., 16,* 1900–1911.

Proctor, F. H., 1989: Numerical simulations of an isolated microburst. Part II: Sensitivity experiments. *J. Atmos. Sci., 46,* 2143–2165.

Ray, P. S., and D. P. Jorgensen, 1988: Uncertainties associated with combining airborne and ground-based Doppler radar data. *J. Atmos. Oceanic Technol., 5,* 177–196.

——, and M. Stephenson, 1990: Assessment of the geometric and temporal errors associated with airborne Doppler radar measurements of a convective storm. *J. Atmos. Oceanic Technol., 7,* 206–217.

——, B. C. Johnson, K. W. Johnson, J. S. Bradberry, J. J. Stephens, K. K. Wagner, R. B. Wilhelmson, and J. B. Klemp, 1981: The morphology of several tornadic storms on 20 May 1977. *J. Atmos. Sci., 38,* 1643–1663.

Raymond, D., 1978: Pressure perturbations in deep convection: An experimental study. *J. Atmos. Sci., 35,* 1704–1711.

Roberts, R. D., and J. W. Wilson, 1995: The genesis of three non-supercell tornadoes observed with dual-Doppler radar. *Mon. Wea. Rev., 123,* 3408–3436.

Rotunno, R., 1984: An investigation of a three-dimensional asymmetric vortex. *J. Atmos. Sci., 41,* 283–298.

——, and J. B. Klemp, 1982: The influence of the shear-induced pressure gradient on thunderstorm motion. *Mon. Wea. Rev., 110,* 136–151.

——, and ——, 1985: On the rotation and propagation of simulated supercell thunderstorms. *J. Atmos. Sci., 42,* 271–292.

Roux, F., 1985: Retrieval of thermodynamic fields from multiple-Doppler radar data using the equations of motion and the thermodynamic equation. *Mon. Wea. Rev., 113,* 2142–2157.

——, 1988: The west African squall line observed on 23 June 1981 during COPT 81: Kinematics and thermodynamics of the convective region. *J. Atmos. Sci., 45,* 406–426.

Schlesinger, R. E., 1980: A three-dimensional numerical model of an isolated thunderstorm. Part II: Dynamics of updraft splitting and mesovortex couplet evolution. *J. Atmos. Sci., 37,* 395–420.

Skolnik, M. I., 1980: *Introduction to Radar Systems.* McGraw Hill, 581 pp.

Smull, B. F., and R. A. Houze, 1987a: Dual-Doppler radar analysis of a mid-latitude squall line with a trailing region of stratiform rain: Radar and satellite observations. *J. Atmos. Sci., 44,* 2128–2148.

——, and ——, 1987b: Rear-inflow in squall lines with trailing stratiform precipitation. *Mon. Wea. Rev., 115,* 2869–2889.

Snow, J. T., 1984: On the formation of particle sheaths in columnar vortices. *J. Atmos. Sci., 41,* 2477–2491.

——, and R. L. Pauley, 1984: On the thermodynamic method for estimating maximum tornado wind speeds. *J. Climate Appl. Meteor., 23,* 1465–1468.

Srivastava, R. C., 1987: A model of intense downdrafts driven by melting and evaporation. *J. Atmos. Sci., 44,* 1752–1773.

Testud, J. P., P. H. Hildebrand, and W.-C. Lee, 1995: A procedure to correct airborne Doppler radar data for navigation errors using the echo returned from the earth's surface. *J. Atmos. Oceanic Technol., 12,* 800–820.

Trapp, R. J., 1999: Observations of nontornadic low-level mesocyclones and attendant tornadogenesis failure during VORTEX. *Mon. Wea. Rev., 127,* 1693–1705.

Ulaby, F. T., R. K. Moore, and A. K. Fung, 1981: *Microwave Remote Sensing.* Vol. I. Addison-Wesley, 456 pp.

Verlinde, J., 1997: Airborne Doppler radar analysis of a TOGA COARE waterspout storm. *Mon. Wea. Rev., 125,* 3008–3017.

Vivekanandan, J., D. S. Zrnic, S. M. Ellis, R. Oye, A. V. Ryzhkov, and J. Straka, 1999: Cloud microphysics retrieval using S-band dual-polarization measurements. *Bull. Amer. Meteor. Soc., 80,* 381–388.

Wakimoto, R. M., and J. W. Wilson, 1989: Non-supercell tornadoes. *Mon. Wea. Rev., 117,* 1113–1140.

——, and J. K. Lew, 1993: Observations of a Florida waterspout during CaPE. *Wea. Forecasting, 8,* 412–423.

——, and N. T. Atkins, 1996: Observations on the origins of rotation: The Newcastle Tornado during VORTEX 94. *Mon. Wea. Rev., 124,* 384–407.

——, and C. Liu, 1998: The Garden City, Kansas, storm during VORTEX 95. Part II: The wall cloud and tornado. *Mon. Wea. Rev., 126,* 393–408.

——, and H. Cai, 2000: Analysis of a nontornadic storm during VORTEX 95. *Mon. Wea. Rev., 128,* 565–592.

——, W. Blier, and C. H. Liu, 1992: The frontal structure of an explosive oceanic cyclone: Airborne radar observations of ERICA IOP 4. *Mon. Wea. Rev., 120,* 1135–1155.

——, W.-C. Lee, H. B. Bluestein, C.-H. Liu, and P. H. Hildebrand, 1996: ELDORA observations during VORTEX 95. *Bull. Amer. Meteor. Soc., 77,* 1465–1481.

——, C. Liu, and H. Cai, 1998: The Garden City, Kansas, storm during VORTEX 95. Part I: Overview of the storm's life cycle and mesocyclogenesis. *Mon. Wea. Rev., 126,* 372–392.

Wilson, J. W., and D. L. Megenhardt, 1997: Thunderstorm initiation, organization, and lifetime associated with Florida boundary layer convergence zones. *Mon. Wea. Rev., 125,* 1507–1525.

——, R. D. Roberts, C. Kessinger, and J. McCarthy, 1984: Microburst wind structure and evaluation of Doppler radar for airport wind shear detection. *J. Climate Appl. Meteor., 23,* 898–915.

Wurman, J., 2002: The multiple-vortex structure of a tornado. *Wea. Forecasting, 17,* 473–505.

——, and S. Gill, 2000: Finescale radar observations of the Dimmitt, Texas (2 June 1995) tornado. *Mon. Wea. Rev., 128,* 2135–2164.

——, J. M. Straka, E. N. Rasmussen, M. Randall, and A. Zahrai, 1997: Design and deployment of a portable, pencil-beam, pulsed, 3-cm Doppler radar. *J. Atmos. Oceanic Technol., 14,* 1502–1512.

Xu, Q., and C.-J. Qiu, 1995: Adjoint-method retrievals of low-altitude wind fields from single-Doppler reflectivity and radial-wind data. *J. Atmos. Oceanic Technol., 12,* 1111–1119.

Yorty, D. P., E. J. Zipser, and S. W. Nesbitt, 2001: Global distribution of extremely intense storms between 36 S and 36 N using evidence from the TRMM radar. Preprints, *30th Int. Conf. on Radar Meteorology,* Munich, Germany, Amer. Meteor. Soc., 334–336.

Ziegler, C. L., E. N. Rasmussen, T. R. Shepherd, A. I. Watson, and J. M. Straka, 2001: The evolution of low-level rotation in the 29 May 1994 Newcastle–Graham, Texas, storm complex during VORTEX. *Mon. Wea. Rev., 129,* 1339–1368.

Zrnic, D. S., D. W. Burgess, and L. Hennington, 1985: Doppler spectra and estimated wind speed of a violent tornado. *J. Climate Appl. Meteor., 24,* 1068–1081.

The David Atlas Symposium at the Orange County Convention Center in Orlando, Florida

Dave Atlas

Dave Atlas and Roger Lhermitte

Bob Serafin and Peter Hildebrand

Dave Atlas

Jenny Sun and Jim Wilson

Speakers preparing for the next session

Live images from CHILL.
Dave Brunkow, Pat Kennedy, and Bob Bowie

Dave Atlas, Rit Carbone, and Bob Serafin

Isztar Zawadzki, Frédéric Fabry,
Wen-Chau Lee, and Ben Jou

Chapter 6

Recent Developments in Observation, Modeling, and Understanding Atmospheric Turbulence and Waves

KENNETH S. GAGE

NOAA/Aeronomy Laboratory, Boulder, Colorado

EARL E. GOSSARD*

NOAA/Environmental Technology Laboratory, Boulder, Colorado

Gage

1. Introduction

This review covers recent developments in turbulence and waves pertinent to radar meteorology. As radar techniques have become more sophisticated and Doppler radar profilers have come into widespread use, more has been learned about the use of radar as a tool in observing atmospheric turbulence and waves. While much of the paper is devoted to recent developments in observations of turbulence and waves, the review also places these developments in a broader context of developments in fluid dynamics that are fundamental to the understanding of atmospheric observations.

Turbulence is one of the most basic processes in fluid dynamics, but it has remained relatively unresolved owing in part to its nonlinear nature and the number of variables involved (Wyngaard 1992). Theoretical progress has relied upon simplifying assumptions concerning homogeneity, stationarity, isotropy, etc., which are often not satisfied in nature. Nevertheless, considerable progress has been made within the framework of applying statistical dynamics to turbulence using simplifying assumptions referred to above.

The past decade has seen much progress in the development of models capable of resolving the instabilities that lead to turbulence in the atmosphere. Direct numerical simulation (DNS) has led to new insight into these processes and is useful in modeling the generation of waves by convection. In addition large eddy simulation (LES) has had an important impact on our ability to simulate the convective boundary layer, which is helping to advance our ability to improve boundary layer parameterizations in global climate models.

The topics covered in this review have received considerable attention in the decade following the publication of *Radar in Meteorology* (Atlas 1990) and our aim is to synthesize some of the most important developments in turbulence and waves that extend topics reported in "Radar research on the atmospheric boundary layer" (Gossard 1990) and "Radar observations of the free atmosphere" (Gage 1990).

The advances reported here have been made by many groups working in different disciplines. The work builds on decades of pioneering research in radar meteorology as well as related disciplines in radio science. Considerable effort was expended in the 1950s and 1960s to understand the role of turbulence in propagation of UHF

* Retired.

radio waves beyond the horizon and radar scattering from the optically clear atmosphere. In many respects the developments reported here have been made possible and motivated by the earlier research of pioneers in the field such as David Atlas.

We begin our review by recounting some of the early developments in the field, which in our opinion have helped to set the stage for more recent developments. We then present a summary of recent advances in our understanding of atmospheric turbulence and our ability to measure it. An important part of our examination of turbulence and waves is the vertical structure associated with it and we will summarize recent developments in observations of the layered structure of the atmosphere. A related topic is stratified turbulence, which has received a fair amount of attention from numerical studies and laboratory research. We conclude our review with a summary of recent developments in our ability to observe atmospheric wave motions using radar and other tools. Considerable progress has been made in identifying the major sources of internal gravity wave motions in the stable atmosphere.

2. Historical background

As mentioned in the introduction, a rich history lies behind the developments in radar probing of the clear atmosphere that have taken place in the past decade. The early history is recorded in several places (see, e.g., Gage and Balsley 1980; Hardy and Gage 1990). Two branches that have led to our current understanding of atmospheric turbulence that affects radio wave propagation and scattering are 1) UHF radio wave propagation beyond the horizon (see, e.g., Du Castel 1966) and 2) Doppler radar studies of the clear atmosphere described in some detail by Hardy and Gage (1990) and Gossard and Strauch (1983). Progress has been accelerated in recent years owing to the widespread use of profilers in research and operations (Rogers et al. 1993; Parsons et al. 1994; Carter et al. 1995; Wilczak et al. 1996; Gage et al. 2002).

Rapid strides were made in the ability to observe the atmosphere using radar during and immediately following World War II. The need to develop reliable over-the-horizon communication systems in the 1950s motivated much research on the mechanisms responsible for troposcatter. Much was learned about the role of turbulence and coherent refractive index gradients in the propagation of radio waves during this period. At the same time radar studies showed that radar returns were common from the optically clear atmosphere (Friend 1949; Atlas 1959). Plank (1956) summarized the early history of radar studies of the clear atmosphere and described several types of angel echoes arising from radar backscattering. At long wavelengths, layer-type (type I) echoes were found associated with stable regions of the atmosphere and are now regarded as being quasi-specular arising from coherent layered structure

in radio refractive index. Another common variety of angels is commonly referred to as point or dot angels. These type-II echoes were observed by radars with wavelengths in the 3–10-cm range. The origin of the type-II angels was hotly debated for a long time and led to an important series of experiments at Wallops Island in the 1960s.

Dave Atlas and colleagues from the Weather Radar Branch at the Air Force Cambridge Research Laboratories (AFCRL) collaborated with Isadore Katz and colleagues from the Johns Hopkins Applied Physics Laboratory and conducted a series of experiments using the powerful radars located at Wallops Island, Virginia. Radars operating at three wavelengths (3.2, 10.7, and 71.5 cm) were utilized in these experiments designed to determine the nature of the clear air echoes.

One of the early results of the Wallops Island research was unambiguous evidence that dot angels were due to insects and birds since these were seen preferentially on the 3- and 10-cm radars (Hardy et al. 1966). However, using these sensitive radars it was also noted that another class of clear air echoes is observed that is associated with refractive index fluctuations due to Bragg scattering as demonstrated with in situ probing of the radar volume by Kropfli et al. (1968). We will return to these issues again in section 4 where recent experiments using dual wavelengths allow unambiguous separation of Rayleigh scattering due to hydrometeors and insects from Bragg scattering due to turbulence.

Atlas and colleagues used the Wallops Island facility to good advantage to study the relationship between clear air echoes seen by the radars and Clear Air turbulence (CAT). During this period a clear association was demonstrated between the CAT observed on the radars and the occurrence of shear flow instabilities in the hydrostatically stable atmosphere commonly referred to as Kelvin–Helmholtz instability (KHI; see, e.g., Atlas et al. 1966a,b). An example of the clear air echoes observed by the Wallops Island radars is shown in Fig. 6.1. The research that was conducted during this period at Wallops Island was especially noteworthy because it combined radar observations at several frequencies with in situ probing using aircraft and helicopters as well as other instrumentation suitable for case studies of CAT episodes.

During the late 1960s Dave Atlas collaborated with Earl Gossard and Juergen Richter in a series of observational studies using frequency-modulated continuous-wave (FM-CW) radar (Richter 1969) capable of very-high-resolution observations of breaking Kelvin–Helmholtz waves (Gossard et al. 1970; Atlas et al. 1970; Metcalf and Atlas 1973). An example of FM-CW radar observation of Kelvin–Helmholtz instability developing over the FM-CW radar is shown in Fig. 6.2. The structure seen in these observations demonstrates that layered fine structure is possible at meter scales within the troposphere. We will return to the subject of layered fine structure in section 5.

FIG. 6.1. A range height indicator (RHI) obtained from the 10.7-cm radar at Wallops Island on 1 Jun 1966 at 1932 UTC. The range marks are at 5 n mi (9.3 km) intervals and the nearly horizontal marker near the top of the photo is at a height of 20 000 ft (6.1 km). The braided structure near the height of 3.5 km occurs in a visually clear atmosphere and is the result of breaking waves in a statically stable layer. The echoes below 1.8 km are due to scattering from convective structures in the boundary layer. After Hicks and Angell (1968).

Following the pioneering work of Atlas and colleagues at Wallops Island, a series of experiments were conducted by Browning and colleagues studying KHI utilizing the 10.7-cm Defford radar in England. Simultaneous radar and aircraft measurements provided important confirming evidence of the link between radar echoes and KHI (Browning 1971).

The early research that established the foundations of radar probing of the clear atmosphere has had important consequences for Doppler radar wind profiling. Unlike most observations taken by conventional scanning radars, Doppler radar profilers observe vertically or nearly vertically. Issues such as turbulent layer thickness and inner scale size are important for wind profiling. Unfortunately, to date there have been few sustained field campaigns in which in situ observations have been made with Doppler radar profilers operating at more than one frequency.

Doppler radar profilers do provide a wealth of observational data relevant to turbulence and waves. They observe continuously over a range of altitudes depending on frequency, power aperture product, and meteorological conditions (Balsley and Gage 1982). Much of the background in this subject has been covered in Gage (1990). The rest of this review will emphasize profiler observations of turbulence and waves utilizing observations made in several field campaigns over the past 15 years. An example of the unique observation of vertical motions made by profilers is shown in Fig. 6.3, which reproduces a multiple height time series of vertical velocities observed using the 50-MHz mesosphere–stratosphere–troposphere (MST) radar at Poker Flat, Alaska, on 12 October 1979. A well-defined monochromatic wave is observed extending over several range gates in unfiltered observations of vertical velocities maximizing in amplitude at 10.7 km. Gage et al. (1981) studied the structure and evolution of the vertical velocities, reflectivity, and spectral width of this event

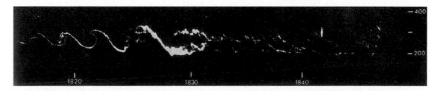

FIG. 6.2. The FM-CW time–height observation of Kelvin–Helmholtz shearing instability made with a vertically pointing, 10-cm wavelength FM-CW radar located near San Diego, California, on 6 Aug 1969. The times are Pacific Standard Time and the height scale is in m. The sequence from left to right, shows the growth of a billow, its breaking, and subsequent decay. After Gossard et al. (1970).

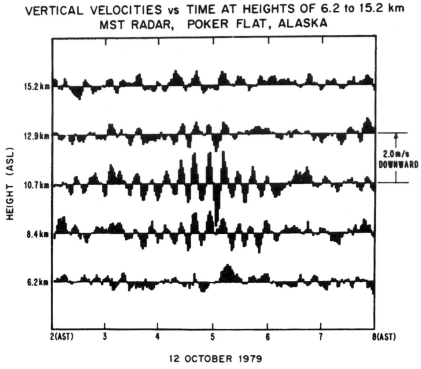

FIG. 6.3. Internal gravity waves seen in vertical velocities observed on the vertically pointing beam of the Poker Flat MST radar on 12 Oct 1979. After Gage et al. (1981).

and concluded that the observations were consistent with a wave modulating the layer in which these observations are embedded.

Vertical velocity fluctuations are ubiquitous in profiler observations and are thought to reflect the presence of a spectrum of waves in the atmosphere. These waves may be thought of as analogous to the Garrett–Munk spectrum of waves in the oceans (Garrett and Munk 1972, 1975; VanZandt 1982; Staquet and Sommeria 2002) and exhibit a fairly universal character. We will consider progress in understanding the spectrum and sources of waves in the atmosphere in sections 9 and 10 of this review.

3. Atmospheric turbulence and waves in the boundary layer and free troposphere

In this section we consider some of the influences of the stability of the atmosphere on the dynamics of turbulence and waves and some of the simplifying assumptions that are made in the study of turbulence and waves in the atmosphere. Turbulence and waves both represent perturbed states of the atmosphere that are subject to the governing equations of fluid mechanics. Several fundamental features distinguish waves from turbulence and it is relatively straightforward to distinguish a monochromatic wave from a chaotic state of fully developed three-dimensional turbulence. In this context a "monochromatic" wave is a disturbance that

is easily distinguished in a time series as having at least several quasi-sinusoidal cycles with a coherent phase relationship evident in spaced observations such as adjacent heights in profiler time series. However, a spectrum of waves is not easily distinguished from turbulence (Dewan 1979). Indeed, in a recent review Staquet and Sommeria (2002) use the term *wave turbulence* to refer to a nearly universal spectrum of incoherent waves resulting from multiple sources.

The most important single feature of waves, which distinguishes them from turbulence, is their propagation. Turbulence does not propagate but rather is generated and dissipated in situ. Turbulence is generally considered to be imbedded in a flow and to satisfy the Taylor transformation between time and space. Small-scale turbulence is dissipative and causes diffusion of mass much more effectively than molecular diffusion. Turbulence is created for the most part by instabilities in the basic flow fields. In the atmosphere, turbulence occurs in hydrostatically stable, neutral, and unstable environments. Buoyancy in hydrostatically unstable fluids leads to vertical motion and convection, which is typically accompanied by turbulence. In hydrostatically stable environments turbulence can be generated in sheared flows when the Richardson number is low enough. Waves cannot propagate in hydrostatically unstable environments.

The temperature lapse rate is the most important factor governing small-scale waves in the atmosphere. For example, a parcel of air displaced upward for whatever reason will tend to oscillate with the Brunt–Väisälä frequency N, defined by $N^2 = (g/\Theta)(d\Theta/dz)$ where Θ is potential temperature and g is acceleration due to gravity. A buoyancy wave or internal gravity wave typically has periods between the Brunt–Väisälä period and the local inertial period defined by $2\pi/2\Omega \sin\phi$; where Ω is the angular velocity of the earth in radians per second and ϕ is latitude. A typical period of buoyancy waves in the free troposphere is 15 min and the inertial period at 45° latitude is close to 17 h. In this review we use the term internal gravity waves to refer to these waves that may have an intrinsic frequency, ω, anywhere in the range $N \geq \omega \geq 2\Omega \sin\phi$. Short-period internal gravity waves have intrinsic periods close to the Brunt–Väisälä frequency and propagate in the vertical plane while low-frequency internal gravity waves possess quasi-horizontal motions and have intrinsic frequencies closer to the inertial frequency. These waves are usually referred to as inertia–gravity waves.

a. Radar backscattering from atmospheric turbulence

Radar backscattering from the optically clear atmosphere arises primarily from Bragg scattering from refractivity turbulence in the inertial subrange. The inertial subrange of homogeneous isotropic turbulence is bounded by an outer scale L_0 and an inner scale l_0. The intensity of atmospheric turbulence is typically expressed in terms of $\varepsilon(\mathrm{m}^2\,\mathrm{s}^{-3})$, the eddy dissipation rate of turbulence. The outer scale of turbulence is often expressed in terms of the Ozmidov length scale as

$$L_0 = 2.3\varepsilon^{1/2}N^{-3/2}. \qquad (6.1)$$

Typical values of L_0 in the free atmosphere are thought to be in the range of 10 to 100 m (Crane 1980). Since the Brunt–Väisälä frequency is directly related to the hydrostatic stability, L_0 is largest in turbulent regions with low hydrostatic stability and smallest in weakly turbulent regions that are hydrostatically very stable.

The inner scale of the inertial subrange is given in terms of η_0 (m), the Kolmogorov microscale

$$\eta_0 = (v^3/\varepsilon)^{1/4}, \qquad (6.2)$$

where $v(\mathrm{m}^2\,\mathrm{s}^{-1})$ is kinematic viscosity ($v = \mu/\rho$, where μ is dynamic viscosity and ρ is atmospheric density). The inner scale is given by Hill and Clifford (1978) as

$$l_0 = 7.4\eta_0. \qquad (6.3)$$

Gossard et al. (1984a) have explored the critical radar wavelength λ_c (m) for the sensitivity of radars to turbulence in the lower troposphere in relation to the Kolmogorov microscale. They define $\lambda_c = 8\pi\eta_0$ (or, equivalently, $\lambda_c = 3.4\, l_0$) by choosing the wavenumber k (radians m^{-1}) at which the spectral power is reduced by viscous effects to half that of the inertial subrange in Hill's spectrum (Hill 1978). The inner scale of the inertial subrange of locally homogeneous, isotropic tur-

bulence depends on the intensity of turbulence. It is small when turbulence intensity is large and largest when turbulence intensity is weak. Typical values of l_0 in the lower atmosphere are on the order of several centimeters, consistent with a critical radar wavelength close to 10 cm. At times when the eddy dissipation rate is small, λ_c becomes larger than 10 cm as was shown by Gossard et al. (1984b). Note that l_0 and λ_c generally increase with altitude in part because of the decrease in atmospheric density ρ. Eaton and Nastrom (1998) provide estimates of inner and outer scales of turbulence in the free atmosphere above White Sands Missile Range, New Mexico.

b. Turbulence and waves in the boundary layer

The atmospheric boundary layer is very important in atmospheric dynamics since it is the region of heat and momentum fluxes between the free atmosphere and the surface. The lowest 100 m or so of the atmospheric boundary layer is referred to as the surface boundary layer. The surface boundary layer has been studied extensively especially under conditions of uniformly flat terrain (Garratt 1992; Kaimal and Finnigan 1994). This is the domain of classic micrometeorology that made rapid advances in the 1950s through the 1970s owing to its accessibility to near surface in situ instruments. Above the surface boundary layer is the deeper atmospheric boundary layer sometimes referred to as the planetary boundary layer. While the surface boundary layer has been extensively studied, there is still much to be learned about the atmospheric boundary layer especially in regions of nonuniform terrain, various states of vegetation and soil moisture, and under varying conditions of atmospheric stability.

Atmospheric boundary layers vary in thickness. The convective boundary layer is heated from below causing thermal instability that leads to convection and turbulence that leads to a mixed layer capped by an inversion. Typically, daytime boundary layers are thicker over land where the surface heating is greatest and thinnest over the ocean where surface heating is minimal. Consequently, boundary layers over land are subject to a more robust diurnal cycle. In a stable boundary layer, such as typically forms at night, turbulence is suppressed and intermittent (Mahrt 1999).

Figure 6.4 shows part of the diurnal cycle of the lower atmosphere under clear air conditions as seen by a vertically pointing 915-MHz profiler at the Flatland Atmospheric Observatory (FAO) in central Illinois (Angevine et al. 1998). The growth and decay of the convective boundary layer can be seen in the time–height cross sections of reflectivity.

Recent research on the atmospheric boundary layer is summarized by Högstrom (1996), Nieuwstadt and Duynkerke (1996), Mahrt (1999), Högstrom et al. (2002). Boundary layer simulation is a very important part of atmospheric numerical models (Holtslag and

Day 266 1995

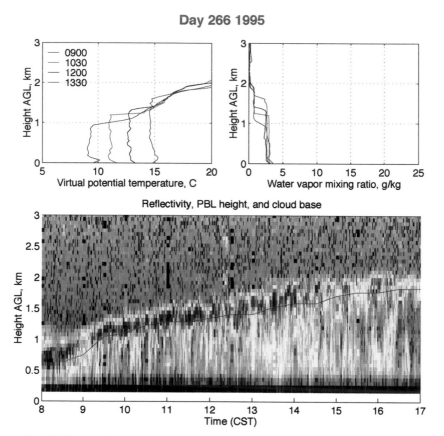

FIG. 6.4. (bottom) Example of reflectivities observed in the cloud-free convective boundary layer at FAO on 23 Sep 1995; (top left) virtual temperature profiles and [COLOR] (top right) vertical profiles of water vapor mixing ratio. After Angevine et al. (1998).

Duynkerke 1998; Stevens et al. 1998). In particular, much attention has been focused on cloudy boundary layers with extensive stratocumulus that develop off the west coast of the Americas. The extensive regions of marine stratocumulus play an important role in the heat balance of the planet. Although the marine boundary layer has received less attention than boundary layers over land surfaces since it is less accessible to observation, it clearly plays an important role in atmospheric dynamics and climate.

A promising new tool for examining the local structure of turbulence in the convective boundary layer is the turbulence eddy profiler (TEP; Mead et al. 1998). The TEP is a volume-imaging radar system that uses digital beam forming techniques to resolve the atmospheric boundary layer. Pollard et al. (2000) report boundary layer measurements obtained using the TEP in Pennsylvania in 1996. The TEP has the potential for providing the kind of detailed observations needed to validate numerical simulations of the boundary layer.

c. Turbulence and waves in the free atmosphere

While the atmospheric boundary layer is mostly turbulent especially during daytime convective conditions,

the free troposphere is almost always hydrostatically stable and is turbulent only in regions of sufficient vertical wind shear or in regions occupied by deep convection and clouds. Kelvin–Helmholtz instability of stratified shear flows is thought to be the primary mechanism for the production of turbulence in a hydrostatically stable environment. Under these circumstances turbulence is typically weak and intermittent being confined to layers of intense local shear. The shear may be organized by larger-scale waves such as inertial waves or in some cases produced by breaking waves. Once formed, the turbulence tends to occupy a horizontal domain much larger than its thickness. Historically, turbulence in the free atmosphere was studied extensively because of its relevance to aviation and electromagnetic wave propagation but it is obviously of great interest owing to its contribution to diffusion and transport.

Fairall et al. (1991) developed a stochastic model using random phase between wavenumber amplitude components and a reverse Fourier transform of an assumed vertical wavenumber spectrum proportional to m^{-3} to simulate profiles of wind, temperature, and Brunt–Väisälä frequency. They found that, assuming layers with $Ri < 0.25$ to be turbulent, layer thickness in the free troposphere is near 35 m and about 25% of the troposphere appears to be actively turbulent.

The layered structure of turbulence in the atmosphere has many important consequences. For example, eddy diffusivity is much greater in the horizontal than it is in the vertical. Furthermore, wind profilers that measure a reflectivity-weighted wind will bias wind measurements toward regions with more intense turbulence and, consequently, higher reflectivities. The consequences of nonuniform distributions of reflectivity on wind measurement are considered in Johnston et al. (2002). They found, for example, that profilers that use large range gates especially at low altitudes often are biased if the measured wind is assigned to the middle of the range gate of the profiler. This happens whenever refractivity gradients are large through the range gate of the profiler as is often the case at low altitudes.

4. Atmospheric turbulence measurement

The dominant scattering mechanism responsible for echoes observed by Doppler radar wind profilers is Bragg scattering (Balsley and Gage 1982). Bragg scattering arises from inhomogeneities in radio refractive index caused by atmospheric turbulence acting on gradients of radio refractive index (Ottersten 1969). Summaries of Bragg scattering as it relates to profiler observations can be found in Gage and Balsley (1980) and Gage (1990). According to the theory developed by Tatarskii (1971) the refractivity turbulence can be parameterized by $C_n^2 (\mathrm{m}^{-2/3})$ and the radar volume reflectivity η (m^{-1}) can be expressed in terms of C_n^2 by

$$\eta = 0.38 C_n^2 \lambda^{-1/3}, \qquad (6.4)$$

where $\lambda(\mathrm{m})$ is the radar wavelength. Here C_n^2 is the refractivity turbulence structure-function parameter (Peltier and Wyngaard 1995) defined for locally homogeneous, isotropic turbulence in the inertial subrange by

$$[n(x_0 + \Delta x) - n(x_0)]^2 = C_n^2 (\Delta x)^{2/3}. \qquad (6.5)$$

In Eq. (6.5) x is a position variable with dimensions of length and n is the radio refractive index defined by

$$n - 1 = 3.73 \times 10^{-1} e/T^2 + 77.6 \times 10^{-6} p/T, \qquad (6.6)$$

where p (hPa) is the atmospheric pressure, e (hPa) is the partial pressure of water vapor, and T (Kelvin) is the absolute temperature.

A recent test of this theory and Eq. (6.4) is reported by Cohn (1994) who used three collocated radars (34, 11.5, and 1.5 cm) at Millstone Hill in Massachusetts to show that the $\lambda^{-1/3}$ wavelength dependence is frequently observed but that typical measurements show a power law that is distributed about $\lambda^{-1/3}$.

Atmospheric turbulence is usually measured in terms of either one of the structure-function parameters, such as the refractivity turbulence structure-function parameter C_n^2 defined in Eq. (6.5), or the eddy dissipation rate that characterizes the intensity of turbulence within the

framework of Kolmogorov theory. The most straightforward in situ determination of the eddy dissipation rate is to measure the spectrum of turbulence within the inertial subrange and determine the best value of epsilon that fits the observed spectrum. This method can be used for aircraft measurements as reported in Chan et al. (1998).

Turbulence measurement in the atmosphere is complicated by intermittency. Intermittency in the atmospheric boundary layer has been studied by Mahrt (1989). Mahrt distinguishes between "global" intermittency and "small-scale" intermittency. Global intermittency involves the familiar patchiness of turbulence that is very pronounced in the free atmosphere. Small-scale intermittency refers to the fact that eddy dissipation rates and other turbulence parameters are highly variable in space and time owing to the influence of the turbulent eddies themselves that tend to create sharp gradients in the flow and regions of high dissipation interspersed with regions of lower dissipation. Consequently, sampling problems arise and it is often difficult to say with confidence that turbulence measurements are representative. For a recent, more general discussion of small-scale intermittency in turbulent fluids see the review by Sreenivasan and Antonia (1997).

For radar measurement of turbulence two approaches have been used (Gage 1990; Cohn 1995). Several authors have used the zeroth moment of the radar spectra under clear air conditions and others have used the spectral width to determine the eddy dissipation rate. However, it is important in making these calculations to be sure that the radar echoes are indeed due to turbulence and not particulate backscatter. A simple method to unambiguously separate turbulent backscatter from particulate backscatter was demonstrated in the Maritime Continent Thunderstorm Experiment (MCTEX) as reported in Gage et al. (1999). The main results are summarized at the end of this section.

a. Turbulence measurement from spectral width

A standard technique for turbulence measurement with Doppler radar has been to use the Doppler spectral width determined from the Doppler spectrum. This technique has been in widespread use (Atlas 1964; Hocking 1985; Gage 1990). Recent results are reported by Cohn (1995), Nastrom (1997), Nastrom and Eaton (1997), Delage et al. (1997), and Rao et al. (2001). Simultaneous radar and in situ measurements have been reported by Dalaudier et al. (1989), Luce et al. (1995, 1996), and Bertin et al. (1997).

In order to use the spectral width method, corrections must be applied to the measured velocity variance to account for shear and beam broadening effects. The technique was first used in the presence of hydrometeors in which the spread of the hydrometeor fall velocities was taken into account. In the absence of hydrometeors, beam broadening and shear broadening are important

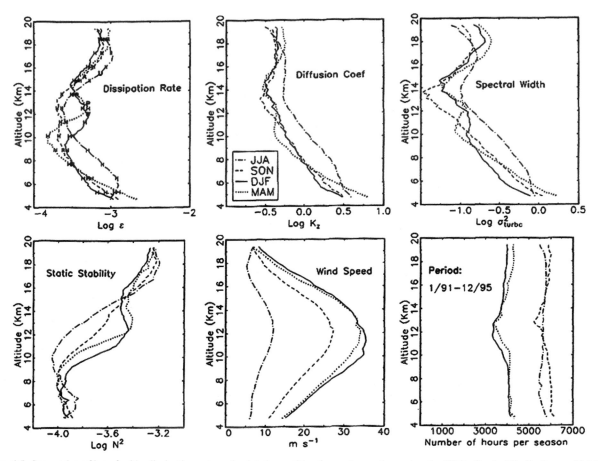

FIG. 6.5. Seasonal profiles of eddy dissipation rate and related quantities from observation using the White Sands Missile Range 50-MHz profiler in New Mexico. After Nastrom and Eaton (1997).

influences. Beam broadening depends on the radar beamwidth and the horizontal wind velocity. It can be minimized by using a radar antenna with a narrow beamwidth. Shear broadening depends on the radar pulse length, zenith angle of the radar beam, and the wind shear in the observing volume. Shear broadening can be minimized by looking vertically and using a short pulse length. In addition, at lower very high frequency (VHF), it is necessary to correct for the influence of specular echoes when looking vertically. Another potential factor influencing the spectral width is the contribution due to short period internal waves, which appears not to be significant in the lower atmosphere (Nastrom and Eaton 1997).

Seasonal profiles of eddy dissipation rate and diffusion coefficients are shown in Fig. 6.5. Typical values of epsilon reported by Nastrom and Eaton (1997) for White Sands Missile Range reproduced in Fig. 6.5 are on the order of 10^{-3} $m^2 s^{-3}$. Cohn (1995) reports values of epsilon in the range of 10^{-5} and 10^{-3} $m^2 s^{-3}$, which were obtained using the Millstone Hill radar in Massachusetts. The MU radar in Japan has also been utilized for the measurement of turbulence parameters in the free atmosphere (Low et al. 1998; Hermawan and Tsuda

1999). Radar measurements of eddy dissipation rates may be on the high side. Perhaps the best method for measuring eddy dissipation rates are with in situ probes capable of recording the spectrum of inertial range turbulence. For example, Chan et al. (1998) report values of eddy dissipation rates observed using the meteorological measurement system (MMS) on the National Aeronautics and Space Administration (NASA) DC-8. Figure 6.6 shows a vertical profile of eddy dissipation rates determined from the MMS over Kansas while ascending through cirrus clouds during the First International Satellite Cloud Climatology Project Regional Experiment (FIRE) I campaign. These values lie in the range of 10^{-7} to 10^{-2} $m^2 s^{-3}$. The lowest value of eddy dissipation rate that can be measured by the MMS is 10^{-7} $m^2 s^{-3}$ and the measured values are thought to be accurate to within 10%. In the free troposphere, of course, eddy dissipation rates are highly variable. It is likely that the radar values are biased by the most active turbulence in layers that must be present within the radar observing volume for measurements to be possible. Aircraft typically sample with 100-m resolution, which may not be adequate for resolving inertial range turbulence within thin layers. Muschinski et al. (2001) report on

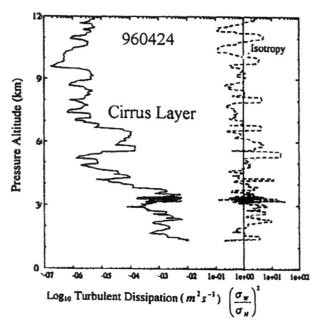

FIG. 6.6. A vertical profile of eddy dissipation measured by aircraft over Kansas. After Chan et al. (1998).

some new developments using kite-, balloon-, and helicopter-borne measurement systems that offer considerable promise for making improved in situ turbulence measurements.

While most of the reported radar measurements have been made in the free troposphere, Jacob-Koaly et al. (2002) have recently reported eddy dissipation rates measured in the boundary layer using spectral width determination from a UHF wind profiler operating in the convective boundary layer during Turbulence Radar Aircraft Cells (TRAC-98). These authors report eddy dissipation rates in the range of 10^{-4} to 3×10^{-3} m^2 s^{-3}, which were comparable to concurrent aircraft measurements. The measured values decreased with altitude away from the surface.

b. Turbulence measurement from backscattered power

Another method for estimating eddy dissipation rates from turbulence measurements is based on relationships that give epsilon in terms of the refractivity turbulence structure-function parameter C_n^2. Of course, C_n^2 is an important turbulence parameter that is well suited to measurement by radar (Van Zandt et al. 1978). Gage (1990) reviewed the C_n^2 measurements that had been made by several Doppler radar profilers mostly examining the free troposphere as well as a theoretical model that was used to calculate C_n^2 based on sounding data. More recent work using this theoretical model applied to observations from the 915-MHz Stapleton profiler is reported in Warnock et al. (1988). Also, in an observational study Frisch et al. (1990) have examined the

variations of C_n^2 between 4 and 18 km in Colorado over a 5-yr period using the Colorado network of VHF wind profilers.

The model of VanZandt et al. (1978) and Warnock and VanZandt (1985) is a statistical model that uses thermodynamic sounding data from a nearby radiosonde to calculate the expected C_n^2. Dalaudier et al. (1989) and Luce et al. (1996) have taken a more direct approach utilizing coordinated high-resolution measurements of temperature and humidity to compare in situ measurements of C_n^2 with radar-observed C_n^2. The original attempt by Dalaudier et al. (1989) led to "puzzling" results due to instrumental problems that were corrected in the later work (Luce et al. 1996). The latest results, reproduced in Fig. 6.7, are based on the assumption that homogeneous isotropic turbulence is responsible for the oblique echoes observed by VHF radars. The observations used by Luce et al. (1996) were taken during the RAdars, SCIdar, BAlloons (RASCIBA90) campaign at Aire sur l'Adour during February–March 1990. Scidar is the term used for turbulence measurement using stellar scintillation detection and ranging (Avila et al. 1997). Comparisons of the reconstructed profiles from in situ measurements with radar observations are shown in Fig. 6.7. The radar observations were taken using the 45-MHz Provence radar located near the launching site for the balloon soundings. Only the oblique (15° off zenith) radar observations are compared to the reconstructed profiles. The comparisons agree quite well in magnitude reproducing some of the same features in vertical structure seen in the radar observations. These results support the hypothesis that the echoes are due to isotropic inertial subrange turbulence.

Distributions of C_n^2 have been determined by Gage et al. (1997) for the lower troposphere in the deep Tropics during the rainy season in the Tiwi Islands near Darwin during MCTEX. The results in Fig. 6.8 are for two collocated profilers at Garden Point, Australia. Figure 6.8 shows the vertical distribution of the composite daily cycle of C_n^2 values. Note the clear diurnal cycle at the lowest levels with largest values of C_n^2 occurring during hours of peak insolation when the convective boundary layer should be well developed. The largest values of C_n^2 are of order 10^{-12} m$^{-2/3}$, which is considerably larger than the midlatitude values from Colorado reported by Warnock et al. (1988). Since C_n^2 does not vary with frequency, the values of C_n^2 should be the same for both radars provided the half-wavelength of each radar lies within the inertial subrange. The bottom panel of Fig. 6.8 shows the differences in C_n^2 values deduced from the two radars. Note that the values of C_n^2 agree well at the lower heights where C_n^2 is largest but that S-band C_n^2 is systematically less at night and at the higher altitudes where turbulence is reduced. We attribute this to the S-band radar having its half-wavelength less than the inner scale of turbulence. This result underscores the advantages of operating radars at more than one wavelength to obtain unambiguous turbulence data.

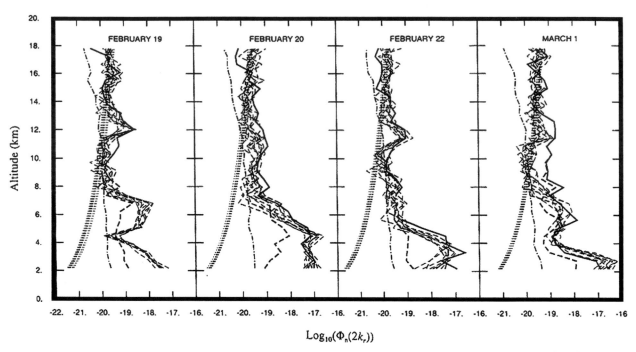

FIG. 6.7. Comparison between radar observations and reconstructions using high-resolution temperature measurements of the 3D spectrum of refractive index fluctuations. The radar detectability threshold is indicated by dotted lines. The noise level of the reconstructions is given as a dash–dotted line. Humidity corrections are included except for 19 Feb and around 6.5 km on 22 Feb. After Luce et al. (1996).

While the 16.5-cm half-wavelength of the UHF profiler will typically lie within the inertial subrange, the 5.3-cm half-wavelength of the S-band profiler may at times be smaller than the inner scale of turbulence. This is most likely to occur when the turbulence is weak.

Gage et al. (1980) developed an approach to estimate eddy dissipation rates from radar measurements of C_n^2. The method has been examined by Cohn (1995) and recently has been critically reviewed by Hocking and Mu (1997), who found the method useful provided care is taken in its application. In a recent study Furumoto and Tsuda (2001) compared the power method with the spectral width method and found that, while the average values were comparable, the two measurements differed in detail. Most of the differences were attributed to the contribution of humidity to the power measurements at lower altitudes.

c. Use of two radars for unambiguous turbulence measurement

The backscatter from a volume filled with spherical raindrops is given by Battan (1973) as

$$\eta = \pi^5 \lambda^{-4} |K|^2 Z, \qquad (6.7)$$

where $|K|^2 \approx 0.93$ for water (K is the complex index of refraction), and Z (mm^6 m^{-3}) is the equivalent radar reflectivity factor defined by

$$Z = \int N(D) D^6 \, dD. \qquad (6.8)$$

In Eq. (6.8), D is drop diameter and $N(D)$ is the drop-size distribution. Provided $D < 0.1 \lambda$, so that the back-scattering from the raindrops is due to Rayleigh scattering, the reflectivity factor is independent of the radar wavelength. Since the volume reflectivity has a λ^{-4} dependence, shorter wavelength radars of similar power aperture products are much more sensitive than longer wavelength radars to Rayleigh scattering.

The radar reflectivity expressed in dBZ$_e$ can be used regardless of whether the echoes arise from rain or snow or from refractivity turbulence structure. Following Battan, Z_e is defined by

$$Ze = \lambda^4 \eta / (\pi^5 |K|^2). \qquad (6.9)$$

While $|K|^2$ for ice differs from water, it is conventional to use the value of $|K|^2$ for water in evaluating Eq. (6.9). Then, for snow, D in Eq. (8) is the melted diameter.

In the event that the echoes arise from Bragg scattering, the relationship between C_n^2 and the equivalent reflectivity in dBZ$_e$ can be obtained from Eqs. (6.4), (6.7), and (6.9). It is given by

$$C_n^2 = 7.49 \times 10^{-16} Z_e \lambda^{-11/3} \quad \text{or} \qquad (6.10)$$

$$Z_e = 1.335 \times 10^{16} \lambda^{11/3} C_n^2 \quad \text{or} \qquad (6.11)$$

$$\text{dBZ}_e = 10 \log C_n^2 + 36.7 \log \lambda + 161.26. \qquad (6.12)$$

For Bragg scattering the difference in dBZ$_e$ observed by two profilers with wavelengths λ_1 and λ_2, respectively, is given by

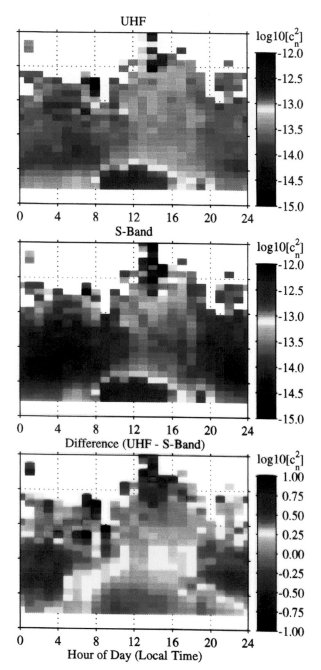

FIG. 6.8. Diurnal variation of refractivity turbulence structure-function parameter observed during MCTEX, Nov–Dec 1995 at Garden Point, Australia. Two profilers are used to screen the data to exclude precipitation echoes or returns from hard targets.

$$(dBZ_e)_{\lambda 1} - (dBZ_e)_{\lambda 2} = 36.7 \log(\lambda_1/\lambda_2), \quad (6.13)$$

provided that both $0.5\lambda_1$ and $0.5\lambda_2$ lie within the inertial subrange of turbulence. The two wavelengths used in the MCTEX campaign were 32.6 cm (920 MHz) and 10.58 cm (2835 MHz). Using these values for the two profiler wavelengths λ_1 and λ_2 in Eq. (6.12) yields an equivalent reflectivity difference of 17.94 dBZ_e. The longer wavelength profiler has the greater equivalent

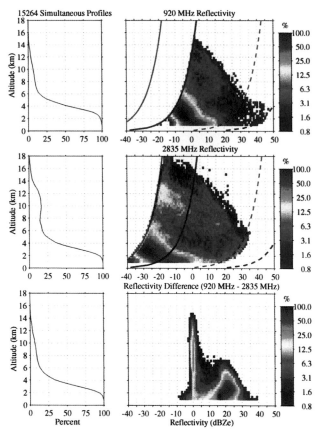

FIG. 6.9. The 2D histograms of reflectivity seen on vertically pointing beams of (top) 920- and (middle) 2835-MHz profilers; (bottom) the 2D histograms for the reflectivity difference. After Gage et al. (1999).

reflectivity for Bragg backscatter as can be seen from Eq. (6.12).

Frequency distributions of equivalent reflectivity and differential reflectivity observed by the two profilers for the entire MCTEX campaign (13 November–7 December 1995) are shown in Fig. 6.9. The panels on the left of Fig. 6.9 show the height distributions of echoes observed by the two profilers and on the bottom left the height distribution of echoes observed simultaneously by both profilers (which closely resembles the height distribution of echoes seen by the 920-MHz profiler). Note that the height distribution of the 2835-MHz echoes substantially exceeds the echoes observed by the 920-MHz profiler in the altitude range 6–12 km. The reason for this is the approximate 20-dB increased sensitivity of the 2835-MHz profiler to hydrometeors.

The top panels on the right of Fig. 6.9 show the frequency distribution of occurrence of equivalent reflectivity versus altitude for the 920- and the 2835-MHz profilers, respectively. The solid blue and red curves are the threshold values of detectability for the 920- and 2835-MHz profilers and the dashed blue and red lines are the threshold values for saturation for the 920- and 2835-MHz profilers, respectively. Note that the distri-

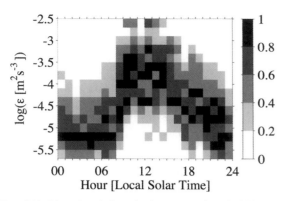

FIG. 6.10. Diurnal variation of ε from a set of vertical histograms showing the number of occurrences of particular values of log (ε) at 1592 m obtained by applying Hill's model to MCTEX dual-profiler observations. After VanZandt et al. (2000).

bution of equivalent reflectivities seen by the 920-MHz profiler at low altitudes is typically 15–20 dBZ_e larger than the corresponding values observed in the same altitude range by the 2835-MHz profiler. These echoes are mostly from Bragg scatter.

The bottom right panel in Fig. 6.9 shows the frequency distribution with altitude of reflectivity difference. Here two populations of echoes are clearly visible. The dominant echoes at the lowest altitudes present most of the time are due to Bragg scatter and the echoes clustered around zero reflectivity difference are due to Rayleigh scatter. It is interesting to note that most of the hydrometeor echoes are distributed in the range of 5–12 km.

Once it can be reasonably assured that turbulence is responsible for the echoes observed by the profiler, it is possible to determine the refractivity turbulence structure-function parameter directly from the measured volume reflectivity using Eq. (6.4). If the 915-MHz profiler is observing turbulence within the inertial subrange and the S-band profiler is observing turbulence in the viscous subrange, however, the reflectivity seen by the S-band profiler will be less than what would be seen if it were observing inertial subrange turbulence. Accordingly, the reflectivity difference between the two profilers will then exceed the theoretical difference of 18 dB if both profilers were sensing inertial range turbulence.

VanZandt et al. (2000) have developed a method to determine eddy dissipation rates from the reflectivities determined from two profilers operating at different frequencies. The method utilizes the Hill model (Hill 1978) for the turbulence spectrum and requires one profiler to operate in the inertial subrange and the second profiler to operate in the viscous subrange of turbulence. Figure 6.10 shows the composite diurnal cycle of the number of occurrences of particular values of log (ε) at 1592 m obtained during the MCTEX campaign. Note that the values of ε fall in the range of 10^{-6} to 10^{-2} m^2 s^{-3},

with the largest values in the middle of the day and the lowest values late at night and in the early morning.

While most of this section has been concerned with turbulence measurements using radar, several recent studies have used lidar to measure turbulence and turbulent processes in the lower troposphere. Banakh et al. (1999) present measurements of eddy dissipation rate in the boundary layer using CW Doppler lidar. Wulfmeyer (1999a,b) reports on investigations of turbulent processes in the lower troposphere using water vapor differential absorption light detecting and ranging laser radar (DIAL) and radar–radio acoustic sounding system (RASS).

d. Turbulence measurements in clouds

During the past decade there has been increasing attention given to measuring turbulence within clouds. For example, Knight and Miller (1993, 1998) and Kollias et al. (2001) have used radar to study the turbulence in fair weather cumuli while Gultepe and Starr (1995) and Gultepe et al. (1995) have used aircraft to examine the turbulence within cirrus clouds. More importantly, much of the work on cloud turbulence has been directed toward an improved understanding of the stratocumulus clouds that form extensive low-level cloud decks over the eastern Pacific. Unlike the convective boundary layer that is heated from below, the turbulence in marine stratocumulus is driven by the longwave radiative cooling at the top of these extensive cloud shields (Dreidonks and Duynkerke 1989).

The observation of turbulence in clouds by radar involves interpretation of the returns in terms of their contributions from Rayleigh and Bragg scattering (White et al. 1996). A multifrequency approach is needed to overcome the ambiguities associated with the two sources of scattering. Knight and Miller (1993, 1998) have examined the earliest radar returns from fair weather cumulus and found that strong mantle echoes occur around the sides and tops of cumulus clouds owing to the mixing of dry and moist air near the cloud boundaries. Their observations were made in Florida during the Convection and Precipitation/Electrification experiment (CaPE) and the Small Cumulus Microphysics Study (SCMS) in July and August 1991 and 1995. Knight and Miller point out that the mantle echoes they observed are similar to those reported by Atlas (1959). Dual-wavelength experiments using 10- and 3-cm wavelength radars demonstrate that most of the early radar returns are due to Bragg scatter. White et al. (1996) reach similar conclusions in a study of shallow trade wind cumuli observed with a 915-MHz wind profiler on the R/V *Moana Wave* during the Tropical Instability Wave Experiment (TIWE) in November–December 1991. By employing a ceilometer, White et al. (1996) conditionally sampled the trade wind cumuli and found that the C_n^2 was considerably enhanced in the presence of the clouds compared to the surrounding clear air.

These authors found C_n^2 to be greater than 10^{-12} m$^{-2/3}$ within the clear marine boundary layer and about five times larger in the presence of clouds. Kollias et al. (2001) have used a 33-cm wind profiler and a 3-mm wavelength cloud radar to study the structure of fair weather cumuli near Miami. They discuss the determination of eddy dissipation rate from spectral width measurements using the vertically pointing cloud radar. While in general it is difficult to retrieve the eddy dissipation rate within the cloud, they found that the eddy dissipation rate within the updraft core was in the range of 10^{-3} m^2 s^{-3} to 4×10^{-3} m^2 s^{-3}. Shear broadening of the vertical velocity makes it difficult to extract the eddy dissipation rate near the cloud boundary. The important role of turbulence on the growth of cloud droplets by condensation and coalescence is discussed by Jonas (1996) and the role of turbulence in entrainment within clouds is reviewed by Telford (1996).

Turbulence in convective precipitating clouds has been examined by Meischner et al. (2001) using C-band scanning Doppler radar during the Lightning Produced NO$_x$ (LINOX) campaign and the European Lightning Nitrogen Oxides (EULINOX) campaign in 1996 and 1998, respectively. During these campaigns, coordinated aircraft flights were made through convective storms that permitted a comparison of eddy dissipation rates measured by radar and aircraft. The radar measurements were made using the spectral width method. Results from the two campaigns showed reasonable agreement between the two platforms. The most intense turbulence was found in regions of strong updrafts. The aircraft measured epsilon in the range of 0.05 to 0.1 m^2 s^{-3} and the radar estimated epsilon of 0.01 m^2 s^{-3}. The radar was found to give lower estimates than the aircraft inside the updraft area. The threshold for detectability for the C-band Doppler weather radar used for this work was found to be about 10^{-3} m^2 s^{-3}.

Marine stratocumulus clouds have been studied extensively during FIRE I and during the Atlantic Stratocumulus Transition Experiment (ASTEX). Turbulence structure observed within marine stratocumulus observed by radar during ASTEX has been reported by Frisch et al. (1995). Using the Environmental Technology Laboratory (ETL) 8-mm and (k$_a$-band) Doppler radar (Kropfli et al. 1995), they measured the vertical-velocity variance within the marine stratocumulus as the clouds advected over the radar located on the island of Porto Santo 880 km west of Casablanca, Morocco. They confined their analysis to reflectivities less than -17 dBZ so that their observations would not be contaminated by the fall velocity of cloud droplets. In the upper part of the cloud they found a maximum vertical-velocity variance greater at night than during the day. These observations show the influence of the radiative cooling at the top of the clouds, which is countered by shortwave heating during daytime. In the upper part of the clouds, downdrafts are more intense than updrafts with less area than the updrafts, consistent with an in-terpretation of upside-down convection driven by cooling at cloud tops. However, in the lower part of the cloud a daytime maximum of vertical-velocity variance was found and the updrafts were more intense than the downdrafts, indicating that daytime heating was driving the convection from below. Kollias and Albrecht (2000) have used a 3-mm cloud radar in a similar way to investigate the turbulence structure in a continental stratocumulus cloud near Miami, Florida. They also found a near-cloud-top maximum in vertical-velocity variance at night. However, during the early morning hours the height of maximum variance descended systematically, indicating turbulence activity descending toward the lower part of the cloud. In addition to the studies on the turbulence structure of marine and continental stratocumulus, Feingold et al. (1999) have considered the dependence of droplet formation and drizzle upon turbulence in stratocumulus.

Cirrus clouds also play an important role in the heat balance of the atmosphere. Thin cirrus is known to preferentially occur near the tropopause especially in the Tropics and thick cirrus anvil shields are dominant features of convectively active regions such as those found over the western Pacific warm pool. Some advances have been made in the past decade in investigating the nature of turbulence in cirrus clouds (Gultepe and Starr 1995; Gultepe et al. 1995). Gultepe and Starr (1995) report measurement of turbulence in cirrus clouds made during FIRE I in October 1986. Their turbulence measurements were made by the National Center for Atmospheric Research (NCAR) *King Air* flying at constant altitude. Eddy dissipation rates were determined using structure functions and spectral methods. Dissipation rates within cirrus were found to be highly variable but exceeded values obtained from adjacent cloud-free regions. Typical values of epsilon within the cirrus were in the range of 10^{-6} to 10^{-3} m^2 s^{-3}, with the largest values measured in the lower conditionally unstable portions of the cloud and the smallest values measured in the highest statically stable portions of the cloud. These results were obtained with a sampling of about 100 m, which may not be adequate for sampling inertial range turbulence under these conditions. Gultepe et al. (1995) present results of an analysis of the dynamical characteristics of cirrus clouds observed in FIRE II over Kansas in November and December 1991 and show evidence for the existence of complex coherent structures possessing vorticity.

5. Finescale structure

VHF radar observations of the free atmosphere at vertical and oblique incidence show a distinct aspect sensitivity and enhanced backscattering at vertical incidence. This "specular" behavior is routinely observed at lower VHF and appears to be well correlated with hydrostatic stability (Gage and Green 1978). Based on these and other similar observations it has been widely

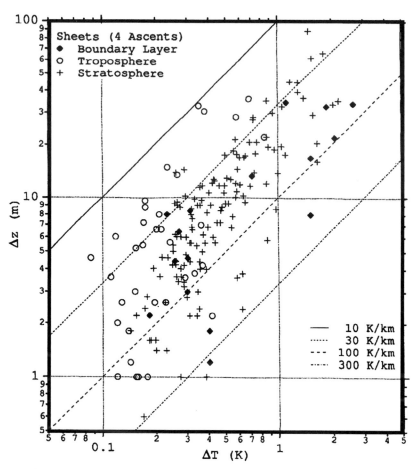

Fig. 6.11. Distribution of sheet thicknesses with temperature difference across the sheet in four balloon ascents using a high-resolution sensor. After Dalaudier et al. (1994).

held that a transversely coherent layered structure is needed to explain the observations but there has been no consensus to date on the physical mechanisms that give rise to this structure. Gage (1990) and Luce et al. (2001a) consider several mechanisms that could give rise to the requisite structure and the observed backscattering.

Several studies have been made in the past decade of the fine structure of the atmosphere using in situ sensors to probe for sheets and layers that might be responsible for the observed echoes. Salathe and Smith (1992) examined the small-scale temperature structure near the tropopause using data recorded on the NCAR *Sabreliner* during the Genesis of Atlantic Lows Experiment (GALE). These authors were able to show the existence of microstructure in the thermal field that became more pronounced in the stratosphere as compared to the troposphere. By comparing with measured vertical velocity fluctuations, Salathe and Smith demonstrated that the microstructure changes in the thermal field were not accompanied by changes in vertical-velocity microstructure, indicating that the temperature microstructure was independent of small-scale turbulence.

Dalaudier et al. (1994) also examined temperature

structure up to 25 km using special balloon-borne fast-response cold wire thermometers. These sensors provide a very-high-resolution (0.2 m) measurement of temperature and give valuable information about the characteristics of sharp temperature gradients in the troposphere and lower stratosphere based on four balloon ascents in France in February 1990. The authors found that the sharp gradients referred to as "sheets" are more numerous in the stratosphere than the troposphere. Confining attention to regions of increasing temperature with height they found 16%–46% of the increase in temperature occurred in sheets within the troposphere and 22%–50% of the increase of temperature in the lower stratosphere was confined to sheets. During these flights, the fraction of the height within these stable regions that were covered by sheets was found to be 2%–14% in the troposphere and 3%–17% in the lower stratosphere. A statistical distribution of the thicknesses and the amplitudes of the sheets is reproduced in Fig. 6.11. While 10 m is a typical thickness for a sheet, many sheets can be found with thicknesses as small as 1 m. The authors go on to calculate the backscattered power from partial reflections that would be observed by a VHF radar viewing this structure at vertical incidence.

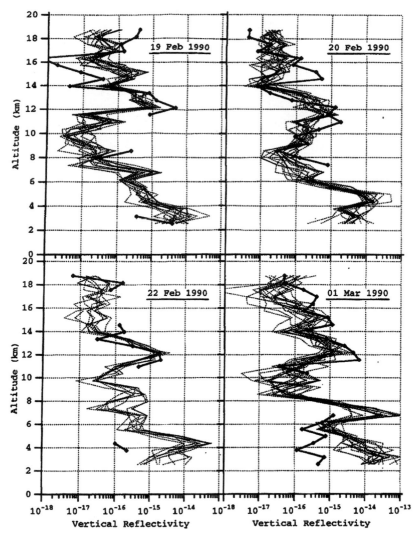

FIG. 6.12. Comparison of vertical reflectivity observed using a VHF radar and the reflectivity calculated from a partial reflections owing to sheets in the temperature field. After Dalaudier et al. (1994).

The results are compared with observations in Fig. 6.12. These results show that the observed in situ layered structure appears to explain much of the VHF radar observations at vertical incidence although the partial reflection model with assumed horizontal layered structure extending over a Fresnel zone is probably too simple to account for all of the observed features (Luce et al. 1995).

Muschinski and Wode (1998) have reported simultaneous high-resolution measurements of the fine structure of coexisting temperature and humidity sheets in the lower troposphere using a helicopter-borne turbulence system (HELIPOD). Their data were taken near 75°N off the east coast of Greenland in October 1995. The observations show structure in the thermal field similar to that reported by Dalaudier et al. (1994) but also provide concurrent high-resolution profiles of humidity mixing ratio and zonal and meridional compo-

nents of horizontal winds as shown in Fig. 6.13. The coexisting layer structure in humidity and temperature is similar to the Boulder Atmospheric Observatory (BAO) tower observations of Gossard et al. (1985). There is a tendency for the base of dry layers to be associated with inversions in temperature although the correspondence in temperature and humidity structure is not one for one.

Balsley et al. (1998) have developed a state-of-the-art kite system for detailed probing of the lower troposphere; Muschinski et al. (2001) present intercomparisons between kite- and balloon-borne instruments and the HELIPOD system. These instruments all provide detailed high-resolution profiles in the lower troposphere and the authors present observations from two field campaigns: the Profiler HELIPOD Intercomparison Experiment (PHELIX), which took place at the Vandenberg Air Force Base in California in November

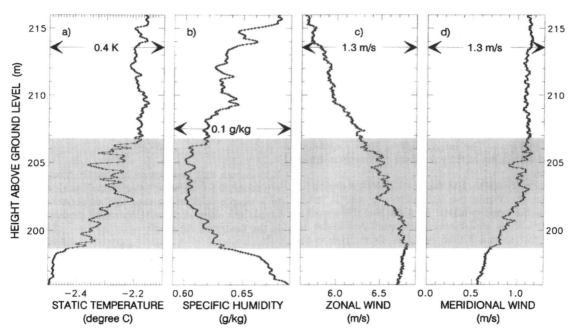

FIG. 6.13. High-resolution vertical profiles of wind temperature and humidity obtained off the east coast of Greenland in October 1995 using a helicopter-borne sounding system (HELIPOD) showing layered structure and coexisting temperature and humidity "sheets." After Muschinski and Wode (1998).

1997, and a second campaign at the White Sands Missile Range in New Mexico in April 1998. These instruments all have the capability of measuring finescale features of the lower atmosphere and precise in situ turbulence measurements for atmospheric dynamics research and for intercomparison with remote sensing observations.

The past decade has seen considerable advances in our ability to observe high-resolution features in the atmosphere using active remote probes. For example, Eaton et al. (1995) presented new results from an FM-CW profiler in New Mexico. An example of the excellent resolution obtainable with this instrument is shown in Fig. 6.14. In addition, several authors have explored new methods for obtaining high-resolution measurements using conventional profiler instruments. Frequency domain interferometry (FDI) has been employed with some success to examine high-resolution features in the lower and middle atmosphere (Kudeki and Stitt 1987; Chilson et al. 1997; Muschinski et al. 1999a). The FDI technique uses two closely spaced frequencies to provide improved vertical resolution. However, the FDI technique assumes a single scattering layer within the pulse volume and does not resolve multiple layers. Based on work of Franke (1990), Palmer et al.

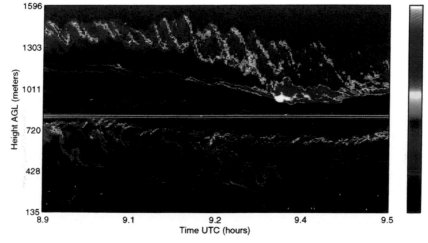

FIG. 6.14. High-resolution reflectivities seen in FM-CW observations at White Sands, New Mexico. After Eaton et al. (1995).

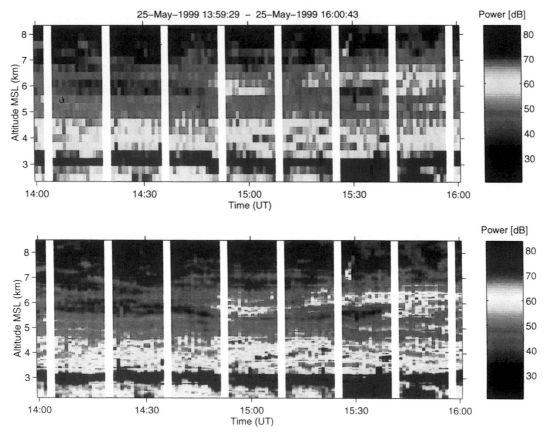

FIG. 6.15. RTI plot of the conventional echo power (top) and the RIM-enhanced echo power (bottom) obtained during SOMARE-99 with the SOUSY VHF radar in northern Germany. After Chilson et al. (2001).

(1998, 1999) have recently developed range imaging (RIM) methods that use more than two frequencies to yield improved vertical resolution when multiple layers occupy the same pulse volume. Palmer et al. (1999) give a detailed explanation of RIM and provide simulations assuming different layer configurations. To be most effective in resolving a multiple-layered structure with n layers, RIM should utilize at least $n + 1$ closely spaced frequencies.

Early RIM measurements have been reported by Palmer et al. (2001) and Chilson et al. (2001) for observations taken by the SOUSY radar in northern Germany. The SOUSY Multifrequency Atmospheric Radar Experiment 1999 (SOMARE-99) provides a demonstration of RIM capabilities. Figure 6.15 compares range time intensity (RTI) plots from the SOUSY radar on 25 May 1999 showing conventional echo power compared to RIM-enhanced echo power. Work of a similar nature in Japan has recently been reported by Luce et al. (2001b,c).

These recent advances in our ability to observe atmospheric fine structure help sharpen our ability to resolve the layered structure of the hydrostatically stable atmosphere. In the coming decade much progress on this subject would be possible in coordinated field campaigns employing the latest in situ and remote probes.

6. Stratified turbulence

It is well known in fluid mechanics that, as turbulence decays in a stably stratified environment, propagating waves are radiated and quasi-horizontal motions possessing vorticity are left behind (Lilly 1983; Hopfinger 1987; Müller et al. 1978; Riley and Lelong 2000). The vorticity-possessing quasi-horizontal motions, often referred to as the "vortical mode," have been reproduced in the laboratory and simulated by direct numerical simulation. This topic has been reviewed recently by Riley and Lelong (2000) who point out that these motions take place under conditions of low Froude number ($F = u'/NL < 1$) where u' is an rms turbulence velocity, L is a length scale of the energy containing motions, and N is the buoyancy frequency introduced in section 3. Riley and Lelong refer to this component of motion as potential vorticity (PV) modes since they possess potential vorticity and the propagating waves do not. Rotation is known to have an important influence on these flows and is parameterized by the Rossby number (u'/fL), where f is half the system rotation rate. Rossby numbers ≥ 1 are relevant to stratified turbulence, which can be regarded as a regime of strong stability and weak rotation.

In the atmosphere this class of motions is thought to

contribute substantially to the mesoscale spectrum of atmospheric motions (Gage 1979; Lilly 1983; Gage and Nastrom 1986). Lilly referred to these motions as stratified turbulence although the same class of motions is often referred to as quasi-two-dimensional turbulence because the vertical motion is suppressed in strongly stable environments and because the motion fields have considerable vertical structure. Their frequency spectrum occupies the same spectral range as internal wave motions, which makes it difficult to differentiate stratified turbulence from a spectrum of waves (e.g., Dewan 1979; VanZandt 1982). Riley et al. (1981), Lilly (1983), and Riley and Lelong (2000) give scaling arguments for two sets of equations of motion governing the wave component and the vortical mode, respectively. In section 10 we will consider the spectrum of internal waves in the atmosphere. In this section we will emphasize stratified turbulence as a possible source of much of the fine structure observed in the atmosphere. It is important to recognize, however, that low-frequency inertial-gravity waves can also produce vertical structure and these will be considered later.

Stratified turbulence has been investigated in the laboratory and by direct numerical simulation. Laboratory studies are necessarily restricted in their size and cannot possibly replicate the range of scales and parameter space covered in the oceanic or atmospheric mesoscale. Nevertheless, considerable insight has been obtained concerning stratified turbulence from laboratory experiments. Several laboratory experiments have simulated some of the fundamental dynamics of stratified turbulence (Itsweire et al. 1986; Maxworthy 1990; Itsweire and Helland 1989; Narimousa et al. 1991; Yap and van Atta 1993). Narimousa et al. (1991) and Yap and van Atta (1993) concentrate on the spectra and energy transfers and show that in the laboratory it is possible to simulate the inverse cascade of stratified turbulence with $k^{-5/3}$ spectra at scales larger than the scale of energy input and steeper spectral slopes at smaller scales. Hopfinger (1987) and Riley and Lelong (2000) concentrate on issues surrounding the collapse of three-dimensional turbulence in strongly stratified flows and its relationship to the residual motions that are left behind in wakes after waves are radiated away. Laboratory experiments demonstrate that a region of strongly stratified turbulence will decay into a field of quasi two-dimensional horizontal motions that possess vertical vorticity with a characteristic collapse of vertical motions, and increase in the horizontal dimension of the layer similar to the dynamics of wake collapse described in Hopfinger (1987).

In addition to the laboratory experiments, considerable progress has been made in the direct numerical simulation of stratified flows. These studies help to interpret laboratory experiments and to understand the energy transfers within the stratified flows and the structures that develop as these flows evolve. One long-standing issue is the difficulty of simulating high–Reynolds number flows, which makes it difficult to simulate many atmospheric flows with high resolution.

Numerical studies to date have focused on decaying turbulence in stratified flows and forced turbulence. Metais and Herring (1989) analyzed decaying turbulence in a strongly stratified flow and found the same general features reported in laboratory studies by Itsweire et al. (1986). Bartello (1995) has examined the interaction of PV and wave modes in decaying turbulence and found them to be closely coupled. Herring and Metais (1989) examined the dynamics of forced, nonrotating, strongly stratified flows with varying amounts of vertical variability. The existence of the vertical variability, which reflects the tendency of layered structure to form in stratified turbulence, was found to inhibit the $k^{-5/3}$ inverse cascade to larger scales of motion. Lelong and Riley (1991) consider wave–vortical mode interactions in strongly stratified flows. Riley and Lelong (2000) point out that the inverse cascade is a result of PV–PV mode interactions and that, in the presence of vertical variability, wave–PV mode interactions dominate the dynamics and result in the cascading of PV mode energy to smaller scales. Metais et al. (1994) considered both two- and three-dimensional forcing with varying degrees of rotation. The results showed that in the presence of rotation the inverse cascade was more likely to develop. Vallis et al. (1997) and Lilly et al. (1998) used low-resolution meteorological models to show that a $k^{-5/3}$ regime develops when convectively driven turbulence is present at the mesoscale with or without rotation. Bartello (2000) suggests that the model atmosphere behaves more two dimensionally in a low-resolution model because of the lack of vertical variability.

Many of the laboratory experiments cited above have been designed primarily to simulate oceanographic structure and dynamics while the numerical studies have been motivated also by the need to improve the understanding of the dynamics of waves and turbulence in the atmosphere. Observational studies addressing the issues discussed in this section have been relatively sparse perhaps owing to the difficulties of unraveling the issues involved. Nevertheless, several studies have been reported that are summarized next.

Masmoudi and Weil (1988) analyzed sodar observations collected from four sites during the MESO-GERS-84 campaign. Gers is a region in southwest France and Meso refers to the mesoscale nature of the campaign. The analysis tested the Taylor hypothesis over the domain of the network of stations and examined structure functions of horizontal wind speed. The spectra of the wind speed followed a $k^{-5/3}$ spectral slope and otherwise appeared consistent with stratified turbulence. Högström et al. (1999) have analyzed the mesoscale velocities over the Baltic Sea also finding consistency with stratified turbulence. Cho et al. (1999a) reported horizontal wavenumber spectra of winds, temperature, and trace gases measured by aircraft during the Pacific Exploratory Mission (PEM). The spectra determined from the PEM flights were similar to those reported by Nastrom and Gage (1985) for the Global Atmospheric Sampling Program (GASP). A detailed analysis of the PEM spectra was reported by Cho et al. (1999b). They tentatively concluded that the observed spectra were likely from several sources including quasi-two-dimensional turbulence. However, in a recent study Cho and Lindborg (2001) and Lindborg and Cho (2001) analyzed data collected from the European MOZAIC program and, using a third-order velocity structure function (Lindborg 1999), could find no support for an inverse cascade even though the analysis yielded spectra with slopes similar to those reported by Nastrom and Gage (1985).

7. Turbulence simulation

Up to this point this review has focused primarily on observations of atmospheric turbulence. While observations provide our ultimate source of knowledge of the dynamics of turbulence in the atmosphere, there are also important contributions to be made by laboratory and numerical simulations. Advances in laboratory simulations that are especially germane to the atmosphere are reviewed in Fernando (1991) and Fernando and Hunt (1996). Advances in numerical simulation of atmospheric turbulence are reviewed by Stevens and Lenschow (2001).

Wyngaard and Peltier (1996) make a useful distinction between modeling and simulation. To these authors *modeling* means ". . .representing turbulence through approximate equations whose solutions have behavioral similarities to turbulence" and *simulation* means ". . .using equations that are derivable from the exact set and, hence remain faithful to the essential physics." DNS is clearly in the latter category whereas LES is a hybrid containing elements of both modeling and simulation.

The last few decades have seen extraordinary advances in our ability to simulate turbulence and the past decade, in particular, has brought together researchers with a common interest in turbulence to examine the role of simulation in connection with experiments and observations. For the first time there seems to be a convergence of interests and capabilities to advance turbulence understanding by coordinating research in numerical simulation with experiments and observations (Stevens and Lenschow 2001). There has already been a dramatic increase in the use of numerical simulation by the atmospheric science community to improve understanding of the structure of the atmospheric boundary layer. While it is beyond the scope of this review to give a comprehensive summary of developments in numerical simulation, it is important to recognize the contributions to turbulence understanding especially in the boundary layer that are being made by this work.

The philosophy of numerical simulation is discussed at length by Stevens and Lenschow (2001). These authors, coin the term *pseudofluid* to draw a clear distinction between the direct numerical simulation of a "real" fluid by solution of the Navier–Stokes equations from the numerical simulation of a pseudofluid by solution of an equation in which the subgrid-scale motions have been parameterized or "filtered."

Atmospheric turbulence spans such a broad range of scales between energy-containing eddies and dissipation length scales that it is very difficult to perform DNSs of atmospheric turbulence especially for high–Reynolds number flows (Mason 1994). For this reason LES methods have been developed that resolve the large-scale energy-containing range of turbulence but parameterize the small-scale less energetic part of the turbulence spectrum (Mason 1994; Saiki et al. 2000; Moeng 1998; Muschinski et al. 1999b; Stevens and Lenschow 2001). LES has a long history of application to the convective boundary layer where the range of turbulent scales of motion are much too large to simulate directly.

a. Direct numerical simulation

Direct solution of the Navier–Stokes equations is the most straightforward approach to numerical simulation of turbulence. DNS can be used whenever the domain of a problem is sufficiently constrained to permit the resolution of all the relevant scales of motion. Realistically this is not possible for many geophysical flows and LES has been used very successfully in these cases. An example of a numerical study using DNS is presented by Coleman et al. (1990). These authors have used DNS to explore the dynamics of the turbulent Ekman boundary layer. The work builds on earlier studies by Faller (1965) and LeMone (1973). Faller drew attention to the inflectional instability as a source of longitudinal roll vortices in the planetary boundary layer and LeMone also studied the longitudinal roll vortices observed in the planetary boundary layer. It is well known that these coherent structures can be produced in a convectively unstable boundary layer with shear. Coleman et al. (1990) were able to show that these coherent structures do not appear to develop in the neutral Ekman boundary layer.

A number of studies of the evolution and breakdown of Kelvin–Helmholtz billows have been pursued with DNS (Fritts et al. 1996b; Palmer et al. 1994, 1996; Werne and Fritts 1999). The work by Fritts and colleagues complements similar studies by Klaassen and Peltier (1991), Caulfield and Peltier (1994), and Afanasyev and Peltier (2001). The DNS simulations are able to reveal in great detail how the Kelvin–Helmholtz instability evolves from a two-dimensional disturbance to fully three-dimensional turbulence. Figure 6.16 shows the development and transition to turbulence in the DNS simulations of Werne and Fritts (1999). Related studies have focused on the dynamics of wave breaking (Fritts et al. 1994, 1996a, 1998; Isler et al. 1994; Andreassen et al. 1994, 1998). The importance of DNS to simula-

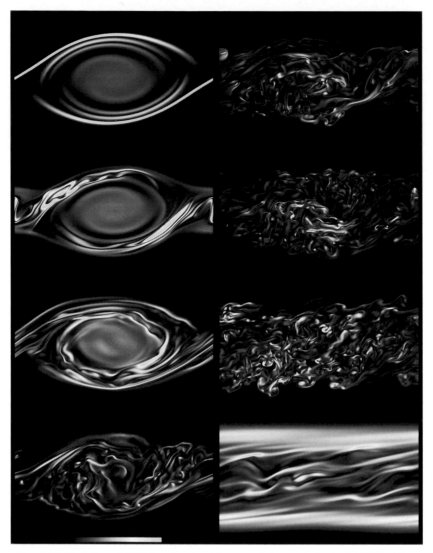

Fig. 6.16. Evolution of KHI showing transition to turbulence obtained with DNS. After Werne and Fritts (1999).

tions of stratified turbulence has already been pointed out in section 6.

Gibson-Wilde et al. (2000) have used DNS to investigate *virtual* VHF radar measurements of turbulence in the mesosphere. The mesospheric turbulence simulated by the DNS was generated by KHI and possessed inertial range characteristics. Their simulations enabled a comparison of turbulence parameters derived from *virtual* Doppler spectra with turbulence parameters calculated directly from the turbulence simulated by the DNS. They found that the energy dissipation rate determined from the *virtual* Doppler spectral widths was systematically smaller than the values determined from the DNS by about a factor of 5. These first DNS simulations of radar measurements show the potential utility of DNS for aiding the interpretion of radar observations.

b. Large eddy simulation

Moeng (1998) presents a recent review of the application of LES to the simulation of atmospheric boundary layers. She points out that the range of scales of interest to the simulation of the atmospheric boundary layer extends from 1 km down to 10^{-3} m. It would take 10^{18} grid points to adequately cover this entire range of scales. Since this is currently well beyond the ability of DNS the methodology of LES has been developed to simulate the dynamics of the resolved scales. For most purposes the largest (resolved) scales are responsible for most of the transport of heat, moisture, and momentum, etc. This is especially true for the convective boundary layer.

LES methodology is reviewed by Lesieur and Metais (1996) and a critical review has been presented by Ma-

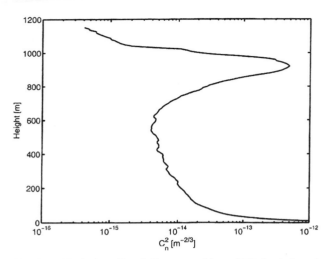

FIG. 6.17. Vertical profile of C_n^2 generated in an LES simulation of the convective boundary layer. After Muschinski et al. (1999b).

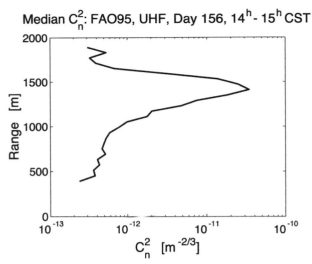

FIG. 6.18. Vertical profile of C_n^2 averaged for 1 h during the afternoon in early June 1995 with the 915-MHz profiler located at the FAO in central Illinois.

son (1994). Lesieur and Metais discuss various methods of treating the unresolved scales of motion. Mason (1994) points out that the development of LES is a natural outgrowth of numerical weather prediction where it is necessary to parameterize subgrid-scale motions. He goes on to show that LES is a powerful tool for turbulence modeling especially for high–Reynolds number flows that are beyond the reach of DNS. He cautions, however, that LES may fail near boundaries and within stable layers where the scales of motion and turbulence are suppressed.

Intercomparisons of different LES models for the convective boundary layer have been published by Nieuwstadt et al. (1993) and for a shear-driven boundary layer by Andrén et al. (1994). These comparisons show generally good agreement but more sensitivity to grid resolution for the neutral boundary layer than the convective boundary layer. Saiki et al. (2000) present recent work on the simulation of the stable boundary layer using LES.

LES has also been used to simulate cloud-topped boundary layers. Simulation of trade wind cumulus clouds by LES is presented in Cuijpers and Duynkerke (1993); simulations of shallow stratocumulus-topped boundary layers are reported in Stevens et al. (1996, 1998). Intercomparison studies of simulations of stratocumulus-topped boundary layers have been reported by Moeng et al. (1996). These studies demonstrate the utility of the LES methodology in improving our understanding of the dynamics of the atmospheric boundary layer.

Another important application of LES germane to this review is the simulation of electromagnetic wave propagation through the turbulent atmosphere. In a pioneering study, Muschinski et al. (1999b) have applied LES to the simulation of wind profiler signals. An example of the simulation of a C_n^2 profile for the convective boundary layer using LES is reproduced in Figure 6.17.

As pointed out by Muschinski et al. (1999b), the resolution of the LES model, which is 8 m in the vertical, 16 m in the horizontal, and 0.8 s, is still too coarse to simulate what the radar sees at the half-wavelength scale of the Bragg component. Nevertheless, in the convective boundary layer the resolution is sufficient to resolve turbulence in the inertial subrange. The Muschinski profile for C_n^2 agrees well in shape to the profile of C_n^2 shown in Fig. 6.18 from the daytime convective boundary layer observed using a 915-MHz profiler at FAO. The top of the convective boundary layer is typically capped by a stable inversion with strong refractivity gradients that produce a strong echo (cf. Fig. 6.4). Similar work using LES as a tool for comparing turbulence measurements by DIAL is reported by Wulfmeyer (1999b). In a recent related study the application of LES to EM wave propagation in the lower atmosphere has been investigated by Wyngaard et al. (2001).

Both DNS and LES are powerful tools in our search to understand and simulate the dynamics of small-scale systems in the atmosphere that often go unresolved in numerical models and inadequately observed in the atmosphere. Nevertheless, as Wyngaard (1998) observes, it is imperative to recognize the need to make new observations for further progress in model development and validation.

8. Observations of waves in the atmosphere

Much progress has been made in the past decade in recognizing various modes of atmospheric waves and identifying their sources. It has been known for some time that the wave motions in the lower atmosphere have a profound influence on the dynamics of the middle and upper atmosphere (Hines 1963; Hodges 1967; Lindzen 1981; McLandress 1998). Indeed, as waves propagate into the middle and upper atmosphere their amplitude increases as atmospheric density decreases. This fact is responsible for the dominant role that waves play in the middle and upper atmosphere. Also, as waves propagate to higher altitudes the waves of short vertical wavelength become unstable and break. The breaking of these waves produces a stress on the winds of the middle and upper atmosphere.

In this section we review briefly advances in our knowledge of the structure and dynamics of waves in the lower atmosphere as gained from observations. Most waves discussed in this section will be internal gravity wave including low-frequency inertia–gravity waves that develop in the rotating stable atmosphere. Internal gravity waves can be found almost everywhere except the convective boundary layer where unstable thermal stratification precludes their existence. Planetary-scale waves and tides will not be considered here. For a comprehensive description of waves in the atmosphere see Gossard and Hooke (1975).

Much has been learned about internal gravity waves from continuous observations of wind profilers. The profilers observe simultaneously and continuously at many heights, which makes them ideal for detecting waves. The short-period internal waves are especially well resolved in vertical-velocity observations, which are routinely made by profilers. Monochromatic waves are often seen in profiler observations at many locations. They appear typically as time series of vertical motions that are in phase over a large height range.

a. Observations of monochromatic waves

Ecklund et al. (1982) reported observations of waves in vertical motions in the lee of the Rocky Mountains in Colorado using wind profiler observations. These authors showed a clear relationship between the magnitude of vertical-velocity fluctuations and the strength of winds over the nearby mountains. Indeed, when the winds blew from the east where the terrain is relatively flat, much of the vertical wind variability ceased. The Flatland Atmospheric Observatory radar was constructed in Illinois to contrast the vertical velocity variability over flat terrain with what was observed in Colorado.

Several studies have been made using profilers to identify the dynamics of internal gravity waves in the lower atmosphere. Three spaced profilers were operated in southern France during the Alpine Experiment (ALPEX; Ecklund et al. 1985). The profilers were located on the southern coast of France in relatively flat terrain downwind of the Alps. During mistral winds, the magnitude of vertical wind variability was clearly seen to be related to the strength of winds and their direction relative to the mountains. The existence of multiple profiler sites, separated by 5–6 km, provided an opportunity to explore the propagation characteristics of the waves. Data collected in the ALPEX campaign were carefully analyzed by Carter et al. (1989). Only a few cases of

monochromatic waves could be identified at each of the sites and cross correlated to yield wave parameters. Typical horizontal wavelengths were 10–20 km and horizontal phase speeds were in the range of 4–15 m s^{-1}.

Balsley and Carter (1989) examined waves associated with topography as seen in vertically directed profiler data from Christmas Island, Kiribati, and Pohnpei, Federated States of Micronesia (FSM). Christmas Island is a very flat atoll located in the Republic of Kiribati south of the Hawaiian Islands and two degrees north of the equator. Pohnpei is located in the FSM in the western Pacific and possesses substantial terrain with a maximum elevation close to 800 m. Balsley and Carter showed vertical motions at Pohnpei that could be related to the island's considerable topography and demonstrated that similar wave motions were lacking at Christmas Island. They also demonstrated a critical layer relationship between the winds aloft and low-level trade winds at Pohnpei that governed whether the waves were present aloft.

Before the advent of profilers, detailed case studies of wave events that reveal their vertical structure were rare. Ralph et al. (1993) analyzed a mesoscale ducted gravity wave observed during FRONTS 84 in southwestern France. Vertical velocities measured by a 50-MHz profiler and surface pressure fluctuations measured by a network of surface pressure stations resolved a trapped 90-min period wave on 19 June 1984 that was found to have an approximate 76-km wavelength. The wave was trapped within a duct formed between the ground and a stable lower-tropospheric layer bounded above by an unstable or near-neutral layer. In addition, there was a critical layer (where the mean wind speed was equal to the ground-relative phase velocity) present near the top of the duct.

Liziola and Balsley (1998) have investigated the occurrence of quasi-horizontally propagating waves in the troposphere using the Piura radar, which is part of the National Oceanic and Atmospheric Administration–Colorado University Trans-Pacific Profiler Network (TPPN). Figure 6.19 contains an example of multi-height time series of vertical velocities observed at Piura on 25 January 1990. In this example the waves have peak amplitude near 1.5 m s^{-1} and period close to 10 min. Note that there is very little or no phase change with altitude, which is typical of the waves seen in the profiler observations. It is very likely that these are trapped waves that are propagating horizontally. Liziola and Balsley (1998) constructed a closely spaced antenna array with spacing of a few hundred meters and were able to determine phase velocities of propagating waves up to a few meters per second. Horizontal wavelengths were found to be in the range of 1–3 km. They also investigated the climatology of the waves and found that they were most common during August when easterlies (from the Andes) were strongest.

The observations of vertical motions using wind profilers have provided much new information on the dy-

Vertical Velocity

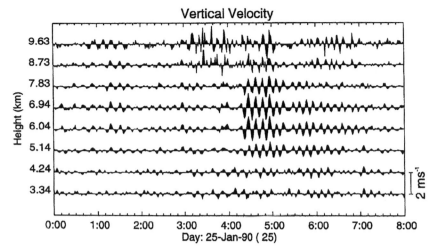

FIG. 6.19. Short-period quasi-sinusoidal internal gravity waves observed 25 Jan 1990 using the 50-MHz profiler at Piura, Peru, near the Andes in northern Peru. After Liziola and Balsley (1990).

namics of wave motions in the atmosphere. The relatively frequent occurrence of quasi-periodic disturbances at locations near significant topography and the relative lack of such quasi-periodic disturbances at places like Christmas Island and at the Flatland Atmospheric Observatory (Nastrom et al. 1990) in central Illinois implicates orography as the cause of these disturbances. These facts suggest the development of trapped nonstationary waves associated with wind flowing over mountains and other complex terrain.

Röttger (2000) has reviewed profiler observations of atmospheric waves over mountainous areas in many locations and has discussed many of the issues concerning the interpretation of observations made by profilers using the Doppler beam swinging (DBS) method. He

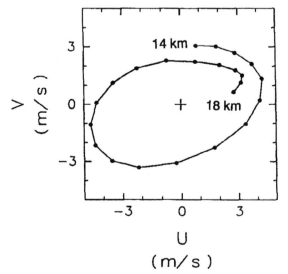

FIG. 6.20. Inertia-gravity wave observed by the MU radar near Shigaraki, Japan. After Sato (1993).

points out that oblique measurements made with the DBS method are hard to interpret and that there are advantages in using spaced antenna (SA) techniques in observing lee waves. Substantial progress in understanding the dynamics of mountain lee waves has been reported in a series of papers using observations from the VHF radar located at Aberystwyth (Mitchell et al. 1994; Prichard et al. 1995; Thomas et al. 1992, 1999; Worthington 1998, 1999a,b; Worthington and Thomas 1996, 1997a,b, 1998; Worthington et al. 1999). These authors have made the most comprehensive observational study of lee-wave dynamics to date.

Other important contributions to our understanding of the dynamics of waves in the vicinity of complex terrain have been made by Ralph et al. (1992, 1997, 1999), Caccia et al. (1997a,b), and Caccia (1998). Nance and Durran (1997, 1998) have modeled the dynamics of trapped mountain lee waves taking into account mean flow variability and non-linear wave dynamics. They have performed two-dimensional wave simulations in idealized time-varying background flows.

In addition to the short-period internal gravity waves discussed above, longer-period inertia-gravity waves have been observed at many locations. The inertia-gravity waves are long-period waves with intrinsic period somewhat shorter than the local inertial period. They are most clearly identified in hodographs of wind soundings, which reveal a characteristic turning of the horizontal wind with height (Cadet and Teitelbaum 1979; Gill 1982; Sato 1989, 1993, 1994; Chan et al. 1991; Danielsen et al. 1991). An example of an inertia–gravity wave observed by the MU radar in Japan is reproduced from Sato (1993) in Fig. 6.20. Figure 6.20 reproduces an inertia-gravity wave observed by the MU radar during the passage of Typhoon Kelly in October 1987. The wave parameters found were a period of 20 h, a vertical wavelength of 2.7 km, and a horizontal wavelength of 300 km. The inertia-gravity waves are often confused with lee waves (Cornish and Larsen 1989; Röttger 2000). Some controversy concerning the origin of these waves (Hines 1989) has been settled by Cho (1995) in favor of inertia-gravity waves. Many studies have shown that inertia-gravity waves are commonly observed in profiler wind soundings at many locations (Cornish and Larsen 1989; Thomas et al. 1992; Sato 1994; Riggin et al. 1995; Yamanaka et al. 1996; Sato et al. 1997).

9. Sources of waves in the atmosphere

Atmospheric waves are preferred modes of atmospheric response to a variety of forcing mechanisms. Until recently very little was known about the sources of the observed waves in the atmosphere except that they were likely to result from flow over complex terrain, Kelvin–Helmholtz instability, convection (especially deep penetrative convection), and geostrophic adjustment in large-scale atmospheric flows.

The role of topography in the generation of gravity waves has been documented by several studies in addition to the work reviewed in the previous section. Nastrom et al. (1987) and Jasperson et al. (1990) examined the relationship of the variance of mesoscale winds measured by commercial aircraft during the Global Atmospheric Sampling Program (GASP) with the underlying surface. The variance was found to be highly correlated with the roughness of the topography. Variance in the troposphere was also related to wind speed. The variance of wind and temperature on scales of 4–80 km was found to be as much as 6 times greater over mountains than over oceans. At larger scales, variance was also greater over mountains compared to oceans but by a smaller amount.

Nastrom and Fritts (1992) extended the earlier work with two case studies aimed at an improved understanding of the linkage between wind variance observed in the GASP flights, underlying terrain, and meteorological conditions. They found that, while in most instances variances were greater over mountains than over plains and oceans, in some cases the variances were not enhanced by topography. Instances of small variances over the mountains were clearly related to regimes of weak wind over the local terrain and high static stability in the lower troposphere that would suppress convection. These results can be compared with the profiler results from Platteville (Ecklund et al. 1982) and Flatland results of Nastrom et al. (1990). For Platteville the vertical velocity variance was directly related to wind speed at 500 hPa and at Flatland the vertical velocity variance was most closely related to stability in the lower troposphere independent of wind speed.

In the aircraft data the variability was either due to waves or stratified turbulence (Nastrom et al. 1987) so a less ambiguous determination of the influence of topography on the generation of gravity waves would be

obtained from an examination of vertical motions. This has been done at many locations using profilers as reviewed in the previous section. Nastrom et al. (1990) investigated sources of gravity wave activity seen in the vertical velocities at Flatland. Generally, the magnitude of vertical velocity variance was found to be less than at Platteville in the lee of the Colorado Rockies. Observations from Flatland have provided an opportunity to investigate the contributions of other sources to gravity wave generation over very flat terrain. The occurrence of large vertical velocity variance in Flatland observations is related to specific events such as frontal passages or thunderstorms. Statistical studies showed variances systematically larger under cloudy skies than under clear skies, suggesting a link with convection. Variances under these disturbed conditions were typically 50% greater than under clear undisturbed conditions where the variance was typically in the range of $100–300 \text{ cm}^2 \text{ s}^{-2}$.

Fritts and Nastrom (1992) used the GASP dataset to investigate the relationship of enhanced wind and temperature variability to fronts, convection, and jet streams. In case studies they were able to relate enhanced variability with one or more sources and found enhancements as large as one to two orders of magnitude extending to scales of 64 km or more. Their results are summarized statistically in Fig. 6.21. Horizontal velocity variances in apparently source-free regions are typically near $0.1–0.4 \text{ m}^2 \text{ s}^{-2}$ and temperature variance is less than 0.4 K^2.

Fritts and Luo (1992) in a theoretical study have examined the excitation of inertia-gravity waves by the geostrophic adjustment process. Their work builds on earlier studies by many authors and concludes that the geostrophic adjustment process is a likely source for low-frequency internal gravity waves.

It is widely recognized that convection is a major source of waves in the atmosphere (Einaudi et al. 1987; Doviak et al. 1991; Rottman et al. 1992; Pfister et al. 1993; Sato et al. 1995; Alexander et al. 2000; Piani et al. 2000). Alexander et al. (1995) and Alexander and Holton (1997) have modeled the gravity wave response above deep convection in a squall-line simulation. Figure 6.22 is taken from Alexander et al. (1995) and shows gravity waves present in the vertical velocity of an anvil structure in a convective system. Observations of anvil structures also show evidence of gravity wave disturbances as shown in Fig. 6.23. Yang and Houze (1995) also show trapped gravity wave modes that are present in a midlatitude squall-line simulations.

Near-surface outflow from deep convective storm systems often creates a solitary wave that can be observed by radar and lidar (Doviak and Ge 1984; Doviak et al. 1989; Fulton et al. 1990; Doviak et al. 1991; Koch et al. 1991). Solitary waves are long nonlinear waves that are trapped by an inversion and propagate intact for long distances (Christie 1989). Trapping mechanisms for the low-level internal gravity waves are discussed by Crook

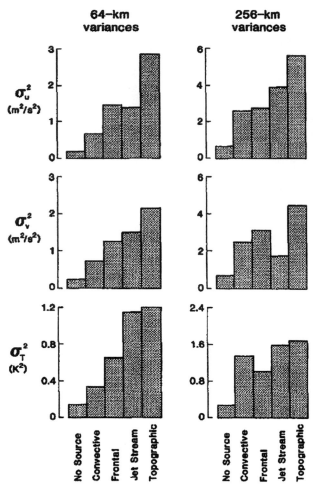

FIG. 6.21. Magnitude of horizontal-velocity variances as a function of source in 64- and 256-km flight legs from commercial aircraft collected during GASP. After Fritts and Nastrom (1992).

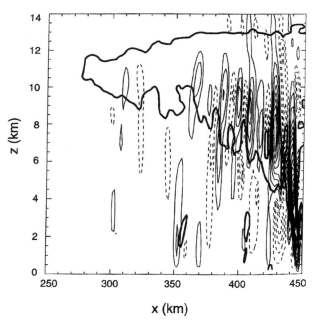

FIG. 6.22. Gravity waves simulated above deep convection in a squall-line simulation. Contours show vertical velocity. After Alexander et al. (1995).

(1988), who points out that while trapping requires an inversion it is aided by opposing winds. The theory for solitary waves in the atmosphere has been reviewed by Rottman and Einaudi (1993). They develop the theory for two classes of solitary waves. The first class of solitary waves is confined to the lowest few kilometers of the troposphere possessing horizontal scales of a few kilometers and phase speeds of the order of 10 m s^{-1}. The second class of solitary waves occupies the entire troposphere and has horizontal scales on the order of 100 km and phase speeds on the order of 25–100 m s^{-1}. The solitary waves may play an important role in the propagation of squall lines (Carbone et al. 1990; Crook et al. 1990; Koch et al. 1993; Trexler and Koch 2000) with important implications for mesoscale predictability (Koch and O'Handley 1997).

Another class of waves known as *convection waves* develops over convective boundary layers in clear weather. While internal gravity waves are precluded from propagating within the hydrostatically unstable atmosphere, *convection waves* often develop in the stable free troposphere above the convective boundary layer. These waves have been seen in aircraft observations (Kuettner et al. 1987; Hauf 1993) and are observed by profilers as reported by Gage et al. (1989) and reproduced in Fig. 6.24. The observations reproduced in Fig. 6.24 were made in Liberal, Kansas, on 29 June 1985 on a clear day during the Preliminary Regional Experiment for Stormscale Operational and Research Meteorology (PRE-STORM) campaign. Liberal, Kansas, is located in a region of very flat terrain far from any mountains. The observed waves appear to fill the entire free troposphere and possess little or no phase progression with altitude, suggesting that these are trapped modes. Clark et al. (1986) have simulated *convection waves* in a model. *Convection waves* appear to develop when the convective boundary layer possesses a corrugated top and winds flow over this surface under favorable synoptic conditions.

As can be seen from the results reviewed in sections 8 and 9, substantial progress has been made in observing and understanding the generation of internal gravity waves in the atmosphere. Recent progress in specifying the climatology of gravity waves in the troposphere and stratosphere has been reported by Murayama et al. (1994), Tsuda et al. (1994a,b), Allen and Vincent (1995), Ogino et al. (1995), Gravilov and Fukao (1999, 2001), Alexander and Vincent (2000), Tsuda et al. (2000), Vincent and Alexander (2000), Dhaka et al. (2001), and Alexander et al. (2002). In an important paper Alexander (1998) has considered the filtering effects of winds and stability on vertically propagating waves in interpreting observations of atmospheric

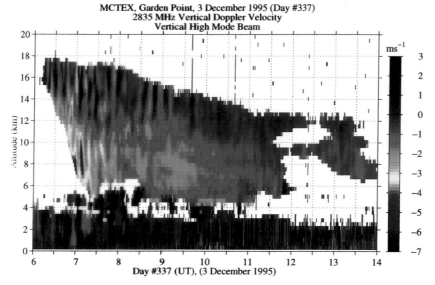

FIG. 6.23. Observations of quasi-periodic fluctuations of vertical velocity observed on 3 Dec 1995 by a vertically pointing 2835-MHz profiler at Garden Point, Australia, during MCTEX. After Ecklund et al. (1999).

waves from diverse sources. Transport of momentum by gravity waves has been investigated by Fritts et al. (1990), Kim and Mahrt (1992), Fritts and VanZandt (1993), Alexander and Pfister (1995), Alexander and Rosenlauf (1996), and Chang et al. (1997). Collectively, the research of the past decade has greatly advanced our ability to parameterize waves in global models (Bacmeister 1993; Alexander and Dunkerton 1999).

10. The spectrum of atmospheric waves

In previous sections we have seen that atmospheric waves can arise from many sources. Of course, we can most easily identify monochromatic waves and other large-amplitude wave disturbances in observations. However, waves are ubiquitous and as they interact they lose their identity with their sources and create a background spectrum of waves that can be nearly universal in character (Staquet and Sommeria 2002). The situation is similar in some respects to what is found in the ocean except that the speed of ocean currents is much smaller than atmospheric winds and the ocean has a definite upper boundary that reflects waves. VanZandt (1982) pointed out that a spectrum of waves similar to the Garrett–Munk spectrum found in the ocean should be present in the atmosphere. In this section we review the empirical evidence supporting the existence of a nearly universal spectrum of waves in the atmosphere. We focus attention here on the spectrum of vertical motions.

With a few notable exceptions, the horizontal winds in the atmosphere are several orders of magnitude larger than vertical velocities. While typical instantaneous vertical motions can be as large as 1 m s^{-1} in the free troposphere, vertical velocity standard deviations are typically at most a few tens of centimeters per second

(Ecklund et al. 1986; VanZandt et al. 1991; Williams et al. 2000) and long-term mean vertical motions averaged over many hours rarely exceed a few centimeters per second.

The spectrum of vertical motions has been considered by Ecklund et al. (1985, 1986) based on observations with vertically directed wind profilers in a number of locations. The characteristic shape of the vertical-velocity spectrum is fairly flat with a peak near the Brunt–Väisälä period. Indeed, Ecklund et al. (1986) showed that the amplitude of the spectrum is less in the stratosphere than it is in the troposphere as expected for internal waves. The spectrum of vertical motions as observed by profilers appears to have a fairly universal shape and amplitude, under undisturbed conditions and far from sources, very similar to what is found in the ocean for the Garrett–Munk spectrum. However, there are substantial differences in the amplitude of the spectrum of vertical motions in the atmosphere near sources of waves such as winds blowing over complex topography.

In order to investigate the influence of complex terrain on vertical motions, the Flatland radar was constructed in central Illinois. At Flatland the vertical-velocity spectrum was found to be very similar to that found in other locations of complex terrain under undisturbed conditions of low winds or winds blowing over flat terrain. Figure 6.25 illustrates the features of the vertical-velocity spectrum. As reported by VanZandt et al. (1991), a flat spectrum is seen at FAO under all wind conditions, although the amplitude of the spectrum does vary with wind speed. VanZandt and colleagues were able to demonstrate using a spectral model that the amplitude variations seen in the Flatland spectra are explicable as-

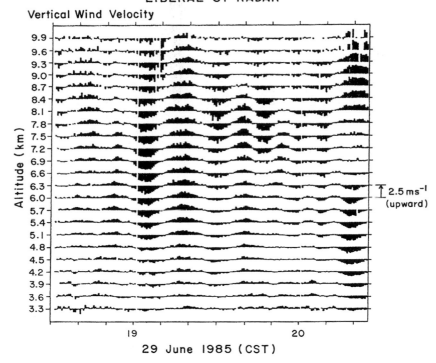

FIG. 6.24. Convection waves observed at Liberal, Kansas, during PRE-STORM on the afternoon of 29 Jun 1985 in multiheight time series of vertical velocities. After Gage et al. (1989).

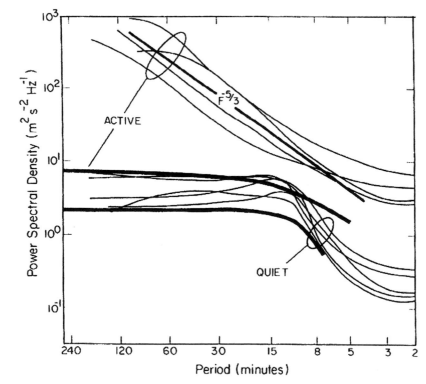

FIG. 6.25. Spectra of vertical velocities observed at several locations including the FAO. After VanZandt et al. (1991).

suming they are indeed a spectrum of waves. This success in modeling the dependence of the spectrum on wind speed by incorporating Doppler shifting effects considerably strengthens the interpretation of the vertical velocity spectrum as a wave spectrum. We consider next the behavior of the vertical velocity spectrum under disturbed conditions near mountains.

Also shown in Fig. 6.25 is the spectrum under disturbed conditions of strong winds blowing over complex terrain. Under these circumstances the spectrum is substantially modified with a greatly enhanced magnitude and a slope approaching $-5/3$. Worthington and Thomas (1996, 1998) have shown that similar spectra are observed during mountain wave events at Aberystwyth. Worthington and Thomas (1996) show that the frequency spectra below a critical layer where turbulence is enhanced shows the $-5/3$ spectral slope while in the same event above the critical level the spectra are flat.

Several authors have considered possible explanations for the occurrence of the $-5/3$ spectral slope in the frequency spectra of vertical velocity during disturbed mountain wave conditions. Gage and Nastrom (1990) hypothesized that quasi-horizontal motions along tilted isentropic surfaces might be responsible for these frequency spectra owing to the fact that the vertical-velocity frequency spectra resembled the horizontal velocity spectra observed at the same time albeit with reduced amplitude. However, Worthington and Thomas argued that at Aberystwyth the frequency spectra are truly vertical motions of the mountain waves and not due to quasi-horizontal motions created by the tilting of isentropic surfaces. Instead, they show that a simple model with a pattern of mountain waves that moves stochastically with respect to the ground produces spectra that resemble the observed frequency spectra. They suggest that a computer model in which the background wind is varied consistent with a $-5/3$ spectral dependence should be used to further examine the spectral dependence of the vertical velocity.

While atmospheric vertical-velocity spectra are almost certainly due to waves, it is not obvious that horizontal-velocity spectra are also due to waves. In a series of papers Nastrom and Gage (1984, 1985) examined the horizontal-velocity spectra measured by commercial aircraft and Gage and Nastrom (1985, 1986) considered the interpretation of the observed spectra. The gravity wave spectral model advanced by VanZandt (1982, 1985) and Scheffler and Liu (1985) provides a way to predict the horizontal-velocity spectrum due to waves when the vertical wave spectrum is known. Gage (1990) compared the observed horizontal-velocity spectra with the horizontal-velocity spectrum due to waves and argued that there was more energy in the observed spectrum than was consistent with the gravity wave model fit to the observed vertical velocity spectrum. Högstrom et al. (1999) also examined the frequency spectrum of vertical velocity from aircraft and compared it to horizontal-velocity spectrum. They concluded that the ver-

tical-velocity spectrum was due to waves and that the horizontal-velocity spectrum was due to quasi-two-dimensional turbulence. This topic was also considered in section 6.

Internal gravity waves that propagate vertically grow in amplitude and eventually break. The waves that possess small vertical scales break at the lowest heights and only the waves with longer vertical wavelengths are able to propagate into the upper atmosphere. Vertical wavenumber spectra have been reported by many authors (Smith et al. 1987; Fritts et al. 1988; Tsuda et al. 1989; VanZandt and Fritts 1989). Example spectra (Fritts et al. 1988) from the MU radar in Japan are shown in Fig. 6.26. Viewed as energy density these spectra fall off as m^{-3}, where m is vertical wavelength. The bottom panels show the same spectra plotted as in energy content form and clearly show dominant vertical scales in the range of 2–3 km.

11. Concluding remarks

In this review we have considered some of the recent advances in our understanding of turbulence and waves in the atmosphere. We have emphasized observations of turbulence and waves that have contributed to an improved understanding of the structure and dynamics of atmospheric turbulence and waves. Considering Dave Atlas's many pioneering contributions to radar studies of the *clear* atmosphere, it is most appropriate to review these advances in a volume that celebrates the many scientific contributions of Dave Atlas.

During the past decade substantial advances have been made in our ability to quantitatively measure the intensity of atmospheric turbulence. For the first time it has been possible to unambiguously determine the magnitude of the refractivity turbulence structure-function parameter and the eddy dissipation rate using radar techniques. While some uncertainties remain, on-balance radar measurements of turbulence compare favorably for the most part with coordinated in situ measurements. The remaining uncertainties are concerned with sampling issues related to intermittency.

One new area for turbulence research that appears particularly promising is the interaction of turbulence in cloud microphysics. Turbulence is an important process in clouds yet it is still difficult to make measurements that would help to improve our understanding of cloud microphysics and our ability to retrieve cloud and precipitation-sized hydrometeor distributions. Some recent advances in measurements of turbulence in clouds suggest that future research in this area could be very timely.

Another area of active research attracting considerable attention is high-resolution measurements using novel radar and in situ techniques. In their recent summary of a turbulence workshop held in Boulder, Colorado, Muschinski and Lenschow (2001) point toward the need for improved understanding of submeter-scale

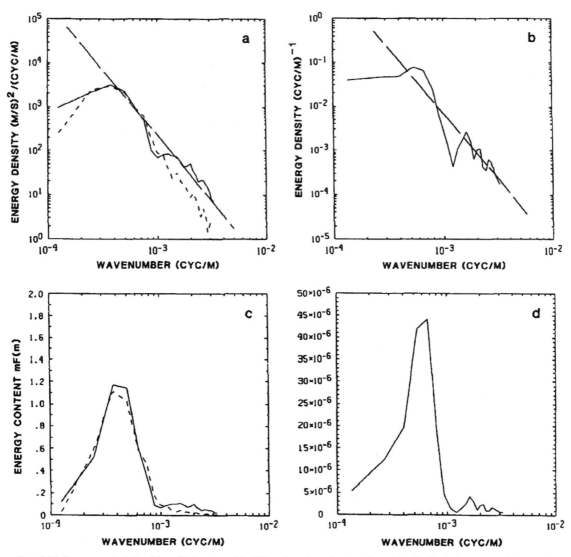

FIG. 6.26. Power spectral densities of (a) eastward (solid) and northward (dashed) radial velocity, (b) normalized temperature between 13 and 20.5 km for 24–25 Oct. The corresponding area-preserving spectra are (c) radial velocity and (d) normalized temperature. Long dashed lines show saturated PSDs. Observations are from the MU radar observatory near Shigaraki, Japan.

turbulence. For example, our lack of understanding of these smallest scales of turbulence has implications for the quality of profiler observations.

Owing to the fact that the analytical description of waves in the atmosphere is more advanced than for atmospheric turbulence and that the new tools such as Doppler radar profilers provide continuous information on velocities simultaneously over a range of altitudes, there have been major advances in our understanding of the role of waves in atmospheric dynamics. Wave drag, for example, is now widely recognized as a fundamental process that must be parameterized in numerical models.

Internal gravity waves are found to be ubiquitous and to conform well to theory. Even the spectrum of waves is understood as a fairly universal spectrum that is robust

at least as can be judged from the vertical-velocity spectrum and the vertical wavenumber spectrum. Other processes such as stratified turbulence are important at least in some places and may play an important role in the creation of layered structure. There is still some ambiguity in the differentiation of waves and turbulence in horizontal-velocity fluctuations.

While major advances have taken place in our ability to simulate turbulence and waves, much remains to be done. Direct numerical simulation is limited in the domain it can be applied to especially in large Reynolds number turbulence. Large eddy simulations have provided great insight into the dynamics of the convective atmospheric boundary layer. Both DNS and LES show considerable potential utility for improved interpretation of radar observations.

Given the attention that mesoscale meteorology and climate is receiving, there has been a decrease in the amount of research being done on turbulence. Yet important problems remain that present significant intellectual challenges and promise to yield substantial useful results using the sophisticated tools that are now becoming available.

Acknowledgments. The authors thank several scientists who provided materials for this review and who discussed recent developments with them. We especially wish to thank Wayne Angevine, Phil Chilson, Wallace Clark, Joe Werne, and Christopher Williams who provided figures used in the review; and Wayne Angevine, Greg Nastrom, Phil Chilson, Ben Balsley, and Andreas Muschinski who discussed recent developments in their research. The authors also thank Debe Fisher for scanning numerous figures and Barbara Herrli for her diligent work on word processing and checking references.

REFERENCES

Afanasyev, Ya. D., and W. R. Peltier, 2001: Numerical simulations of internal gravity wave breaking in the middle atmosphere: The influence of dispersion and three-dimensionalization. *J. Atmos. Sci.,* **58,** 132–153.

Alexander, M. J., 1998: Interpretations of observed climatological patterns in stratospheric gravity wave variance. *J. Geophys. Res.,* **103,** 8627–8640.

——, and L. Pfister, 1995: Gravity wave momentum flux in the lower stratosphere over convection. *Geophys. Res. Lett.,* **22,** 2029–2032.

——, and K. H. Rosenlauf, 1996: Nonstationary gravity wave forcing of the stratospheric zonal mean wind. *J. Geophys. Res.,* **101,** 23 465–23 474.

——, and J. R. Holton, 1997: A model study of zonal forcing in the equatorial stratosphere by convectively induced gravity waves. *J. Atmos. Sci.,* **54,** 408–419.

——, and T. J. Dunkerton, 1999: A spectral parameterization of mean-flow forcing due to breaking gravity waves. *J. Atmos. Sci.,* **56,** 4167–4182.

——, and R. A. Vincent, 2000: Gravity waves in the tropical lower stratosphere: A model study of seasonal and interannual variability. *J. Geophys. Res.,* **105,** 17 983–17 993.

——, J. R. Holton, and D. R. Durran, 1995: The gravity wave response above deep convection in a squall line simulation. *J. Atmos. Sci.,* **52,** 2212–2226.

——, J. H. Beres, and L. Pfister, 2000: Tropical stratospheric gravity wave activity and relationships to clouds. *J. Geophys. Res.,* **105,** 22 299–22 309.

——, T. Tsuda, and R. A. Vincent, 2002: Latitudinal variations of observed gravity waves with short vertical wavelengths. *J. Atmos. Sci.,* **59,** 1394–1404.

Allen, S., and R. Vincent, 1995: Gravity wave activity in the lower atmosphere: Seasonal and latitudinal variations. *J. Geophys. Res.,* **100,** 1327–1350.

Andreassen, Ø., C. E. Wasberg, D. C. Fritts, and J. R. Isler, 1994: Gravity wave breaking in two and three dimensions. 1. Model description and comparison of two dimensional evolutions. *J. Geophys. Res.,* **99,** 8095–8108.

——, P. O. Hvidsten, D. C. Fritts, and S. Arndt, 1998: Vorticity dynamics in a breaking internal gravity wave. Part 1. Initial instability evolution. *J. Fluid Mech.,* **367,** 27–46.

Andrén, A., A. Brown, J. Graf, P. J. Mason, C.-H. Moeng, F. T. M. Nieuwstadt, and U. Schumann, 1994: Large eddy simulation of

a neutrally stratified boundary layer: A comparison of four computer codes. *Quart. J. Roy. Meteor. Soc.,* **120,** 1457–1484.

Angevine, W. M., A. W. Grimsdell, L. M. Hartten, and A. C. Delany, 1998: The Flatland boundary layer experiments. *Bull. Amer. Meteor. Soc.,* **79,** 419–431.

Atlas, D., 1959: Meteorological "angel" echoes. *J. Meteor.,* **16,** 6–11.

——, 1964: Advances in radar meteorology. *Advances in Geophysics,* Vol. 10, Academic Press, 317–478.

——, Ed.,1990: *Radar in Meteorology.* Amer. Meteor. Soc., 806 pp.

——, K. R. Hardy, K. M. Glover, I. Katz, and T. G. Konrad, 1966a: Tropopause detected by radar. *Science,* **153,** 1110–1112.

——, ——, and K. Naito, 1966b: Optimizing the radar detection of clear air turbulence. *J. Appl. Meteor.,* **5,** 450–460.

——, J. I. Metcalf, J. H. Richter, and E. E. Gossard, 1970: The "birth" of CAT and microscale turbulence. *J. Atmos. Sci.,* **27,** 903–913.

Avila, R., J., Vernin, and E. Masciadri, 1997: Whole atmospheric-turbulence profiling with scidar. *Appl. Opt.,* **36,** 7898–7905.

Bacmeister, J. T., 1993: Mountain wave drag in the stratosphere and mesosphere inferred from observed winds and a simple mountain-wave parameterization scheme. *J. Atmos. Sci.,* **50,** 377–399.

Balsley, B. B., and K. S. Gage, 1982: On the use of radar for operational wind profiling. *Bull. Amer. Meteor. Soc.,* **63,** 1009–1018.

——, and D. A. Carter, 1989: Mountain waves in the tropical Pacific atmosphere: A comparison of vertical wind fluctuations over Pohnpei and Christmas Island using VHF wind profilers. *J. Atmos. Sci.,* **46,** 2698–2715.

——, M. L. Jensen, and R. G. Frehlich, 1998: The use of state-of-the-art kites for profiling the lower atmosphere. *Bound.-Layer Meteor.,* **87,** 1–25.

Banakh, V. A., I. N. Smalikho, F. Kopp, and C. Werner, 1999: Measurements of turbulence energy dissipation rate with a CW Doppler lidar in the atmospheric boundary layer. *J. Atmos. Oceanic Technol.,* **16,** 1044–1061.

Bartello, P., 1995: Geostrophic adjustment and inverse cascades in rotating stratified turbulence. *J. Atmos. Sci.,* **52,** 4410–4428.

——, 2000: Potential vorticity, resonance and dissipation in rotating, convective turbulence. *Fluid Mech. Astrophys. Geophys.,* **8,** 309–322.

Battan, L. J., 1973: *Radar Observation of the Atmosphere.* University of Chicago Press, 324 pp.

Bertin, F., J. Barat, and R. Wilson, 1997: Energy dissipation rates, eddy diffusivity, and the Prandtl number: An *in situ* experimental approach and its consequences on radar estimate of turbulent parameters. *Radio Sci.,* **32,** 791–804.

Browning, K., 1971: Structure of the atmosphere in the vicinity of large-amplitude Kelvin–Helmholtz billows. *Quart. J. Roy. Meteor. Soc.,* **97,** 283–299.

Caccia, J. L., 1998: Lee wave vertical structure monitoring using height–time analysis of VHF ST radar vertical velocity data. *J. Appl. Meteor.,* **37,** 530–543.

——, B. Benech, and V. Klaus, 1997a: Space–time description of nonstationary trapped lee waves using ST radars, aircraft, and constant volume balloons during the PYREX experiment. *J. Atmos. Sci.,* **54,** 1821–1833.

——, M. Crochet, and K. Saada, 1997b: ST radar evaluation of the standard deviation of the air vertical velocity perturbed by the local orography. *J. Atmos. Solar Terr. Phys.,* **59,** 1127–1131.

Cadet, D., and H. Teitelbaum, 1979: Observational evidence of internal inertia-gravity waves in the tropical stratosphere. *J. Atmos. Sci.,* **36,** 892–907.

Carbone, R. E., J. E. Conway, N. A. Crook, and M. W. Moncrieff, 1990: The generation and propagation of a nocturnal squall line. Part I: Observations and implications for mesoscale predictability. *Mon. Wea. Rev.,* **118,** 26–49.

Carter, D. A., B. B. Balsley, W. L. Ecklund, K. S. Gage, R. Garello, and M. Crochet, 1989: Investigations of internal gravity waves using three vertically directed closely spaced wind profilers. *J. Geophys. Res.,* **94,** 8633–8642.

——, K. S. Gage, W. L. Ecklund, W. M. Angevine, P. E. Johnston, A. C. Riddle, J. Wilson, and C. R. Williams, 1995: Developments in lower tropospheric wind profiling at the NOAA Aeronomy Laboratory. *Radio Sci.,* **30,** 977–1001.

Caulfield, C. P., and W. R. Peltier, 1994: Three dimensionalization of the stratified mixing layer. *Phys. Fluids,* **6,** 3803–3805.

Chan, K. R., S. G. Scott, S. W. Bowen, S. E. Gaines, E. F. Danielsen, and L. Pfister, 1991: Horizontal wind fluctuations in the stratosphere during large-scale cyclogenesis. *J. Geophys. Res.,* **96,** 17 425–17 432.

——, J. Dean-Day, S. W. Bowen, and T. P. Bui, 1998: Turbulence measurements by the DC-8 meteorological measurement system. *Geophys. Res. Lett.,* **25,** 1355–1358.

Chang, J. L., S. K. Avery, A. C. Riddle, S. E. Palo, and K. S. Gage, 1997: First results of tropospheric gravity wave momentum flux measurements over Christmas Island. *Radio Sci.,* **32,** 727–748.

Chilson, P. B., A. Muschinski, and G. Schmidt, 1997: First observations of Kelvin–Helmholtz billows in an upper level jet using VHF frequency domain interferometry. *Radio Sci.,* **32,** 1149–1160.

——, R. D. Palmer, A. Muschinski, D. A. Hooper, G. Schmidt, and H. Steinhagen, 2001: SOMARE-99: A demonstrational field campaign for ultrahigh-resolution VHF atmospheric profiling using frequency diversity. *Radio Sci.,* **36,** 695–707.

Cho, J. Y. N., 1995: Inertio-gravity wave parameter estimation from cross-spectral analysis. *J. Geophys. Res.,* **100,** 18 727–18 737.

——, and E. Lindborg, 2001: Horizontal velocity structure functions in the upper troposphere and lower stratosphere. 1. Observations. *J. Geophys. Res.,* **106,** 10 223–10 232.

Cho, J. Y. N., and Coauthors, 1999a: Horizontal wavenumber spectra of winds, temperature, and trace gases during Pacific Exploratory Missions. Part I: Climatology. *J. Geophys. Res.,* **104,** 5697–5716.

Cho, J. Y. N., R. E. Newell, and J. D. Barrick, 1999b: Horizontal wavenumber spectra of winds, temperature, and trace gases during the Pacific Exploratory Missions Part 2: Gravity waves, quasi-two-dimensional turbulence, and vortical modes. *J. Geophys. Res.,* **104,** 16 297–16 308.

Christie, D. R., 1989: Long nonlinear waves in the lower atmosphere. *J. Atmos. Sci.,* **46,** 1462–1490.

Clark, T. L., T. Hauf, and J. P. Kuettner, 1986: Convectively forced internal gravity waves: Results from two-dimensional numerical experiments. Quart. *J. Roy. Meteor. Soc.,* **112,** 899–925.

Cohn, S. A., 1994: Investigations of the wavelength dependence of radar backscatter from atmospheric turbulence. *J. Atmos. Oceanic Technol.,* **11,** 225–238.

——, 1995: Radar measurements of turbulent eddy dissipation rate in the troposphere: A comparison of techniques. *J. Atmos. Oceanic Technol.,* **12,** 85–95.

Coleman, G. N., J. H. Ferziger, and P. R. Spalart, 1990: A numerical study of the turbulent Ekman layer. *J. Fluid Mech.,* **213,** 313–348.

Cornish, C. R., and M. F. Larsen, 1989: Observations of low-frequency gravity waves in the lower stratosphere over Arecibo. *J. Atmos. Sci.,* **46,** 2428–2439.

Crane, R. K., 1980: A review of radar observations of turbulence in the lower stratosphere. *Radio Sci.,* **15,** 177–193.

Crook, N. A., 1988: Trapping of low-level internal gravity waves. *J. Atmos. Sci.,* **45,** 1533–1541.

——, R. E. Carbone. M. W. Moncrieff, and J. W. Conway, 1990: The generation and propagation of a nocturnal squall line. Part II: Numerical simulations. *Mon. Wea. Rev.,* **118,** 50–65.

Cuijpers, J. W. M., and P. G. Duynkerke, 1993: Large eddy simulation of trade wind cumulus clouds. *J. Atmos. Sci.,* **50,** 3894–3908.

Dalaudier, F., M. Crochet, and J. Vernin, 1989: Direct comparison between *in situ* and radar measurements of temperature fluctuation spectra: A puzzling result. *Radio Sci.,* **24,** 311–324.

——, C. Sidi, M. Crochet, and J. Vernin, 1994: Direct evidence of "sheets" in the atmospheric temperature field. *J. Atmos. Sc.,* **51,** 237–248.

Danielsen, E. F., R. S. Hipskind, W. L. Starr, J. F. Vedder, S. E. Gaines,

D. Kley, and K. K. Kelly, 1991: Irreversible transport in the stratosphere by internal waves of short vertical wavelength. *J. Geophys. Res.,* **96,** 17 433–17 452.

Delage, D., R. Roca, F. Bertin, J. Delcourt, A. Cremieu, M. Massebeuf, R. Ney, and P. van Velthoven, 1997: A consistency check of three radar methods for monitoring eddy diffusion and energy dissipation rates through the tropopause. *Radio Sci.,* **32,** 757–767.

Dewan, E. M., 1979: Stratospheric spectra resembling turbulence. *Science,* **204,** 832–835.

Dhaka, S. K., P. K. Devrajan, Y. Shibagaki, R. K. Choudhary, and S. Fukao, 2001: Indian MST radar observations of gravity wave activities associated with tropical convection. *J. Atmos. Solar Terr. Phys.,* **63,** 1631–1642.

Doviak, R. J. and R. S. Ge, 1984: An atmospheric solitary gust observed with a Doppler radar, a tall tower, and a surface network. *J. Atmos. Sci.,* **41,** 2259–2573.

——, K. W. Thomas, and D. R. Christie, 1989: The wavefront shape, position, and evolution of a great solitary wave of translation. *IEEE Trans. Geosci. Remote Sens.,* **27,** 658–665.

——, S. S. Chen, and D. R. Christie, 1991: A thunderstorm-generated solitary wave observation compared with theory for nonlinear waves in a sheared atmosphere. *J. Atmos. Sci.,* **48,** 87–111.

Dreidonks, A. G. M., and P. G. Duynkerke, 1989: Current problems in the stratocumulus-topped atmospheric boundary layer. *Bound.-Layer Meteor.,* **46,** 275–304.

Du Castel, F., 1966: *Tropospheric Radiowave Propagation Beyond the Horizon.* Pergamon Press, 236 pp.

Eaton, F. D., and G. D. Nastrom, 1998: Preliminary estimates of vertical profiles of inner and outer scales from White Sands Missile Range, New Mexico. *Radio Sci.,* **33,** 895–903.

——, S. A. McLaughlin, and J.R. Hines, 1995: A new frequency modulated continuous wave radar for studying planetary boundry layer morphology. *Radio Sci.,* **30,** 75–88.

Ecklund, W. L., K. S. Gage, B. B. Balsley, R. G. Strauch, and J. L. Green, 1982: Vertical wind variability observed by VHF radar in the lee of the Colorado Rockies. *Mon. Wea. Rev.,* **110,** 1451–1457.

——, B. B. Balsley, D. A. Carter, A. C. Riddle, M. Crochet, and R. Garello, 1985: Observations of vertical motions in the troposphere and lower stratosphere using three closely spaced ST radars. *Radio Sci.,* **20,** 1196–1206.

——, K. S. Gage, G. D. Nastrom, and B. B. Balsley, 1986: A preliminary climatology of atmospheric vertical velocity spectrum observed by clear-air Doppler radar. *J. Climate Appl. Meteor.,* **25,** 885–892.

——, C. R. Williams, P. E. Johnston, and K. S. Gage, 1999: A3-GHz profiler for precipitating cloud studies. *J. Atmos Oceanic Technol.,* **16,** 309–322.

Einaudi, F., W. L. Clark, D. Fua, J. L. Green, and T. E. VanZandt, 1987: Gravity waves and convection in Colorado during July 1983. *J. Atmos. Sci.,* **44,** 1534–1553.

Fairall, C. W., A. B. White, and D. W. Thomson, 1991: A stochastic model of gravity-wave-induced clear-air turbulence. *J. Atmos. Sci.,* **48,** 1771–1790.

Faller, A. J., 1965: Large eddies in the atmospheric boundary layer and their possible role in the formation of cloud roles. *J. Atmos Sci.,* **22,** 176–184.

Feingold, G., A. S. Frisch, B. Stevents, and W. R. Cotton, 1999: On the relationship among cloud turbulence, droplet formation and drizzle as viewed by Doppler radar, microwave radiometer and lidar. *J. Geophys. Res.,* **104,** 22 195–22 203.

Fernando, H. J. S., 1991: Turbulent mixing in stratified fluids. *Annu. Rev. Fluid Mech.,* **23,** 455–493.

——, and J. C. R. Hunt, 1996: Some aspects of turbulence and mixing in stably stratified layers. *Dyn. Atmos. Oceans,* **23,** 35–62.

Franke, S. J., 1990: Pulse compression and frequency domain interferometry with a frequency hopped MST radar. *Radio Sci.,* **25,** 565–574.

Friend, A. W., 1949: Theory and practice of tropospheric sounding by radar. *Proc. IRE,* **37,** 116–138.

Frisch, A. S., B. L. Weber, D. B. Wuertz, R. G. Strauch, and D. A. Merritt, 1990: The variations of C_n^2 between 4 and 18 km above sea level as measured over 5 years. *J. Appl. Meteor.,* **29,** 645–651.

——, D. H. Lenschow, C. W. Fairall, W. H. Shubert, and J. S. Gibson, 1995: Doppler radar measurements of turbulence marine stratiform clouds during ASTEX. *J. Atmos. Sci.,* **52,** 2800–2808.

Fritts, D. C., and Z. Luo, 1992: Gravity wave excitation by geostrophic adjustments of the jet stream. Part I: Two-dimensional forcing. *J. Atmos. Sci.,* **49,** 681–697.

——, and G. D. Nastrom, 1992: Sources of mesoscale variability of gravity waves. Part II: Frontal, convective, and jet stream excitation. *J. Atmos. Sci.,* **49,** 111–127.

——, and T. E. VanZandt, 1993: Spectral estimates of gravity wave energy and momentum fluxes. Part I: Energy dissipation, acceleration, and constraints. *J. Atmos. Sci.,* **50,** 3685–3694.

——, T. Tsuda, T. Sato, S. Fukao, and S. Kato, 1988: Observational evidence of a saturated gravity wave spectrum in the troposphere and lower stratosphere. *J. Atmos. Sci.,* **45,** 1741–1759.

——, ——, T. E. VanZandt, S. A. Smith, T. Sato, S. Fukao, and S. Kato, 1990: Studies of velocity fluctuations in the lower atmosphere using the MU Radar. Part II: Momentum fluxes and energy densities. *J. Atmos. Sci.,* **47,** 51–66.

——, J. R. Isler, and Ø. Andreassen, 1994: Gravity wave breaking in two and three dimensions. 2: Three-dimensional evolution and instability structure. *J. Geophys. Res.,* **99,** 8109–8123.

——, J. F. Garten, and Ø. Andreassen, 1996a: Wave breaking and transition to turbulence in stratified shear flows. *J. Atmos. Sci.,* **53,** 1057–1085.

——, T. L. Palmer, Ø. Andreassen, and I. Lie., 1996b: Evolution and breakdown of Kelvin–Helmholtz billows in stratified compressible flows. Part I: Comparison of two- and three-dimensional flow. *J. Atmos. Sci.,* **53,** 3173–3191.

——, S. Arendt, and Ø. Andreassen, 1998: Vorticity dynamics in a breaking internal gravity wave. Part 2. Vortex interactions and transition to turbulence. *J. Fluid Mech.,* **367,** 47–63.

Fulton, R., D. S. Zrnic, and R. J. Doviak, 1990: Initiation of a solitary wave family in the demise of a nocturnal thunderstorm density current. *J. Atmos Sci.,* **47,** 319–337.

Furomoto, J., and T. Tsuda, 2001: Characteristics of energy dissipation rate and effect of humidity on turbulence echo power revealed by MU radar-RASS measurements. *J. Atmos. Solar Terr. Phys.,* **63,** 289–294.

Gage, K. S., 1979: Evidence for a $k^{-5/3}$ law inertial range in mesoscale two-dimensional turbulence. *J. Atmos. Sci.,* **36,** 1950–1954.

——, 1990: Radar observations of the free atmosphere: Structure and dynamics. *Radar in Meteorology,* D. Atlas, Ed., Amer. Meteor. Soc., 534–565.

——, and J. L. Green, 1978: Evidence for specular reflection from monostatic VHF radar observations of the stratosphere. *Radio Sci.,* **13,** 991–1001.

——, and B. B. Balsley, 1980: On the scattering and reflection mechanisms contributing to clear air radar echoes from the troposphere, stratosphere, and mesosphere. *Radio Sci.,* **15,** 243–257.

——, and G. D. Nastrom, 1985: On the spectrum of atmospheric velocity fluctuations seen by ST/MST radar and their interpretation. *Radio Sci.,* **20,** 1339–1347.

——, and ——, 1986: Theoretical interpretation of atmospheric wavenumber spectra of wind and temperature observed by commercial aircraft during GASP. *J. Atmos. Sci.,* **43,** 729–740.

—— and ——, 1990: A simple model for the enhanced frequency spectrum of vertical velocity based on tilting of atmospheric layers by lee waves. *Radio Sci.,* **25,** 1049–1056.

——, T. E. VanZandt, and J. L. Green, 1980: Use of Doppler radar for the measurement of atmospheric turbulence parameters from the intensity of clear-air echoes. *Radio Sci.,* **15,** 407–416.

——, D. A. Carter, and W. L. Ecklund, 1981: The effect of gravity waves on specular echoes observed by the Poker Flat MST radar. *Geophys. Res. Lett.,* **8,** 599–602.

——, W. L. Ecklund, and D. A. Carter, 1989: Convection waves observed using a VHF wind-profiling Doppler radar during the PRESTORM experiment. Preprints, *24th Conf. on Radar Meteorology,* Tallahassee, FL, Amer. Meteor. Soc., 705–708.

——, C. R. Williams, P. E. Johnston, and W. L. Ecklund, 1997: Unambiguous refractivity turbulence measurements using UHF and S-band profilers. Preprints, *28th Conf. on Radar Meteorology,* Austin, TX, Amer. Meteor. Soc., 139–140.

——, ——, W. L. Ecklund, and P. E. Johnston, 1999: Use of two profilers during MCTEX for unambiguous identification of Bragg scattering and Rayleigh scattering. *J. Atmos. Sci.,* **56,** 3679–3691.

——, ——, W. L. Clark, P. E., Johnston, and D. A. Carter, 2002: Profiler contributions to Tropical Rainfall Measuring Mission (TRMM) ground validation field campaigns. *J. Atmos. Oceanic Technol.,* **19,** 843–863.

Garratt, J. R., 1992: *The Atmospheric Boundary Layer.* Cambridge University Press, 316 pp.

Garrett, C., and W. Munk, 1972: Space–time scales of internal waves. *Geophys. Fluid Dyn.,* **2,** 225–264.

——, and ——, 1975: Space–time scales of internal waves: A progress report. *J. Geophys. Res.,* **80,** 291–297.

Gavrilov, N. M., and S. Fukao, 1999: A comparison of seasonal variations of gravity wave intensity observed by the MU radar with a theoretical model. *J. Atmos. Sci.,* **56,** 3485–3494.

——, and ——, 2001: Hydrodynamic tropospheric wave sources and their role in gravity wave climatology of the upper atmosphere from the MU radar observations. *J. Atmos. Sol. Terr. Phys.,* **63,** 931–943.

Gibson-Wilde, D., J. Werne, D. Fritts, and R. Hill, 2000: Direct numerical simulation of VHF radar measurements of turbulence in the mesosphere. *Radio Sci.,* **35,** 783–798.

Gill, A., 1982: *Atmosphere–Ocean Dynamics.* Academic Press, 662 pp.

Gossard, E. E., 1990: Radar research on the atmospheric boundary layer. *Radar in Meteorology,* D. Atlas, Ed., Amer. Meteor. Soc., 477–527.

——, and W. H. Hooke, 1975: *Waves in the Atmosphere: Atmospheric Infrasound and Gravity Waves: Their Generation and Propagation.* Elsevier Scientific, 456 pp.

——, and R. G. Strauch, 1983: *Radar Observation of Clear Air and Clouds.* Elsevier, 280 pp.

——, J. H. Richter, and D. Atlas, 1970: Internal waves in the atmosphere from high resolution radar measurements. *J. Geophys. Res.,* **75,** 903–913.

——, R. B. Chadwick, T. R. Detman, and J. E. Gaynor, 1984a: Capability of surface-based clear-air Doppler radar for monitoring meteorological structure of elevated layers. *J. Climate Appl. Meteor.,* **23,** 474–485.

——, W. D. Neff, R. J. Zamora, and J. E. Gaynor, 1984b: The fine structure of elevated refractive layers: Implications for over-the-horizon propagation and remote sounding systems. *Radio Sci.,* **19,** 1523–1533.

——, J. E. Gaynor, R. J. Zamora, and W. D. Neff, 1985: Fine structure of elevated stable layers observed by sounder and *in situ* tower sensors. *J. Atmos. Sci.,* **42,** 2156–2169.

Gultepe, I., and D. O'C. Starr, 1995: Dynamical structure and turbulence in cirrus clouds: Aircraft observations during FIRE. *J. Atmos. Sci.,* **52,** 4159–4182.

——, ——, A. J. Heymsfield, T. Uttal, T. P. Ackerman, and D. L. Westphal, 1995: Dynamical characteristics of cirrus clouds from aircraft and radar observations in micro- and meso-γ scales. *J. Atmos. Sci.,* **52,** 4060–4078.

Hardy, K. R., and K. S. Gage, 1990: The history of radar studies of the clear atmosphere. *Radar in Meteorology,* D. Atlas, Ed., Amer. Meteor. Soc., 130–142.

——, D. Atlas, and K. M. Glover, 1966: Multiwavelength backscatter from the clear atmosphere. *J. Geophys. Res.,* **71,** 1537–1552.

Hauf, T., 1993: Aircraft observation of convection waves over southern Germany—A case study. *Mon. Wea. Rev.,* **121,** 3282–3290.

Hermawan, E., and T. Tsuda, 1999: Estimation of turbulence energy dissipation rate and vertical eddy diffusivity with the MU radar RASS. *J. Atmos. Sol. Terr. Phys.,* **61,** 1123–1130.

Herring, J. R., and O. Metais, 1989: Numerical experiments in forced stably stratified turbulence. *J. Fluid Mech.,* **202,** 97–115.

Hicks, J. J., and J. K. Angell, 1968: Radar observations of breaking gravitational waves in the visually clear atmosphere. *J. Appl. Meteor.,* **7,** 114–121.

Hill, R. J., 1978: Spectra of fluctuations in refractivity, temperature, humidity and the temperature humidity cospectrum in the inertial and dissipation ranges. *Radio Sci.,* **13,** 953–961.

——, and S. F. Clifford, 1978: Modified spectrum of atmospheric temperature fluctuations and its application to optical propagation. *J. Opt. Soc. Amer.,* **68,** 892–899.

Hines, C. O., 1963: The upper atmosphere in motion. *Quart. J. Roy. Meteor. Soc.,* **89,** 1–42.

——, 1989: Tropopausal mountain waves over Arecibo: A case study. *J. Atmos. Sci.,* **46,** 476–488.

Hocking, W. K., 1985: Measurement of turbulent eddy dissipation rates in the middle atmosphere by radar techniques: A review. *Radio Sci.,* **20,** 1403–1422.

——, and P. K. L. Mu, 1997: Upper and middle troposphere kinetic energy dissipation rates from measurements of C_n^2—Review of theories, *in-situ* investigations, and experimental studies using the Buckland Park atmospheric radar in Australia. *J. Atmos. Sol. Terr. Phys.,* **59,** 1779–1803.

Hodges, R. R., Jr., 1967: Generation of turbulence in the upper atmosphere by internal gravity waves. *J. Geophys. Res.,* **72,** 3455–3458.

Högstrom, U., 1996: Review of some basic characteristics of the atmospheric surface layer. *Bound.-Layer Meteor.,* **78,** 215–246.

——, A.-S. Smedman, and H. Bergstrom, 1999: A case study of two-dimensional stratified turbulence. *J. Atmos. Sci.,* **56,** 959–976.

——, J. C. R. Hunt, and A.-S. Smedman, 2002: Theory and measurements for turbulence spectra and variances in the atmospheric surface layer. *Bound.-Layer Meteor.,* **103,** 101–124.

Holtslag, A. A. M., and P. G. Duynkerke, Eds., 1998: *Clear and Cloudy Boundary Layers.* Royal Netherlands Academy of Arts and Sciences, 372 pp.

Hopfinger, E. J., 1987: Turbulence in stratified fluids: A review. *J. Geophys. Res.,* **92,** 5287–5303.

Isler, J. R., D. C. Fritts, Ø. Andreassen, and C. E. Wasberg, 1994: Gravity wave breaking in two and three dimensions. 3: Vortex breakdown and transition to isotropy. *J. Geophys. Res.,* **99,** 8125–8137.

Itsweire, E. C., and K. N. Helland, 1989: Spectra and energy transfer in stably stratified turbulence. *J. Fluid Mech.,* **207,** 419–452.

——, ——, and C. W. Van Atta, 1986: The evolution of grid generated turbulence in a stably stratified fluid. *J. Fluid Mech.,* **162,** 299–338.

Jacoby-Koaly, S., B. Campistron, S. Bernard, B. Benech, F. Ardhuin-Girard, J. Dessens, E. Dupont, and B. Carissimo, 2002: Turbulent dissipation rates in the boundary layer via UHF wind profiler Doppler spectral width measurements. *Bound.-Layer Meteor.,* **103,** 361–389.

Jasperson, W. H., G. D. Nastrom, and D. C. Fritts, 1990: Further study of the terrain effects on the mesoscale spectrum of atmospheric motions. *J. Atmos. Sci.,* **47,** 979–987.

Johnston, P. E., L. M. Hartten, C. H. Love, D. A. Carter, and K. S. Gage, 2002: Range errors in wind profiling caused by strong refractivity gradients. *J. Atmos. Oceanic Technol.,* **19,** 934–953.

Jonas, P. R., 1996: Turbulence and cloud microphysics. *Atmos. Res.,* **40,** 283–306.

Kaimal, J. C., and J. J. Finnigan, 1994: *Atmospheric Boundary Layer Flows.* Oxford University Press, 289 pp.

Kim, J., and L. Mahrt, 1992: Momentum transport by gravity waves. *J. Atmos. Sci.,* **49,** 735–748.

Klaassen, G. P., and W. R. Peltier, 1991: The influence of stratification on secondary instability in free shear layers. *J. Fluid Mech.,* **227,** 71–106.

Knight, C. A., and L. J. Miller, 1993: First radar echoes from cumulus clouds. *Bull. Amer. Meteor. Soc.,* **74,** 179–188.

——, and ——, 1998: Early radar echoes from small, warm cumulus: Bragg and hydrometeor scattering. *J. Atmos. Sci.,* **55,** 2974–2992.

Koch, S. E., and C. O'Handley, 1997: Operational forecasting and detection of mesoscale gravity waves. *Wea. Forecasting,* **12,** 253–281.

——, P. B. Dorian, R. Ferrare, S. H. Melfi, W. C. Skillman, and D. Whiteman, 1991: Structure of an internal bore and dissipating gravity current as revealed by Raman lidar. *Mon. Wea. Rev.,* **119,** 857–887.

——, F. Einaudi, P. B. Doran, S. Lang, and G. M. Heymsfield, 1993: A mesoscale gravity-wave event observed during CCOPE. Part IV: Stability analysis and Doppler-derived wave vertical structure. *Mon. Wea. Rev.,* **121,** 2483–2510.

Kollidas, P., and B. Albrecht, 2000: The turbulence structure in a continental stratocumulus cloud from millimetre-wavelength radar observations. *J. Atmos. Sci.,* **57,** 2417–2434.

——, ——, R. Lhermitte, and A. Savtchenko, 2001: Radar observations of updrafts, downdrafts, and turbulence in fair weather cumuli. *J. Atmos. Sci.,* **58,** 1750–1766.

Kropfli, R. A., I. Katz, T. G. Konrad, and E. B. Dobson, 1968: Simultaneous radar reflectivity measurements and refractive index spectra in the clear atmosphere. *Radio Sci.,* **3,** 991–994.

——, and Coauthors, 1995: Cloud physics studies with 8 mm wavelength radar. *Atmos. Res.,* **35,** 299–313.

Kudeki, E., and G. R. Stitt, 1987: Frequency domain interferometry: A high resolution radar technique for studies of atmospheric turbulence. *Geophys. Res. Lett.,* **14,** 198–201.

Kuettner, J. P., P. A. Hildebrand, and T. L. Clark, 1987: Convection waves: Observations of gravity wave systems over convectively active boundary layers. *Quart. J. Roy. Meteor. Soc.,* **113,** 445–467.

Lelong, M. P., and J. J. Riley, 1991: Internal wave–vortical mode interactions in strongly stratified flows. *J. Fluid Mech.,* **232,** 1–19.

LeMone, M. A., 1973: The structure and dynamics of horizontal roll vortices in the planetary boundary layer. *J. Atmos. Sci.,* **30,** 1077–1091.

Lesieur, M., and O. Metais, 1996: New trends in large-eddy simulations of turbulence. *Annu. Rev. Fluid Mech.,* **28,** 45–82.

Lilly, D. K., 1983: Stratified turbulence and the mesoscale variability of the atmosphere. *J. Atmos. Sci.,* **40,** 749–761.

——, G. Bassett, K. Drogemeier, and P. Bartello, 1998: Stratified turbulence in the atmospheric mesoscales. *Theor. Comput. Fluid Dyn.,* **11,** 139–154.

Lindborg, E., 1999: Can the atmospheric kinetic energy spectrum be explained by two-dimensional turbulence? *J. Fluid Mech.,* **388,** 259–288.

——, and J. Y. N. Cho, 2001: Horizontal velocity structure functions in the upper troposphere and lower stratosphere, 2. Theoretical considerations. *J. Geophys. Res.,* **106,** 10 233–10 241.

Lindzen, R. S., 1981: Turbulence and stress owing to gravity wave and tidal breakdown. *J. Geophys. Res.,* **86,** 9707–9714.

Liziola, L. E., and B. B. Balsley, 1998: Studies of quasi-sinusoidal horizontally propagating gravity waves in the troposphere using the Piura ST wind profiler. *J. Geophys. Res.,* **103,** 8641–8650.

Low, D. J., T. Adachi, and T. Tsuda, 1998: MU radar–RASS measurements of tropospheric turbulence parameters. *Meteor. Z.,* **7,** 345–354.

Luce, H., M. Crochet, F. Dalaudier, and C. Sidi, 1995: Interpretation of VHF ST radar vertical echoes from *in situ* temperature sheet observations. *Radio Sci.,* **34,** 1077–1083.

——, F. Dalaudier, M. Crochet, and C. Sidi, 1996: Direct comparison between *in situ* and VHF oblique radar measurements of refractive index spectra: A new successful attempt. *Radio Sci.,* **31,** 1487–1500.

——, M. Crochet, and F. Dalaudier, 2001a: Temperature sheets and aspect sensitive radar echoes. *Ann. Geophys.,* **19,** 899–920.

——, M. Yamamoto, S. Fukao, and M. Crochet, 2001b: Extended radar observations with the frequency domain interferometric imaging (FII) technique. *J. Atmos. Sol.-Terr. Phys.,* **63,** 1033–1041.

——, ——, ——, D. Helal, and M. Crochet, 2001c: A frequency domain radar interferometric imaging (FII) technique based on high resolution methods. *J. Atmos. Sol.-Terr. Phys.,* **63,** 221–234.

Mahrt, L., 1989: Intermittency in atmospheric turbulence. *J. Atmos. Sci.,* **46,** 79–95.

——, 1999: Stratified atmospheric boundary layers. *Bound.-Layer Meteor.,* **90,** 375–396.

Masmoudi, M., and A. Weil, 1988: Atmospheric mesoscale spectra and structure functions of mean horizontal velocity fluctuations measured with a Doppler sodar network. *J. Appl. Meteor.,* **27,** 864–873.

Mason, P. J., 1994: Large eddy simulation: A critical review of the technique. *Quart. J. Roy. Meteor. Soc.,* **120,** 1–26.

Maxworthy, T., 1990: The dynamics of two dimensional turbulence. *The Physical Oceanography of Sea Straits,* L. J. Pratt, Ed., Kluwer Academic, 567–574.

McLandress, C., 1998: On the importance of gravity waves in the middle atmosphere and their parameterization in general circulation models. *J. Atmos. Sol.-Terr. Phys.,* **60,** 1357–1383.

Mead, J. B., G. Hopcraft, S. J. Frasier, B. D. Pollard, C. D. Cherry, D. H. Schaubert, and R. E. McIntosh, 1998: A volume-imaging radar wind profiler for atmospheric boundary layer turbulence studies. *J. Atmos. Oceanic Technol.,* **15,** 849–859.

Meischner, P., R. Baumann, H. Höller, and T. Jank, 2001: Eddy dissipation rates in thunderstorms estimated by Doppler radar in relation to aircraft in situ measurements. *J. Atmos. Oceanic Technol.,* **18,** 1609–1627.

Metais, O., and J. R. Herring, 1989: Numerical simulations of freely evolving turbulence in stably stratified fluids. *J. Fluid Mech.,* **202,** 117–148.

——, J. J. Riley, and M. Lesieur, 1994: Inverse cascade in stably stratified rotating turbulence. *Dyn. Atmos. Oceans,* **23,** 193–203.

Metcalf, J., and D. Atlas, 1973: Microscale ordered motions and atmospheric structure associated with thin echo layers in stably stratified zones. *Bound.-Layer Meteor.,* **4,** 7–35.

Mitchell, N. J., L. Thomas, and I. T. Pritchard, 1994: Gravity waves in the stratosphere and troposphere observed by lidar and MST radar. *J. Atmos. Terr. Phys.,* **56,** 939–947.

Moeng, C.-H., 1998: Large eddy simulation of atmospheric boundary layers. *Clear and Cloudy Boundary Layers,* A. A. M. Holtslag and P. G. Duynkerke, Eds., Royal Netherlands Academy of Sciences, 67–83.

——, and Coauthors, 1996: Simulations of a stratocumulus-topped planetary boundary layer: Intercomparisons among different numerical codes. *Bull. Amer. Meteor. Soc.,* **77,** 261–278.

Müller, P., D. J. Olbers, and J. Willebrand, 1978: The IWEX spectrum. *J. Geophys. Res.,* **83,** 479–500.

Murayama, Y., T. Tsuda, and S. Fukao, 1994: Seasonal variation of gravity wave activity in the lower atmosphere observed with the MU radar. *J. Geophys. Res.,* **99,** 23 057–23 069.

Muschinski, A., and C. Wode, 1998: First *in situ* evidence for coexisting submeter temperature and humidity sheets in the lower free troposphere. *J. Atmos. Sci.,* **55,** 2893–2906.

——, and D. H. Lenschow, 2001: Future directions for research on meter- and submeter-scale atmospheric turbulence. *Bull. Amer. Meteor. Soc.,* **82,** 2831–2843.

——, P. B. Chilson, S. Kern, J. Nielinger, G. Schmidt, and T. Prenosil, 1999a: First frequency domain interferometry observations of large-scale vertical motion in the atmosphere. *J. Atmos. Sci.,* **56,** 1248–1258.

——, P. P. Sullivan, D. B. Wuertz, R. J. Hill, S. A. Cohn, D. H. Lenschow, and R. J. Doviak, 1999b: First synthesis of wind-profiler signals on the basis of large-eddy simulation data. *Radio Sci.,* **34,** 1437–1459.

——, R. Frehlich, M. Jensen, R. Hugo, A. Hoff, F. Eaton, and B. Balsley, 2001: Fine scale measurements of turbulence in the lower troposphere: An intercomparison between a kite- and balloon-borne and a helicopter-borne measurement system. *Bound.-Layer Meteor.,* **98,** 219–250.

Nance, L. B., and D. R. Durran, 1997: A modelling study of nonstationary trapped mountain lee waves. Part I: Mean-flow variability. *J. Atmos. Sci.,* **54,** 2275–2291.

——, and ——, 1998: A modeling study of nonstationary trapped mountain lee waves. Part II: Nonlinearity. *J. Atmos. Sci.,* **55,** 1429–1445.

Narimousa, S., T. Maxworthy, and G. R. Spedding, 1991: Experiments on the structure of forced, quasi-two-dimensional turbulence. *J. Fluid Mech.,* **223,** 113–133.

Nastrom, G. D., 1997: Doppler radar spectral width broadening due to beamwidth and wind shear. *Ann. Geophys.,* **15,** 786–796.

——, and K. S. Gage, 1984: Kinetic energy spectrum of large- and mesoscale atmospheric processes. *Nature,* **310,** 36–38.

——, and ——, 1985: A climatology of atmospheric wavenumber spectra observed by commercial aircraft. *J. Atmos. Sci.,* **42,** 950–960.

——, and D. C. Fritts, 1992: Sources of mesoscale variability of gravity waves. Part I: Topographic excitation. *J. Atmos. Sci.,* **49,** 101–110.

——, and F. D. Eaton, 1997: Turbulence eddy dissipation rates from radar observations at 5–20 km at White Sands Missile Range, NM. *J. Geophys. Res.,* **102,** 19 495–19 505.

——, D. C. Fritts, and K. S. Gage, 1987: An investigation of terrain effects on the mesoscale spectrum of atmospheric motions. *J. Atmos. Sci.,* **44,** 3087–3096.

——, M. R. Peterson, J. L. Green, K. S. Gage, and T. E. VanZandt, 1990: Sources of gravity wave activity seen in the vertical velocities observed by the flatland VHF radar. *J. Appl. Meteor.,* **29,** 783–792.

Nieuwstadt, F. T. M., and P. G. Duynkerke, 1996: Turbulence in the atmospheric boundary layer. *Atmos. Res.,* **40,** 111–142.

——, P. J. Mason, C.-H. Moeng, and U. Schumann, 1993: Large eddy simulation of the convective boundary layer: A comparison of four computer codes. *Turbulent Shear Flows 8,* F. Ourst, Ed., Springer Verlag, 343–367.

Ogino, S., M. D. Yamanaka, and S. Fukao, 1995: Meridional variation of lower stratospheric gravity wave activity: A quick look at Hakuho-maru J-COARE cruise rawinsonde data. *J. Meteor. Soc. Japan,* **73,** 407–413.

Ottersten, H., 1969: Atmospheric structure and radar backscattering, in clear air. *Radio Sci.,* **4,** 1179–1193.

Palmer, R. D., S. Gopalam, and T.-Y. Yu, 1998: Coherent radar imaging using Capon's method. *Radio Sci.,* **33,** 1585–1598.

——, T.-Y. Yu, and P. B. Chilson, 1999: Range imaging using frequency diversity. *Radio Sci.,* **34,** 1485–1496.

——, P. H. Chilson, A. Muschinski, G. Schmidt, T.-Y. Yu, and H. Steinhagen, 2001: SOMARE-99: Observations of tropospheric scattering layers using multiple frequency range imaging. *Radio Sci.,* **36,** 681–693.

Palmer, T. L., D. C. Fritts, Ø. Andreassen, and I. Lie, 1994: Three-dimensional evolution of Kelvin–Helmholtz billows in stratified compressible flow. *Geophys. Res. Lett.,* **21,** 2287–2290.

——, and ——, 1996: Evolution and breakdown of Kelvin–Helmholtz billows in stratified compressible flows. Part II: Instability structure, evolution, and energetics. *J. Atmos. Sci.,* **53,** 3192–3212.

Parsons, D., and Coauthors, 1994: The Integrated Sounding System: Description and preliminary observations from TOGA COARE. *Bull. Amer. Meteor. Soc.,* **75,** 553–567.

Peltier, L. J., and J. C. Wyngaard, 1995: Structure–function parameters in the convective boundary layer from large-eddy simulation. *J. Atmos. Sci.,* **52,** 3641–3660.

Pfister, L., S. Scott, M. Lowenstein, S. Bowen, and M. Legg, 1993:

Mesoscale disturbances in the tropical stratosphere excited by convection: Observations and effects on the stratospheric momentum budget. *J. Atmos. Sci.,* **50,** 1058–1075.

Piani, C., D. Durran, M. J. Alexander, and J. R. Holton, 2000: A numerical study of three-dimensional gravity waves triggered by deep tropical convection and their role in the dynamics of the QBO. *J. Atmos. Sci.,* **57,** 3689–3702.

Plank, V. G., 1956: A meteorological study of radar angels. Geophys. Res. Pap., 52, AFCRC, 117 pp.

Pollard, B. D., S. Khanna, S. J. Frasier, J. C. Wyngaard, D. W. Thomson, and R. E. McIntosh, 2000: Local structure of the convective boundary layer from a volume-imaging radar. *J. Atmos. Sci.,* **57,** 2281–2296.

Prichard, I. T., L. Thomas, and R. M. Worthington, 1995: The characteristics of mountain waves observed by radar near the west coast of Wales. *Ann. Geophys.,* **13,** 757–767.

Ralph, F. M., M. Crochet, and V. Venkateswaran, 1992: A study of mountain lee waves using clear-air radar. *Quart. J. Roy. Meteor. Soc.,* **118,** 597–627.

——, ——, and ——, 1993: Observations of a mesoscale ducted gravity wave. *J. Atmos. Sci.,* **50,** 3277–3291.

——, P. J. Neiman, T. L. Keller, D. Levinson, and L. Fedor, 1997: Observation, simulations, and analysis of nonstationary trapped lee waves. *J. Atmos. Sci.,* **54,** 1308–1333.

——, ——, and ——, 1999: Deep-tropospheric gravity waves created by leeside cold fronts. *J. Atmos. Sci.,* **56,** 2986–3009.

Rao, D. N., T. N. Rao, M. Venkataratnam, P. Srinivasulu, and P. B. Rao, 2001: Diurnal and seasonal variability of turbulence parameters observed with Indian mesosphere-stratosphere-troposphere radar. *Radio Sci.,* **36,** 1439–1457.

Richter, J. H., 1969: High-resolution tropospheric radar sounding. *Radio Sci.,* **4,** 1260–1268.

Riggin, D., D. C. Fritts, C. D. Fawcett, and E. Kudeki, 1995: Observations of inertia-gravity wave motions in the stratosphere over Jicamarca, Peru. *Geophys. Res. Lett.,* **22,** 3239–3242.

Riley, J. J., and M.-P. Lelong, 2000: Fluid motions in the presence of strong stable stratification. *Annu. Rev. Fluid Mech.,* **32,** 623–657.

——, R. W. Metcalfe, and M. A. Weissman, 1981: Direct numerical simulations of homogeneous turbulence in density stratified fluids. *Nonlinear Properties of Internal Waves,* B. J. West, Ed., La Jolla Institute, AIP Conf. Proc. 76, 79–112.

Rogers, R. R., W. L. Ecklund, D. A. Carter, K. S. Gage, and S. A. Ethier, 1993: Research applications of a boundary layer wind profiler. *Bull. Amer. Meteor. Soc.,* **74,** 567–580.

Röttger, J., 2000: ST radar observations of atmospheric waves over mountainous areas: A review. *Ann. Geophys.,* **18,** 750–765.

Rottman, J. W., and F. Einaudi, 1993: Solitary waves in the atmosphere. *J. Atmos. Sci.,* **50,** 2116–2136.

——, ——, S. E. Koch, and W. L. Clark, 1992: A case study of penetrative convection and gravity waves over PROFS Mesonetwork on 23 July 1983. *Meteor. Atmos. Phys.,* **47,** 205–227.

Saiki, E. M., C.-H. Moeng, and P. P. Sullivan, 2000: Large-eddy simulation of the stably stratified planetary boundary layer. *Bound.-Layer Meteor.,* **95,** 1–30.

Salathe, E. P., Jr., and R. B. Smith, 1992: *In situ* observations of temperature microstructure above and below the tropopause. *J. Atmos. Sci.,* **49,** 2032–2036.

Sato, K., 1989: Inertial gravity wave associated with a synoptic scale pressure trough observed by the MU radar. *J. Meteor. Soc. Japan,* **67,** 325–334.

——, 1993: Small-scale wind disturbances observed by the MU radar during the passage of Typhoon Kelly. *J. Atmos. Sci.,* **50,** 518–537.

——, 1994: A statistical study of the structure, saturation and sources of inertio-gravity waves in the lower stratosphere observed with the MU radar. *J. Atmos. Terr. Phys.,* **56,** 755–774.

——, H. Hashiguchi, and S. Fukao, 1995: Gravity waves and turbulence associated with cumulus convection observed with the UHF/VHF clear-air Doppler radars. *J. Geophys. Res.,* **100,** 7111–7119.

——, D. J. O'Sullivan, and T. J. Dunkerton, 1997: Low-frequency inertia-gravity waves in the stratosphere revealed by three-week continuous observation with the MU radar. *Geophys. Res. Lett.,* **24,** 1739–1742.

Scheffler, A. O., and C. H. Liu, 1985: Observation of gravity wave spectra in the atmosphere using MST radars. *Radio Sci.,* **20,** 1309–1322.

Smith, S. A., D. C. Fritts, and T. E. VanZandt, 1987: Evidence of a saturation spectrum of atmospheric gravity waves. *J. Atmos. Sci.,* **44,** 1404–1410.

Sreenivasan, K. R., and R. A. Antonia, 1997: The phenomenology of small-scale turbulence. *Annu. Rev. Fluid Mech.,* **29,** 435–472.

Staquet, C., and J. Sommeria, 2002: Internal gravity waves: From instabilities to turbulence. *Annu. Rev. Fluid Mech.,* **34,** 559–593.

Stevens, B., and D. H. Lenschow, 2001: Observations, experiments and large-eddy simulation. *Bull. Amer. Meteor. Soc.,* **82,** 283–294.

——, G. Feingold, W. R. Cotton, and R. L. Walko, 1996: Elements of the microphysical structure of numerically simulated nonprecipitating stratocumulus. *J. Atmos. Sci.,* **53,** 980–1007.

——, W. R. Cotton, G. Feingold, and C.-H. Moeng, 1998: Large-eddy simulations of strongly precipitating, shallow, stratocumulus-topped boundary layers. *J. Atmos. Sci.,* **55,** 3616–3638.

Tatarskii, V. I., 1971: *The Effects of the Turbulent Atmosphere on Wave Propagation.* Israel Program for Scientific Translations, 472 pp. [NTIS TT 68-50464.]

Telford, J. W., 1996: Clouds with turbulence: The role of entrainment. *Atmos. Res.,* **40,** 261–282.

Thomas, L., T. Pritchard, and I. Astin, 1992: Inertia-gravity waves in the troposphere and lower stratosphere. *Ann. Geophys.,* **10,** 690–697.

——, R. M. Worthington, and A. J. McDonald, 1999: Inertia-gravity waves in the troposphere and lower stratosphere associated with a jet stream exit region. *Ann. Geophys.,* **17,** 115–121.

Trexler, C. M., and S. E. Koch, 2000: The life cycle of a mesoscale gravity wave observed by a network of Doppler wind profilers. *Mon. Wea. Rev.,* **128,** 2423–2466.

Tsuda, T., T. Inoue, D. C. Fritts, T. E. VanZandt, S. Kato, T. Sato, and S. Fukao, 1989: MST radar observations of a saturated gravity wave spectrum. *J. Atmos. Sci.,* **46,** 2440–2447.

——, Y. Murayama, T. Nakamura, R. A. Vincent, A. H. Manson, C. E. Meek, and R. L. Wilson, 1994a: Variations of the gravity wave characteristics with height, season and latitude revealed by comparative observations. *J. Atmos. Terr. Phys.,* **56,** 555–568.

——, ——, H. Wiryosumarto, S.-W. B. Harijono, and S. Kato, 1994b: Radiosonde observations of equatorial atmosphere dynamics over Indonesia. 2: Characteristics of gravity waves. *J. Geophys. Res.,* **99,** 10 507–10 516.

——, M. Nishida, C. Rocker, and R. H. Ware, 2000: A global morphology of gravity wave activity in the stratosphere revealed by the GPS occultation data (GPS/MET). *J. Geophys. Res.,* **105,** 7257–7274.

Vallis, G. K., G. J. Shuts, and M. E. B. Gray, 1997: Balanced mesoscale motion and stratified turbulence forced by convection. *Quart. J. Roy. Meteor. Soc.,* **123,** 1621–1652.

VanZandt, T. E., 1982: A universal spectrum of buoyancy waves in the atmosphere. *Geophys. Res. Lett.,* **9,** 575–578.

——, 1985: A model for gravity wave spectra observed by Doppler sounding system. *Radio Sci.,* **20,** 1323–1330.

——, and D. C. Fritts, 1989: A theory of enhanced saturation of the gravity wave spectrum due to increases in atmospheric stability. *Pure Appl. Geophys.,* **130,** 399–420.

——, J. L. Green, K. S. Gage, and W. L. Clark, 1978: Vertical profiles of refractivity turbulence structure constant: Comparison of observations from the Sunset Radar with a new theoretical model. *Radio Sci.,* **13,** 819–829.

——, G. D. Nastrom, and J. L. Green, 1991: Frequency spectra of vertical velocity from Flatland VHF radar data. *J. Geophys. Res.,* **96,** 2845–2855.

——, W. L. Clark, K. S. Gage, C. R. Williams, and W. L. Ecklund, 2000: A dual wavelength radar technique for measuring the turbulent energy dissipation rate ε. *Geophys. Res. Lett.,* **27,** 2537–2540.

Vincent, R. A., and M. J. Alexander, 2000: Gravity waves in the tropical lower stratosphere: An observational study of seasonal and internal variability. *J. Geophys. Res.,* **105,** 17 971–17 982.

Warnock, J. M., and T. E. VanZandt, 1985: A statistical model to estimate refractivity turbulence structure constant C_n^2 in the free atmosphere. NOAA Tech. Memo. ERL AL-10, Boulder, Colorado, 175 pp.

——, N. Signups, and R. G. Strauch, 1988: Comparison between height profiles of C_n^2 measured by the Stapleton UHF clear-air Doppler radar and Model Calculations. Preprints, *Eighth Symp. on Turbulence and Diffusion,* San Diego, CA, Amer. Meteor. Soc., 267–270.

Werne, J., and D. C. Fritts, 1999: Stratified shear turbulence: Evolution and statistics. *Geophys. Res. Lett.,* **26,** 439–442.

White, A. B., C. W. Fairall, A. S. Frisch, B. W. Orr, and J. B. Snider, 1996: Recent radar measurements of turbulence and microphysical parameters in marine boundary layer clouds. *Atmos. Res.,* **40,** 177–221.

Wilczak, J. M., E. E. Gossard, W. D. Neff, and W. L. Bernhard, 1996: Ground-based remote sensing of the atmospheric boundary layer: 25 years of progress. *Bound.-Layer Meteor.,* **78,** 321–349.

Williams, C. R., W. L. Ecklund, P. E. Johnston, and K. S. Gage, 2000: Cluster analysis techniques to separate air motion and hydrometeors in vertical incident profiler observations. *J. Atmos. Oceanic Technol.,* **17,** 949–962.

Worthington, R. M., 1998: Tropopausal turbulence caused by the breaking of mountain waves. *J. Atmos. Sol.-Terr. Phys.,* **60,** 1543–1547.

——, 1999a: Alignment of mountain wave patterns above Wales: A VHF radar study during 1990–1998. *J. Geophys. Res.,* **104,** 9199–9212.

——, 1999b: Calculating the azimuth of mountain waves, using the effect of tilted fine-scale stable layers on VHF radar echoes. *Ann. Geophys.,* **17,** 257–272.

——, and L. Thomas, 1996: Radar measurements of critical-layer absorption in mountain waves. *Quart. J. Roy. Meteor. Soc.,* **122,** 1263–1282.

——, and ——, 1997a: Impact of the tropopause on the upward propagation of mountain waves. *Geophys. Res. Lett.,* **24,** 1071–1074.

——, and ——, 1997b: Long period gravity waves and associated VHF radar echoes. *Ann. Geophys.,* **15,** 813–822.

——, and ——, 1998: The frequency spectrum of mountain waves. *Quart. J. Roy. Meteor. Soc.,* **124,** 687–703.

——, R. D. Palmer, and S. Fukao, 1999: An investigation of tilted aspect sensitive scatters in the lower atmosphere using the MU and Aberystwyth VHF radars. *Radio Sci.,* **34,** 413–426.

Wulfmeyer, V., 1999a: Investigation of turbulent processes in the lower troposphere with water vapor DIAL and Radar–RASS. *J. Atmos. Sci.,* **56,** 1055–1076.

——, 1999b: Investigations of humidity skewness and variance profiles in the convective boundary layer and comparison of the latter with large eddy simulation results. *J. Atmos. Sci.,* **56,** 1077–1087.

Wyngaard, J. C., 1992: Atmospheric turbulence. *Annu. Rev. Fluid Mech.,* **24,** 205–233.

——, 1998: Boundary-layer modelling: History, philosophy and sociology. *Clear and Cloudy Boundary Layers,* A. A. M. Holtslag and P. G. Duynkerke, Eds., Royal Netherlands Academy of Sciences, 325–332.

——, and L. J. Peltier, 1996: Experimental micrometeorology in an era of turbulence simulation. *Bound.-Layer Meteor.,* **78,** 71–86.

——, N. Seaman, S. J. Kimmel, M. Otte, X. Di, and K. Gilbert, 2001: Concepts, observations, and simulation of refractive index turbulence in the lower atmosphere. *Radio Sci.,* **36,** 643–670.

Yamanaka, M. D., S. Ogino, S. Kondo, T. Shimomai, Y. Shibagaki, Y. Maekawa, and I. Takayabu, 1996: Inertio-gravity waves and subtropical multiple tropopauses: Vertical wavenumber spectra of wind and temperature observed by the MU radar, radiosondes and operational rawinsonde network. *J. Atmos. Terr. Phys.,* **58,** 785–805.

Yang, M.-J., and R. A. Houze Jr., 1995: Multicell squall-line structure as a manifestation of vertically trapped gravity waves. *Mon. Wea. Rev.,* **123,** 641–661.

Yap, C. T., and C. W. van Atta, 1993: Experimental studies of the development of quasi-two-dimensional turbulence in a stably stratified fluid. *Dyn. Atmos. Oceans,* **19,** 289–323.

Chapter 7

The Assimilation of Radar Data for Weather Prediction[*]

JUANZHEN SUN AND JAMES W. WILSON

National Center for Atmospheric Research, Boulder, Colorado

Sun   Wilson

1. Introduction

Since the early years of radar meteorology, scientists have given a good deal of attention to the development of techniques for deducing the unobserved meteorological fields from Doppler radar observations. For example, Lhermitte and Atlas (1961) and Atlas (1964) described the velocity-azimuth display (VAD) method for determining a horizontally averaged wind profile from single-Doppler radial velocity observations. Lhermitte (1970) developed a method to derive the detailed horizontal wind field from dual-Doppler radial velocity observations. These early research activities were largely driven by the desire to understand the dynamical processes of convective systems in the atmosphere. In the late 1980s, however, scientists began to recognize that radar observations would also play an important role in future mesoscale and convective-scale data assimilation systems for short-term weather forecasting. This scientific vision was gained, in large part, as a result of the planned implementation of the Weather Surveillance Radar-1988 Doppler (WSR-88D) network in the United States and the rapid increase in computer power. With these increased resources, it appeared that running numerical weather prediction (NWP) models with resolutions that are able to resolve mesoscale and cloud-

scale features was within reach. On account of the considerable societal and economic impact of the operational storm-scale NWP and the scientific and technological advancement, Lilly (1990) argued that it was time for convective-storm research scientists to apply their knowledge and to show whether storm-scale NWP was a realistic goal. Since the WSR-88D network is the major observing system capable of sampling the four-dimensional structure of storm-scale flows, a key scientific and technical challenge is whether these observations can be used to initialize high-resolution storm prediction models.

Doppler radar measures radial wind speed, reflectivity, and variance of the velocity spectrum in volumes containing a sufficiently high concentration of scatterers. The WSR-88D radars are able to observe precipitation at least within 200–250 km of each radar as well as boundary layer flows by clear air returns (e.g., insects) within 50–120 km. These measurements have high spatial and temporal resolution (1 km × 1° for reflectivity and 0.25 km × 1° for radial velocity every 5–10 min; Klazura and Imy 1993) and good accuracy (better than 1 m s^{-1} random error for radial velocity), but they are limited to the radial velocity and reflectivity. The variables that are essential for initializing a convective-scale numerical model have to be retrieved. Although there existed retrieval techniques for deducing the unobserved meteorological fields from Doppler radar observations prior to the last decade, these tech-

* The National Center for Atmospheric Research is sponsored by the National Science Foundation.

niques either required dual-Doppler observations or they were incapable of obtaining high-resolution fields from a single-Doppler radar. For instance, the VAD technique (Lhermitte and Atlas 1961) can only derive the averaged wind over a specified radius of the radar (typically 5–25 km) and the tracking radar echoes by correlation (TREC; Rinehart 1979; Tuttle and Foote 1990) technique is able to obtain the wind field with a spatial resolution at best 6 km $\times$ 8°. Besides the poor resolution, these techniques suffer from other limitations and are clearly not adequate for the purpose of initializing a cloud-scale numerical model. For instance, when TREC is applied to storm echoes, it tracks the propagation of the storm rather than the air motion. It was, therefore, well recognized that new single-Doppler wind retrieval techniques need to be developed in order to produce a high-resolution analysis that can be used for model initialization.

The single-Doppler wind retrieval is the first step in achieving the goal of assimilating radar observations into a numerical model. The object of atmospheric data assimilation is to produce a physically consistent four-dimensional representation of the state of the atmosphere by continuously merging observations with a numerical model. Data assimilation usually consists of two main components: analysis and a short forecast to prepare the first guess or background field for the next cycle. Some of the data assimilation systems that will be reviewed in this paper only include an analysis step at the time of this writing. In spite of that, we will still refer to them by the word "assimilation" because we believe they will evolve into a complete data assimilation system through continuous development in the near future.

Although the research activities on the assimilation of radar data were spurred by the motivation of producing explicit storm-scale numerical weather prediction, analyses produced through this process can be used in other ways. For example, the low-level wind analysis field obtained by assimilating radar data and surface data into a Boussinesq dynamical model has been used in a knowledge-based thunderstorm nowcasting system (Sun and Crook 2001a). Furthermore, assimilation of radar and other type of mesoscale data collected in research field experiments can be applied to diagnostic studies of convective systems.

In this paper, a number of data assimilation techniques that show potential for the meso- and convective-scale are presented, progress on the assimilation of radar data for weather prediction is described, and future challenges are discussed. The data assimilation techniques are introduced in section 2. In section 3, single-Doppler radar retrieval and radar data assimilation on the convective scale are reviewed and some of the results are presented. The research on assimilation of radar data on the mesoscale is reviewed in section 4. In the last section, the future of radar data assimilation is discussed.

2. General description of data assimilation techniques

The goal of atmospheric data assimilation is to produce a physically consistent estimate of the atmospheric flow on a regular grid using all the available information. The available information generally consists of the observations from a heterogeneous array of in situ and remote instruments sampled imperfectly and irregularly in space and time, the physical laws governing the flow, and any prior knowledge about the flow. Since data from these sources of information have uncertainties, and ideal assimilation system should produce not only the estimate of the flow but also its uncertainty originating from the uncertainties of the various sources of information.

If all the available information and the associated uncertainty are given, an idealized equation for finding the optimal estimate of the atmospheric flow can be derived based on Bayesian probabilistic arguments (Lorenc 1986). However, since the atmospheric data assimilation problems usually involve complex, nonlinear numerical models with large dimension, simplifications must be sought in practice. Different assimilation techniques are the results of different approximations involved in the simplification. In this section, we introduce three data assimilation techniques that hold most promise to the problem of assimilation of radar data: the three-dimensional variational technique (3D-VAR), the ensemble Kalman filter (EnKF), and the four-dimensional variational technique (4D-VAR). Our focus will be on explaining the concept of these techniques. Readers are referred to Tarantola (1987), Talagrand (1997), and Cohn (1997) for the underlying fundamentals and mathematical derivation of each technique.

a. Three-dimensional variational technique

3D-VAR is a technique that attempts to fit a first guess field through a cost function to the observations. The first guess field is usually referred to as a background field. It can be, for instance, the result of a numerical model prediction or of a climatological analysis. With the background knowledge and the observations as well as estimates of their uncertainties, a 3D-VAR problem can be defined and an optimal analysis can be found by the variational calculus. The variational calculus involves the determination of stationary points of an integral expression known as a functional or cost function. The cost function in 3D-VAR measures the misfit of the analysis vector $\mathbf{x}$ to the background $\mathbf{x}_b$ and to the observation $\mathbf{y}$. It is defined by

$$J = (\mathbf{x} - \mathbf{x}_b)^\mathrm{T}\mathbf{B}^{-1}(\mathbf{x} - \mathbf{x}_b)$$
$$+ (\mathbf{y} - \mathbf{H}\mathbf{x})^\mathrm{T}\mathbf{O}^{-1}(\mathbf{y} - \mathbf{H}\mathbf{x}). \quad (7.1)$$

The precision of the background and the observation data is represented by $\mathbf{B}^{-1}$ and $\mathbf{O}^{-1}$, where $\mathbf{B}$ and $\mathbf{O}$ are background and observation error covariance matrices, respectively. An error covariance matrix contains the statistics of error magnitude and correlation between different measurements. The definition of these matrices will be further explained later in this section. Here T denotes the matrix transpose. Since **x** and **y** can be different variables and on different grids, the observation operator **H** is used to represent an analytical function that relates the model variables to the observation variables (for instance, the relation between u, v, and w and the radial velocity v_r from radar observations) and a transformation between the different grid meshes by an interpolation scheme.

The optimal analysis is found by minimizing the cost function (7.1). The solution procedure of 3D-VAR is described in appendix A. In the cost function (7.1), the inverse of the error covariance matrix **B** measures the precision of the background information. Similarly, the inverse of **O** measures the precision of the observations. A common practice in atmospheric data assimilation is to assume the observation errors are uncorrelated in space such that the matrix **O** has a simple diagonal form. Most of the operational optimal interpolation (OI) and 3D-VAR systems use forecast fields from a numerical model as the background information. The error covariance of the forecast background **B** is usually modeled based on some simple hypotheses on the shape and spatial extension of the presumed covariance functions (see, e.g., Parrish and Derber 1992; Daley 1992; Hayden and Purser 1995).

The 3D-VAR technique belongs to a class of static data assimilation methods because the temporal variability of the atmospheric state is not taken into account. A separate step is required to merge the observations with a dynamical model. After each analysis, the forecast model is integrated forward, starting from the analysis initial conditions, until the next observation time. The forecast at this time is taken as the background field and is then combined with the new observations using an analysis algorithm. This process forms the basis of sequential data assimilation. In sequential assimilation, each new piece of observation is used for correcting the latest estimate, so the best fit of the model solution to observations is achieved at the end of the assimilation period through the propagation of information from the past in a sequential manner.

b. Ensemble Kalman filter

The ensemble Kalman filter approach is similar to the 3D-VAR technique except that it provides a way to estimate the evolution of forecast background error (which is assumed constant in 3D-VAR) using a numerical model. The evolution of the atmosphere is governed by the physical laws, which are available in practice in the form of a numerical model. The numerical model can be used as another source of information in data assimilation. Kalman filter and 4D-VAR approaches are the two techniques capable of accounting for the evolution of the atmosphere represented by a numerical model. The Kalman filtering provides a systematic way to estimate the evolution of the forecast background error covariance matrix assuming a system whose state evolves linearly with time. In brief, the Kalman filtering technique uses Eqs. (7.A3)–(7.A4) in appendix A as the basis of the analysis, but instead of specifying a modeled covariance matrix **B** as in OI and 3D-VAR, a rigorous calculation of the forecast background error covariance matrix **O** is given by applying a linear model that represents the evolution of the atmosphere.

Although the Kalman filter provides an elegant way to calculate and propagate the background covariance error, one major difficulty in applications to meteorological problems is its numerical cost. Various simplifications have been sought in order to reduce the computational cost (see, e.g., Cohn and Parrish 1991; Dee 1991, Bouttier 1994). In recent years, a number of studies have considered the use of an ensemble to estimate the forecast error covariance (see, e.g., Evensen 1994; Houtekamer and Mitchell 1998; Hamill and Snyder 2000), which is referred to as ensemble Kalman filter. The EnKF attempts to estimate the forecast background error covariance by randomly sampling a limited number of forecast realizations. The premise is that the random sampling is able to determine the forecast background error covariance more realistically than the analytical modeling techniques as used in 3D-VAR. One advantage of the EnKF technique is that an estimate of the analysis error covariance can be computed from the ensemble while it is more difficult to obtain in variational techniques.

Data assimilation using the EnKF technique follows the sequential procedure described in section 2a. The difference is that n (the number of ensemble members) analyses, instead of one as in 3D-VAR, are performed with each new set of observations. Correspondingly, n forecasts are made and these n forecasts are used to compute the background error covariance matrix **B**. As a result, not only the forecast background (given by the mean of the n ensemble forecasts) but also the error covariance are propagated by the numerical model.

c. Four-dimensional variational technique

Another technique that allows the incorporation of a numerical model is the four-dimensional variational data assimilation. 4D-VAR can be considered as an extension of 3D-VAR in such a way that all observations distributed within a time window $[t_0, t_N]$ are taken into account. The 4D-VAR approach therefore provides a way to globally adjust a model to observations distributed in time. A numerical model that is supposed to represent the evolution of the estimate vector **x** is used as a priori information. Assume that the nonlinear model equation is given in the form

$$\frac{\partial \mathbf{x}}{\partial t} = F(\mathbf{x}), \qquad (7.2)$$

where $\mathbf{x}$ denotes the model state and F stands for all the mathematical functions involved in the nonlinear model. If the observational error covariance $\mathbf{O}$ is independent of time, the 3D cost function (7.1) can be extended to

$$J = (\mathbf{x}_0 - \mathbf{x}_b)^{\mathrm{T}} \mathbf{B}^{-1} (\mathbf{x}_0 - \mathbf{x}_b)$$

$$+ \int_{t_0}^{t_N} (\mathbf{y} - \mathbf{Hx})^{\mathrm{T}} \mathbf{O}^{-1} (\mathbf{y} - \mathbf{Hx}) \, dt. \qquad (7.3)$$

The variable $\mathbf{x}_0$ represents the model state at initial time. The model states at successive times are linked by the model equation (7.2). Minimizing the cost function (7.3) under the constraint (7.2) will produce the best linear unbiased estimate at any time from all the available observations within the time window $[t_0, t_N]$ (Talagrand 1997).

Since the state variables at successive time steps are related through the dynamical relation (7.2), it is easy to see that the control variable in the cost function (7.3) is the initial state $\mathbf{x}_0$. To iteratively solve the minimization problem (7.2)–(7.3), the gradient of the cost function with respect to $\mathbf{x}_0 (\partial J/\partial x_{0i}, i = 1 \ldots, n$, where n is the dimension of the control variable $\mathbf{x}$) needs to be determined. Although it is possible to find this gradient by sequentially perturbing each of the initial state components, this is too costly for practical implementation; it requires integrating the model over the assimilation period as many times as there are control variable components. The adjoint method provides an effective way for calculating the gradient at a cost that is usually a few times the cost of one model integration for each iteration in the minimization procedure. The derivation of the adjoint equation associated with the nonlinear equation (7.2) is shown in Appendix B. The minimization procedure is also explained in this appendix.

It should be noted that the solution nonuniqueness for the minimization of the cost function (7.2)–(7.3) can be a potential problem in 4D-VAR (Li 1991). The possibility of the solution nonuniqueness increases when a long assimilation window is used and the data do not contain adequate information. Incorporation of additional physical constraints such as spatial or temporal smoothness constraint can help alleviate the problem.

Over the past two decades, considerable effort has been devoted to the field of 4D-VAR data assimilation through the use of adjoint model. Experiments have been conducted using research as well as operational models (see, e.g., Lewis and Derber 1985; Talagrand and Courtier 1987; Navon et al. 1992; Zupanski 1993; Zou and Kuo 1996). Recent operational implementation of the European Centre for Medium-Range Weather Forecasts (ECMWF) incremental 4D-VAR system (a simplified version of the full 4D-VAR) has shown prom-

ising performance (Rabier et al. 2000; Mahfouf and Rabier 2000; Klinker et al. 2000).

The 4D-VAR technique represents another form of data assimilation: variational assimilation, which has some fundamental differences from the sequential assimilation. Unlike the sequential assimilation, which finds the best analysis at a single observational time, the variational assimilation seeks an optimal fit of the model solution to observations over an assimilation period by adjusting the estimate states in this period simultaneously. In this global adjustment, a numerical model is used to link the state of the atmosphere at different times. The analysis over the assimilation period is influenced by all of the observations distributed in time. The information is propagated both from the past into the future and from the future into the past. A plot (Fig. 7.1) is used to illustrate the concept of the 4D-VAR technique. Also depicted in Fig. 7.1 is the illustration of a sequential-type technique. It can be shown that, in the case of a perfect linear model, if the forecast error covariance is propagated by the Kalman filter formulation, the sequential and the variational estimates converge to the same solution at the end of the estimation period (see, e.g., Talagrand 1997).

Table 7.1 lists and compares the techniques introduced in this section. The OI technique is also included in the table since the technique has been used widely in the past for operational numerical models.

3. Assimilation of radar data at convective scale

We divide our discussion on the assimilation of radar data into convective scale and mesoscale. The convective scale is discussed in this section and the mesoscale in the next section. There are a number of ways to define these scales (Emanuel 1984) based on empirical, theoretical, or practical considerations. Our definition in this paper is mainly from a practical approach. We refer to a convective-scale system as any phenomenon too small to be observed by the conventional network of in situ observations, which means a spatial scale of less than 100 km and a timescale of less than half a day. Another consideration is that the large-scale forcing is not significant. Mesoscale, on the other hand, is referred to those systems that can be discerned by the radar network as well as the conventional in situ network and satellites. In addition, the large-scale forcing plays an important role in these systems.

Although the fundamental principles in convective-scale data assimilation are the same as in large-scale data assimilation, their emphases and challenges are different, due to differences in the phenomenon of interest and in the observing systems. The main observing system for the convective scale is a radar network. As already mentioned, Doppler radars provide spatially and temporally high-resolution measurements, but these measurements are limited to the radial velocity and reflectivity, which are not direct model variables. The var-

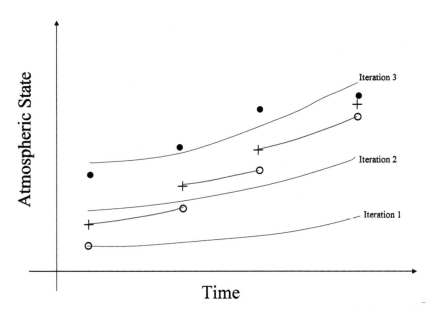

FIG. 7.1. Illustration of the variational assimilation and sequential assimilation by assuming a single point of the atmospheric state varying with time. The black dots denote observations at different times, the circles are the forecast background, and the crosses show the analysis. The red lines indicate trajectories obtained at different iterations from a variational assimilation and the blue lines indicate the forecast trajectories in a sequential assimilation algorithm.

iables that are essential for initializing a convective-scale numerical model have to be retrieved. On the large-scale, however, the main observing system is the radiosonde network, which provides measurements of all variables, except for vertical velocity, needed for initialization of a forecast model. These measurements have a spatial resolution much poorer than that in a numerical model. Consequently, the major challenge in the convective-scale data assimilation is to retrieve the unobserved model variables, while optimal interpolation is the main objective for the large scale. Another difference between the convective-scale and large-scale data assimilation is the phenomenon of interest. The convective-scale weather systems have high temporal variability and no simple balances are known to exist on this scale while various balances (e.g., the geostrophic balance) can be used in large-scale data assimilation. As a result of the aforementioned differences, methods that have shown good performance for the large-scale may not be suitable for the convective scale.

Another point worth mentioning is that the convective-scale data assimilation and numerical weather prediction are still in their infancy. Since the statistics need-

ed in some of the data assimilation techniques require the availability of an operational numerical forecast system and a system that is capable of explicitly predicting convections at the scale comparable with radar observations is not yet available, simplifications and assumptions must be made on these statistics.

In this section, we describe some of the single-Doppler radar retrieval and assimilation techniques that were developed over the last 10 years or so. These techniques are reviewed in three sections: single-Doppler wind retrieval, variational assimilation, and sequential assimilation. Analysis results from some of the data assimilation systems will be shown. Forecast experiments initialized by radar data from some case studies will also be described and their results will be discussed. Our review focuses on the assimilation and forecast using radar data from single-Doppler observations because the operational network can not provide dual-Doppler coverage. One of the early studies on initialization and prediction of convection using Doppler radar data was conducted by Lin et al. (1993) using multiple-Doppler observations. Their work, needless to say, also made im-

TABLE 7.1. Comparison of data assimilation techniques.

Technique	Dynamic model	Global fitting	Time-varying error statistics	Analysis error covariance	Computation cost
OI	No	No	No	No	Low
3D-VAR	No	Yes (3D)	No	No	Low
EnKF	Yes	No	Yes	Yes	High
4D-VAR	Yes	Yes (4D)	No	No	High

portant contributions to the initialization and prediction of convective storms.

a. Single-Doppler wind retrieval

If dual-Doppler radial velocity observations are available, the three-dimensional wind can be easily derived through the use of the continuity equation. The spacing of operational Doppler radars (e.g., WSR-88D network) is not good enough to provide dual-Doppler coverage; hence, retrievals need to rely on data from single-Doppler observations. In the early 1960s, Lhermitte and Atlas (1961), and Atlas (1964) showed that, provided the fields of wind and precipitation fall speed are horizontally homogeneous, these fields could be determined by the VAD technique. Under the constant wind and fall speed assumption, the mean radial velocity can be written as a sine function of azimuth angle at a given elevation angle and a constant altitude. The amplitude and phase of this sine curve are measures of the speed and direction of the wind at the sampled altitude, and the displacement of the entire sine curve from zero velocity is a measure of the precipitation fall speed. Caton (1963), Harrold (1966), and Browning and Wexler (1968) generalized the idea of Lhermitte and Atlas (1961) and Atlas (1964) by including the determination of the divergence and deformation through harmonic analysis.

The VAD technique has been used widely both in research and operation to estimate the horizontal wind as a function of height near a Doppler radar. However, the VAD analysis is unable to provide high-resolution analysis that is needed in initializing storm-scale numerical prediction models. A basic retrieval problem facing convective-scale scientists is to determine the detailed 3D boundary layer wind using single-Doppler clear air observations. The low-level wind retrieval has received a good deal of attention not only because it is simpler than a full parameter retrieval that involves microphysics, but also because boundary layer wind convergence plays an important role in convection initiation (Wilson and Schreiber 1986). Techniques for retrieving thermodynamical and microphysical fields from dual-Doppler synthesized winds were developed in the past (Gal-Chen 1978; Ziegler 1985; Roux 1985). Therefore, if the 3D wind fields can be determined from single-Doppler observations, the other fields can be retrieved by applying these techniques.

Clear air scatterers in the boundary layer during the warm season ($>10°C$) are primarily from insects that are carried by the winds. With the exception of occasional periods of bird migrations, the velocity of these scatters accurately portrays the wind velocity (Wilson et al. 1994). When the reflectivity signal is due primarily to clear air scatterers as distinct from precipitation, then the reflectivity conservation equation should, presumably, be valid without the source term. This assumption forms the basis for a number of single-Doppler wind

retrieval techniques (Qiu and Xu 1992; Laroche and Zawadzki 1994; Shapiro et al. 1995; Zhang and Gal-Chen 1996; Liou 1999; Gao et al. 1999). Most of these techniques have been tested using real single-Doppler observations and have demonstrated that the low-level wind can be retrieved with acceptable accuracy when compared with dual-Doppler analysis.

The 3D-VAR system developed by Gao et al. (1999, 2001) is perhaps most promising among the single-Doppler wind retrieval algorithms, due partly to its ability to include various constraints and to deal with data voids by including a background term, and partly to its better compatibility with large-scale data assimilation systems. The 3D-VAR technique attempts to minimize a cost function in the form

$$J = J_b + J_o + J_p, \qquad (7.4)$$

where J_b represents the background term that measures the distance between the estimate and a specified background. The variable J_o denotes the observation term that measures the difference between the estimated radial velocity and the observed radial velocity. The last term J_p represents the enforced constraints. In their system, Gao et al. (2001) used three constraints: reflectivity conservation, anelastic mass continuity, and second-order spatial smoothness. The reflectivity conservation is a prognostic equation in which the temporal and spatial derivatives are computed using reflectivity observations while the velocity fields, eddy viscosity, and the source term are determined by the 3D-VAR analysis.

The 3D-VAR system was tested on both model-simulated data and real data. Figure 7.2 shows the retrieved wind vector and vertical velocity contours in a vertical cross section through a supercell storm that occurred at Arcadia, Oklahoma. The results using observations from two radars are plotted in Fig. 7.2a and from one radar in Fig. 7.2b. It is shown that the configuration of the storm is well retrieved when observations from only a single radar are used, but the strength of the updraft is much weaker compared to the experiment that used two radars.

b. Variational assimilation

The variational assimilation technique was applied to convective-scale data assimilation a few years after the adjoint technique was introduced to the meteorology and oceanography community. Wolfsberg (1987), Sun et al. (1991), and Kapitza (1991) were the first to test the 4D-VAR technique on the single-Doppler retrieval problem using simulated data. Since then, the technique has been used to retrieve wind and thermodynamic field from real data (Sun and Crook 1994) and to obtain microphysical quantities (Verlinde and Cotton 1993; Sun and Crook 1997, 1998; Wu et al. 2000) with simulated and real data. Recently, the Variational Doppler Radar Analysis System (VDRAS) developed at the National Center for Atmospheric Research (NCAR) has been implemented

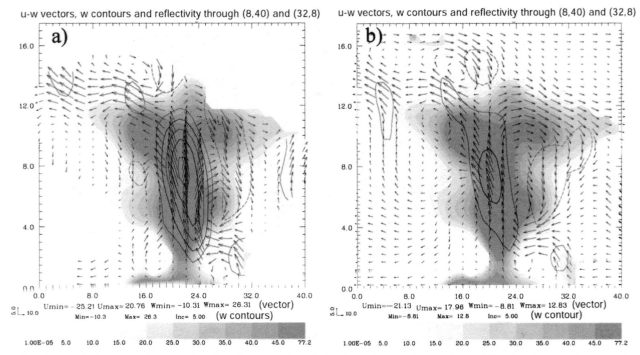

u-w vectors, w contours and reflectivity through (8,40) and (32,8)

a)

Umin= −25.21 Umax= 20.76 Wmin= −10.31 Wmax= 26.31 (vector)

Min= −10.3 Max= 26.3 Inc= 5.00 (w contours)

1.00E−05 5.0 10.0 15.0 20.0 25.0 30.0 35.0 40.0 45.0 77.2

u-w vectors, w contours and reflectivity through (8,40) and (32,8)

b)

Umin= −21.13 Umax= 17.96 Wmin= −8.81 Wmax= 12.83 (vector)

Min= −8.81 Max= 12.8 Inc= 5.00 (w contour)

1.00E−05 5.0 10.0 15.0 20.0 25.0 30.0 35.0 40.0 45.0 77.2

FIG. 7.2. Vertical cross section of retrieved wind vector and vertical velocity (contours) using the 3D-VAR Doppler analysis method for the Arcadia, Oklahoma, 17 May 1981 tornadic storm. The plots are vertical cross sections from the experiment (a) using dual-Doppler radars and (b) using single-Doppler radar. The shaded area is reflectivity. The axis label has the unit of km and the velocity is in the unit of m s^{-1}. Adapted from Gao et al. (2001).

to run operationally in a few field demonstration programs (Sun and Crook 2001a; Crook and Sun 2002) to produce a low-level wind and temperature analysis from single- or dual-Doppler observations.

The main components of VDRAS include a cloud-scale numerical model, the adjoint of the numerical model, a cost function, a minimization algorithm, specifications of weighting coefficients, and a simple background error algorithm. For details on each of these components, readers are referred to Sun and Crook (1997, 2001a). The numerical model used to represent the convective-scale motion is anelastic with Kessler-type warm rain microphysical parameterization. A detailed description of the numerical model can be found in Sun and Crook (1997). There are seven prognostic equations: one each for the three velocity components u, v, and w, the liquid water potential temperature θ_l, the water vapor mixing ratio q_v, the total water mixing ratio q_t, and the rainwater mixing ratio q_r. The pressure p is diagnosed through a Poisson equation. The temperature T and the cloud water mixing ratio q_c are diagnosed from the prognostic variable by assuming that all vapor in excess of the saturation value is converted to cloud water. The lateral boundary conditions of the numerical model are open, such that the inflow is prescribed and the outflow is extrapolated using the closest inner two grid points. The top and bottom boundary conditions are set to zero for vertical velocity, and all

other variables are defined such that their normal derivatives vanish.

By fitting the model to observations over a specified time period, a set of optimal initial conditions of the constraining numerical model can be obtained. The cost function has the same form as that in 3D-VAR given by (7.4). The general form of the background term J_b is given by the first term in the cost function (7.3). Although a common practice in data assimilation is to use the numerical forecast from the previous analysis–forecast cycle as the background, in our system a different approach is taken. A background analysis is obtained using all available observations or estimates other than the radar observation. In the past, some or all of these observations—upper-air sounding, profiler, surface mesonet, and VAD wind estimate—have been used in VDRAS to define a mesoscale background through a mesoscale analysis procedure. When profilers and surface mesonet are not available, an upper-air sounding and the VAD wind are combined to define the background.

The second term J_o in Eq. (7.4) represents the discrepancy of the analysis from the radar observations. Its general form is given by the second term in Eq. (7.3). When radar data are used in the cost function, under the assumption that the errors in the observations are not correlated, the observation term in Eq. (7.3) becomes

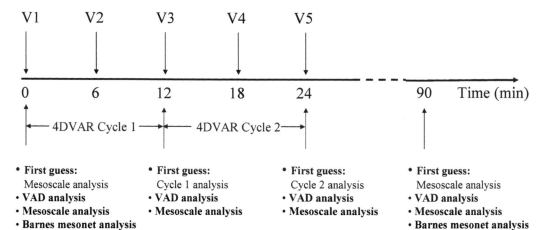

FIG. 7.3. Illustration of 4D-VAR cycles. The radar volumetric input data are shown by the arrows above the time axis; the first guess and the analyses performed at the beginning of each cycle are listed. The Barnes mesonet analysis is performed every 90 min. The mesoscale analysis is an analysis that blends the mesonet analysis and the VAD analysis [see Sun and Crook (2000) for details].

$$J_o = \sum_{\sigma, t} \{ \eta_v [F(v_r) - v_r^o]^2 + \eta_Z [F(Z) - Z^o]^2 \}, \quad (7.5)$$

where v_r^o and Z^o represent the input radial velocity and reflectivity, and v_r and Z are their model counterparts, respectively. The model radial velocity is not a predicted variable but can be computed from the velocity components (u, v, w). The operator F stands for the function that transforms a variable from its model grid to the input data grid. Its expression can be found in Sun and Crook (2001a). The variables η_v and η_Z are weighting coefficients for radial velocity and reflectivity, respectively, and they are specified based on the typical scales of these fields. The summation in Eq. (7.5) is over σ and t, which stand for the spatial and temporal extents of the assimilation window.

As already mentioned, the radar reflectivity measures the concentration of the precipitating hydrometeors when there is precipitation. In the clear air, the reflectivity measures the concentration of small particles such as insects that exist in the boundary layer. For the clear air reflectivity, a conservation equation is used to model the evolution of the reflectivity. For precipitation particles, the reflectivity is converted to the rainwater mixing ratio q_r using a z–q_r relation. In the latter case, the second term in (7.5) is formulated by the difference between the model rainwater and the derived rainwater from the observed reflectivity.

The third term J_p in Eq. (7.4) is the spatial and temporal smoothness penalty term, which takes the following form:

$$J_p = \sum_{\sigma, t, i, j} \alpha_{1i} \left(\frac{\partial A_j}{\partial x_i} \right)^2 + \alpha_{2i} \left(\frac{\partial^2 A_j}{\partial x_i^2} \right)^2 + \alpha_{3i} \left(\frac{\partial A_j}{\partial t} \right)^2$$
$$+ \alpha_{4i} \left(\frac{\partial^2 A_j}{\partial t^2} \right)^2, \quad (7.6)$$

where A_j represents any of the model-dependent vari-

ables and x_i represents the spatial dimension (x, y, z). The weighting coefficients in the penalty terms are determined in a trial-and-error fashion. Discussion on the determination of these coefficients can be found in Sun and Crook (2001a).

A continuous cycling procedure is employed in VDRAS. The length of the assimilation window in each analysis cycle is usually set to 12 min, consisting of three radar volumes in WSR-88D storm mode or two radar volumes in clear air mode. The storm mode covers 14 elevation angles in 5–6 min, and a radar is put into this mode when precipitation is present or expected within the scan region of the radar. The clear air mode covers five elevation angles in 10 min. This mode is used to detect early formation of precipitation and boundary layer discontinuities, and to obtain vertical wind profiles. The antenna scans more slowly and more averaging is done to improve detection of the weaker reflectivities from the "optically clear" air. An optimal trajectory is obtained from each cycle using data within the assimilation window. Only the analysis fields at the final time of the assimilation window are written to disk and displayed. These analysis fields are also used as first guess fields in the next analysis cycle. Figure 7.3 shows an example of how the analysis proceeds in time assuming observations from a WSR-88D radar are used and the radar is operating in storm mode (assuming each volume takes 6 min to complete). If the analysis system does not find any previous analysis within a specified window (24 min, as in the case of the 4D-VAR cycle 1 in Fig. 7.3), the mesoscale analysis is used as the first guess.

It should be noted that, in the operational boundary layer VDRAS, each 3D radar data volume is treated as if the data are collected at a single time in spite of the fact that each measurement corresponds to an observation time and an entire volume scan usually takes 5–

10 min. VDRAS is able to sequentially assimilate the volumetric data by ingesting only a portion of the data at each time step. Sun and Crook (1998) found differences as large as 30% in the retrieved fields between two experiments with and without a time field designating the observation time of each data sample. The radar volume is treated as a snapshot in the operational application of the boundary layer VDRAS due mainly to the efficiency consideration since a longer assimilation window is required to account for the time difference in a data volume.

VDRAS has been used for two major applications. One is to provide low-level wind and temperature analysis using mainly clear air returns in the boundary layer. In this application the numerical model is reduced to a dry Boussinesq system and only the lower atmosphere (usually below 3 km) is covered. This we refer to as boundary layer VDRAS. The second application of VDRAS is to initialize a storm prediction model by retrieving the velocity, thermodynamical, and microphysical variables using hydrometeor radar returns and clear air returns in the low level, such that an explicit numerical prediction of thunderstorms can be made. Both of these applications are discussed below.

1) APPLICATIONS OF BOUNDARY LAYER VDRAS

Boundary layer VDRAS was implemented in a real-time setting at the Washington–Baltimore National Weather Service Forecast Office in Sterling, Virginia, in 1998 and 1999 (Sun and Crook 2001a). It was also implemented at the Bureau of Meteorology in Sydney, Australia, as part of the Sydney 2000 Forecast Demonstration Project (FDP; Keenan et al. 2001) sanctioned by the World Weather Research Program that included the period of the 2000 Sydney Summer Olympics (Crook and Sun 2002). In these field tests VDRAS assimilated data from one or two radars as well as observations from surface mesonet and wind profilers. For both installations the VDRAS wind information was available for forecaster viewing and was input into the NCAR Thunderstorm Nowcasting System called the "Auto-nowcaster (ANC)" (Roberts et al. 1999; Wilson et al. 1998). As shown schematically in Fig. 7.4, the boundary layer VDRAS includes a data ingest, a quality control and data preprocessing step, the 4D-VAR assimilation, and Display by Configurable Interactive Data Display (CIDD) developed at NCAR. The output of VDRAS along with other predictor fields goes into the ANC to produce 0-1 hour nowcast.

In the summer of 2001, the system was run to assimilate data from the Denver WSR-88D (radar station KFTG) and, for the first time, from the Denver Terminal Doppler Weather Radar (TDWR) located about 7 km south of the KFTG radar. Since these two radars are closely located, dual-Doppler coverage provides little dual-Doppler velocity information. However, the TDWR radar helped improve the analysis due to the

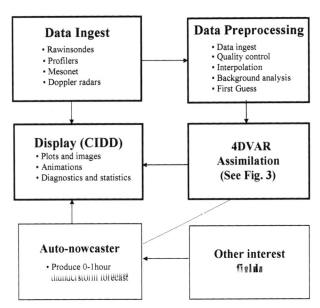

FIG. 7.4. The principal elements of the operational boundary layer VDRAS.

high quality of the data and its additional scan levels in the lower atmosphere. Figure 7.5 compares the wind analysis without and with the TDWR radar. In Fig. 7.5a, the retrieved horizontal wind vectors and the horizontal convergence at the lowest model level from the KFTG radar are displayed along with the KFTG reflectivity at the 0.5 elevation angle. The corresponding retrieval in Fig. 7.5b includes the TDWR Doppler velocities and reflectivity. It is clearly shown that when the TDWR data are included in the assimilation, an apparent more realistic wind field and stronger divergence field is obtained. Furthermore, the cold air surging from a new cell that developed north of the main convection region is well retrieved when the TDWR radar is included.

VDRAS wind retrievals are an integral part of the ANC. The ANC is an automated system for producing 30- and 60-min nowcasts of convective-storm location and intensity. A unique aspect of the ANC is its ability to initiate new storms and grow and dissipate existing storms. The ANC ingests a variety of data, including radar, satellite, lightning, surface, upper-air observations, and VDRAS boundary layer wind fields. Analysis algorithms operate on these data to perform a variety of functions that include data quality control, feature detection, tracking, and characterization. The outputs from these algorithms are used to define predictor fields. Examples of predictor fields are extrapolated echoes, presence of a boundary, boundary collision, storm motion relative to boundary motion, and satellite-observed cumulus or cumulus congestus. Functions are applied to each predictor field to create a field of thunderstorm likelihood fields. The likelihood numbers range from −1 (most unlikely) to +1 (most likely). The thunderstorm likelihood fields are weighted and summed to create a combined likelihood field. A threshold value is

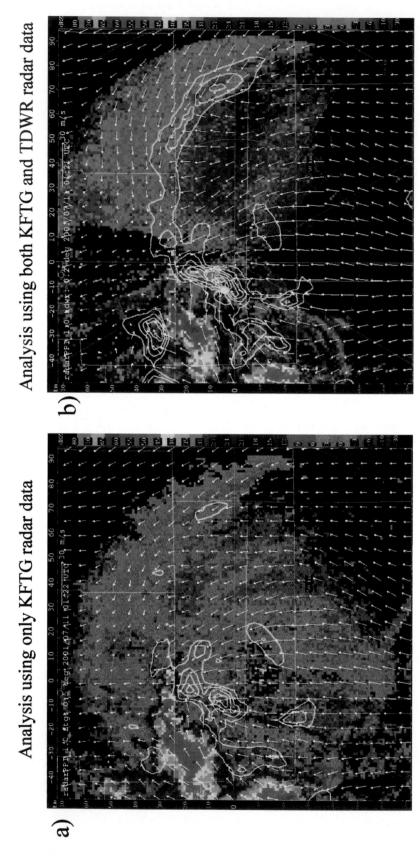

FIG. 7.5. Retrieved horizontal wind vector and convergence field (contours) at $Z = 0.18$ km. The contour interval is 0.4×10^{-3} s^{-1} starting from 0.4×10^{-3} s^{-1}. (a) Plotted results are from the experiment using only the KFTG radar observations, and the overlaid reflectivity is at the 0.5 elevation angle from the KFTG radar. (b) Plotted results are from the experiment using both KFTG and TDWR radar observations, and the overlaid reflectivity is at the 0.2 elevation angle from the TDWR radar.

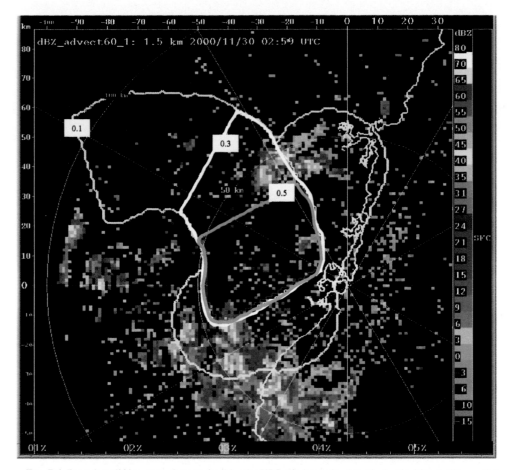

FIG. 7.6. Boundary lifting area characterized by the VDRAS-determined vertical velocities (yellow = 0.1 m s⁻¹, White = 0.3 m s⁻¹, and cyan = 0.5 m s⁻¹) overlaid on the extrapolated radar reflectivity field. The lifting area, vertical velocities, and reflectivity echoes are 60-min extrapolations of existing conditions at 0259 UTC. The data are for 30 Nov 2000 from the Sydney 2000 FDP. Range rings are at 50-km intervals.

selected to convert this final likelihood field into a categorical nowcast of a thunderstorm. The functions for converting to thunderstorm likelihood and weight to apply to each predictor field are based on research experiments (Wilson and Mueller 1993).

The initiation and growth of convective storms by the ANC is highly dependent on the detection and characterization of boundary layer convergence lines and is based on scientific studies that show that storm initiation and growth are often favored along boundary layer convergence lines (Purdom 1973, 1976, 1982; Wilson and Schreiber 1986; Carbone et al. 1990; Fankhauser et al. 1995; Koch and Ray 1997). The strength and depth of the vertical velocity associated with the boundary is often an indicator of the likelihood of convection and its intensity (Foote 1984; Crook and Klemp 2000).

One of the predictor fields called "MaxW" is based on VDRAS vertical velocities. The term MaxW arises because the maximum vertical velocities at 1-km height within circles spaced linearly along the boundary are obtained from VDRAS at the time the nowcast is prepared. The nominal radius of the circles is 5 km with spacing between centers of 10 km. The purpose of this field is to characterize the strength of the convergence line. The convergence line is extrapolated to its nowcast position at 30 and 60 min based on its past motion. It is assumed that the magnitude of the vertical velocities at nowcast time will move with the boundary without change.

Figure 7.6 is an example of how the MaxW field is used by the ANC to improve 60-min nowcasts. The example is from the Sydney 2000 FDP on 30 November 2000. In this case a large area of storm initiation and growth was associated with the interaction of the sea-breeze front and a gust front. Figure 7.6 shows a "band-aid" shaped feature (yellow contour) that is called the boundary lifting area. It is ~35 km wide about the boundary and has been characterized by three contours of MaxW (0.1, 0.3, and 0.5 m s⁻¹). These contours show the 60-min extrapolated position. As discussed above, they were extrapolated with the boundary motion. Also shown in Fig. 7.6 is the extrapolated position of radar reflectivity echoes. Note that the high MaxW values are mostly not located over the extrapolated echoes. Prior

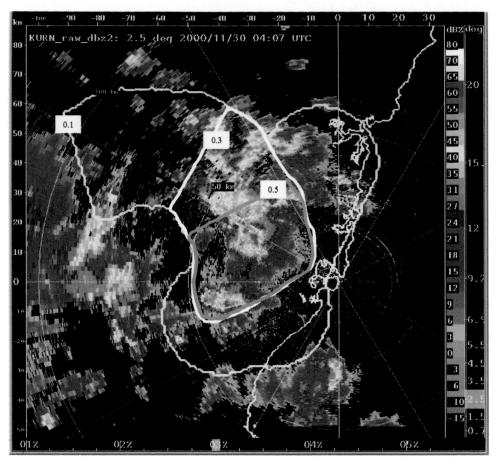

FIG. 7.7. Same as Fig. 7.6 except the verifying echo at 0407 UTC is shown instead of the 60-min extrapolated echo. It is evident that significant storm initiation took place in the area of high vertical velocities.

to initiation the area of maximum convergence was not associated with the existing echoes. Figure 7.7, which shows the actual reflectivity at verification time, indicates that extensive storm initiation has taken place in the region of MaxW values >0.3 m s^{-1}, where only a small echo had been extrapolated. Such high correlation of vertical velocity and storm presence and strength are not always observed. Other factors that can be significant contributors are stability (Crook 1996; Weckwerth 2000), the relative motion between the storms and boundary (Weisman and Klemp 1986; Wilson and Megenhardt 1997), and the horizontal-vorticity balance across boundaries (Rotunno et al. 1988; Weisman and Klemp 1986).

Further evidence of the utility of the VDRAS winds to improve the nowcasts for the 30 November case is illustrated in Fig. 7.8 where the time history of the Critical Success index (Donaldson et al. 1975) for three nowcasting techniques is compared. The nowcasts are for 60 min and for reflectivity >34 dBZ. The forecasts are verified on a 1-km grid with no credit given if the nowcast is in error by as little as 1 km, thus the low value of the skill scores. The first technique (labeled extrapolation) is that by simple extrapolation of existing

radar echoes based on their past motion. The second technique (labeled "only bdry") nowcasts reflectivity >34 dBZ everywhere within the boundary lifting zone. The third technique (labeled MaxW) nowcasts reflectivity >34 dBZ everywhere within the lifting area where MaxW is >0.35 m s^{-1}. It is apparent from Fig. 7.8 hat nowcasting storms to cover the entire boundary lifting area generally provided higher skill scores than the echo extrapolation technique. In addition the skill was generally increased further by only nowcasting storms where the vertical velocity in the lifting area was >0.35 m s^{-1}. Such improvement is not realized for all cases because of other circumstances already listed above. However, research in progress indicates that nowcasts that utilize boundary characterization information are generally superior to echo extrapolation–only techniques.

2) VDRAS INITILIZATION FOR EXPLICIT FORECAST OF THUNDERSTORMS

While the boundary layer VDRAS has demonstrated some success when applied to real-time retrieval of the low-level wind and temperature, the full VDRAS with

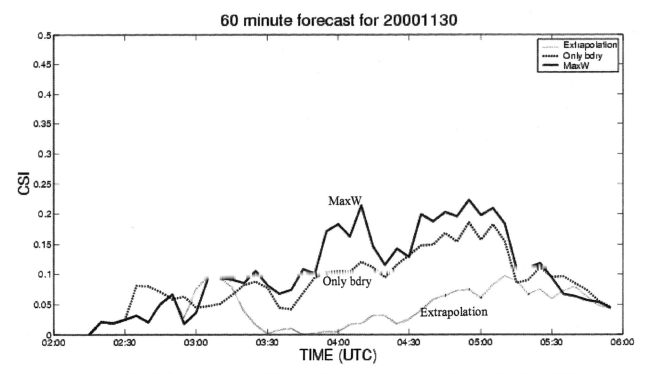

FIG. 7.8. Time series of the Critical Success Index (CSI) for the 60-min nowcasts of radar reflectivity >34 dBZ on 30 Nov (same case as Fig. 7.6). Three nowcast techniques are presented: 1) extrapolation, which is linear extrapolation of radar echoes based on their past motion; 2) only bdry, which is nowcasting echo >34 dBZ within the entire boundary lifting area; and 3) MaxW, which nowcasts reflectivity everywhere with the boundary lifting area where the vertical velocity is >0.

microphysical retrieval has mainly been used in case studies for initialization and prediction of thunderstorms. In the boundary layer VDRAS, our focus is to obtain the flow structure in the boundary layer where a radar receives returns from clear air particles. When the full VDRAS is used, the emphasis of the data assimilation is to provide internal structure of the thunderstorms observed through the hydrometeor returns. The model vertical domain extends to a height of 16 km, in contrast to only a few kilometers in the boundary layer VDRAS. Greater challenge is present in the full VDRAS because there are more unknown variables to be determined and the model can be less reliable due to uncertainties often present in the microphysical parameterization.

The first case study using the VDRAS with the microphysical processes was conducted by Sun and Crook (1998) on an airmass storm observed during the Convection and Precipitation/Electrification Experiment (CAPE). Experiments using observations from both dual-Doppler radars and single-Doppler radar were conducted and the results did not show significant difference. The retrieved thermodynamical and microphysical fields were verified against aircraft observations. Figure 7.9 shows a comparison between the Wyoming *King Air* N2UW observation and the retrieval at 3.0 km above ground along an aircraft track using dual-Doppler observations. The retrieval is seen to have fairly good

agreement with the observation. VDRAS was also used by Wu et al. (2000) in a study of dynamical and microphysical retrieval of an observed hailstorm case from the Microburst and Severe Thunderstorm (MIST) dual-Doppler observations. In that study, the VDRAS-retrieved fields were compared with dual-Doppler analyzed fields. VDRAS reproduced the main features of the hailstorm although the intensities were weaker. Warner et al. (2000) applied VDRAS to the initialization and forecasting of the flash flood case of 13 July 1996 that occurred near Buffalo Creek, Colorado, using the Denver WSR-88D level-II data. Figure 7.10 is a plot of the reflectivity field at the initialization time and for 30- and 60-min forecast periods, compared with the observed reflectivity for this case. The threat score for the 1-h forecast shown in Fig. 11 indicates that the numerical forecast significantly improves over persistence and extrapolation. The forecast storm was able to last more than 2 h, but its correlation with the observed storm decreased rapidly after 2 h.

A recent study by Crook and Sun (2001) showed that forecasts initialized by radar data through VDRAS were sensitive to the specification of the storm environment. In their study using supercell storm data observed by a single-Doppler radar, they have found that the strength of the storm at the end of the 2-h forecast depended very much on the low-level moisture and the mean wind direction. To further investigate the sensitivity of fore-

Florida Air-Mass Storm

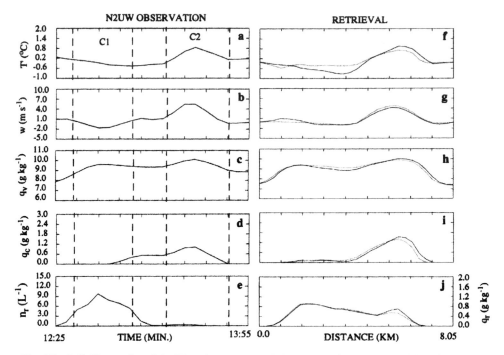

FIG. 7.9. (left) Time series of the Wyoming *King Air* N2UW observations (at $z = 3.0$ km) for a Florida airmass storm during CaPE. The displayed fields are (a) perturbation temperature, (b) vertical velocity, (c) water vapor mixing ratio, (d) mixing ratio of cloud droplets, and (e) number of concentration of large drops. (right) The retrieved fields along the same track. The same fields are displayed as in the left panels, except in the last panel where the retrieved rainwater mixing ratio is shown. The solid line and the dashed line represent results from two slightly different experiments; (left) The dashed lines indicate the boundaries of two convective cells marked by C1 and C2. From Sun and Crook (1997).

casting supercell storms with respect to data availability and quality, Sun and Crook (2001b) performed assimilation and forecast experiments using data from a simulated supercell storm. Using radial velocity and rainwater data from the control simulation as "observations," a series of assimilation and forecast experiments were conducted by degrading the radial velocity and rainwater data and the environment sounding data in various ways. The model initialization was performed at both the cumulus stage and growth stage, so the sensitivity experiments were conducted at both stages. Their main findings from this study are as follows: 1) the forecast is most sensitive to the accuracy of the low-level moisture from the environmental sounding and the availability of the radar radial velocity observations; and 2) at both stages, assimilation of radial velocity observations plays an important role for the subsequent forecast, but the impact of the low-level moisture is more significant at the cumulus stage than at the growth stage. Figure 12 shows the rainwater correlation coefficient between the sensitivity experiment and the control simulation for four experiments, two performed at each of the cumulus stage and the growth stage. The solid curve shows the result from the experiment in which the low-

level water vapor is reduced by 1 g kg⁻¹ and the dotted curve from the experiment that uses only reflectivity observations. The other two curves display results from similar experiments but at the growth stage.

c. Sequential assimilation

The 4D-VAR assimilation attempts to determine the best analysis for all the prognostic model variables by combining the observations and the numerical model in a single analysis step. Furthermore, since the evolution of the atmospheric state is taken into account by incorporating the numerical model into the analysis, each 4D-VAR assimilation cycle produces an analysis that fits the observations over a period of time. In contrast, a number of other techniques that have been developed over the last decade for radar data assimilation require multiple analysis steps to produce the initial conditions of a numerical model at a single time. In addition, the merging of the analysis with the numerical model is performed in a separate step by integrating the model forward, starting from the analysis and using the forecast as the first guess for next analysis. These systems are similar because of their sequential nature, but the

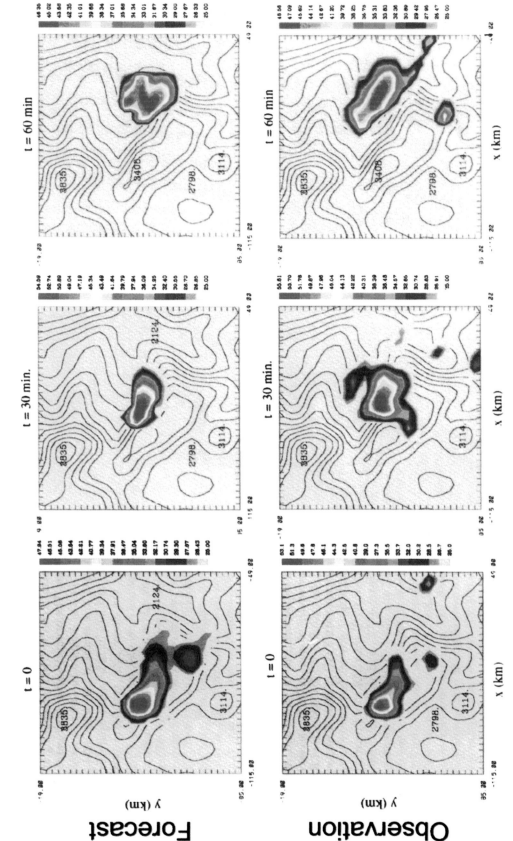

FIG. 7.10. Numerical model forecast (upper) and verifying radar reflectivity (lower) of the Buffalo Creek flash flood that occurred in the Colorado mountains. Forecast periods are $t = 0$, 30, and 60 min. The initial conditions for the model forecast are from VDRAS. The contours show the surface elevation.

Skill Score for Buffalo Creek Storm

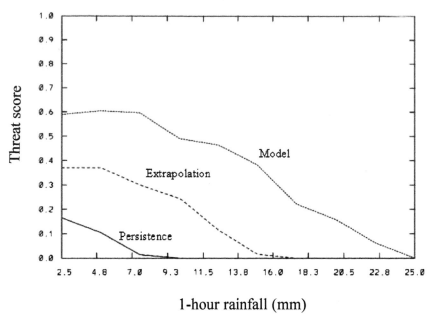

FIG. 7.11. Threat score for the Buffalo Creek storm from the 1-h model forecast shown in Fig. 7.10 and compared with extrapolation and persistence forecasts.

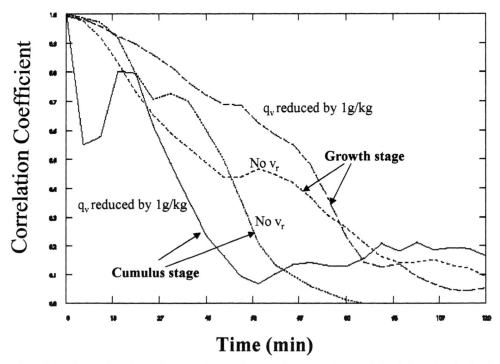

FIG. 7.12. Time series of correlation coefficient between forecast and control simulation of a simulated supercell storm. Two sensitivity experiments are shown at each of the cumulus stage and the growth stage: one with reduced low-level moisture in the mean profile and the other with no radial velocity observations.

analysis procedure in each system is quite different. In this section, we briefly review these systems.

The first sequential assimilation system was developed by Liou (1990). In his study, the continuous updating technique of Charney et al. (1969) was blended with a diagnostic pressure and temperature retrieval technique (Gal-Chen 1978) using a cloud-scale numerical model. Experiments were conducted on model-generated data of a thermal bubble convection in which the velocity component u (instead of radial velocity) was assumed to be observed. It was shown that the assimilation run converged toward the control run after a period of around 20 min. The results from this study was encouraging because it demonstrated that it was possible to obtain the flow fields by combining information from a single-Doppler radar, a numerical model, and the thermodynamic retrieval technique. The shortcoming of this technique is that the radial velocity cannot be directly ingested because it is not a model variable. Thus, when this technique is applied to real data, a single-Doppler retrieval scheme must be incorporated. At the Center for Analysis and Prediction of Storm (CAPS), a sequential system was developed and applied to real data. This system consists of the single-Doppler velocity retrieval (SDVR) scheme developed by Shapiro et al. (1995), a variational velocity adjustment to blend with the background wind, a thermodynamic retrieval, and a moisture specification step. By successively applying these techniques, the initial conditions are specified and the numerical model is integrated forward until the next observation becomes available.

The CAPS sequential system for radar data assimilation was implemented into the Advance Regional Prediction System (ARPS) model in a case study of a supercell storm that occurred in Arcadia, Oklahoma (Weygandt et al. 2002a,b). The observed radial velocity from a nearby Doppler radar was used as input to the SDVR scheme to determine the 3D wind within the radar data coverage region. The retrieved velocity field was then blended, through variational adjustment, with a horizontally homogeneous wind provided by a nearby sounding. The temperature and pressure perturbation was obtained by applying the thermodynamical retrieval technique, and the moisture field was specified based on the radar-observed reflectivity by assuming that regions where the rainwater exceeded 0.1 g kg^{-1} and the retrieved vertical velocity exceeded 3 m s^{-1} were saturated. Using the initial conditions provided in this manner, the ARPS model was integrated forward and the prediction was verified against a dual-Doppler analysis. Figure 7.13 shows the rainwater correlation between the prediction and the analysis for a period of about 45 min. In Fig. 7.14b, the 26-min forecast field of the rainwater mixing ratio and wind vector at the height of 2.25 km is shown and compared with the dual-Doppler analysis field (Fig. 7.14a). The general storm evolution and motion are reasonably well predicted for about 30 min. However, the predicted storm weakens too rapidly com-

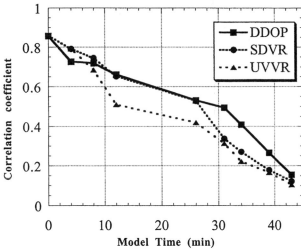

3-D rainwater correlation

FIG. 7.13. Time series of correlation coefficients between the verifying analysis and various prediction cases. DDOP represents the case starting from a dual-Doppler analysis, SDVR from the single-Doppler analysis, and UVVR from a simplified version of SDVR (from Weygandt et al. 2002b).

pared with the observations. The authors speculated that the inability of retrieving the low-level cold pool by this technique might have been the cause of the rapid decaying of the predicted storm.

Montmerle et al. (2001) presented another sequential assimilation system that successively employed the 3D wind retrieval technique of Laroche and Zawadzki (1994), a variational thermodynamical retrieval technique, and a moisture specification method in the attempt to obtain the initial conditions using bistatic Doppler radar data for the simulation of a midlatitude convective storm. Their assimilation and forecast experiments have shown that the simulated shapes and intensities of the precipitation cells were of the same order as those observed during the 30 min of the integration, but the model solution diverged from the observations beyond the 30-min period. They suggested that this divergence was mainly due to the model's inability to represent the stratiform part of the storm.

More recently, the EnKF technique has been applied to the convective scale by Snyder et al. (2001). Initial tests on a simulated supercell storm indicated that the filter was stable and well behaved for an ensemble of 50 members. Further study using simulated data and real data will be conducted in the near future (C. Snyder and F. Zhang 2002, personal communication). Although the EnKF remains largely unproven for radar data assimilation, it has a number of appealing properties. One of them is that it provides a more dynamically consistent way of combining numerical model and observations compared to other sequential assimilation algorithms discussed in this section.

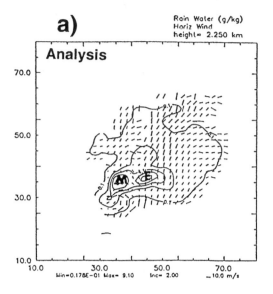

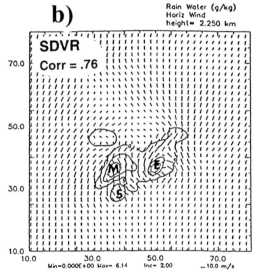

FIG. 7.14. Low-level ($z = 2.25$ km) rainwater and storm-relative wind vectors for (a) the dual-Doppler analysis and (b) from a 26-min prediction using the sequential assimilation technique of CAPS. Contours are in 2 g kg^{-1} increments starting with 0.1 g kg^{-1}. The centers of the three main cells are marked. From Weygandt et al. (2002b).

4. Assimilation of radar data at mesoscale

Assimilation of radar data into mesoscale numerical models presents unique challenges. Because mesoscale phenomena are often closely associated with large-scale forcings, radar data alone are insufficient to initialize a mesoscale model; hence, it is necessary to combine observations from large-scale observing networks. If a sophisticated scheme based on the 4D-VAR or EnKF technique is used, assimilating full volumes of radar data into a high-resolution mesoscale model can result in high computational cost. In the last few years, study has begun to explore the feasibility of incorporating radar observations into mesoscale data assimilation systems and to assess their impact on forecast of mesoscale

weather systems. In this section, we provide a brief review on the use of radar data in mesoscale models.

a. Assimilation of radar estimate rainfall

In a study of a prefrontal rainband with embedded convection, Guo et al. (2000) assimilated the hourly rainfall data generated by merging 15 WSR-88D radar precipitation estimates with approximately 500 rain gauges over the Kansas–Oklahoma area. The Mesoscale Model version 5 (MM5) 4D-VAR system with a horizontal resolution of 20 km was used for the assimilation. The cost function associated with the rainfall estimates was defined by

$$\mathbf{J}_r = \sum_{0 \le k \le N} [\mathbf{R}\mathbf{x}_k - \mathbf{R}_k^{obs}]^T \mathbf{W}_r [\mathbf{R}\mathbf{x}_k - \mathbf{R}_k^{obs}], \quad (7.7)$$

where k is the temporal index, $\mathbf{R}^{obs}$ represents the observed rainfall and x the model variables, R denotes the operator that calculates the model-predicted rainfall, and $\mathbf{W}_r$ is the diagonal weighting matrix for rainfall data. The 4D-VAR approach clearly has an advantage in assimilating rainfall data over some of the traditional techniques because it does not require a priori diagnostic adjustment on moisture, divergence, heating rate, etc., as in some of the traditional techniques (e.g., Wang and Warner 1988; Carr and Baldwin 1991; Krishnamurti et al. 1991). Rather, the rainfall data are directly assimilated into the model along with other observations. Guo et al. (2000) found that assimilating the rainfall data improved the precipitation amount and pattern and had a significant impact on the temperature retrieval during the assimilation period. Since no forecasting experiments were performed, it was not known whether the improved analysis could result in any positive impact on the subsequent forecast. In a more recent study by Zupanski et al. (2002), the 4-km hourly precipitation analysis from both radar and gauge measurements produced at the National Centers for Environmental Prediction (NCEP; Baldwin and Mitchell 1996) were assimilated into the NCEP Eta model with a horizontal model resolution of 32 km. They examined two cases: one is the Great Plains tornado outbreak of 3 May 1999 and the other is the East Coast blizzard of 2000. It was found that the impact of the precipitation data is insignificant on the former case, but substantial on the latter case. They speculated that the reason for the small impact on the tornado outbreak case was that the precipitation was sporadic during the assimilation period. However, in the blizzard case, a large and continuous area of precipitation was observed and assimilated.

Recently, the assimilation of rainfall estimates was also tested in the operational Eta Data Assimilation System (EDAS). Lin et al. (2001) reported that the hourly rainfall analysis similar to that used in Zupanski et al. (2002) was assimilated into the Eta model by the EDAS system. The model's latent heating, moisture, and cloud water fields were adjusted to make them more consistent with the precipitation analysis. They found that the assimilation of the precipitation analysis significantly improved the model's precipitation field during EDAS and, as a result, had a positive impact on the soil moisture field. Additionally, the assimilation often had a positive impact on the first 6-h forecast of the precipitation.

Although some positive impact has been found in assimilating rainfall estimates, Guo et al. (2000) have pointed out that the assimilation was sensitive to the specification of the weighting W_r. A large range of variation can be found in the specification of the weighting from different studies, suggesting further investigation on the error variance of the estimate rainfall should be pursued.

b. Assimilation of VAD wind

A standard output from the WSR-88D network is the VAD wind data. These VAD wind data have a horizontal resolution higher than the routine radiosonde data in the United States, so the dataset represents a potentially valuable source of information for mesoscale data assimilation. Michelson and Seaman (2000) examined the quality of the WSR-88D VAD winds using radiosonde measurements that were taken at sites with collocated WSR-88D radar sites in a case study of a high-ozone episode. They found that the mean wind speed and direction errors computed over 90 pairs of measurements were very low, but the rms errors for speed and direction were much greater, indicating that there is much scatter in the data. This result suggests that quality control is very important before the VAD winds can be used in data assimilation schemes. After removing unreliable VAD wind data, Michelson and Seaman (2000) conducted experiments to assess the impact of the these data by the Newtonian relaxation technique, or "nudging" (Stauffer and Seaman 1990). Evaluation using independent data demonstrated that the assimilation of VAD wind data significantly reduced model wind error, especially below 2 km, where the wind data were most numerous and the VAD winds were more reliable.

c. Assimilation of WSR-88D radial velocity and reflectivity data

Assimilation of WSR-88D level-II radial wind and reflectivity data into the MM5 mesoscale model was attempted by Xu et al. (2001). A snowstorm near New York City in 1997 was used in the case study. As a first step, Observing System Simulation Experiments (OSSEs) were conducted using simulated data from MM5. The MM5–4D-VAR was applied to assimilate radial velocity and rainwater mixing ratio data extracted from a control simulation. The OSSEs showed that the assimilation system was unable to accurately retrieve the unobserved model variables. For example, the rms error in the temperature field was reduced only by 30% after 50 iterations. However, the encouraging result was that

even though the error reduction was small, the prediction starting from the initial conditions obtained from the MM5–4D-VAR assimilation showed substantial improvement over directly inserting the observations into the model.

5. Future

Clearly, the assimilation of radar data for weather prediction is still in its infancy. There are far more open questions than solved ones. However, the encouraging results obtained in the last decades warrant further research efforts. With the steady increase in computing power, the operational models will be able to run at grid spacings below 10 km in the near future. Such resolution opens the possibility of using radar data for initialization of the explicit structure of thunderstorms in operational numerical weather forecast. We envision that the research activity in the assimilation of radar data will grow substantially in the next decade. Among many important research directions, we speculate as follows.

1) *Further development of data assimilation techniques that are suitable for convective-scale weather phenomena.* The 4D-VAR technique has demonstrated its ability in retrieving the wind, thermodynamical, and microphysical fields in a dynamically consistent way so as to provide numerical forecasts of storms that were able to last longer than the sequential systems currently available. Compared to any sequential assimilation system, a clear advantage of the 4D-VAR technique is its ability to process more information by including all the data observed at a specified time window at each data assimilation cycle. However, since it treats the forecast model as a deterministic system, uncertainties in the observation and the model can limit the utility of the system. An ensemble-based system, such as the EnKF, has the promise to provide the initial condition with an estimate of its maximum likelihood. Although the EnKF technique has a host of issues that need to be explored and addressed, it should certainly be examined for data assimilation on the convective scale, given the intermittent nature of many of the convective-scale phenomena. It is possible that the best system, even though it is currently impractical due to the tremendous computation cost, could be integration of the EnKF and 4D-VAR, in which the EnKF provides the background statistics information for the 4D-VAR assimilation.

2) *Integration of radar data with other types of data.* The coverage of radar data is limited to the region where either hydrometeors or boundary layer reflectors exist. Furthermore, due to the increasing height of radar beams with distance from the radar and earth curvature effects, boundary layer wind data are usually confined to radar ranges within 100 km. Radar data alone cannot capture the full picture of the flow

structure associated with a storm. Other data sources, for example, from aircraft, wind profiler, and surface mesonet, can be used to define an environmental background and supplement the information near the surface. Since data from these sources have an average resolution much coarser than that of the radar data and are expected to have larger decorrelation length, simultaneous assimilation of all the data types requires a spatially adaptive error correlation function. If an appropriate background error covariance can be modeled, simultaneous assimilation of these different types of data would be able to retain more information than the two-step procedure currently used in VDRAS.

3) *Investigation of radar data impact on the mesoscale.* It is not known yet whether the high-resolution radial velocity observations from radar will benefit the mesoscale NWP, specifically with resolutions coarser than 10 km. A recent study (M. Xu 2002, personal communication) using MM5–4D-VAR system indicated that the assimilation of radial velocity data had a negative impact on the precipitation forecast compared to an experiment that assimilated reflectivity data alone. We believe that research on the impact of assimilating radar data into numerical models to forecast mesoscale weather systems, especially those strongly governed by large-scale forcing, such as convective rainbands embedded in hurricane circulations or tropical cyclones and snowbands associated with cold fronts, is lacking and should be expanded in the near future. One of the challenges is the modeling of background error statistics. The algorithms that are currently used in operational centers are not suitable for convective mesoscale systems. More detailed statistics based on geographical location or weather system type might be necessary for mesoscale applications.

4) *Operational implementation of convective-scale data assimilation and weather prediction.* The most benefit the Doppler radar can provide is likely in the convective scale. Given that most weather systems on the convective scale are often driven by highly local effects, it is probably most economic to conduct the convective-scale NWP, and its data assimilation, in a distributed and "on-demand" manner (Droegemeier 2000). That means local forecast offices guide the execution of their own "customized" versions of a unified model [for instance, the Weather Research and Forecasting (WRF) model] and target areas of particular active weather. This scenario might prove effective for both the prediction and the data assimilation. Due to the different emphases in the convective-scale data assimilation from that in the large scale, it is a formidable task to design an assimilation system that is able to integrate radar data and data from large-scale observing networks in a dynamically consistent manner and to retain the high-resolution information contained in the radar

observations. For example, the hydrostatic balance equation that is often used in many of the meso- or large-scale data assimilation systems is clearly not suitable for the convective scale. Therefore, we propose that the convective-scale data assimilation using radar data and the subsequent prediction are conducted as a separate process. This process is run on demand and uses the coarser-resolution model output to provide background information and boundary conditions. The data assimilation scheme in this fine-resolution model run can be different from that in the coarser resolution.

5) *Data quality control and estimate of error statistics.* Automated data quality control is one of the key components required in radar data assimilation. The WSR-88D radars are equipped with some quality control algorithms, for instance, the ground clutter removal algorithm. However, contaminated data often exist due to bird migrations, velocity folding, anomalous propagation, and sea clutter. Estimate of error statistics for the observed radial velocity and the derived rainwater mixing ratio, or estimate of rainfall from the reflectivity observations is another area that demands a good deal of attention. Active involvement by the experts in radar observation and signal processing is crucial in order to make significant progress in this area.

Acknowledgments. Many of the NCAR's activities reported here were supported by the FAA Aviation Weather Research Program to whom we are grateful. We are also grateful to Drs. Gao and Weygandt for providing us the electronic data of their figures. This research is partially supported by the National Science Foundation through an Interagency Agreement in response to requirements and funding by the Federal Aviation Administration's Aviation Weather Development Program. The views expressed are those of the authors and do not necessarily represent the official policy or position of the U.S. government.

APPENDIX A

3D-VAR Solution Procedure

The stationary point $\mathbf{x}_a$ of the unknown variable (or control variable) $\mathbf{x}$ can be obtained by taking the first-order variation of J with respect to $\mathbf{x}$ and setting it to zero. Starting from Eq. (7.1), the first-order variation can be written as

$$\delta J = 2\mathbf{B}^{-1}(\mathbf{x} - \mathbf{x}_b)\delta\mathbf{x} - \mathbf{H}^\mathsf{T}\mathbf{O}^{-1}(\mathbf{y} - \mathbf{Hx})\delta\mathbf{x}. \quad (7.A1)$$

At the stationary point $\mathbf{x}_a$ the variation of the cost function is zero, that is,

$$\mathbf{B}^{-1}(\mathbf{x}_a - \mathbf{x}_b) - \mathbf{H}^\mathsf{T}\mathbf{O}^{-1}(\mathbf{y} - \mathbf{Hx}_a) = 0, \quad (7.A2)$$

which yields the best analysis $\mathbf{x}_a$:

$$\mathbf{x}_a = \mathbf{x}_b + \mathbf{W}(\mathbf{y} - \mathbf{H}\mathbf{x}_b), \qquad (7.A3)$$

where $\mathbf{W}$ is given by

$$\mathbf{W} = \mathbf{B}\mathbf{H}^\mathrm{T}(\mathbf{H}\mathbf{B}\mathbf{H}^\mathrm{T} + \mathbf{O})^{-1}. \qquad (7.A4)$$

The vector $\mathbf{y} - \mathbf{H}\mathbf{x}_b$ in Eq. (7.A3) is often referred to as the innovation vector and the matrix $\mathbf{W}$ is called the gain matrix. Equation (7.A3) shows that the analysis is the sum of the prior information and a correction term, which is proportional to the difference between the new information brought in by the observation and what is known from the prior information. When $\mathbf{W}$ is computed according to Eq. (7.A4), the analysis produced is optimal in the sense that it gives the minimum variance. However, directly solving Eqs. (7.A3)–(7.A4) is not an easy task due to the large dimension of the matrices that need to be stored and inverted. The full matrix of $\mathbf{B}$ has the dimension of $n \times n$ where n is the number of the control variables defined as the multiplication of the number of the grid points and the number of the model variables. Therefore, simplifications need to be made in order to reduce the computational cost for operational implementation. For instance, one way to simplify the problem is through data selection; that is, one can assume only observations located in the vicinity of a given model grid point have influence on the analysis at that point. The solution procedure of directly solving the linear system (7.A3)–(7.A4) under certain simplifications is used in the OI (Gandin 1963; Daley 1991), a technique widely applied in meteorology before the 3D-VAR technique was introduced. In contrast with the OI, the 3D-VAR assimilation technique takes a different computational approach to determine the best analysis defined by Eqs. (7.A3)–(7.A4). The minimum point of the cost function J is obtained approximately by iteratively minimizing the cost function (7.1). The commonly used algorithm is a variant based on the conjugate gradient method (Gill et al. 1981). In the iterative procedure, the cost function and its gradient are computed at each iteration and used to define the best descending direction. As a result, the large computation involved in solving Eqs. (7.A3)–(7.A4) is avoided. Since in the 3D-VAR technique the optimal analysis is achieved by iteratively approaching the minimum of the cost function (7.1), no data selection (e.g., by defining a radius of influence) is required so a global fitting (i.e., all observations have influence on the analysis of a single grid point) can be sought.

APPENDIX B

Adjoint Equation

The adjoint equation can be obtained by using the technique of Lagrange multiplier (see, e.g., Thacker and Long 1988; Sun et al. 1991). First, a Lagrangian function is defined so that the constrained minimization problem (7.2)–(7.3) is reduced to an unconstrained minimization problem:

$$\mathbf{L} = \mathbf{J} + \int_{t_0}^{t_N} \mathbf{\Lambda}^\mathrm{T}\left[\frac{\partial \mathbf{x}}{\partial t} - \mathbf{F}(\mathbf{x})\right] dt, \qquad (7.B1)$$

where $\mathbf{J}$ is given by Eq. (7.3) and $\mathbf{\Lambda}$ denotes the Lagrange multiplier (also known as the adjoint variable). Then the first-order variation of $\mathbf{L}$ with respect to $\mathbf{x}$ and $\mathbf{\Lambda}$ is taken:

$$\delta\mathbf{L} = \delta\mathbf{J} + \int_{t_0}^{t_N} \delta\mathbf{\Lambda}^\mathrm{T}\left[\frac{\partial \mathbf{x}}{\partial t} - \mathbf{F}(\mathbf{x})\right] dt$$
$$+ \int_{t_0}^{t_N} \mathbf{\Lambda}^\mathrm{T}\left[\frac{\partial \delta\mathbf{x}}{\partial t} - \mathbf{F}'(\mathbf{x})\delta\mathbf{x}\right] dt. \qquad (7.B2)$$

Applying Eq. (7.3), the variation of $\mathbf{J}$ can be expressed as

$$\delta\mathbf{J} = 2\mathbf{B}^{-1}(\mathbf{x}_0 - \mathbf{x}_b)\delta\mathbf{x}_0$$
$$- \int_{t_0}^{t_N} 2\mathbf{H}'\mathbf{O}^{-1}(\mathbf{y} - \mathbf{H}\mathbf{x})\,\delta\mathbf{x}. \qquad (7.B3)$$

In Eqs. (7.B2)–(7.B3), $\mathbf{F}'$ and $\mathbf{H}'$ are the Jacobians of the respective nonlinear model and observation operator $\mathbf{F}$ and $\mathbf{H}$. The last term in Eq. (7.B2) can be expanded using integration by parts:

$$\int_{t_0}^{t_N} \mathbf{\Lambda}^\mathrm{T}\left[\frac{\partial \delta\mathbf{x}}{\partial t} - \mathbf{F}'(\mathbf{x})\delta\mathbf{x}\right] dt$$
$$= \mathbf{\Lambda}_N^\mathrm{T}\delta\mathbf{x}_N - \mathbf{\Lambda}_0^\mathrm{T}\delta\mathbf{x}_0 - \int_{t_0}^{t_N} \frac{\partial \mathbf{\Lambda}^\mathrm{T}}{\partial t}\delta\mathbf{x}\,dt$$
$$- \int_{t_0}^{t_N} \mathbf{\Lambda}^\mathrm{T}\mathbf{F}(\mathbf{x})\delta\mathbf{x}\,dt. \qquad (7.B4)$$

Note that we have used the notation $\mathbf{\Lambda}_0$ and $\mathbf{\Lambda}_N$ to represent $\mathbf{\Lambda}|_{t_0}$ and $\mathbf{\Lambda}|_{t_N}$, respectively, and $\mathbf{x}_0$ to represent $\mathbf{x}|_{t_0}$. Now substituting Eqs. (7.B3) and (7.B4) into Eq. (7.B2) and setting $\delta\mathbf{L}$ to zero results in the following Euler Lagrange equations:

$$\frac{\partial \mathbf{\Lambda}}{\partial t} + \mathbf{F}'^\mathrm{T}(\mathbf{x})\mathbf{\Lambda} + 2\mathbf{H}^\mathrm{T}\mathbf{O}^{-1}(\mathbf{y} - \mathbf{H}\mathbf{x}) = 0, \quad (7.B5)$$

$$\mathbf{\Lambda}_0 - 2\mathbf{B}^{-1}(\mathbf{x}_0 - \mathbf{x}_b) = 0, \qquad (7.B6)$$

$$\mathbf{\Lambda}_N = 0, \qquad (7.B7)$$

$$\frac{\partial \mathbf{x}}{\partial t} - \mathbf{F}(\mathbf{x}) = 0. \qquad (7.B8)$$

Equations (7.B5)–(7.B7) are the so-called adjoint equations. Integration of the adjoint equation (7.B5) backward in time, starting from the condition (7.B7) will result in the required gradient of the cost function. This can be verified by substituting Eqs. (7.B4)–(7.B5) and (7.B7)–(7.B8) into Eq. (7.B2) and letting $\delta\mathbf{L}$ vanish. The resulting equation is

$$\delta\mathbf{J} = \mathbf{\Lambda}_0^{\mathrm{T}}\delta\mathbf{x}_0, \qquad (7.\mathrm{B}9)$$

which implies that

$$\frac{\partial\mathbf{J}}{\partial\mathbf{x}_0} = \mathbf{\Lambda}_0. \qquad (7.\mathrm{B}10)$$

Thus, the adjoint variable at the initial time is equal to the gradient of the cost function $\mathbf{J}$ with respect to the initial condition $\mathbf{x}_0$ of the model state variable.

The minimization procedure is described as follows. The process is started with a first guess estimate of the initial conditions. The numerical model is then integrated forward and the output is used in the computation of the cost function and stored for the adjoint backward integration. The adjoint backward integration yields the gradient of the cost function. With the information of the cost function and its gradient, the minimization algorithm finds an improved estimate of the initial conditions. The process is repeated until the cost function's reduction is no longer significant, which typically requires 50–100 iterations.

REFERENCES

Atlas, D., 1964: Advances in radar meteorology. *Advances in Geophysics,* Vol. 10, Academic Press, 317–478.

Baldwin, M. E., and K. E. Mitchell. 1996: The NCEP hourly multisensor U.S. precipitation analysis. Preprints, *11th Conf. on Numerical Weather Prediction,* Norfolk, VA, Amer. Meteor. Soc., J95–96.

Bouttier, F., 1994: A dynamical estimation of forecast error covariances in an assimilation system. *Mon. Wea. Rev.,* **122,** 2376–2390.

Browning, K. A., and R. Wexler, 1968: The determination of kinematic properties of a wind field using Doppler radar. *J. Appl. Meteor.,* **7,** 105–113.

Carbone, R. G., and Coauthors, 1990: Convective dynamics: Panel report. *Radar in Meteorology,* D. Atlas, Ed., Amer. Meteor. Soc., 391–400.

Carr, F. H., and M. Baldwin, 1991: Incorporation of observed precipitation estimates during the initialization of synoptic and mesoscale storms. Preprints, *First Int. Symp. on Winter Storms,* New Orleans, LA, Amer. Meteor., Soc., 71–75.

Caton, P. A. F., 1963: Wind measurement by Doppler radar. *Meteor. Mag.,* **92,** 213–222.

Charney, J., M. Halem, and R. Jastrow, 1969: Use of incomplete historical data to infer the present state of the atmosphere. *J. Atmos. Sci.,* **26,** 1160–1163.

Cohn, S. E., 1997: An introduction to estimation theory. *J. Meteor. Soc. Japan,* **75,** 257–288.

——, and D. F. Parrish, 1991: The behavior of forecast error covariances for a Kalman filter in two dimensions. *Mon. Wea. Rev.,* **119,** 1757–1785.

Crook, N. A., 1996: Sensitivity of moist convection forced by boundary layer processes to low-level thermodynamic fields. *Mon. Wea. Rev.,* **124,** 1767–1785.

——, and J. B. Klemp, 2000: Lifting by convergence lines. *J. Atmos. Sci.,* **57,** 873–890.

——, and J. Sun, 2001: Assimilation and forecasting experiments on supercell storms. Part II: Experiments with WSR-88D data. Preprints, *14th Conf. on Numerical Weather Prediction,* Ft. Lauderdale, FL, Amer. Meteor. Soc., 147–150.

——, and J. Sun, 2002: Assimilating radar, surface, and profiler data for the Sydney 2000 Forecast Demonstration Project. *J. Atmos. Oceanic Technol.,* **19,** 888–898.

Daley, R., 1991: *Atmospheric Data Analysis.* Cambridge Atmospheric and Space Science Series, Cambridge University Press. 457 pp.

——, 1992: Estimating model-error covariances for application to atmospheric data assimilation. *Mon. Wea. Rev.,* **120,** 1735–1746.

Dee, D. P., 1991: Simplification of the Kalman filter for meteorological data assimilation. *Quant. J. Roy. Meteor. Soc.,* **117,** 365–384.

Donaldson, R. J., R. M. Dyer, and M. J. Kraus, 1975: An objective evaluation of techniques for predicting severe western events. Preprints, *Ninth Conf. on Severe Local Storms,* Norman, OK, Amer. Meteor. Soc., 321–326.

Droegemeier, K. K., 2000: The numerical prediction of thunderstorms: Challenges, potential benefits and results from real-time operational tests. *WMO Bull.,* **46,** 324–336.

Emanuel, K., 1984: What does "mesoscale" mean? *Dynamics of Mesoscale Weather Systems,* J. B. Klemp, Ed., NCAR, 2–12.

Evensen, G., 1994: Sequential data assimilation with a nonlinear quasi-geostrophic model using Monte Carlo methods to forecast error statistics. *J. Geophys. Res.,* **99** (C5), 10 143–10 162.

Fankhauser, J. C., N. A. Crook, J. Tuttle, L. J. Miller, and C. G. Wade, 1995: Initiation of deep convection along boundary layer convergence lines in a semitropical environment. *Mon. Wea. Rev.,* **123,** 291–313.

Foote, G. B., 1984: Study of hail growth utilizing observed storm conditions. *J. Climate Appl. Meteor.,* **23,** 84–101.

Gal-Chen, T., 1978: A method for the initialization of the anelastic equations: Implications for matching models with observations. *Mon. Wea. Rev.,* **106,** 587–606.

Gandin, L. S., 1963: *Objective Analysis of Meteorological Fields* (in Russian). Gidrometeorizdar, 238 pp. (English translation by Israel Program for Scientific Translations, 1965, 242 pp.)

Gao, J., M. Xue, A. Shapiro, and K. K. Droegemeier, 1999: 3D variational wind retrievals from single-Doppler radar. Preprints, *29th Int. Conf. on Radar Meteorology,* Montreal, QC, Canada, Amer. Meteor. Soc., 12–16.

——, ——, K. K. Droegemeier, and A. Shapiro, 2001: A 3-D variational method for single-Doppler velocity retrieval applied to a supercell storm case. Preprints, *30th Int. Conf. on Radar Meteorology,* Munich, Germany, Amer. Meteor. Soc., 456–458.

Gill. P. E. Gill. P. E., W. Murray, and M. H. Wright, 1981: *Practical Optimization.* Academic Press, 401 pp.

Guo, Y., Y.-H. Kuo, J. Dudhia, and D. Parson, 2000: Four-dimensional variational data assimilation of heterogeneous mesoscale observations for a strong convective case. *Mon. Wea. Rev.,* **128,** 619–643.

Hamill, T. M., and C. Snyder, 2000: A hybrid ensemble Kalman filter–3D-variational analysis scheme. *Mon. Wea. Rev.,* **128,** 2905–2919.

Harrold, T. W., 1966: Measurement of horizontal convergence in precipitation using a Doppler radar. *Quart. J. Roy. Meteor. Soc.,* **92,** 31–40.

Hayden, C. M., and R. J. Purser, 1995: Recursive filter objective analysis of meteorological fields: Applications to NESDIS operational processing. *J. Appl. Meteor.,* **34,** 3–15.

Houtekamer, P. L., and H. L. Mitchell, 1998: Data assimilation using an ensemble Kalman filter technique. *Mon. Wea. Rev.,* **126,** 796–811.

Kapitza, H., 1991: Numerical experiments with the adjoint of a nonhydrostatic mesoscale model. *Mon. Wea. Rev.,* **119,** 2993–3011.

Keenan, T., and Coauthors, 2001: The World Weather Research Programme (WWRP) Sydney 2000 Forecast Demonstration Project: Overview. Preprints, *30th Int. Conf. on Radar Meteorology,* Munich, Germany, Amer. Meteor. Soc., 474–476.

Klazura, G. E., and D. Imy, 1993: A description of the initial set of analysis products available from the NEXRAD WSR-88D system. *Bull. Amer. Meteor. Soc.,* **74,** 1293–1310.

Klinker, E., F. Rabier, G. Kelly, and J.-F. Mahfouf, 2000: The ECMWF operational implementation of four-dimensional variational assimilation. Part III: Experimental results and diagnostics with operational configuration. *Quart. J. Roy. Meteor. Soc.,* **126,** 1143–1170.

Koch, S. E., and C. A. Ray, 1997: Mesoanalysis of summertime convergence zones in central and eastern North Carolina. *Wea. Forecasting,* **12,** 56–77.

Krishnamurti, T. N., J. Xue, H. S. Bedi, K. Ingles, and D. Oosterhof, 1991: Physical initialization for numerical weather prediction over the Tropics. *Tellus,* **43,** 53–81.

Laroche, S., and I. Zawadzki, 1994: A variational analysis method for the retrieval of three-dimensional wind field from single-Doppler data. *J. Atmos. Sci.,* **51,** 2664–2682.

Lewis, J. M., and J. C. Derber, 1985: The use of adjoint equation to solve a variational adjustment problem with advective constraints. *Tellus,* **37A,** 309–322.

Lhermitte, R. M., 1970: Dual-Doppler observation of convective storm circulation. *Proc. 14th Radar Meteorology Conf.,* Tucson, AZ, Amer. Meteor. Soc., 139–144.

——, and D. Atlas, 1961: Precipitation motion by pulse Doppler radar. *Proc. 9th Weather Radar Conf.,* Kansas City, MO, Amer. Meteor. Soc., 218–223.

Li, Y., 1991: A note on the uniqueness problem of variational adjustment approach to four-dimensional data assimilation. *J. Meteor. Soc. Japan,* **69,** 581–595.

Lilly, D. K., 1990: Numerical prediction of thunderstorm—Has its time come? *Quart. J. Roy. Meteor. Soc.,* **116,** 779–798.

Lin, Y., P. S. Ray, and K. W. Johnson, 1993: Initialization of a modeled convective storm using Doppler radar-derived fields. *Mon. Wea. Rev.,* **121,** 2757–2775.

——, M. E. Baldwin, K. E. Mitchell, E. Rogers, and G. J. DiMego, 2001: Spring 2001 changes to Eta analysis and forecast system: Assimilation of observed precipitation data. Preprints, *14th Conf. on Numerical Weather Prediction,* Fort Lauderdale, FL, Amer. Meteor. Soc., J92–95.

Liou, Y.-C., 1990: Retrieval of three-dimensional wind and temperature fields from one component wind data by using the four-dimensional data assimilation technique. M.S. thesis, Dept. of Meteorology, University of Oklahoma, 112 pp.

——, 1999: Single radar recovery of cross-beam wind components using a modified moving frame of reference technique. *J. Atmos. Oceanic Technol.,* **16,** 1003–1016.

Lorenc, A. C., 1986: Analysis methods for numerical weather prediction. *Quart. J. Roy. Meteor. Soc.,* **112,** 1177–1194.

Mahfouf, J.-F., and F. Rabier, 2000: The ECMWF operational implementation of four-dimensional variational assimilation. Part II: Experimental results with improved physics. *Quart. J. Roy. Meteor. Soc.,* **126,** 1171–1190.

Michelson, S. A., and N. L. Seaman, 2000: Assimilation of NEXRAD-VAD winds in summertime meteorological simulations over the northeastern United States. *J. Appl. Meteor.,* **39,** 367–383.

Montmerle, T., A. Caya, and I. Zawadzki, 2001: Simulation of a midlatitude convective storm initialized with bistatic Doppler radar data. *Mon. Wea. Rev.,* **129,** 1949–1967.

Navon, I. M., X. Zou, J. Derber, and J. Sela, 1992: Variational data assimilation with an adiabatic version of the NMC spectral model. *Mon. Wea. Rev.,* **120,** 1433–1446.

Parrish, D., and J. Derber, 1992: The National Meteorological Center's spectral statistical-interpolation analysis system. *Mon. Wea. Rev.,* **120,** 1747–1763.

Purdom, J. F. W., 1973: Satellite imagery and the mesoscale convective forecast problem. Preprints, *Eighth Conf. on Severe Local Storms,* Denver, CO, Amer. Meteor. Soc., 244–251.

——, 1976: Some uses of high resolution GOES imagery in the mesoscale forecasting of convection and its behavior. *Mon. Wea. Rev.,* **104,** 1474–1483.

——, 1982: Subjective interpretations of geostationary satellite data for nowcasting. *Nowcasting,* K. Browning, Ed., Academic Press, 149–166.

Qiu, C., and Q. Xu, 1992: A simple adjoint method of wind analysis for single-Doppler data. *J. Atmos. Oceanic Technol.,* **9,** 588–598.

Rabier, F., H. Jarvinen, E. Klinker, J.-F. Mahfouf, and A. Simmons, 2000: The ECMWF operational implementation of four-dimensional variational assimilation. Part I: Experimental results with simplified physics. *Quart. J. Roy. Meteor. Soc.,* **126,** 1143–1170.

Rinehart, R. E., 1979: Internal storm motions from a single non-Doppler weather radar. Tech. Note NCAR/TN-146 + STR, 262 pp.

Roberts, R., T. Saxon, C. Mueller, J. Wilson, A. Crook, J. Sun, and S. Henry, 1999: Operational application and use of NCAR's thunderstorm nowcasting system. Preprints, *15th Int. Conf. on Interactive Information and Processing Systems for Meteorology, Oceanography, and Hydrology,* Dallas, TX, Amer. Meteor. Soc., 158–161.

Rotunno, R., J. B. Klemp, and M. L. Weisman, 1988: A theory for strong, long-lived squall lines. *J. Atmos. Sci.,* **45,** 463–485.

Roux, F., 1985: Retrieval of thermodynamic fields from multiple Doppler radar data, using the equations of motion and the thermodynamic equation. *Mon. Wea. Rev.,* **113,** 2142–2157.

Shapiro, S., S. Ellis, and J. Shaw, 1995: Single-Doppler velocity retrievals with Phoenix II data: Clear air and microburst wind retrievals in the planetary boundary layer. *J. Atmos. Sci.,* **52,** 1265–1287.

Snyder, C., F. Zhang, J. Sun, and A. Crook, 2001: Tests of an ensemble Kalman filter at convective scales. Preprints, *14th Conf. on Numerical Weather Prediction,* Fort Lauderdale, FL, Amer. Meteor. Soc., 444–446.

Stauffer, D. R., and N. L. Seaman, 1990: Use of four-dimensional data assimilation in a limited-area mesoscale model. Part I: Experiments with synoptic-scale data. *Mon. Wea. Rev.,* **118,** 1250–1277.

Sun, J. D., and N. A. Crook, 1994: Wind and thermodynamic retrieval from single-Doppler measurements of a gust front observed during Phoenix II. *Mon. Wea. Rev.,* **122,** 1075–1091.

——, and ——, 1997: Dynamical and microphysical retrieval from Doppler radar observations using a cloud model and its adjoint. Part I: Model development and simulated data experiments. *J. Atmos. Sci.,* **54,** 1642–1661.

——, and ——, 1998: Dynamical and microphysical retrieval from Doppler radar observations using a cloud model and its adjoint. Part II: Retrieval experiments of an observed Florida convective storm. *J. Atmos. Sci.,* **55,** 835–852.

——, and ——, 2001a: Assimilation and forecasting experiments on supercell storms. Part I: Experiments with simulated data. Preprints, *14th Conf. on Numerical Weather Prediction,* Ft. Lauderdale, FL, Amer. Meteor. Soc., 142–146.

——, and ——, 2001b: Real-time low-level wind and temperature analysis using single WSR-88D data. *Wea. Forecasting,* **16,** 117–132.

——, W. Flicker, and D. K. Lilly, 1991: Recovery of three-dimensional wind and temperature fields from single-Doppler radar data. *J. Atmos. Sci.,* **48,** 876–890.

Talagrand, O, 1997: Assimilation of observations: An introduction. *J. Meteor. Soc. Japan,* **75,** 191–209.

——, and P. Courtier, 1987: Variational assimilation of meteorological observations with the adjoint vorticity equation—Part I. Theory. *Quart. J. Roy. Meteor. Soc.* **113,** 1311–1328.

Tarantola, A., 1987: *Inverse Problem Theory.* Elsevier, 613 pp.

Thacker, W. C., and R. B. Long, 1988: Fitting dynamics to data. *J. Geophys. Res.,* **93,** 1227–1240.

Tuttle, J. D., and G. B. Foote, 1990: Determination of the boundary layer airflow from a single Doppler radar. *J. Atmos. Oceanic Technol.,* **7,** 218–232.

Verlinde, J., and W. R. Cotton, 1993: Fitting microphysical observations of nonsteady convective clouds to a numerical model: An application of the adjoint technique of data assimilation to a kinematic model. *Mon. Wea. Rev.,* **121,** 2776–2793.

Wang, W., and T. T. Warner, 1988: Use of four-dimensional data assimilation by Newtonian relaxation and latent heat forcing to improve a mesoscale model precipitation forecast: A case study. *Mon. Wea. Rev.,* **116,** 2593–2613.

Warner, T. T., E. E. Brandes, C. K. Mueller, J. Sun, and D. N. Yates, 2000: Prediction of a flash flood in complex terrain. Part I: A comparison of rainfall estimates from radar, and very-short-range rainfall simulations from a dynamic model and an automated algorithmic system. *J. Appl. Meteor.,* **39,** 815–825.

Weckwerth, T. M., 2000: The effect of small-scale moisture variability on thunderstorm initiation. *Mon. Wea. Rev.,* **128,** 4017–4030.

Weisman, M. L., and J. B. Klemp, 1986: Characteristics of isolated convective storms. *Mesoscale Meteorology and Forecasting,* P. S. Ray, Ed., Amer. Meteor. Soc., 331–358.

Weygandt, S., A. Shapiro, and K. Droegemier, 2002a: Retrieval of model initial fields from single-Doppler observations of a supercell thunderstorm. Part I: Single-Doppler velocity retrieval. *Mon. Wea. Rev.,* **130,** 433–453.

——, ——, and ——, 2002b: Retrieval of model initial fields from single-Doppler observations of a supercell thunderstorm. Part II: Thermodynamic retrieval and numerical prediction. *Mon. Wea. Rev.,* **130,** 454–476.

Wilson, J. W., and W. E. Schreiber, 1986: Initiation of convective storms at radar-observed boundary-layer convergence lines. *Mon. Wea. Rev.,* **114,** 2516–2536.

——, and C. K. Mueller, 1993: Nowcasts of thunderstorm initiation and evolution. *Wea. Forecasting,* **8,** 113–131.

——, and D. L. Megenhardt, 1997: Thunderstorm initiation, organization, and lifetime associated with Florida boundary convergence lines. *Mon. Wea. Rev.,* **125,** 1507–1525.

——, T. M. Weckwerth, J. Vivekanandan, R. M. Wakimoto, and R. W. Russell, 1994: Boundary layer clear-air radar echoes: Origin of echoes and accuracy of derived winds. *J. Atmos. Oceanic Technol.,* **11,** 1184–1206.

——, N. A. Crook, C. K. Mueller, J. Sun, and M. Dixon, 1998: Nowcasting thunderstorms: A status report. *Bull. Amer. Meteor. Soc.,* **79,** 2079–2099.

Wolfsberg, D., 1987: Retrieval of three-dimensional wind and temperature fields from single-Doppler radar data. CIMMS Rep. 84, 91 pp. [Available from Cooperative Institute for Mesoscale Meteorological Studies, 401 East Boyd, 100 East Boyd St., Norman, OK 73019.]

Wu, B., J. Verlinde, and J. Sun, 2000: Dynamical and microphysical retrieval from Doppler radar observations of a deep convective cloud. *J. Atmos. Sci.,* **57,** 262–283.

Xu, M., N. A. Crook, J. Sun, and R. Rasmussen, 2001: Assimilation of radar data for 1–4 hour snowband forecasting using a mesoscale model. Preprints, *14th Conf. on Numerical Weather Prediction,* Fort Lauderdale, FL, Amer. Meteor. Soc., 283–286.

Zhang, J., and T. Gal-Chen, 1996: Single-Doppler wind retrieval in the moving frame of reference. *J. Atmos. Sci.,* **53,** 2609–2623.

Ziegler, C., 1985: Retrieval of thermal and microphysical variables in observed convective storms. Part 1: Model development and preliminary testing. *J. Atmos. Sci.,* **42,** 1487–1509.

Zou, X., and Y. Kuo, 1996: Rainfall assimilation through an optimal control of initial and boundary conditions in a limited-area mesoscale model. *Mon. Wea. Rev.,* **124,** 2859–2882.

Zupanski, D., M. Zupanski, E. Rogers, D. F. Parrish, and G. J. DiMego, 2002: Fine-resolution 4DVAR data assimilation for the Great Plain tornado outbreak of 3 May 1999. *Wea. Forecasting,* **17,** 506–525.

Zupanski, M., 1993: Regional four-dimensional variational data assimilation in a quasi-operational forecasting environment. *Mon. Wea. Rev.,* **121,** 2396–2408.

Chapter 8

Innovative Signal Utilization and Processing[*]

FRÉDÉRIC FABRY

J. S. Marshall Radar Observatory, McGill University, Montreal, Quebec, Canada

R. JEFFREY KEELER

Research and Technology Facility, National Center for Atmospheric Research, Boulder, Colorado

Fabry   **Keeler**

1. What is signal processing?

Shrouded behind jargon and understood only by a small community, "signal processing" in radar meteorology seems to have a certain mystique associated with it. Its roots are embedded in the branches of mathematics and engineering that deal with the processing of time series data, in our case from received radar echoes. These weak radar echoes contain not only returns from weather (e.g., precipitation), but also a variety of artifacts that we wish were not present (e.g., ground clutter) and distract us from our primary goal of measuring precipitation rate, wind speed, turbulence, or any variety of more specialized atmospheric characteristics. Minimally, we wish to measure the intensity or power of the signal (the moment zero of the Doppler spectrum), its mean Doppler frequency (the first moment), and its variability in velocity (the second central moment) within many thousands of range cells per second.

Before delving into details, let us first consider the radar electric fields and waves to be processed. After completing the journey from the transmitter to the scat-

tering targets and back to the receiver, the radar signals must be converted into usable information. Signal processing encompasses all of the techniques and computations required to transform the raw radar measurements, namely, amplitude (or power) and angular phase of the received wave at one or a few polarization states, into parameters that may be interpreted physically. This transformation involves two key aspects that are not totally independent of each other: the suppression of data artifacts and the extraction of as much information as possible from the time series of received power and phase. With the rapid increase in computing power experienced since radar signals were first digitized in the early 1970s, our ability to extract information from the digital samples has been continuously expanding. This paper attempts to provide a brief overview of some of these recent innovative developments and speculates on what the future may hold.

By design, signal-processing hardware and software are partly dictated by the specifications of the radar hardware. As a result, signal-processing code will vary considerably from one type of radar to another to the point that an upgrade in hardware usually implies an upgrade in the signal processing. Because they shield the user from hardware perceived to be complex, signal-

* The National Center for Atmospheric Research is sponsored by the National Science Foundation.

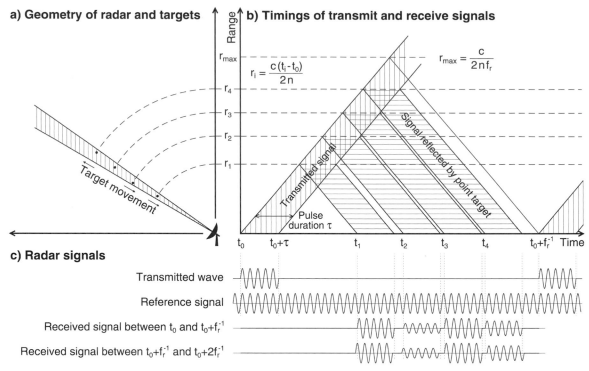

a) Geometry of radar and targets

b) Timings of transmit and receive signals

$$r_i = \frac{c(t_i - t_0)}{2n}$$

$$r_{max} = \frac{c}{2nf_r}$$

Pulse duration τ

c) Radar signals

Transmitted wave

Reference signal

Received signal between t_0 and $t_0 + f_r^{-1}$

Received signal between $t_0 + f_r^{-1}$ and $t_0 + 2f_r^{-1}$

FIG. 8.1. Pictorial description of the nature of radar signals for single point targets. (a) Geometry of the problem with the radar at the origin, the beam axis stripped vertically, and four point targets, the two nearest ones approaching from the radar and the farthest one receding from it. (b) Time-range plot of the transmitted pulse of duration τ (stripped vertically) and the reflected signals from each point target (stripped horizontally). In combination with the speed of light in vacuum c and the index of refraction n of the medium through which radar waves travel, the pulse repetition frequency f_r determines the maximum range r_{max} up to which echoes can be observed without range folding. (c) Illustration of the transmitted wave, the reference signal with respect to which the phase of targets will be computed, and the received signals as a function of time (or range) from two successive radar pulses. The strength of the signal returned for each target depends on target size, range (not illustrated here), and position within the beam; it typically remains constant in time unless the radar beam pointing direction changes. However, the phase of the signal will change from pulse to pulse if the target moves away or toward the radar.

processing algorithms are often inadequately understood by meteorologists who neither feel the need nor have the inclination to understand the frequently detailed mathematics inherent in them. In its essence, signal processing always attempts to maximize the amount of meteorologically relevant information one can extract from radar returns. However, to better understand and be conversant in signal processing, one must first appreciate the nature of the radar signal itself.

2. The radar signal from weather targets

Most weather radars illuminate their target volumes by sending high-power short pulses of microwaves at a single wavelength, such as near 10.5 cm for the Weather Surveillance Radar (WSR-88D) and other S-band radars or 5.5 cm for C-band radars. If the microwave pulse encounters discontinuities in the refractive index of the medium, such as an air–raindrop interface, a small portion of the pulse will be reflected. Returns from all these refractive index discontinuities are then detected and amplified by the radar receiver. For each transmitter pulse and at each range gate, the receiver system determines the amount of power received from targets and

the phase of the returned wave with respect to the transmitter signal. The time series of power and phase at a given polarization, for example, $P_{HH}(r, t_i)$, $\varphi_{HH}(r, t_i)$ for the power and phase of targets at range r and at times $(t_1, t_2, \ldots t_n)$ for a radar at horizontal polarization, form the digital data that will then be treated by our signal-processing system.

a. Radar signal from a single point target

Let us consider four point targets at different positions within the radar beam (Fig. 8.1). If the radar illuminates these targets by sending a pulse of duration τ, each of these point targets will reflect a small fraction of that pulse for the same duration. If the targets are far enough from each other, that is, if their range difference to the radar exceeds $(c\tau)/(2n)$, where c is the speed of light in vacuum and n is the refractive index of the medium, their signals will not overlap. The strength of the signal returned to the radar will vary from target to target, but it will remain constant in time if the target's apparent shape and illumination by the radar do not change. The phase of the target depends on the time of travel of the radar pulse modulo $1/f$, where f is the microwave fre-

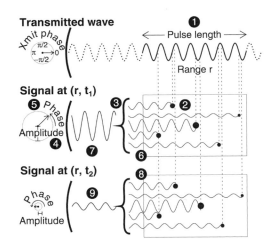

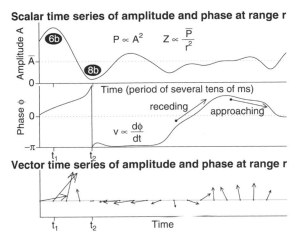

FIG. 8.2. Illustration of the processes (left) leading to the weather radar signals observed at a given range r (right). Transmitted single-frequency radar pulses (1) illuminate a volume of space. Each target in that volume, represented by circles of various sizes, will scatter a portion of the wave back to the receiver, the amplitude and phase of the return depending on the size and the position of the targets (2). The echoes of all the targets will interfere and combine with one another (3), resulting in a single return wave back to the receiver. That wave has two main properties, an amplitude A (4), and a phase φ (5) with respect to the transmit wave. The radar return can be expressed as a complex number $A \exp(j\varphi) = I + jQ$ and illustrated by a vector (5). If targets move in time (note the shifts in their position between t_2 and t_1), the phase of their individual return will change. If the returns of many targets interfere constructively (6), the resulting signal will be strong (7); if they interfere destructively (8), the resulting signal will be weak (9). The radar signal amplitude hence fluctuates in time as targets move with respect to each other in and out of spatial configurations where constructive (6b) and destructive (8b) interference dominates the scattered wave. Because of these fluctuations, it is necessary to observe echoes for a certain amount of time (tens to hundreds of milliseconds) before one can obtain a good reflectivity measurement.

quency of the radar pulse in hertz. Any small displacement of the target toward or away from the radar will result in a significant change in phase: at S-band, a 1-mm radial displacement will cause a phase change of about 7°, while most coherent radars can resolve phase to well within 1°. This property is used to measure the radial or Doppler velocity of targets by observing the change of the phase of targets from pulse to pulse.

b. Radar signal from a distributed target

The situation becomes somewhat more complex when multiple targets, or a distributed target such as precipitation, are observed by radar (Fig. 8.2). In such cases, the strength of the echo not only depends on the properties of individual point targets, but also on their relative position, as the returns from all point targets will interfere constructively or destructively with one another depending on the distribution of targets with respect to the phase of the incoming wave. At one extreme, if the positions of point targets with respect to the radar wave are such that targets are found in *equal number for all values of the phase of the transmitted radar pulse*, then for each target scattering a wave of phase φ there will be one and only one target of phase $\varphi + \pi$ whose out-of-phase signal will cancel the signal from the first target. In such cases, considerable destructive interference will occur and the signal from the target will be weak. At the other extreme, if point targets tend to be clustered in a narrow range of values of the transmitted pulse phase, their returns will constructively interfere

with each other and the signal from the target will be strong. In the case of precipitation, targets get reshuffled in position by turbulence, wind shear within the beam, and by their different fall speeds. Hence, they find themselves alternately in configurations where constructive interference dominates and in other configurations where destructive interference dominates. The radar echo of precipitation therefore fluctuates considerably in intensity with time. For the same reason, the rate of change of echo phase with time caused by the radial velocities of this specific arrangement of targets will also fluctuate, but to a smaller extent, and the effect on the measured velocity is typically much smaller than on reflectivity. Instead of a single velocity, the targets have a distribution of radial velocities with respect to the radar. In general, for convenience, one often assumes that the distribution of radial velocities can be approximated by a Gaussian distribution with a mean velocity v_{DOP} and a spectrum width, or width of the distribution, σ_v. Often, the distribution of velocities will be considerably more complex because of the presence of different types of targets going at different velocities such as ground versus precipitation echoes, or clear air versus precipitation targets at vertical incidence. In such cases, the distribution of target velocity may have to be described by a detailed Doppler spectrum, or periodogram, to get an accurate description.

Because of echo fluctuation, any single measurement of reflectivity may bear little resemblance to the mean reflectivity that would be observed by averaging the signals from all possible combinations of target positions (Marshall and Hitschfeld 1953). For example, single samples are exponentially distributed with the mean and standard deviation being equal to the true mean power; consequently, the standard deviation of a power estimate from N independent samples is proportional to $N^{-1/2}$. But as more measurements are taken, the echo intensity will evolve in time as targets reshuffle in position within the sampling volume. The time to independence is defined as the time required for targets to move enough with respect to each other that a new measurement of the target reflectivity will be statistically independent from the previous one. This time to independence will increase with wavelength λ and decrease with the width of the velocity distribution σ_v according to $\tau_{\mathrm{indep}} = \lambda/(4\pi^{1/2}\sigma_v)$. For microwave weather surveillance radars and normal shear and turbulence within the pulse volume, it is of the order of 10 ms. To obtain a good estimate of mean reflectivity, it is necessary to get many independent estimates because of the fluctuating nature of radar echoes from precipitation. Presently, this fact is the main constraint for determining the time that radars will point in the same direction: one must wait long enough to obtain a sufficient number of independent measurements to achieve the needed accuracy in reflectivity. For example, about 20 independent estimates are required to achieve a 1-dB accuracy in a point measurement. Simultaneously, the averaging of many independent noise power estimates improves our measurement of the background noise. If the mean noise power is better characterized, it becomes easier to detect weak echoes and an effective increase in the radar sensitivity is achieved by subtracting the noise-only from the signal-plus-noise measurement.

c. Measurement ambiguities

In addition to the accuracy issues mentioned above, the measurements of reflectivity and Doppler velocity can also suffer from three primary sources of ambiguity. First, precipitation and other weather targets may not necessarily generate the only echo being observed. In any volume scan, one will find near-stationary targets such as echoes from the ground, as well as moving targets such as airplanes, boats, chaff, birds, and insects. While some of these targets may have meteorological value such as insect echoes acting as wind field tracers that allow detection of convergence lines, many more do not. Most of these unwanted echoes will not cause significant problems to an experienced user, but they will ruin most attempts to use the data quantitatively through automated algorithms. Hence, they must somehow be suppressed or removed.

The two other sources of ambiguities are related to the choice of the radar pulse repetition frequency f_r. On one end, if f_r is too high, the next radar pulse may be fired before all significant echoes from the previous pulse have reached the receiver. Under such circumstances, distant echoes at ranges greater than r_{max} (Fig.

8.1) will be range folded and mapped at the wrong range. On the other end, if one reduces f_r too much, it is possible that the change in the phase of targets between successive pulses will exceed $\pm 180°$. This will occur if the magnitude of the radial velocity v_{DOP} exceeds $f_r \lambda/4$, where λ is the radar wavelength. If this happens, a large phase change will be indistinguishable from a smaller one 360° away, and velocity aliasing occurs.

d. The goals of signal processing

Since signal processing seeks to transform the digitized radar signals into meteorologically useful measurements, most of the work in the field focuses on three main activities:

1) increasing the accuracy and sensitivity of the measurements for a given sampling time (or decreasing the sampling time for a given accuracy) by maximizing the number of independent samples measured per unit time,
2) eliminating the measurement ambiguities using a variety of approaches each dedicated to a specific ambiguity, and
3) measuring new parameters that may be of meteorological interest.

We shall address these signal processing goals in the following sections.

3. Traditional radar meteorology signal processing

Before we describe a few of what may be considered "innovative" signal-processing techniques that address the three goals above, let us first briefly describe some standard, or traditional, signal-processing methods widely used today.

a. Pulse-pair processing and the correlation function

Traditional signal-processing techniques generally follow one of the following two sets of approaches: frequency-domain or time-domain processing. In frequency-domain processing, batches of pulses (frequently 2^n) are processed together, computing the Doppler power spectrum in the frequency domain using Fourier analysis. Traditional Fourier signal analysis assumes the radar echo to be composed of a sum of sinusoids (the Fourier components) containing the elements of the fluctuating radar signal and the contaminating artifacts. In time-domain processing, computations may be performed on the same batch of pulses using autocorrelation techniques or by averaging calculations done on individual pulses (to compute power), adjacent "pulse pairs" (to compute mean Doppler velocity and spectrum width), or sets of n pulses to perform filtering of unwanted echoes (such as zero-frequency ground echoes).

Pulse-pair processing consists of taking the complex

time series of amplitude and phase of radar signals and computing the complex correlation function $R(m)$ of that data sequence with a shifted (or lagged by m positions) replica of itself. In mathematical form, the autocorrelation value at lag m can be written as

$$R(m) = \sum [A_i \exp(j\varphi_i)][A_{i-m} \exp(-j\varphi_{i-m})]$$
$$= \sum A_i A_{i-m} \exp[j(\varphi_i - \varphi_{i-m})], \qquad (8.1)$$

where A_i and φ_i are the radar-echo amplitude and phase for pulse i, and j is the square root of -1. For lag $m = 0$, $R(0) = \sum A_i^2$, or the sum of the received power from each pulse. For a lag $m = 1$, the argument of the complex number $R(1)$ reduces to the mean power-weighted phase difference between successive pulses, from which the mean Doppler velocity may be computed (Woodman and Hagfors 1969; Rummler 1968). The pulse-pair estimators have been employed in radar meteorology since the beginnings of Doppler processing, primarily because they rapidly process the very large number of sampled time-domain radar echoes. Keeler and Passarelli (1990) describe these traditional techniques, their origins, and their evolution.

The radar signal detected is a combination of the echo from various scatterers, generally referred to as "signal," and thermal noise originating from the atmosphere and from the receiver system itself, as well as other electrical interference, generally referred to as "noise." The ratio of these two is the signal-to-noise ratio (SNR). If plotted in vector form, the signal vector will rotate at a more or less steady rate (see Fig. 8.2, bottom right, where the Doppler velocity is close to zero) and slowly fluctuate in amplitude, while the noise vector will have a random direction and rapid fluctuation in intensity such that there will be no correlation from one pulse to the next. If we compute $R(1)$, the contribution of the noise will remain small as each individual $A_i A_{i-1}$ $\exp[j(\varphi_i - \varphi_{i-1})]$ vector for noise will point in a different direction since there is no correlation in the phase of successive pulses. For signal, each $A_i A_{i-1} \exp[j(\varphi_i - \varphi_{i-1})]$ vector will point more or less in the same direction, as $\varphi_i - \varphi_{i-1}$ does not vary greatly from pulse to pulse. Noise power will contribute significantly only to $R(0)$. Furthermore, as lag m is increased, the magnitude of $R(m)$ will slowly decrease as the range of values ($\varphi_i - \varphi_{i-m}$) can take increases, while for signals coming from a single type of targets, for example, not contaminated by clutter, ($\varphi_i - \varphi_{i-m}$) $\approx m$ ($\varphi_i - \varphi_{i-1}$). The result of the computation of $R(m)$ for various m is a corkscrew-shaped helix depicting the complex autocorrelation function (Fig. 8.3). The size of the helix at lag 0 gives the received signal power P_s plus the delta function denoting the noise power P_n; the rate of rotation of the corkscrew is related to the Doppler velocity; and the decreasing amplitude with lag reflects the spectrum width or increasing independence between radar samples at these lags.

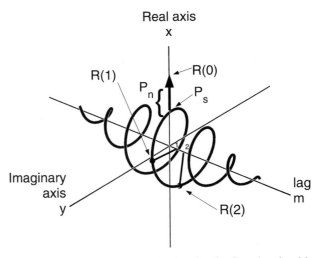

FIG. 8.3. Complex autocorrelation function for Gaussian signal in white noise as function of lag. Here P_s is the signal power and P_n is the noise power. The ratio P_s/P_n is the SNR of this particular set of received echo samples. Autocorrelation values at lag 0, 1, and 2 are shown. The Doppler spectrum is the magnitude-squared Fourier transform of this function. The derivative at $R(0)$ is directly proportional to first moment of Doppler spectrum (i.e., the pulse-pair velocity estimator). From Keeler and Passarelli (1990).

b. Artifact removal

Artifacts in the data, such as ground targets or interference, have correlation functions with specific characteristics, ground clutter having a near-zero Doppler and narrow spectral width, or slow decorrelation with increasing lag. If clutter is mixed with signal, the correlation functions of the two sources of echo add together. The time-domain methods described above used to obtain the spectral moment parameters Z, v_{DOP}, σ_v from the autocorrelation function do nothing to suppress artifacts or noise or to improve the accuracy of the measurement; they are only an expedient method of estimating the (frequently contaminated) meteorological parameters of the radar echo through the spectral moments. However, one may design 1) time-domain digital filters (Oppenheim and Schafer 1989; Press et al. 1992) to remove artifacts (and unfortunately some signal) having specific characteristic frequencies, or 2) frequency-domain filters by computing the Fourier spectrum and cutting or reducing those frequencies containing undesired echoes. It is clear that spectrum, or frequency domain, processing permits an additional degree of freedom in the frequency dimension for potential suppression of these artifacts and improvement of the meteorological estimates. In the past, spectral processing has been computationally demanding because of the large number of radar echoes to be processed, but this is no longer the case. We shall address spectral processing in some detail in the next section. Some data artifacts, such as point target aircraft or tall towers, are typically removed by ad hoc detection and suppression techniques and interpolating the meteorological data across the data gap.

c. *Digital filtering*

Digital filters are simply numerical filters that operate in the time domain and pass or suppress different groups of frequencies within the desired frequency range. They are most frequently employed in the pulse-time domain for processing the (I, Q) "baseband," or Doppler domain centered at zero frequency, signatures that we covered before. However, they are also seeing new applications in the range-time domain for processing "passband," or the intermediate frequency domain signal from which the raw (I, Q) data are derived. Presently, the primary applications of digital filtering techniques are ground clutter suppression filters in the first case and digital "matched filter" receiver implementations in the second case.

Traditional techniques of noise suppression and removal include intermediate frequency (IF) matched filtering (analog in most radars or digital in radars employing a digital receiver) and simple data thresholding based on signal power or some other signal quality indicator. Ideal filters by design (or definition) maximize the SNR of the filtered signal and are matched to the transmit waveform. However, there may be more compelling reasons to use other nonmatched filtering techniques when advanced waveforms or processing algorithms are used. Digital receiver filters have advantages over analog matched filters in that the numerical coefficients do not drift as they do when analog components age and the filter may be designed precisely to completely cancel any hardware imbalance characteristic of analog filtering of the complex waveforms.

d. *Digital thresholding*

Because both the signal and the noise are fluctuating, radar signal parameter estimation has a statistical nature. When all the signal processing has been completed and the final meteorological estimated parameters are to be detected or displayed, one frequently wishes to set thresholds so that only the accurately estimated parameters are used for further data processing. Simple power thresholding is often used effectively. However, since both the fluctuating character of the signal and its SNR contribute to the final accuracy of the estimates, it may be desirable to use a joint signal quality indicator as a detection threshold. Often, the signal quality indicator is based on a ratio of $|R(1)|$ and $R(0)$ (Keeler and Passarelli 1990), as a fast decorrelation between pulses is indicative of either strong noise or difficult-to-process signal.

4. Trends in radar meteorology signal processing

Innovative signal-processing techniques abound (Table 8.1). The literature in the past 40 or more years has described procedures for processing complex, fluctuating signals in various "optimal" fashions for a variety of astronomical, geophysical, oceanic, and atmospheric applications. Because of the meteorological radar community's desire for real time processing of an extremely large number of measurement volumes (10 000–100 000 each second) only a few of these have been applied to date. In this section, we address our interpretation of innovative techniques and applications according to the previously identified signal-processing goals.

a. *Information extraction—Sensitivity and accuracy*

1) SPECTRAL PROCESSING

In time-domain processing we make assumptions about symmetric Gaussian-shaped spectra, white noise, lack of spectral artifacts, and reasonably long dwell times in computing spectral moments from which we estimate the meteorological base data. In reality, we have asymmetric weather spectra, various spectral artifacts (caused by ground/sea clutter, birds, airplanes, and assorted other point targets), and sometimes nonwhite noise (typically caused by filters in the receiver chain). By adding another degree of freedom, the frequency dimension, to the processing domain, the ability to discriminate and measure characteristics of weak or ambiguous signals is greatly enhanced. For example, radars using time-domain processing can sometimes detect the rotation of a small tornado vortex but only at short ranges where the beam resolution is adequate. However, with spectral processing, it is plausible that detection of a dealiased bimodal spectrum, when corroborated with other supporting data, may be adequate to extend the range of tornado detection to much longer ranges than presently attainable. Furthermore, a wide variety of artifacts can be readily suppressed in the frequency domain using simple detection and removal techniques or by using spectral filtering and reconstruction techniques, as we shall see later.

There is great interest in techniques that would allow radars to sample the atmosphere at a faster rate. Anticipated future problems inherent in short-dwell-time (rapid scan) radar systems include how to accurately estimate spectra from only a short time sample. Classic radar systems theory tells us that spectral resolution is inversely proportional to the radar dwell time. However, modern or advanced spectrum processing techniques may permit using short-dwell-time sampled data appropriately averaged in range (or some other domain) to obtain similar or better resolution spectra for estimating spectral moments. Maximum likelihood and maximum entropy spectrum analyses (Burg 1972) are techniques of the general class of autoregressive (AR) spectrum estimators whereby the observed data are modeled as all-pole filtered white noise instead of a weighted sum of sinusoids according to the Fourier model. In standard data models, the autocorrelation function of the ob-

TABLE 8.1. Main goals of signal processing, and some of the approaches and techniques used to achieve them. In the techniques, SP refers to a technique purely based on signal processing, HS refers to a technique that needs both additional hardware and a new signal processing to be applied, while the number–letter combination links the techniques to the goals and approaches they are based on.

Goals	Approaches	Techniques
1. Reduce sampling time and/or increase accuracy	a. Obtain a greater number of samples in the same time to increase accuracy b. Increase the independence of samples during the sampling time. c. Use innovative methods to parameterize the spectral distribution of echoes.	- Transmit pulse with frequency or phase shifting several times within the pulse: pulse coding (1a; HS) - Oversample and artificially decorrelate (whiten) samples using digital inverse filters (1a; HS) - Use multiple transmit frequencies (1b; HS) - Rotate through a few pointing directions within the decorrelation time by electronic beam steering on phased-array systems (1b; HS) - Use autoregressive techniques to characterize echoes; Figs. 8.4–8.5 (1c; SP)
2. Eliminate ambiguities	a. Eliminate range folding b. Elimiate velocity aliasing c. Identify or eliminate unwanted targets using their signal characteristics	- Multiple pulse repetition frequency (2a, 2b; HS) - Change the transmit phase from pulse to pulse to make second trip targets incoherent (noiselike): phase coding (2a; HS) - Digital filtering of zero-velocity echoes (2c; SP) - Digital adaptive/predictive filtering of all signal but the one originating from weather (2c; SP)
3. Measure new parameters	a. Process the existing signal more intelligently b. Add receivers at other locations to obtain information from another geometry. c. Add hardware and software to transmit, receive, and process data at more frequencies and/or more polarization states	- Make use of nonmeteorological signals: refractivity measurements using ground targets; Fig. 8.6 (3a; SP) - Make use of data obtained from additional antenna and receivers to get tangential wind information; Fig. 8.7 (3b; HS) - Make use of data at multiple radar bands to obtain attenuation-related parameters; liquid water, hail, etc. (3c; HS) - Make use of data at multiple polarizations to obtain target shape information; Fig. 8.8 (3c; HS)

served data may be directly computed to as many lags as desired; however, for short dwell times, the longer lags have an unacceptably high variance. In the autoregressive models, the first several correlation lags may be computed the normal way, but the autocorrelation function may be extended to larger lags making minimal assumptions on its form by invoking the "maximum entropy" constraint. A variety of AR spectrum estimation techniques exist: Yule–Walker, Burg, MUSIC, eigenvalue decomposition, maximum likelihood, and others. Figure 8.4 shows a sample of the Burg AR spectrum estimator compared with the Fourier periodogram estimator [the common spectrum computed by fast Fourier transform (FFT)] from a 128-point I/Q dataset taken in Norman, Oklahoma, on the local WSR-88D in May 1997. Note the general agreement of the AR maximum entropy estimator to the envelope (or the mean value) of the Fourier spectrum estimator. As the number of computed correlation lags (the order of the AR spectrum estimate) increases, the approximation to the Fourier estimate becomes closer as both spectra models describe the same physical process.

Even very-short-dwell-time radar observations may be used to estimate high-resolution Doppler spectra if estimates from multiple nearby cells are available that span the desired spatial resolution. If multiple-resolution volumes in range, azimuth, or elevation are available, the actual measured correlation values may be averaged,

and the resulting spectrum estimate will be representative of the full measured volume. Figure 8.5 compares the FFT periodogram with eight Burg maximum entropy spectrum estimates taken from contiguous 16-sample segments of the original 128-point data interval. Note the variability of each maximum entropy spectrum estimate, which may reflect the true variability of the short data samples or the variability of the spectrum estimator itself—likely both. Keeler and Passarelli (1990) and Bringi and Chandrasekar (2001) explore the application of the classic Fourier transform and the advanced AR techniques to spectrum estimation for meteorological measurements.

2) ADAPTIVE DIGITAL FILTERING

The digital matched filter has been identified as a standard IF processing technique to improve the radar sensitivity to weak signals by suppressing the noise spectrum components outside the received signal spectral region. We may extend this digital filtering concept to make the filter evolve in time to adapt to a changing situation. Let us suppose that we want to design an innovative time-variable signal tracking filter to suppress noise or transient interference from the background of a radar echo whose meteorological parameters change with time, say, as the radar antenna rotates. We can easily design an adaptive digital filter that tracks

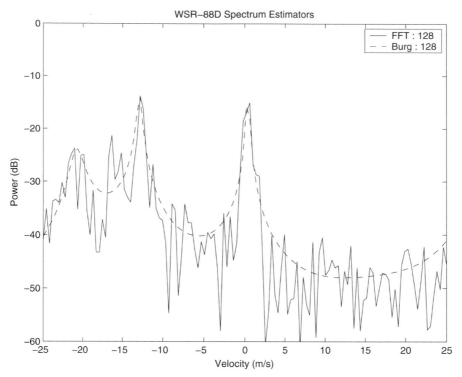

Fig. 8.4. Power spectrum estimates of WSR-88D I/Q data taken from station KOUN on 2 May 1997 of small storm cell located about 50 km from radar. The standard FFT periodogram spectra of 128 samples (solid line) shows ground clutter at 0-m s^{-1} velocity, precipitation spectra near -15-m s^{-1} velocity, and possibly some spectral artifacts. The fourth-order Burg AR maximum entropy spectra of the same 128-point dataset matches the essence of the periodogram spectral density. A higher-order AR spectrum would more closely approximate the Fourier spectrum.

the center frequency and width of the precipitation or clear air signal, thereby enhancing the signal while simultaneously suppressing the power from everywhere else in the Doppler spectrum (Widrow and Stearns 1985). Employing such adaptive digital filtering techniques, we improve the accuracy of the meteorological signal parameters. Many other time-variable digital filtering applications have evolved in other fields as signal-processing chips have become smaller and more powerful, for example, noise-canceling headphones and medical devices that remove ambient noise, geophysical prospecting systems that remove subsurface reflections, and delay distortion filters in our computer modems that allow fast digital transmission on voice-grade telephone systems.

3) INDEPENDENT SAMPLES—ACCURACY AND UPDATE RATE

Accuracy is improved by averaging independent samples of the same random process over different domains. Any domain that increases the number of independent samples will increase the accuracy of any parameter. For example, by using "time diversity" and sampling the received power over increasingly long time periods (even noncontiguous periods), the mean reflectivity of

the echo fluctuating in intensity can be determined more accurately. Similarly, using "frequency diversity" to measure the fluctuating power at several appropriately separated frequencies can improve the mean reflectivity estimate. "Spatial diversity" techniques using adequately separated multiple antennas also yield independent measurements since the echo fluctuations are independent. "Polarization diversity" measurements to increase accuracy are not necessarily an option for distributed targets since the individual scatterers are generally similarly located at all polarizations and the echoes are highly correlated. Nevertheless, it is clear that polarimetric differences contain new information essential to improved precipitation estimation and should be used separately.

Various other techniques are available for increasing independence; some are quite inexpensive in terms of processing power, bandwidth, or simply dollar cost. Pulse compression waveforms have been successfully explored for increasing the independent measurement rate using small range spacing in weather radars (Keeler and Hwang 1995; Mudukutore et al. 1998). However, range sidelobes are difficult to suppress when the return signal is Doppler shifted, even using shaped nonlinear FM waveforms and sophisticated inverse filtering techniques. The same goal of more independent samples can

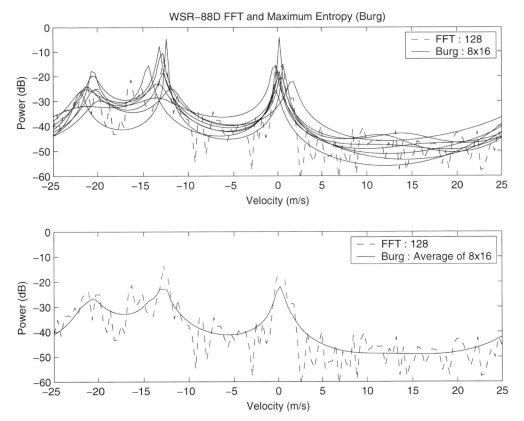

FIG. 8.5. (top) Eight individual fourth-order Burg maximum entropy spectra computed from contiguous sets of 16 samples each taken from the original 128-sample dataset; FFT periodogram of 128 samples is shown in dashed line. (bottom) These maximum entropy spectra may be averaged in various ways to obtain a range averaged short dwell time spectrum estimate; e.g., eight subrange gated intervals could be sampled for only 16 pulses, and averaged. The average AR spectrum in the bottom figure is a simple log domain average of the eight individual spectra.

be achieved by transmitting shorter simple pulses, but the peak transmit power must be raised to maintain sensitivity. Extending the coded pulse length will retain the sensitivity of each pulse compression measurement at high range resolution with a proportional increase of the average transmit power. Hence, the pulse compression process is lossless. However, a significant cost occurs because the limited microwave bandwidth of the radar must be enlarged to accommodate the greater measurement accuracy. Of course, one could equally well decrease the dwell time of each beam while maintaining the original accuracy if so desired (Keeler and Frush 1983a; Strauch 1988). This is the basic concept of the rapid-scan radar technique (Keeler and Frush 1983b): using larger bandwidth to increase volume scan rate (or volume update rate) while retaining the full radar sensitivity to clear air echoes.

Recently, Torres and Zrnic (2001) have proposed to simply oversample the received echo of a standard single-frequency pulse waveform at rates a few times higher than the nominal signal bandwidth of the simple transmit pulse and to whiten or decorrelate these samples with an inverse filter of the same type used for pulse compression filtering. This innovative technique allows

independence gained from the propagating pulse into new scattering regions to be recovered. The inverse filtering operation then effectively yields an increased number of independent samples (and, consequently, accuracy or scan rate). Koivunen and Kostinski (1999) have also recently studied this whitening technique to produced a larger number of independent estimates. This technique should be explored more fully since no additional bandwidth is required in this microwave bandwidth-scarce world and the range sidelobes may be constrained as desired depending on the design of the whitening filter. However, it appears that the cost of obtaining independent samples is a reduced sensitivity of each new estimate. Thus, the pulse oversampling technique trades off SNR for independent samples that may be used to improve accuracy or decrease dwell time. Keeler and Griffiths (1977) applied this identical technique (the oversampled pulse followed by a whitening filter) to acoustic radar processing by implementing an adaptive one-step linear prediction filter that continuously tracked the Doppler shifted acoustic pulse echo received from the atmospheric scatterers. This specific processing technique may be applied equally well to Doppler lidar; the main difference is that in the sodar and lidar cases, the processing to recover meteorological parameters is performed in range time, whereas in the radar case, the parameter estimation processing is performed with data from successive pulses.

b. Artifact suppression

The quality of meteorological radar data has always been a concern for detailed scientific analysis. In the past, ad hoc techniques have been applied to individual algorithms in research and operational radar product generation. Any improvement in front-end processing to yield higher data quality will show dividends in many other areas, particularly assimilation of radar observations into numerical forecasting models.

Spectral processing is an important tool for detecting and removing artifacts that previously have generated gross estimation errors in time-domain processing. Consequently, the meteorological parameters may be estimated with fewer anomalies, less ambiguity, and higher accuracy. For example, sea clutter is difficult to remove from weather echoes using conventional time-domain processing techniques since its mean velocity depends on ocean currents frequently unknown. By transforming the radar echo into the frequency domain and invoking continuity constraints, one may selectively identify, track, and separate the weather spectrum components from the sea echo spectrum components. Similarly, other artifacts that provide unique spectral signatures may be processed using high-level identification parameters to detect and suppress or ignore these artifacts. Furthermore, overlaid echoes arising from precipitation at ranges from multiple trips might be individually tracked and separated—an impossible task with simple pulse-pair processing.

Range folding and velocity dealiasing have always been a challenge in radar meteorology. Second trip echoes fold into the first trip and obscure weak boundary layer and precipitation features; ground clutter and first trip echoes obscure desired second trip measurements; high-velocity measurements alias to unrealistically low values of the wrong sign. Various proposals over the years have addressed these data quality problems. Currently, the WSR-88D Radar Operations Center in Norman, Oklahoma, is sponsoring a program to suppress second trip echoes using phase coding (Sachidananda and Zrnic 1999). The phase coding system is expected to allow separation of first and second trip echoes having up to 40-dB power ratio at nominal spectrum width values. Tests have shown that about 80% of the present overlaid echo, identified by the "purple haze" censoring in the WSR-88D, can be recovered after phase code pulsing and spectral filtering are applied. Furthermore, using a staggered pulse repetition time (PRT) pulsing technique, the WSR-88D is expected to generate dealiased velocities up to ± 50 m s^{-1} while simultaneously suppressing ground clutter in the frequency domain employing an innovative sparse matrix processing tech-

nique (Sachidananda and Zrnic 2000). This technique solves a long-standing problem of how to precisely suppress clutter sampled with nonuniformly spaced pulses. However, this is just the first step. New techniques are being continuously proposed that need evaluation and field validation. Pirttilä and Lehtinen (2001) are presently testing a multiple PRT pulsing and processing technique that simultaneously separates range folded echoes and dealiases velocities.

Many old but promising and innovative signal-processing techniques need re-evaluation and validation. For example, orthogonal waveforms using dual polarization (Doviak and Sirmans 1973; Pazmany et al. 1999), spaced frequencies, or coded waveforms will allow separation of two or more trips of range folded data. Linear predictive filtering may be used to suppress sea clutter, normal and anomalous propagated ground clutter, and point targets artifacts. By accurately predicting future samples of the highly correlated clutter returns, one is left with the less-correlated weather echo and white noise. In this manner, the velocity of the coherent interferer need not be known to perform its suppression. Utilization of adaptive filters in processing of active meteorological sensors has only been marginally explored yet appears to have much to offer, especially in artifact identification and removal. These approaches consist of signal-processing algorithms that directly modify the I/Q samples from which the base data spectral moment estimates are made. Any attempt to improve data quality by this root level signal processing must be fully validated so as not to destroy the existing data so familar to the research and operational communities.

c. Measurements of new parameters

When a radar transmits at a single frequency and single polarization, one may recover the simple spectral moment data, for example, target strength and Doppler velocity information. To estimate additional parameters, one must either process data in unusual, innovative ways, or add hardware components to measure new quantities that provide additional information.

1) INNOVATIVE PROCESSING—REFRACTIVITY

Almost all weather radar data processing relies on the tacit assumption that the target of interest is precipitation, and the other echoes are clutter of one sort or another that must be suppressed. Yet radar information is considerably richer: insect concentrations provide wind tracers to yield information on convergence lines; variable ground echo coverage tell us about changes in the vertical profiles of temperature and moisture in the lower troposphere; etc. Specialized processing is required in order to obtain and use these ancillary sources for new meteorological information.

An example of such a technique is the measurement of surface refractivity maps, or the near-surface refractive index of air, using stable ground targets (Fabry et al. 1997). As the refractive index n of air changes, the time of travel of radar waves between the antenna and a fixed ground target varies slightly, and this results in a change in the measured echo phase to that ground target. That variation is large enough to be detected and accurately measured by carefully selecting and processing the proper targets (Fig. 8.6). These data can then be used to measure the spatial and temporal variations of n with time.

2) INNOVATIVE SENSORS—MULTIPLE-ANTENNA RADAR

If one adds new sensor hardware, such as additional receivers or transmitter–receiver systems, additional datasets can be obtained that, in combination with the original set, can be used to measure additional parameters, since new possibilities for data processing now exist.

For example, data from additional antenna–receiver systems will be similar to the original data in terms of being time series of power and phase, but they will be obtained from a slightly different geometry. As a result, the received power and phase time series will be different from that of the original receiver and will provide complementary information that can be used by specialized signal-processing approaches. If the two, or several, receiver systems are widely separated (by several kilometers), one can measure two, or several, Doppler velocity components. These Doppler velocities can then be combined to compute vector wind velocity over three-dimensional volume of space (Fig. 8.7). Although the multistatic radar concept developed by Wurman et al. (1993) is appealing, it is essential to take into account the considerable sidelobe contamination that may occur as a result of the use of a low-gain, wide beam receiving antenna needed to cover a large sector (de Elía and Zawadzki 2000).

If, instead of being widely separated, the additional receiving systems are within meters of each other, the normal processing of these data will yield similar reflectivity and Doppler velocity. However, the time series of power and phase themselves will be different, though somewhat correlated at a certain time lag. Using an interferometry-based technique, the shape of the correlation functions between the various signals can then be used to determine the magnitude and direction of the tangential winds (Cohn et al. 2001). While this technique works well with radars with large beamwidths that dwell over several seconds in a given direction, such as wind profilers, it may not be practical for scanning radars. However, overlapping or adjacent subarrays of a phased-array radar antenna implementing an electronic step scan sequence are being investigated to obtain the full three-dimensional vector wind field similar to the wind profiler application.

3) MULTIPLE-FREQUENCY RADAR

Adding samples at additional frequencies to standard single-frequency radars will provide new information. If the new frequencies are within a few percent of the primary radar frequency, the propagation properties of the signal and the scattering properties of the atmospheric targets will be similar; as a result, the additional indepedent samples and power will improve the accuracy and sensitivity of the measurement as previously described under the guise of frequency diversity. However, if additional phase processing is performed on the different frequency returns, enhanced range resolution is possible. The technique termed "range imaging" aligns the phases such that constructive interference occurs at subrange resolution intervals (Palmer et al. 1999). In this manner independent samples are increased by effectively compressing the pulse energy into a subrange interval.

If new transceivers having very different frequencies are added to the primary radar, as in a different wavelength band, the scattering properties of targets and probably also the propagation properties of the atmosphere will be greatly different. Depending on whether the effect of propagation or scattering properties is (believed to be) dominant, different types of multiwavelength signatures can be sought. Scattering properties will change if targets behave as non-Rayleigh scatterers for at least one of the wavelengths; these approaches have been proposed to look for signatures of large hydrometeors such as hail using dual-wavelength S- and X-band radars (Carbone et al. 1973). In the last 15 years this line of work has faded on ground-based radars in favor of target recognition based on dual-polarization approaches, but it is still an active subject of research for planned spaceborne weather radars such as the Global Precipitation Mission.

More recently, focus has shifted toward higher-frequency transceiver pairs for the detection of propagation signatures, primarily the measurement of the differential attenuation caused by liquid water (Gosset and Sauvageot 1992). The interest in this technique is driven by its potential for the detection of weather conditions leading to in-flight icing (Pazmany et al. 2001). Unless more than two wavelengths are used, accurate measurement of the differential attenuation is hampered by the wavelength-dependent characteristics of non-Rayleigh targets. Using a similar differential attenuation approach, it is also conceivable to design a three-frequency radar system capable of measuring range-resolved water vapor concentration using one frequency in the 22-GHz water vapor absorption band and two on either side to compensate for scattering and liquid water attenuation effects.

4) DUAL-POLARIZATION MEASUREMENTS

Increasingly, radars have the possibility of measuring the returns of targets at more than one polarization state.

At calibration time

A) Get time series of phase maps in constant n regime $(n = n_{ref})$

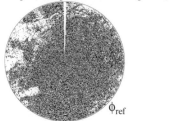

ϕ_{ref}

Average them together; this is the map of reference phase ϕ_{ref}

B1) Use the average SNR to identify echoes from sidelobes

B2) Measure a time series of scan-to-scan phase differences

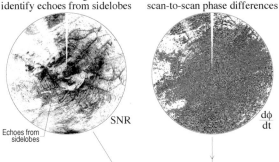

Echoes from sidelobes

SNR

$\dfrac{d\phi}{dt}$

B3) Compute temporal coherence of $d\phi/dt$ from the ratio of lag-1 and lag-0 time correlation

And use the sidelobe detection to mask out sidelobe contamination

Notes:

Calibration steps A and B do not need to be done for the same time sequence. Step A requires very uniform n for a few scans. Step B needs a few hours of data without weather echoes, preferably with some wind and without major changes in n near the surface.

Changes in n over a short time period Δt can be computed by replacing ϕ_{ref} by $\phi(t - \Delta t)$ at step IV.

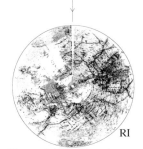

RI

The result is the ground target reliability index map

In real-time

I) Measure the phase of targets and other parameters (SNR, v, σ)

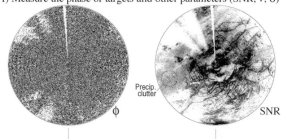

Precip. clutter

ϕ

SNR

II) Use SNR, v, σ to select ground targets

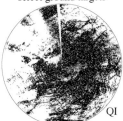

QI

This is the target quality index

IV) Subtract reference phase

III) Take minimum of QI and RI to make the weighting function

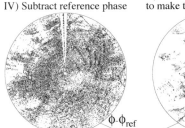

$\phi - \phi_{ref}$

WF

V) Smooth phase difference over small areas using the weighting function

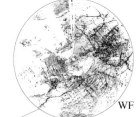

$\phi - \phi_{ref}$ (smoothed)

VI) Derive path values of n from slopes of phase difference

VII) Transform path values of refractive index into a field

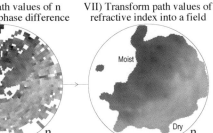

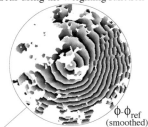

n

Moist

Dry

n

FIG. 8.6. Illustrated flowchart of the processing steps required to obtain refractive index measurements from radar data. Fixed ground targets, as opposed to oscillating ones covered by vegetation, are identified by their slow and coherent phase change in time over a calibration period (left). In parallel, the average value of target phase φ_{ref} is computed for $n = n_{ref}$. In real time (right), ground targets are separated from weather clutter based on their SNR, velocity, and spectrum width data. The current value of target phase is then subtracted from that at calibration time for valid targets and smoothed; $d(\varphi - \varphi_{ref})/dr$ is then computed and converted to $(n - n_{ref})$.

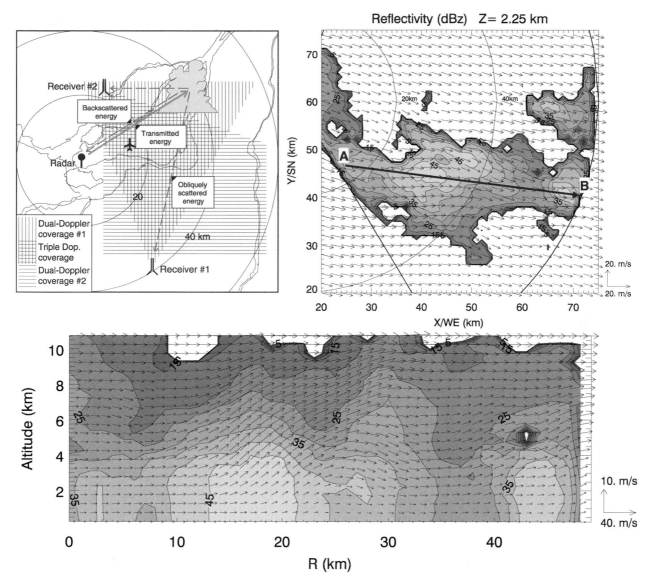

FIG. 8.7. (top left) Configuration and concept behind the Doppler multistatic radar network operated by McGill University. Radar pulses sent by the McGill radar are scattered by meteorological targets such as precipitation from a storm. Some of this energy comes back to the main radar site, providing information on the strength and radial velocity of targets, while some other portion is scattered obliquely to additional receivers from which other components of the wind can be obtained. The measurements from all the receivers are then combined to determine the 3D wind. (top right) Horizontal section of radar reflectivity and horizontal winds of a convective line within coverage of the multistatic radar network. Winds within the storm are derived from the radar data while winds outside are set based on a synoptic-scale analysis. (bottom) Vertical section through the convective line along the line A-B. In this image, the winds being plotted are the projection of the 3D wind on the plane defined by the cross section. Vertical motion is observed over two intensifying cells (kilometers 18 and 44) while descending motion occurs over a decaying cell (kilometer 29). From Fabry and Zawadzki (2001).

Often, the two polarization states used are horizontally polarized (H) and vertically polarized (V) waves to obtain some information on the shape of the targets observed. For each transmitted pulse in one polarization state, say H, one can obtain the power and phase of the returns at H [$P_{HH}(r, t_i)$, $\varphi_{HH}(r, t_i)$] or at V [$P_{HV}(r, t_i)$, $\varphi_{HV}(r, t_i)$], respectively called the copolar and the cross-polar data. If the same is done for transmission at vertical polarization, we end up with four distinct time series of power and phase that can be processed to obtain additional information on target properties. Examples of parameters that can be computed include the ratio P_{HH} over P_{VV} (Z_{DR}), the cross-correlation of the HH and VV time series (ρ_{HV} or ρ_{CO}), the mean phase difference between the HH and VV time series (φ_{DP}), and the ratio of the cross-polar and the copolar signal (L_{DR}), to name a few. Figure 8.8 depicts polarimetric radar waves and their joint processing to derive new information. Bringi and Chandrasekar (2001) provide a multitude of details on polarimetric weather radars.

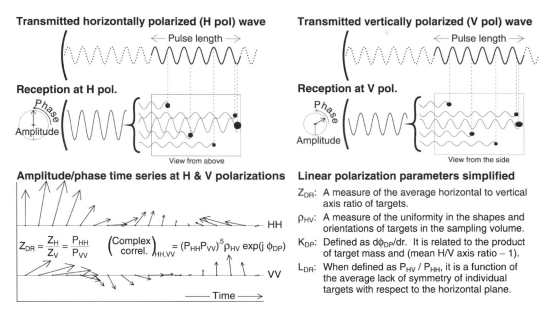

Fig. 8.8. Illustration of the type of information available using horizontal and vertical polarization data, including the process (top) leading to the weather radar signals observed at a given range r (bottom left) at horizontal (HH) and vertical (VV) polarization, and a simplified description of common parameters derived from dual-polarization signals (bottom right). If the radar observes rain, the larger raindrops will look larger at horizontal polarization than at vertical polarization. As a result, the time series of target amplitudes will be different for the two polarizations, not only by an average factor $(Z_{DR})^{0.5}$, but also in the time sequence of relative amplitude (and phase) from pulse to pulse, resulting in a slight decorrelation between the two signals. In addition, if there is considerable rain in the path from the radar to the target, the horizontally polarized wave will be somewhat delayed (and hence compressed) compared to the vertically polarized one. This will result in an average difference in phase φ_{DP} between the time series of the two signals, even in the absence of scattering phase delay that is significant only at the smaller wavelengths.

Dual-polarization parameters provide information on the shapes and the variability in the shapes of targets in the sampling volume as well as differential propagation information. This information, in combination with other radar parameters (Z, v_{DOP}, σ_v, etc.), can be used to determine, to a large extent, the nature of the meteorological targets being observed: rain, melting snow, hail, ground echoes, insects, among others (Vivekanandan et al. 1999). From there, further data processing may be performed to extract meteorological data by suppressing artifacts unambiguously identified by polarimetric data or by using continuity of fields to estimate parameters more accurately. Fuzzy logic, neural networks, and wavelets all have innovative applications in this area as well in the detailed signal-processing algorithms. However, it is beyond the scope of this paper to go into such details.

Polarimetric information can also be used to improve rainfall estimates by separating rainfall signatures from other targets and to provide additional information on the shape of raindrops, and hence more accurate drop-size distributions. The usefulness of the latter information will depend on the climatic regime, as raindrop shape information will be useful only up to the radar range where only raindrops and no ice particles are observed. This range varies with melting-level height and, hence, surface temperature.

5. New hardware and the expanding role of signal processing

There has always been talk of digital receivers that remove spectral artifacts, digital waveform generators that generate highly structured pulse compression waveforms, digital phased-array radars systems that scan instantaneously in random directions, digital beam-forming arrays that simultaneously generate multiple beams, and full vector wind measurement radars not constrained to radial Doppler estimates. These radar components that rely heavily on digital signal-processing techniques may have been innovative at one time and to some extent still are today, but they are rapidly becoming standard radar system technologies that provide essential meteorological information. Admittedly, some of these concepts find sole application in vertical-pointing wind profilers that have significantly reduced processing requirements from scanning weather radars. Typically in profiling radars, the base data rate is 10 000 times slower than scanning weather radars, which permits time periods necessary to achieve these sophisticated and innovative signal-processing techniques. However, as digital computing hardware and software continue to evolve at the present rapid pace and decreasing cost, we may expect banks of general purpose parallel computing elements to replace cards of special digital signal-processing chips produced at much lower volume and higher cost. Compact personal computer boards using standard software, fast data distribution networks, and as much parallel processing as is required appear ready to dominate the radar signal-processing architectures in the near future.

6. Conclusions

As this quick overview has illustrated, signal processing still offers many new techniques to improve the extraction of information from radar echoes. Constant improvements in radar and digital processing hardware allow the implementation of new algorithms, which in turn breed innovation and sophistication in new applications and yet additional new and innovative processing techniques.

At the same time, the gap has never been wider between what signal-processing specialists can offer and what the weather radar community is using. While this paper has focused on innovative techniques to extract information from radar data now and in the future, radar meteorologists have yet to constructively use all the present measurements. For example, spectrum width is computed by most radars, including the WSR-88D, but has found limited operational applications. Besides the signal-processing jargon barrier, one of the problems lies in the fact that radar already provides so much data, of which so little is used quantitatively, that meteorologists do not feel compelled to ask for more. As a result, with the exception of artifact removal and other data quality applications, the signal-processing community receives little guidance in developing new capabilities. Instead of serving the radar community with new concepts to solve high-priority meteorological measurement problems, many signal-processing engineers find themselves trying to guess what meteorologists want or need, often just experimenting with new ideas to see how well they work and then moving on.

It is imperative that the two solitudes find a way to communicate, with engineers gaining an appreciation of the meteorologically relevant information and meteorologists understanding more signal processing to steer the engineers' expertise in the right direction. If the limited quantitative use of radar is really the limitation, engineers should take initial steps to assist the atmospheric scientists make better use of the presently available data by understanding its application in, for example, assimilating radar data into numerical models (Sun and Wilson 2003). When real communication begins and these basic needs are met, additional relevant signal-processing techniques will be applied and new ones will evolve. Only by coordinating efforts between the digital signal-processing technologists and the atmospheric scientists will innovative signal-processing developments relevant to the radar meteorology community occur. Otherwise, the guessing game may continue for a long time.

Acknowledgments. The authors would like to thank the assistance of NCAR staff member Greg Meymaris for making several figures and Mary Landahl for help with references. We also pull our hats off to Rit Carbone and Roger Wakimoto for organizing the symposium from which this volume is derived.

REFERENCES

Bringi, V. N., and V. Chandrasekar, 2001: *Polarimetric Doppler Weather Radar.* Cambridge University Press, 662 pp.

Burg, J. P., 1972: The relationship between maximum entropy spectra and maximum likelihood spectra. *Geophysics,* **37,** 375–376.

Carbone, R. E., D. Atlas, P. Eccles, R. Fetter, and E. Mueller, 1973: Dual-wavelength radar hail detection. *Bull. Amer. Meteor. Soc.,* **54,** 921–924.

Cohn, S. A., W. O. J. Brown, C. L. Martin, M. S. Susedik, G. Maclean, and D. B. Parsons, 2001: Clear air boundary layer spaced antenna wind measurement with the Multiple Antenna Profiler (MAPR). *Ann. Geophys.,* **19,** 845–854.

de Elia, R., and I. Zawadzki, 2000: Sidelobe contamination in bistatic radars. *J. Atmos. Oceanic Technol.,* **17,** 1313–1329.

Doviak, R. J., and D. Sirmans, 1973: Doppler radar with polarization diversity. *J. Atmos. Sci.,* **30,** 737–738.

Fabry, F., and I. Zawadzki, 2001: New observational technologies: Scientific and societal impacts. *Meteorology at the Millennium,* R. P. Pearce, Ed., Academic Press, 72–82.

——, C. Frush, I. Zawadzki, and A. Kilambi, 1997: On the extraction of near-surface index of refraction using radar phase measurements from ground targets. *J. Atmos. Oceanic Technol.,* **14,** 978–987.

Gosset, M., and H. Sauvageot, 1992: A dual-wavelength radar method for ice–water characterization in mixed-phase clouds. *J. Atmos. Oceanic Technol.,* **9,** 538–547.

Keeler, R. J., and L. J. Griffiths, 1977: Acoustic Doppler extraction by adaptive linear prediction filtering. *J. Acoust. Soc. Amer.,* **61,** 1218–1227.

——, and C. L. Frush, 1983a: Coherent wideband processing of distributed targets. *Proc. Int. Geoscience and Remote Sensing Symp.,* San Francisco, CA, IEEE/URSI, 3.1–3.5.

——, and ——, 1983b: Rapid scan Doppler radar development considerations. Preprints, *21st Conf. on Radar Meteorology,* Edmonton, AB, Canada, Amer. Meteor. Soc., 284–290.

——, and R. E. Passarelli, 1990: Signal processing for atmospheric radars. *Radar in Meteorology,* D. Atlas, Ed., Amer. Meteor. Soc., 199–229.

——, and C. A. Hwang, 1995: Pulse compression for weather radar. *Proc. Int. Radar Conf.,* Washington, DC, IEEE, 1–7.

Koivunen, A. C., and A. B. Kostinski, 1999: The feasibility of data whitening to improve performance of weather radar. *J. Appl. Meteor.,* **38,** 741–749.

Marshall, J. S., and W. Hitschfeld, 1953: Interpretation of the fluctuating echo from randomly distributed scatterers. *Can. J. Phys.,* **31,** 962–994.

Mudukutore, A. S., V. Chandrasekar, and R. J. Keeler, 1998: Pulse compression for weather radars. *IEEE Trans. Geosci. Remote Sens.,* **36,** 125–142.

Oppenheim, A. V., and R. W. Schafer, 1989: *Discrete Time Signal Processing.* Prentice Hall, 879 pp.

Palmer, R. D., T. Y. Yu, and P. B. Chilson, 1999: Range imaging using frequency diversity. *Radio Sci.,* **34,** 1485–1496.

Pazmany, A. L., J. C. Galloway, J. B. Mead, I. Popstefanija, R. E. McIntosh, and H. Bluestein, 1999: Polarization diversity pulse-pair technique for millimeter-wave Doppler radar measurements of severe storm features. *J. Atmos. Oceanic Technol.,* **16,** 1900–1911.

——, J. B. Mead, S. M. Sekelsky, and D. J. McLaughlin, 2001: Multi-frequency radar estimation of cloud and precipitation properties using an artificial neural network. Preprints, *30th Int. Conf. on Radar Meteorology,* Munich, Germany, Amer. Meteor. Soc., 154–156.

Pirttilä, J., and M. Lehtinen, 2001: Solving the range-Doppler dilemma with the SMPRF pulse code. Preprints, *30th Int. Conf. on Radar Meteorology* Munich, Germany, Amer. Meteor. Soc., 322–324.

Press, W. H., S. A. Teukolsky, W. T. Vetterling, and B. P. Flannery, 1992: *Numerical Recipes in C: The Art of Scientific Computing.* Cambridge University Press, 994 pp.

Rummler, W. D., 1968: Introduction of a new estimator for velocity spectral parameters. Tech. Memo. mm-68-4121-5, Bell Telephone Laboratories, Whippany, NJ, 24 pp.

Sachidananda, M., and D. S. Zrnic, 1999: Systematic phase codes for resolving range overlaid signals in a Doppler weather radar. *J. Atmos. Oceanic Technol.,* **16,** 1351–1363.

——, and ——, 2000: Clutter filtering and spectral moment estimation for Doppler weather radars using staggered pulse repetition time (PRT). *J. Atmos. Oceanic Technol.,* **17,** 323–331.

Strauch, R. G., 1988: A modulation waveform for short-dwell-time meteorological Doppler radars. *J. Atmos. Oceanic Technol.,* **5,** 512–520.

Sun, J., and J. Wilson, 2003: The assimilation of radar data for weather prediction. *Radar and Atmospheric Science: A Collection of Essays in Honor of David Atlas, Meteor. Monogr.,* No. 52, Amer. Meteor. Soc., 175–198.

Torres, S. M., and D. S. Zrnic, 2001: Optimum processing in range to improve estimates of Doppler and polarimetric variables. Preprints, *30th Int. Conf. on Radar Meteorology,* Munich, Germany, Amer. Meteor. Soc., 325–327.

Vivekanandan, J., D. S. Zrnic, S. M. Ellis, R. Oye, A. V. Ryzhkov, and J. Straka, 1999: Cloud microphysics retrieval using S-band dual polarization radar measurements. *Bull. Amer. Meteor. Soc.,* **80,** 381–388.

Widrow, B., and S. D. Stearns, 1985: *Adaptive Signal Processing.* Prentice Hall, 528 pp.

Woodman, R. F., and T. Hagfors, 1969: Methods for the measurement of vertical ionospheric motions near the magnetic equator by incoherent scattering. *J. Geophys. Res.,* **74** (A5), 1205–1212.

Wurman, J., S. Heckman, and D. Boccippio, 1993: A bistatic multiple-Doppler network. *J. Appl. Meteor.,* **32,** 1802–1814.

Chapter 9

Global and Local Precipitation Measurements by Radar

V. CHANDRASEKAR

Colorado State University, Fort Collins, Colorado

R. MENEGHINI

NASA/Goddard Space Flight Center, Greenbelt, Maryland

I. ZAWADZKI

Department of Atmospheric and Oceanic Sciences, McGill University, Montreal, Quebec, Canada

Chandrasekar   **Zawadzki**

1. Introduction

The detection and measurement of precipitation by radar has been pursued since its introduction as a meteorological tool. The main advantage of using radar for precipitation estimation is that measurements can be made over large areas, with either fairly high temporal and spatial resolution or extensive spatial coverage (about 10 000 km² for ground-based radars and an order of magnitude more for space-based radars). To sample the area covered by a typical ground-based radar, substituting each radar spatial sample with a rain gauge, would require about a quarter-million gauges. Using a similar analogy to space radars, nearly one-half-million gauges would be required per orbit. Since the radar transmitter and receiver normally use the same antenna (monostatic operation), the measurements are sent to a central location at the speed of light by "natural wireless networks." In addition, radars can provide fairly rapid updates of the three-dimensional structure of precipitation. Because of these advantages, radar measurements of precipitation have enjoyed widespread use for meteorological applications, independent of the accuracy or the type of algorithm used to derive precipitation estimates.

Approaches to rainfall measurement can be broadly classified into 1) physically based and 2) statistical–engineering based. Physically based rainfall algorithms, as defined here, rely on physical models of the rain medium without feedback from ground observations, whereas statistical–engineering solutions rely on modifications to the algorithm based on the volumetric structure of radar echoes or on the information from gauge observations. Physically based approaches attempt to solve the inverse electromagnetic problem of obtaining resolution-volume-averaged precipitation estimate from radar backscatter and forward scatter measurements such as reflectivity Z, differential reflectivity Z_{dr}, specific differential propagation phase K_{dp}, or specific attenuation (A). Engineering solutions, on the other hand, seek the best possible estimate of rainfall on the ground, using some feedback mechanism such as gauge data, recognizing that the radar measurements are made aloft. Though not stated in this form, this fundamental distinction was recognized by Zawadzki (1984). Both physically based techniques and engineering solutions have their role in precipitation measurements. Physically based techniques, such as the polarimetric radar measurements, can distinguish rain from frozen hydrometeors such as hail or graupel and from nonmeteorological echoes. Discrimination among hydrometeors is valuable not only for precipitation physics, but it is an important step prior to application of precipitation algorithms. Engineering techniques focus primarily on accurate estimation of rainfall or snowfall on the ground. These range from simple techniques such as tuning the algorithm coefficients with season or with radar range to more sophisticated approaches such as the derivation of nonparametric Z–R relations and the use of neural networks.

Areal rainfall estimates using propagation measurements such as K_{dp} have shown great success recently. Similarly, physically based approaches such as dual-polarization radar estimates of rainfall have been adapted to statistical/engineering techniques such as probability-matched methods. In addition, polarimetric algorithms can be used to determine the amount of rain in a rain–hail mixture. A class of hybrid procedures are evolving that combine the advantages of physically based and statistical/engineering solutions.

Ground-based weather radars have excellent temporal resolution but are usually limited to land-based deployment and do not cover all the land surface of the earth. Space-borne radar is complementary to the ground radar in that it can provide global coverage but is severely limited in its ability to monitor the temporal evolution of precipitation systems. Although most current space-based algorithms are physically based, statistical methods hold great promise, particularly in the estimation of rainfall over large space–timescales. Spaceborne methods of rain estimation differ significantly from ground-based approaches, primarily because of differences in the radar systems deployed on these platforms. The first major difference is the operating frequency. The current space radar for measuring precipitation, the Tropical Rainfall Measuring Mission Precipitation Radar (TRMM PR), operates at 13.8 GHz. This and the proposed space-borne radars at frequencies of 13.6, 35, and 95 GHz are significantly higher than S-band or C-band radars typically used on the ground. While the use of higher frequencies yields adequate resolution from space for a modest antenna size, attenuation correction must be done before parameters of the rainfall can be estimated. In addition, strong backscattering from the surface of the earth limits the lowest altitude at which rainfall estimates can be computed. Comparisons of radar reflectivity factors and rainfall rates derived from space- and ground-based radars suggest that a single-wavelength space-borne radar can yield reasonably accurate estimates of these quantities. Next-generation space-borne sensors will include a dual wavelength and a millimeter cloud radar. These instruments are expected to improve global estimates of precipitation and cloud

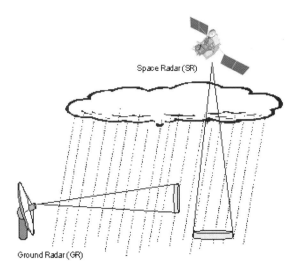

FIG. 9.1. Geometry of space- and ground-based radars. The shaded areas show the resolution volumes.

parameters and will constitute an integral part of our rain and cloud monitoring capabilities.

The potential applications of quantitative radar precipitation measurements are very broad: from hydrology, agriculture, and forestry through water cycle and water resources studies, to nowcasting and finally numerical modeling validation and data assimilation. The aim of radar meteorology is to establish the error structure of quantitative radar precipitation measurements and their time- and space-scale dependence.

This paper provides a summary of some ground- and space-based radar precipitation estimation techniques. In addition, the paper will emphasize the most critical practical elements of "rainfall estimation" that impact the measurement accuracy. We also attempt to provide some insight into new instruments and techniques that are expected to improve our understanding of local and global precipitation. The paper is organized as follows. Section 2 defines the radar basis for characterizing precipitation for both ground and spaceborne radars. The practical issues facing radar measurements of precipitation are discussed extensively, with section 3 focusing on reflectivity-based estimates, section 4 on dual-polarization techniques and section 5 on spaceborne measurements. The paper will conclude with some projections on the future of rainfall estimation from radars.

2. Radar basis for characterizing precipitation

a. Background

The scattering of electromagnetic waves by precipitation and their propagation through precipitation media form the basis of radar-based characterization of precipitation. Figure 9.1 shows the geometry of the scattering volume, within the precipitation medium, for ground and space-borne radars. Consider a two-port antenna system that transmits and receives two states of polarization. Though the polarization state associated with the two ports can be general, for simplicity assume horizontal–vertical (H–V) polarization states. Under this condition the received voltages at the antenna ports due to a single precipitation particle described by scattering matrix **S** can be expressed as

$$\begin{bmatrix} V_H \\ V_V \end{bmatrix} = \frac{\lambda G}{4\pi r^2} \textbf{TST} \begin{bmatrix} M_H \\ M_V \end{bmatrix} \tag{9.1}$$

(Bringi and Chandrasekar 2001), where λ is the operating wavelength, G is the gain of the antenna in the direction of the precipitation particle, M_H, M_V are the input amplitudes to the antenna, and **T** is the transmis-

sion matrix of the precipitation medium. Thus, the fundamental physical equation governing the transmission and reception of electromagnetic waves by a radar includes the propagation through, and backscatter by, precipitation. Depending on the type of design and deployment that is used, such as frequency of operation and whether space or ground based, algorithms are developed to convert the received voltages to characteristics of the precipitation. The precipitation medium on average has a diagonal propagation matrix for H–V polarization in most cases, particularly in rain (Bringi and Chandrasekar 2001) and (9.1) can be simplified under that condition to

$$\begin{bmatrix} V_H \\ V_V \end{bmatrix} = \frac{\lambda G}{4\pi r^2} \begin{bmatrix} S_{HH}\exp(-j2k_{eff}^H r) & S_{HV}\exp[-j(k_{eff}^H + k_{eff}^V)r] \\ S_{VH}\exp[-j(k_{eff}^H + k_{eff}^V)r] & S_{VV}\exp(-j2k_{eff}^V r) \end{bmatrix} \begin{bmatrix} M_H \\ M_V \end{bmatrix}, \tag{9.2}$$

where k_{eff}^H, k_{eff}^V are the effective propagation constants that depend on the forward scatter properties of the precipitation medium, while S_{HH}, S_{VV} and $S_{HV}(=S_{VH})$ are the elements of the backscatter matrix of the single scatterer.

The propagation constants for horizontal and vertical polarizations, k_{eff}^H, k_{eff}^V, can be expressed in terms of real and imaginary parts as

$$k_{eff}^H = k_{re}^H - jk_{im}^H \tag{9.3a}$$

$$k_{eff}^V = k_{re}^V - jk_{im}^V, \tag{9.3b}$$

where k_{re} determines the phase and k_{im} determines the attenuation of the electromagnetic wave due to propagation.

The voltages received (or powers) at the radar are composed of contributions from all the precipitation particles in the resolution volume. Therefore, it is useful to work with radar backscatter cross sections per unit volume that can be related to microphysical properties of precipitation. The volumetric radar cross sections at horizontal (H) and vertical (V) polarizations are given by

$$\eta_{HH} = \int N(D)4\pi|S_{HH}|^2\, dD, \tag{9.4a}$$

$$\eta_{VV} = \int N(D)4\pi|S_{VV}|^2\, dD, \tag{9.4b}$$

where $N(D)$ is the particle size distribution and D is the equivolume diameter. In the Rayleigh limit where the scatterers are spherical and much smaller than the radar wavelength, the volumetric radar cross section reduces to

$$\eta_{ray} = \frac{\pi^5}{\lambda^4}|K_p|^2 \int N(D)D^6\, dD, \tag{9.5}$$

where $|K_p|^2 = |(\varepsilon_r - 1)^2/(\varepsilon_r + 2)|$; ε_r is the complex

dielectric constant of the precipitation particle. The reflectivity factor Z is defined as

$$Z = \int_D D^6 N(D)\, dD. \tag{9.6}$$

When the Rayleigh scattering approximation is not valid, the reflectivity factor is given by the more general expression

$$Z_{H,V} = \frac{\lambda^4}{\pi^5|K_p|^2} \int \sigma_{H,V}(D)N(D)\, dD, \tag{9.7}$$

where the radar cross section $\sigma_{H,V}$ is given by $\sigma_{H,V} = 4\pi|S_{HH,VV}|^2$.

With the assumption of a homogeneous propagation medium, the received power at each polarization state is decreased by the two-way path attenuation in accordance with the relations

$$P_{rec}^H = P_{rec}^{H,int}e^{-4k_{im}^H r} \tag{9.8a}$$

$$P_{rec}^V = P_{rec}^{V,int}e^{-4k_{im}^V r}, \tag{9.8b}$$

where $P_{rec}^{H,int}$, $P_{rec}^{V,int}$ are the intrinsic backscattered powers in the absence of attenuation.

The cross correlation between V_H, V_V yields

$$\frac{\langle V_H^* V_V \rangle}{[\langle |V_H|^2 \rangle \langle |V_V|^2 \rangle]^{1/2}} = |\rho_{co}|e^{j\psi_{dp}}, \tag{9.9}$$

where ρ_{co} is the copolar correlation between S_{HH} and S_{VV} computed over the particle size distributions.

The propagation constants can be related to the forward scatter properties by (van de Hulst 1981)

$$k_{\text{eff}}^{\text{H}} = k_{\text{re}}^{\text{H}} - jk_{\text{im}}^{\text{H}} = k_o + \frac{2\pi}{k_o}\langle f_{\text{HH}}\rangle \quad (9.10a)$$

$$k_{\text{eff}}^{\text{V}} = k_{\text{im}} - jk_{\text{im}}^{\text{V}} = k_o + \frac{2\pi}{k_o}\langle f_{\text{VV}}\rangle, \quad (9.10b)$$

where $f_{\text{HH,VV}}$ are the forward scattering amplitudes for H and V polarizations.

The received voltage is determined by the backscattering properties of the hydrometeors in the resolution volume as well as by the forward scattering properties of the particles over the propagation path from the radar to the resolution volume. From the received voltages, the reflectivity factor (Z), differential reflectivity (Z_{dr}), differential propagation phase (ϕ_{dp}), and specific attenuation can be inferred. All of the above electromagnetic quantities are directly related to microphysical properties of precipitation, which are used in quantitative precipitation estimation (QPE).

The most commonly used measurement for precipitation estimation is the reflectivity factor Z. In practice it has to be measured at a specific polarization state. Historically, the polarization state at which reflectivity is measured was ignored leading to inaccuracies of the order of a few decibels. For ground-based linear polarization radars, the most commonly used polarization state is horizontal. The next important issue to be understood from a radar perspective is the frequency of operation. Most operational ground-based radars are at S (3GHz) or C (5.5 GHz) band. However, due to practical considerations space-borne radars operate at higher frequencies such as K_u band (14 GHz) or higher. Attenuation is usually negligible at S band, but not at C band (5.5 GHz) or higher. In spaceborne systems, attenuation measurements are used in deriving rainfall estimates as discussed later in this section.

The measurements of precipitation described above can be further elaborated upon for rain. The distributions of raindrop sizes and shapes form the building blocks for deriving physically based rain-rate algorithms. Although practical considerations may be just as important, the physical approach provides guidance in developing algorithms for rainfall estimation. The raindrop size distribution (DSD) describes the probability density distribution function of raindrops and can be expressed as

$$N(D) = n_c f_D(D) \text{ m}^{-3} \text{ mm}^{-1}, \quad (9.11)$$

where $N(D)$ is the number of raindrops per unit volume per unit size interval (D to $D + \Delta D$), n_c is the concentration, and $f_D(D)$ is the probability density function. It should be noted that any function used to describe $N(D)$, when integrated over the DSD, must yield the concentration to qualify as a DSD function. This property is a direct result of the fact that any probability density function should integrate to unity. It can be seen that $f_D(D)$ has units of inverse size. Several parametric forms of DSD have been used in the literature including exponential, lognormal, and gamma. A gamma form of

the DSD with three parameters, namely, the median drop diameter D_o, the scaling constant N_w or N_0, and the shape parameter μ, is widely used (Ulbrich 1983).

The mean shape of raindrops plays an important role in the description of dual polarization observations. The equilibrium shape of a raindrop is determined by a balance of forces on the interface involving hydrostatic, surface tension, and aerodynamic forces. It is common to approximate this shape by an oblate spheroid, defined by its axis ratio r ($r = a/b$, where b, a are the semiminor and semimajor axes, respectively). The simplest model describes the shape–size relation as a linear approximation,

$$r = 1 - \beta D, \quad (9.12)$$

where β is the slope of the shape–size relation. Pruppacher and Beard (1971) estimated β to be 0.062 from perturbation models. Subsequently, Beard and Chuang (1987) and Andsager et al. (1999) derived more elaborate expressions for shape–size relations. Chandrasekar et al. (1988) and Bringi et al. (1998) measured the mean shape of raindrops over Colorado and Florida using airborne 2D-P probes and found that the results are in general agreement with theoretical models.

Based on the size and shape distributions of raindrops the various polarimetric radar observables can be calculated. The most commonly used polarimetric measurements for rainfall estimation are the reflectivity factor (say, at horizontal polarization), the differential reflectivity, and the specific differential phase. For rain, Z_{dr} and K_{dp} can be written as integrals of backscatter and forward scatter amplitudes over the DSD. Here Z_{dr} can be written as (Bringi and Chandrasekar 2001)

$$Z_{\text{dr}} = 10 \log_{10}\left[\frac{\int \sigma_{\text{HH}}(D)N(D)\,dD}{\int \sigma_{\text{VV}}(D)N(D)\,dD}\right] \text{ dB}, \quad (9.13)$$

where $\sigma_{\text{HH}} = 4\pi|\mathbf{S}_{\text{HH}}|^2$ and $\sigma_{\text{VV}} = 4\pi|\mathbf{S}_{\text{VV}}|^2$ are the radar cross sections at horizontal and vertical polarization, respectively. Also,

$$K_{\text{dp}} = \frac{180\lambda}{\pi}\int \text{Re}[f_{\text{H}}(D) - f_{\text{V}}(D)]N(D)\,dD \text{ }^\circ\text{km}^{-1}, \quad (9.14)$$

where f_{H}, f_{V} are the forward scattering amplitudes at horizontal and vertical polarization, respectively. Rain rate is the volume flux of water per unit area. The still-air rain rate is given by

$$R = 0.6\pi \times 10^{-3}\int v(D)D^3N(D)\,dD \text{ mm h}^{-1}, \quad (9.15)$$

where $v(D)$ is the terminal fall velocity of raindrops, which depends on parameters such as air density, size, and shape of raindrops. However, for algorithmic purposes the terminal fall velocity can be approximated by a power-law expression of the form

$$v(D) = \alpha D^\beta \text{ m s}^{-1}. \qquad (9.16)$$

It should be noted here that air density or altitude adjustment factor could be introduced to modify (9.16) (Foote and Du Toit 1969; Beard 1985).

The extinction cross section of particles determines the power loss suffered by the incident wave due to absorption and scattering. For Rayleigh scattering, where the radar wavelength is much larger than the particle sizes, the absorption cross section of a raindrop is proportional to its volume. Moreover, in the Rayleigh regime, the impact of scattering on attenuation can be neglected so that the absorption and extinction cross sections are approximately equal. The specific attenuation can be expressed in terms of extinction cross section as (Bringi and Chandrasekar 2001)

$$\gamma = 4.343 \times 10^3 \int \sigma_{\text{ext}}(D)N(D) \, dD \text{ dB km}^{-1}. \qquad (9.17a)$$

The specific attenuation is nearly proportional to the product of water content and the imaginary part of the dielectric constant (ε_λ') of water. The imaginary part of the dielectric constant varies with frequency and temperature. Below about 35 GHz the extinction cross section can be approximated as a power law in the diameter D so that the formula above becomes

$$\gamma \cong 4.343 \times 10^3 C_\lambda \int D^n N(D) \, dD \text{ dB km}^{-1}, \qquad (9.17b)$$

where C_λ, n are functions of temperature and frequency.

b. Precipitation algorithms for ground radar

Both from earth and space, the Z–R algorithm is the most widely used precipitation estimation technique. Initially it was used as an engineering solution to the problem of estimation of rainfall from radar. However, over time, various authors have tried to provide a physical basis for Z–R relation (see, e.g., Atlas et al. 1999; Rosenfeld and Ulbrich 2003). One such approach is the concept of normalized drop size distributions (Sekhon and Srivastava 1971; Willis 1984; Testud et al. 2001; Bringi and Chandrasekar 2001). Using the normalized DSD, Z and R can be related as

$$\frac{Z}{N_w} = a\left(\frac{R}{N_w}\right)^b, \qquad (9.18a)$$

where b is approximately 1.5 and N_w is the normalizing constant defined as the intercept of an equivalent exponential distribution with the same water content W. Equation (9.18) indicates that most of the fluctuations

in Z–R relations can be ascribed to variability in N_w. Zawadzki (1984) has argued that measurement errors can dominate the process of deriving a ground rain rate from Z. Bringi et al. (2003) have studied the variability of N_w over different climatic regimes and conclude that there are significant systematic variations of N_w that could account for changes in the Z–R relations. Nevertheless, it is clear that attention must be paid to all aspects of the rainfall retrieval problem.

Dual-polarization measurement brought in a new aspect to the precipitation measurement problem. Initially the Z_{dr} measurement was proposed as a method to estimate the DSD parameters which, in turn, could be used to derive more accurate rainfall rates (Seliga and Bringi 1976). However, over time, direct estimates of rainfall rate using Z_H and Z_{dr} were developed without the intermediate step of estimating the DSD. Specific differential propagation phase K_{dp} (Seliga and Bringi 1978; Sachidananda and Zrnic 1986; Chandrasekar et al. 1990) has some attractive properties such as immunity to calibration errors, and partial beam blocking, as well as the ability to measure rain in a rain–hail mixture. The technique is particularly useful for the estimation of moderate to heavy rain rates. The physical basis of rainfall estimates from polarimetric radar measurements can be traced to the behavior of these measurements, specifically Z_{dr} and K_{dp}. In Eq. (9.14), K_{dp} is defined as the ensemble average of the difference in forward scatter amplitudes of hydrometers in the propagation medium. The integral in (9.14) can be simplified for rain under a low-frequency approximation to (Bringi and Chandrasekar 2001)

$$K_{\text{dp}} = \left(\frac{2\pi}{\lambda}\right) \times 10^{-3} \, CW(1 - \bar{r}_m), \qquad (9.18b)$$

where C is a dimensional constant that is approximately equal to 3.75, $\bar{r}_m$ is the mass-weighted means axis ratio, and W is the rainwater content. Thus, microphysically K_{dp} is proportional to the rainwater content multiplied by a function of mean raindrop shapes (Jameson 1985). Seliga and Bringi (1976) showed that Z_{dr} is a direct estimate of the median drop diameter D_o (or the mass-weighted mean diameter D_m). Here D_m can be related to Z_{dr} in a power law of the form

$$D_m = a(Z_{\text{dr}})^b. \qquad (9.19a)$$

Considering the expressions for Z and R given by (9.6) and (9.15), the ratio of Z/R can be parameterized to (Bringi and Chandrasekar 2001)

$$\frac{Z}{R} \cong \alpha D_m^{2.33}. \qquad (9.19b)$$

This parameterization when combined with the estimates of D_m in terms of Z_{dr} yields

$$R = \alpha Z_h Z_{\text{dr}}^{-2.33b}, \qquad (9.19c)$$

where α is a scaling constant. Thus, the above simplistic forms of rainfall estimates using Z_H and Z_{dr} or K_{dp} can be optimized by considering power-law estimates of the form

$$R = C_R Z_H^a Z_{dr}^b K_{dp}^c, \qquad (9.19d)$$

where C_R, a, b, c are empirical constants. In the above estimator, if only one or two of the measurements (among Z_H, Z_{dr}, and K_{dp}) are used, the rest of the exponents assume the value zero. In the last decade several algorithms have been proposed to estimate rainfall using Z_H, Z_{dr}, and K_{dp}. These are summarized below.

i) *The R (Z_H, Z_{dr} algorithm.* The rainfall algorithm using Z_H and Z_{dr} is of the form

$$R(Z_H, Z_{dr}) = C_1 Z_H^{a_1} \xi_{dr}^{b_1}, \qquad (9.20a)$$

where

$$Z_{dr}(dB) = 10 \log_{10} \xi_{dr} \quad \text{or}$$

$$R(Z_H, Z_{dr}) = C_1 Z_H^{a_1} 10^{0.1 b_1 Z_{dr}}. \qquad (9.20b)$$

ii) *The R (K_{dp}) algorithm.* This algorithm is of the form

$$R(K_{dp}) = C_2 K_{dp}^{b_2} \quad \text{at a given frequency, or} \qquad (9.21a)$$

$$R(K_{dp}) = 129 \left(\frac{K_{dp}}{f} \right)^{b_2} \qquad (9.21b)$$

as a function of radar transmit frequency (expressed in GHz). The frequency scaling is valid up to 13 GHz (Bringi and Chandrasekar 2001).

iii) *The R (K_{dp}, Z_{dr}) algorithm.* The rainfall algorithm using K_{dp} and Z_{dr} is of the form

$$R(K_{dp}, Z_{dr}) = C_3 K_{dp}^{a_3} \xi_{dr}^{b_3} \quad \text{or} \qquad (9.22a)$$

$$= C_3 K_{dp}^{a_3} 10^{0.1 b_3 Z_{dr}}. \qquad (9.22b)$$

These algorithms and their error structure are summarized in Bringi and Chandrasekar (2001).

c. Single wavelength methods for space-borne radar

The fundamental work on single attenuating-wavelength methods was done nearly five decades ago by Hitschfeld and Bordan (1954). By solving the first-order differential equation associated with the radar equation, they concluded that a weather radar that operates at a single attenuating wavelength is unreliable for estimating the rain rate because imprecise knowledge of the radar calibration and inevitable errors in the Z–R and γ–Z relations result in rapid growth in the estimation error. A way of mitigating the problem was to bound the solution by measurements from a rain gauge or some other point measurement of rain rate. In the late 1970s and early 1980s researchers recognized that the Hitschfeld–Bordan equation also could be made stable by imposing a constraint of path attenuation by using the surface as a reference target. Since that time a number of modifications, improvements, and tests with experimental data have been carried out.

It should be noted that there are a number of single attenuating-wavelength methods that are not discussed here. Although the mirror-image method (Meneghini and Atlas 1986) and the dual-beam technique (Testud and Amayenc 1989) can be applied to single-wavelength radar data, they differ fundamentally from the methods described below. Also omitted from the discussion is the iterative method (Hildebrand 1978) and the slope method of Klett (1981). Combined radar–radiometer algorithms are also beyond the scope of our discussion.

To simplify the equations it is convenient to define an apparent or measured reflectivity factor, Z_m, a quantity directly proportional to the radar return power. This can be expressed in terms of the true reflectivity factor Z by

$$Z_m(r) = Z(r) \exp \left[-0.2 \ln 10 \int_0^r \gamma(s) \, ds \right]$$

$$= Z(r) A(r), \qquad (9.23)$$

where the specific attenuation γ (dB km^{-1}) can be written as the sum of contributions from the hydrometeors, cloud water, and water vapor. If it is assumed that the attenuation from the hydrometeors dominates, then γ can be written as a function of the reflectivity factor in the form of a power law $\gamma = \alpha Z^\beta$. Substituting this into (9.23), differentiating the result with respect to the range r and letting $Z^\beta = u^{-1}$ gives the first-order linear equation

$$du/dr + fu + q\alpha = 0, \qquad (9.24)$$

where $f = \beta d/dr(\ln Z_m)$ and $q = 0.2 \ln 10 \beta$.

The solution of (9.24) that satisfies the initial value problem $u(0) = 1/Z_m^\beta(0) = 1/Z^\beta(0)$ is

$$u(r) = \left\{ u(0) - q \int_0^r \alpha(s) \exp \left[\int_0^s f(u) \, du \right] ds \right\}$$

$$\times \exp \left[-\int_0^r f(u) \, du \right]. \qquad (9.25)$$

Noting that $\exp[\int_0^s f(u) \, du] = (Z_m(s)/Z_m(0))^\beta$, the initial value or Hitschfeld–Bordan solution becomes

$$Z_{HB}(r) = Z_m(r)/[1 - S(r)]^{1/\beta}, \qquad (9.26)$$

where

$$S(r) = q \int_0^r \alpha(s) Z_m^\beta(s) \, ds. \qquad (9.27)$$

The final value solution was first derived by Marzoug and Amayenc (1991). Although their solution is expressed in terms of the specific attenuation γ, the solution also can be written in terms of the reflectivity factor Z (Iguchi and Meneghini 1994). For the final value problem, we impose the condition that at the range gate just above the surface, $r = r_s$, u satisfy the following equation:

$$u(r_s) = A^\beta(r_s)Z_m^{-\beta}(r_s). \quad (9.28)$$

The solution of (9.24) that satisfies (9.28) can be written as

$$Z_{fv}(r) = Z_m(r)/[A^\beta(r_s) + S(r_s) - S(r)]^{1/\beta}. \quad (9.29)$$

The estimate of reflectivity factor given by (9.29) at range r depends only on measurements of Z_m and estimates of α and β at ranges of r and greater. In contrast, the Hitschfeld–Bordan estimate, and the α- and C-adjustment methods described below, depends on data from the storm top ($r = 0$) down to r that usually includes ice and mixed-phase precipitation where α and β are not easily specified. The other advantage of the final value solution is that, unlike the Hitschfeld–Bordan solution, the denominator always remains greater than 0 and, indeed, can get no smaller than the estimate of the path attenuation, $A(r_s)$. However, as with all methods that rely on an independent estimate of the path attenuation, the reliability of the solution is closely tied to the accuracy with which $A(r_s)$ can be determined.

Two other attenuation-correction estimates can be derived by noting that if an independent estimate of path attenuation is available, $A(r_s)$, and if the initial value is to be satisfied, then the following equation must hold:

$$1 - q \int_0^{r_s} \alpha(s)Z_m^\beta(s)\, ds = [A(r_s)]^\beta. \quad (9.30)$$

There are two ways to adjust the parameters on the left-hand side of (9.30) so that the equality is satisfied. In the first, we replace $\alpha(r)$ by $\alpha(r)\varepsilon$. Solving for ε and replacing α by $\alpha\varepsilon$ in (9.26) yields the α-adjustment solution (Meneghini et al. 1983):

$$Z_\alpha(r) = Z_m(r)/[1 - \varepsilon S(r)]^{1/\beta}, \quad (9.31)$$

where

$$\varepsilon = [1 - A^\beta(r_s)]/S(r_s). \quad (9.32)$$

Notice that $Z_\alpha(0) = Z_{HB}(0) = Z_m(0) = Z(0)$; that is, like the Hitschfeld–Bordan, but unlike the final value solution, the α-adjustment solution at $r = 0$ is equal to the measured reflectivity factor. Furthermore, at the storm top, where the attenuation is zero, the measured reflectivity factor and the reflectivity factor are equal, as can be seen from (9.23). At the gate just above the surface, $Z_\alpha(r_s) = Z_{fv}(r_s) = Z_m(r_s)A^{-1/\beta}(r_s)$ so that the α-adjustment and the final value solutions agree at the far range gate. Moreover, if $A(r_s)$ is equal to the true path attenuation factor, then these solutions yield the reflectivity factor since

$$Z_m(r_s)A^{-1/\beta}(r_s) = Z(r_s). \quad (9.33a)$$

A second way to achieve equality in (9.30) is to replace $Z_m(r)$ by $Z_m(r)\delta$. Proceeding as before, and solving for δ, we obtain

$$Z_C(r) = Z_m(r)/[A^\beta(r_s)\varepsilon^{-1} + S(r_s) - S(r)]^{1/\beta}, \quad (9.33b)$$

where ε is given by (9.32). It is not difficult to show that $Z_C(r) = \varepsilon^{1/\beta}Z_\alpha(r)$. In general, ε differs from unity so that Z_c and Z_α are not equal. Moreover, as Z_α satisfies the "natural" boundary conditions at 0 and r_s, Z_C generally will not. An advantage of the C-adjustment method is that it is independent of errors in the radar constant; however, for a well-calibrated radar, the Z_α or Z_{fv} estimates are preferable.

In contrast to most single-wavelength methods, the objective of the approach by Kozu and Nakamura (1991) and Kozu et al. (1991) is to estimate the DSD along the path. However, because a single-wavelength radar does not provide sufficient information to extract a two-parameter DSD at each range gate, the parameter N_w is kept fixed while the median mass diameter D_0 is estimated at each gate. This assumption is justified if N_0 tends to be fairly stable in space and time. As in the other methods mentioned here, the path attenuation $A(r_s)$ is derived either from the surface reference method or from a radiometer-derived brightness temperature along the radar path.

It was recognized early in the development of single-wavelength radar methods that the estimation of rain rate over the full range of intensities would require a Hitschfeld–Bordan correction at light rain rates and an alternative method such as the surface reference technique (SRT) at moderate and high rain rates. Iguchi and Meneghini (1994) proposed several hybrid algorithms that provide a transition between the two solutions. More recently, Iguchi et al. (2000) have described the operational TRMM radar algorithm along with a refinement of the hybrid solution. Here, we outline the idea in its simplest form where the factor ε, defined by (9.32), is modified from its nominal value, say ε_0, to a new value ε, where

$$\varepsilon = 1 + \omega(\varepsilon_0 - 1). \quad (9.34)$$

Substituting this into the α-adjustment estimate yields one possible hybrid method:

$$Z_H(r) = Z_m(r)[1 - \{1 + \omega(\varepsilon_0 - 1)\}S(r)]^{-1/\beta}. \quad (9.35)$$

The factor ω specifies the relative weighting of the Hitschfeld–Bordan and SRT estimates of path attenuation. Note that when $\omega = 1$, and $\omega = 0$, $Z_H(r)$ reduces to the alpha-adjusted and Hitschfeld–Bordan estimates, respectively.

d. Dual-wavelength methods for spaceborne radar

Austin (1947) considered the use of the radar reflectivity factor at one wavelength with relative attenuation at another to deduce raindrop sizes. Atlas (1954) suggested a similar method, which would use dual-wavelength radar to measure cloud liquid water content. One of the primary thrusts of dual-wavelength research has been its application to the problem of hail detection (Atlas and Ludlum 1961; Srivastava and Jameson 1977;

Eccles 1979). An important offshoot of this type of method is that in the absence of hail the attenuation by rain can be estimated.

For the determination of the rain rate, the basic principle of the standard dual-wavelength method is to estimate the path attenuation using the radar equation and the ratio of return powers at the two wavelengths (Eccles and Mueller 1971; Berjulev and Kostarev 1974). The rain rate and liquid water content can be estimated over the interval by means of empirical laws that relate these quantities to the attenuation. To derive the basic equations for this method, we introduce the following definition for the ratio of the radar return powers over the interval $[r_k, r_j]$ $(r_j > r_k)$ at wavelengths λ_1 and $\lambda_2 (\lambda_1 < \lambda_2)$, $\Gamma_P(r_j, r_k; \lambda_1, \lambda_2)$:

$$\Gamma_P(r_j, r_k; \lambda_1, \lambda_2)$$
$$= P(\lambda_1, r_j)P(\lambda_2, r_k)/[P(\lambda_1, r_k)P(\lambda_2, r_j)]. \quad (9.36)$$

A similar quantity can be defined for the ratio of the radar reflectivity factors Z as

$$\Gamma_Z(r_j, r_k; \lambda_1, \lambda_2)$$
$$= Z(\lambda_1, r_j)Z(\lambda_2, r_k)/[Z(\lambda_1, r_k)Z(\lambda_2, r_j)]. \quad (9.37)$$

Use of the basic radar equation shows that Γ_P and Γ_Z are related by the following equation:

$$\Gamma_P(r_j, r_k; \lambda_1, \lambda_2)$$
$$= \Gamma_Z(r_j, r_k; \lambda_1, \lambda_2)$$
$$\times \exp\left\{-0.2 \ln 10 \int_{r_k}^{r_j} [k(\lambda_1, s) - k(\lambda_2, s)] \, ds\right\}$$
$$(9.38)$$

where $\int_{r_k}^{r_j} [k(\lambda_1, s) - k(\lambda_2, s] \, ds$ is the differential attenuation over the interval $[r_k, r_j]$. To estimate this quantity as a function of the power ratio Γ_P requires an estimate of Γ_Z. There are two assumptions under which $\Gamma_Z = 1$: in the first, the rain rate is taken to be uniform over the interval so that $Z(\lambda_i, r_k) = Z(\lambda_i, r_j)$, $i = 1, 2$; in the second, less restrictive, assumption, the reflectivity factor is taken to be wavelength independent (Rayleigh approximation) so that $Z(\lambda_1, r_1) = Z(\lambda_2, r_1)$, $1 = k$ or j. When either of these assumptions is used, the estimated value of Γ_Z is unity and the differential attenuation is obtained from (9.38):

$$\int_{r_k}^{r_j} [k(\lambda_1, s) - k(\lambda_2, s)] \, ds$$
$$= -5 \log_{10} \Gamma_P(r_j, r_k; \lambda_1, \lambda_2). \quad (9.39)$$

For many dual-wavelength datasets, for example, (3 cm, 0.86 cm) or (10 cm, 3 cm), the attenuation at the longer wavelength is much smaller than that at the shorter wavelength so that (9.39) provides an estimate of the interval attenuation at λ_1. The corresponding rain rate

R is then estimated from power laws relating $[k(\lambda_1) - k(\lambda_2)]$ or $k(\lambda_1)$ to R.

As compared with the usual Z–R method at a single wavelength, the standard dual-wavelength approach represents an improvement in that the estimate of R is independent of the radar calibration. Furthermore, the estimate tends to be less sensitive to errors in the drop size distribution, a reflection of the insensitivity in the k–R law to variations in the DSD (Atlas and Ulbrich 1977). Nevertheless, the method is limited in several respects. One of the more severe problems is one of dynamic range. For a 3 cm–10 cm wavelength combination and under conditions of light rain rate, the method must be applied over a fairly long interval to attain sufficient sensitivity, thereby degrading the effective resolution. On the other hand, if the wavelengths are decreased (i.e., the attenuation is increased) to provide good sensitivity, the major difficulty is the loss of signal at the shorter wavelength. A second difficulty is the large numbers of independent samples needed to reduce the variance in the power ratio estimate, (9.36). One other serious source of error is caused by mismatches in the antenna patterns at the two wavelengths. For the detection of hail the number of misses and false alarms can be significant even for minor misalignments in the main and sidelobe patterns (Rinehart and Tuttle 1984). In a rain-only medium the errors are not expected to be as severe, although their magnitude will depend on the spatial inhomogeneity of the rain. A final source of error arises from non-Rayleigh scattering. While this error is usually negligible for lower frequencies and lighter rain rates, it can become an important source of error for typical spaceborne radar frequencies.

The following provides a brief summary of some other dual-wavelength techniques that have been proposed. Over an interval of uniform rain rate, Goldhirsh and Katz (1974) showed that two parameters of a mean DSD could be deduced for both attenuating/nonattenuating and dual-attenuating wavelength combinations. The technique of Fujita (1983) uses a ratio of powers at adjacent range gates and a nonlinear least mean squares technique to estimate rain rate. Meneghini and Nakamura (1990) analyzed a dual-wavelength method based on the Hitschfeld–Bordan equation. A somewhat similar approach was proposed by Marzoug and Amayenc (1994) where the authors find that the method is able to correct for multiple scaling errors such as in α and the radar constant, whereas the single-wavelength methods are able to correct fully for only one type of error. As a consequence, the dual-wavelength algorithm exhibits better performance than the single-wavelength counterparts over the common range of detectable rain rates (Testud et al. 1992). As with all such dual-wavelength methods, however, it must be supplemented with a single-wavelength approach at the light and heavy rain rates.

Integral or differential equations for the DSD parameters $N_0(r)$ and $D_0(r)$ can be derived in several ways

(Meneghini et al. 1992, 1997; Iguchi and Meneghini 1995). A particularly simple way is to express 10 $\log[Z(\lambda_1, r)/Z(\lambda_2, r)]$ and 10 $\log[Z(\lambda_1, r)]$ in terms of parameters of DSD and the measured reflectivity factors. A final value form of the equations is appropriate to cases where independent estimates of the path-integrated attenuation (PIA) are available at both frequencies. The path attenuations can be estimated using the surface reference technique or the standard dual-wavelength technique, (9.39), applied over the entire path along with an empirical relation that provides the attenuation at one wavelength from an estimate of the differential attenuation.

Many of the approaches outlined here have been used in the analysis of airborne dual-wavelength radar and radiometer data and most are candidate algorithms for the dual-wavelength radar proposed for the Global Precipitation Mission satellite. Nevertheless, extensive validation of the techniques, particularly those that provide parameters of the drop size distribution, has not occurred. This constitutes one of the primary challenges for future dual-wavelength airborne and ground-based experiments.

3. Practical issues of precipitation estimation using reflectivity factor

The measured radar reflectivity factor Z and rain rate R are physically linked by the DSD. Thus, if radar measures Z sufficiently close to the ground so that precipitation intensity does not change over this height, the only uncertainty in the transformation from Z to R arises from the DSD variability. These DSD fluctuations limit the measurement accuracy to ~30%–40%, if a single climatological Z–R relationship is used. This accuracy can be improved significantly if the Z–R relationship is changed in accordance with precipitation type. Similar statements can be made about multiparameter radar algorithms. Under ideal conditions a polarimetric rainfall algorithm that uses a combination of reflectivity Z and differential reflectivity Z_{dr} or a combination of K_{dp} and Z_{dr} can estimate rainfall to an accuracy of about 10%–15%. The estimates of accuracy based on physical considerations are not valid under various practical limitations as described below. Three factors limit the lowest height of the radar measurement: earth curvature, blockage by the landscape, and ground-clutter contamination. Between higher altitudes and ground, the rain intensity can change due to further growth or evaporation. If the measurement is above the 0°C isotherm the phase of the precipitation changes as well. The contamination by the bright band and the change in the vertical profile of reflectivity introduce uncertainties that rapidly dominate the errors in radar precipitation estimates. Several methodologies have been proposed to solve this problem such as climatological corrections, range-dependent probability matching, and neural network–based rainfall estimates. A comprehensive evaluation of the errors associated with many of these techniques is yet to be made. A serious drawback is the lack of good ground truth. Gauge measurements, besides having their own limitations, are made at scales very different from those of radar. Because it is impossible to achieve a perfect matching between the radar and gauge measurements, radar measurements must be compared to a cluster of closely spaced gauges (Zawadzki 1975). However, such datasets are rarely available.

The radar reflectivity factor is derived from the power backscattered from the radar resolution volume. However, only a fraction of this power comes from rain that is close enough to the ground to be considered hydrological data. The majority of the backscattered power arrives from rain sufficiently far above the ground that a downward extrapolation is mandatory. To circumvent this problem, the concept of an optimal surface precipitation map derived from radar data can be implemented. There are a number of ways in which this can be achieved. Bolen et al. (1998) used Colorado State University–(CSU–CHILL) data to track the vertical evolution of rainfall fields. Bolen et al. (1998) used the K_{dp} rather than the Z field because K_{dp} is more closely related to rainfall rate than the reflectivity factor. Using the fine-resolution data from the CSU–CHILL radar, they derived an optimal surface on the ground corresponding to the radar data aloft. Their analysis demonstrates the concept of vertical evolution of the rainfall and its importance in determining rainfall estimates at the surface.

The practical problem of operational precipitation estimates involves a number of steps that must be undertaken with care. Assuming that the measurable parameters are well calibrated, the elimination of returns from nonmeteorological targets, foremost ground clutter, is the first challenge. Zero velocity notch filtering during signal processing can mitigate the problem. The effectiveness of this will vary according to the quality of the radar transmitter and the strength of the clutter. In any case, avoidance of the remaining ground echoes at the data processing stage will still be necessary. Algorithms based on Doppler velocity, vertical echo structure, and horizontal echo structure (in that order of importance) can be made quite effective in real-time detection of contamination by ground echoes, or those resulting from anomalous propagation. This technique can detect close to 90% of the contamination. Although not yet extensively evaluated, target identification by polarimetric data is still more effective, detecting nearly 100% of the contamination. Because polarimetric signatures of ground clutter are so clear, they are also more straightforward to implement in software. Voids created by avoidance of the contaminated data can be treated by horizontal interpolation (if the void is of small area) or by vertical extrapolation of noncontaminated data taken at higher elevations. This is similar to the procedure used to extrapolate data at far ranges.

Next in importance is the contamination of surface rainfall by melting particles (snow, graupel, or hail).

These regions of extremely high reflectivity can be particularly pernicious, leading to flood warnings in light rain. The level of the 0°C isotherm can be obtained from soundings, model outputs, and more directly, from the vertical structure of polarimetric radar data themselves. Using all these sources of information is, of course, even better. During frontal passages, when the height of the bright band may change rapidly, radar information is crucial. With a 1° beamwidth, the signature of the bright band can be clearly detected out to ranges of 50–70 km, depending on the thickness and intensity of the bright band. For brightband detection polarimetric radar data are extremely useful, particularly in the presence of wet hail or graupel above or below the melting level. In such cases, polarimetry may be the only adequate detection technique available. There are several strategies that can be used to avoid brightband contamination. The simplest, perhaps, is to substitute the information below (if possible) or the reflectivity in snow above and extrapolate the measurement to ground.

Zones of partial beam blocking must be identified for each antenna elevation and a compensation factor applied to the data. This is particularly critical in complex terrain. The question arises as to whether it is better to use nonblocked data from higher elevation beams or compensate for the blocked fraction of the lowest beam. Although this was not fully evaluated, given the large errors associated with extrapolation to ground, it is probably better to use the lowest elevation possible. Stability of the antenna elevation angles is particularly critical here. Precipitation measurement techniques based on differential phase shift, independent of blocking, can be potentially applied in such blocking situations. Attenuation in rain and on the radome severely limits the QPE capability of radars operating at C band. Polarization diversity may provide a solution to the problem of attenuation in rain but the radome attenuation should be corrected for using properties of the radome material and empirical approximations. It should be noted here that wet radome attenuation will vary with azimuth and elevation and can reach levels as high as 10 dB.

Once the decontamination and correction of data are completed, extrapolation to ground must be done. The broadening of the radar beam with range, the minimum beam elevation angle, and the earth curvature all serve to increase the distance between the radar scattering volume and surface. As this distance increases with range, it is often referred to as the range effect. Here the removal of the bias associated with the vertical profile of reflectivity (VPR) is the primary goal. However, the VPR is highly variable in time and space so that the VPR determined at short ranges may be quite different from the one at far ranges. There is a risk of introducing errors larger than the intended correction if the wrong vertical profile is taken. A conservative approach is preferable. This is again complicated when virga is present. These overhangs may extend over very large areas and if not identified as such could lead to overestimation of

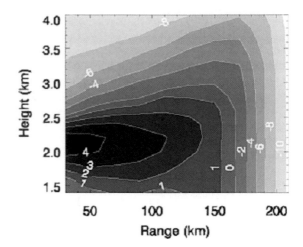

FIG. 9.2. Reflectivity bias (in dBZ) due to range effect obtained by projection of actual data at the 20–40-km range to further ranges by simulation of radar measurements. A bias of 2 dBZ means that 2 dB must be subtracted from the measurement to obtain Z at the surface.

rainfall. At short ranges, where voids are created by removal of ground contaminated data, the nearby VPR is sufficient for an effective extrapolation. At long ranges, and in spite of all the mentioned pitfalls, extrapolated data will be useful for some hydrological applications.

Figure 9.2 shows in a height–range display the bias with respect to low-level data associated with measurements at increasing height and broader beam, that is, the range effect. It was obtained by projecting volume scans of radar data obtained at ranges 20–40 km to farther ranges by simulating the radar measurement process (convolution with a Gaussian beam and post detection integration over 1° azimuth). This figure illustrates the effect for a particular meteorological situation: stratiform rain with a melting layer at 2.2 km and echo tops close to 8 km. The bias line of 2 dBZ indicates a clear effect of the brightband contamination out to 140 km. Figure 9.2 illustrates that the contamination extends well beyond the range of unambiguous detection of the bright band by morphology-based algorithms. It is also clear that the bias correction is critical. Beyond about 180 km the bias rapidly increases. The specific values in Fig. 9.2 will change with type of precipitation. Biases may be less severe in convective rain than in stratiform situations since the convective VPR is, on the average, uniform over much greater vertical extent (Fabry and Zawadzki 1995, and references therein). The parameters that will be most important in determining the values in Fig. 9.2 are the height and intensity of the bright band in stratiform precipitation and the echo top in both stratiform and convective regions. Thus, sets of these diagrams for the various values of the expected parameters will be needed for the VPR bias correction. The use of data from the entire radar network, rather than a single radar, will be helpful in determining the vertical profile of reflectivity. In this manner, a better real-time correc-

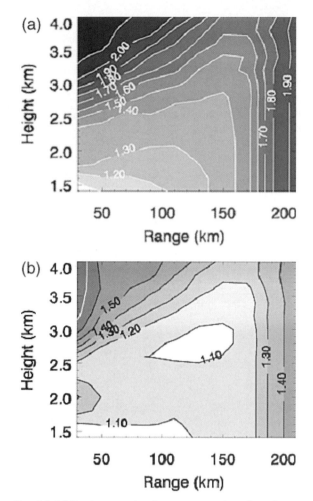

FIG. 9.3. (a) Random error in rain rate expressed as a factor between the reflectivity measurements projected to a given range and the reference field of data at the 20–40-km range transformed using $Z = 200R^{1.6}$. The resolution of the reference field was degraded to far range (3 km $\times$ 3 km) for the error computation. (b) Same as (a) but at a resolution of 17 km $\times$ 17 km. For low antenna elevations and ranges shorter than 180 km the random error is reduced to less than 20% (factor $<$ 1.2).

tion can be applied for radars detecting precipitation at far ranges only. The extrapolation to ground introduces a random error as well as a bias. To eliminate the random error, averaging of the data will be necessary and this will reduce the temporal and/or spatial resolution. For the same simulation shown in Fig. 9.2, Fig. 9.3a shows the random error of the extrapolation to the reference level by comparing the bias-corrected data at a given range to the short-range reference field. These are point-wise errors and diminish with area integration. In strong convection, where the space variability is greater, although the bias due to the range effect may be small or negligible, the random errors will be appreciably more severe. In Fig. 9.3b the same random errors are shown for an area average of 17 km $\times$ 17 km. The random error is reduced greatly at the expense of a drastic reduction in spatial resolution. Time averaging will further

reduce the error. Note that this is the limit of the precision that can be expected from an extrapolation of the VPR. In fact, errors will be appreciably larger because the actual vertical profile of reflectivity at the far ranges will not be the one used for the VPR correction. In summary, the problem of VPR correction needs proper evaluation.

a. Validation

Validation of radar estimates of precipitation can be done at various stages. The effectiveness of extrapolation to ground and other corrections on the equivalent reflectivity factor can be evaluated by comparison with another overlapping radar providing data close enough to the ground to be considered as reference. Overall validation must be done with gauges. The problem here is the comparison of two estimates based on measurements at very different scales. Gauges are affected by instrumental errors and sampling errors (the so-called representativeness errors), the latter being quite severe (see, e.g., Zawadzki 1975). Thus, in comparing radar and gauges the separation of errors is crucial. Dense gauge networks are required for the validation of radar estimates. Consequently, experimental sites for the different climatological regions are required.

A more direct hydrological validation by coupling radar data and hydrological models applied to instrumented basins is less stringent but responds to the ultimate practical need. If hydrological validation is done with radar data processed progressively closer to the optimal surface precipitation map, it would be ultimately more convincing than the validation with gauges.

b. Determination of the radar rainfall relationship from observations

The variability of drop size distributions (DSDs) and the variability of the Z–R relationship have been extensively studied. However, a good deal of confusion still remains. The observed frequent deficit of small drops that lead to the gamma distribution model of DSDs with large shape parameter or μ (Ulbrich 1983) is potentially likely to be the result of the shortcomings of a disdrometer rather than reality. Other instruments, although showing departures from an exponential behavior, do not indicate that a gamma distribution model explains most of the variability in DSDs in comparison to the exponential model. Parameters of the DSD derived from experimental Z–R relationship should be considered with suspicion since the Z–R relationships depend strongly on the instrument and the method by which they were obtained and uncertainty in the data.

A more fundamental question is the uncertainty related to the enormous difference in the sampling volume of disdrometric measurements and the radar. Meteorological fields have variability at all scales that make scale matching critical in general. To illustrate this, the

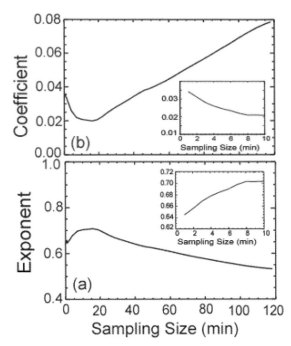

FIG. 9.4. Effect on the R–Z relationship of time averaging a sequence of DSDs: (a) change in the coefficient of the R–Z relationship with the number of averaged 1-min DSDs; (b) same for the exponent.

effect of time averaging on the Z–R relationship obtained by linear regression on logarithms of Z and R derived from disdrometric measurements is shown in Fig. 9.4. Although time and space averages are not identical, the effects of averaging should be similar. Some of the change with averaging in the relationship between rain rate and reflectivity in Fig. 9.4 is due to simple statistical effects: as the scatter decreases with averaging, the regression leads to a different relationship. But it is obvious that by increasing the averaging domain the small-scale variability will be eliminated, which in turn will change the relationship between the moments. Mixing different distributions also leads to different relationships. In spite of these shortcomings disdrometric Z–R relationships are our best guides to their variability due to changes in DSDs. Certainly, averages of several disdrometers deployed over an area comparable to the radar measurement cell would lead to more representative information.

The Z–R relationships can also be obtained by a direct radar–gauge comparison. This can be done in two ways: by regression of synchronous radar and gauge measurements at very short ranges and by matching probabilities of long-term measurements of radar Z at very short ranges and rain gauge R (Calheiros and Zawadzki 1987; Rosenfeld et al. 1994). In both cases the problem of differences in the sampled scales must be carefully accounted for. It should be noted that probability matching leads to a biased Z–R relationship (compared to the relationship obtained by regression) since the method does not distinguish between dependent and independent variables. DSDs obtained from Doppler spectra of vertically pointing radars avoid the problems of sample volume matching and are very desirable. The possibility of deriving Z–R relationship from a polarimetric radar data is a very attractive option and it certainly should be fully exploited in the future (Bringi et al. 2002). For many applications a classification of Z–R relationship according the type of precipitation could be sufficient; however, a simple convective-stratiform-drizzle separation perhaps falls short of the need. This is particularly true for stratiform precipitation where the microphysical processes are highly variable, depending on the depth of snow growth, its habit, and the degree of riming.

Alternative approaches have been suggested for developing observation-based mapping from radar observations to rainfall on the ground. One such procedure is the neural network technique. The neural network technique (similar to probability matching) provides a mechanism to build a nonparametric relation between surface rainfall and the vertical profile of reflectivity aloft. The concept of using neural networks for radar rainfall estimates was first presented by Xiao and Chandrasekar (1995). Since then the methodology has been refined to the point that it has been tested over a large network of gauges for daily applications. Liu et al. (2001) present a methodology whereby they build a neural network using radar and gauge network data up to a certain date in order to estimate rainfall for the next day. At the end of the next day the neural network is modified to include information from gauges to estimate rainfall the following day. Figure 9.5 shows the schematic diagram of the neural network–based rainfall estimate. This procedure is essentially a gauge adjustment, and the neural networks should just be seen as a mechanism to assimilate the large amount of gauge data. Based on analysis of several months of data per year over a couple of years, Liu et. al. (2001) found that the adaptive neural network scheme performs as well or better than the "best Z–R relation," which can be derived only after the fact. This procedure is very promising and must be thoroughly evaluated.

4. Practical aspects and recent advances in precipitation estimation by dual-polarization techniques

Most of the practical issues associated with reflectivity-based rainfall estimates also apply to polarimetric radar observations. However, there are some unique advantages and issues specific to polarimetric radars that will be discussed here. The various rainfall algorithms discussed in section 2 use different combination of Z_H, Z_{dr} and K_{dp}. Each measurement comes with its own advantage and peculiarities, which can be attributed to both microphysical characteristics and measurement issues. Any algorithm involving Z needs accurate measurement of the radar constant. Here Z_{dr} is a relative power measurement and it can be measured relatively

DNN - Dynamic Neural Network

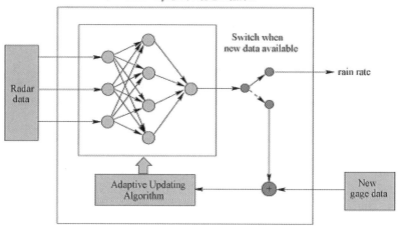

FIG. 9.5. Structure of the adaptive neural network for rain-rate estimation.

accurately without being affected by absolute calibration errors (though such accuracies must be realized in operational radars). Algorithms that use K_{dp} have several advantages that originate from K_{dp} being derived from phase measurements; for example, it is unaffected by absolute calibration error and attenuation and unbiased if rain is mixed with spherical hail. However, K_{dp} is relatively noisy at low rain rates; K_{dp} is derived as range derivative of Φ_{dp} and, in practice, is estimated over a finite path. Therefore, there is a trade-off between accuracy and range resolution of K_{dp}-based algorithms.

The above-mentioned characteristics of the polarimetric radar variables affect the regions over which these algorithms can be applied. Using S-band radars as reference, Chandrasekar et al. (1990) showed that for light rain Z–R relations perform the best, while in moderate and heavy rainfall dual-polarization algorithms provide better rainfall estimates. Dual-polarization radar algorithms have also been applied in a statistical context such as the probability matching technique whereby an optimal rain-rate relation is derived by matching cumulative distributions of rainfall rate (Gorgucci et al. 1995). There are two other recent developments in polarimetric radar rainfall estimates, which include the areal rainfall estimation techniques and the correction for the effective mean shape of rain drops. These topics are discussed in the following.

As discussed in section 3, verification of instantaneous point rainfall estimates has always been a problem. Radar and rain gauge go through fundamentally different processes in measuring rainfall. Radars observe instantaneous snapshots of the precipitation medium and then obtain rainfall rate using an algorithm. Rain gauges, on the other hand, accumulate rain or count raindrops over time and then convert that information to average rain rate. Some modern disdrometers measure rainfall rate by definition (although not without errors). As mentioned in the previous section, areal estimates of rainfall are always more meaningful for comparison

in that they reduce the inherent sampling errors between radar and gauge measurements. Propagation differential phase and cumulative attenuation are path-integrated measurements. As such, they provide a natural integration of rain rate over the radial dimension; azimutual integration is achieved by antenna scanning. This technique has been used to obtain areal rainfall estimates over small basins very successfully (Raghavan and Chandrasekar 1994; Ryzhkov et al. 2000; Bringi et al. 2001). The most recent results of Bringi et al. (2001) near Darwin, Australia, showed that the areal rainfall estimates obtained over a 10 km × 10 km area, compared to areal estimates from a dense gauge network, had a bias of 5% and normalized error of 14%. If such results can be repeated at other locations, then this approach holds great promise.

Another important recent development is the radar-based estimate of mean axis ratio of raindrops. All polarimetric techniques critically depend on the mean shape of raindrops. Among the measurements, K_{dp}-based estimates of rainfall are affected the most. Raindrop oscillations change the mean shape–size relation of raindrops (Andsager et al. 1999). These effects introduce large errors in K_{dp}-based rainfall estimates if equilibrium axis ratio models are used (Keenen et al. 2001). Gorgucci et al. (2000, 2001) derived a solution to this problem where the variation of mean shape–size relation, (parameterized by its slope β) is inferred from the data and accounted for in the algorithms.

Attenuation correction and radar calibration are two other areas where polarimetric radar observations have made a substantial contribution (Gorgucci et al. 1999). While these issues are important for all radar applications, they are particularly critical in rainfall estimation. The topics of attenuation correction and application of self consistency to radar calibration have been discussed in several articles (Holt 1988; Chandrasekar et al. 1990; Bringi et al. 1990; Scarchilli et al. 1996). The fundamental principle of attenuation correction can be recognized by noting that both K_{dp} and attenuation are proportional to the rainwater content (Bringi and Chandrasekar 2001). In simple terms, A is proportional to K_{dp} for the frequency ranges from 5 to 19 GHz. More recently, the initially proposed attenuation correction procedures have been further refined to make them applicable in a wide variety of practical situations. [See, e.g., Section 7.4 of Bringi and Chandrasekar (2001) and the references contained therein.] Attenuation correction is an important first step for use of any radar reflectivity data at frequencies C band and higher. The attenuation correction problem has been addressed using polarization diversity fairly well in rain. However, in mixed-phase precipitation or melting hydrometeor regions, additional research, in conjunction with hydrometer classification work, is needed.

Data quality and hydrometeor classification are two areas where polarimetric radar techniques will have significant impact, irrespective of the type of algorithm used for precipitation. The contribution of polarimetric radar measurements to inferring data quality is likely to be a major practical advantage. Hydrometeor preclassification will help in the choice of the proper algorithm, instead of a "blind" application of a standard rain algorithm. For example, if polarimetric radar observations show a region of hail or rain mixed with hail, the user will be forewarned from applying the usual Z–R relation to that region. Polarimetric data are extremely effective in detecting and mitigating contamination of hydrological information content in radar data.

5. Practical aspects and recent advances in spaceborne precipitation estimation

November 2002 marks the fifth anniversary of the launch of the TRMM satellite and with it the first spaceborne precipitation radar (Kozu et al. 2001; Kummerow et al. 1998; Simpson et al. 1996). Work has already begun on the TRMM follow-on satellite, the Global Precipitation Mission (GPM) with a tentative launch date of 2007. Plans for this mission call for a dual-frequency precipitation radar (DPR), the first of its kind, with frequencies of 13.6 and 35 GHz. With a projected launch date of 2004, the CloudSat and ESSP3 satellites will carry, respectively, a 95-GHz cloud radar and dual-wavelength lidar for cloud remote sensing at nadir (Kollias et al. 1999; L'Ecuyer and Stephens 2002). When viewed against the backdrop of decades of weather satellites without radar or lidar, the question arises as to why now, or perhaps, why not earlier?

In looking for reasons, perhaps foremost is the growing concern with climate, and the need to understand in detail the coupling between oceans and atmosphere evident in the ENSO phenomena. The notion that global warming would be accompanied by an increased incidence of severe weather events suggests that we have the observational capability to test both hypotheses. In any event, there seems to be general agreement that our observations of precipitation are inadequate both spatially and temporally over much of the globe and that the sparseness of data is a hindrance to assimilating rain rate into forecast and climate models. However, the increased interest in weather and climate does not alone account for the change in philosophy with respect to space-borne weather radar since a sufficient number of microwave radiometers can achieve the desired space–time sampling. What makes the radar unique is its ability to provide the 3D structure of the precipitation along with improved estimates of rain rate and rain type (convective and stratiform) over land and ocean. With the addition of a second wavelength, the radar should yield information on the snow and raindrop size distribution, and better discrimination between regions of snow, rain, and mixed-phase precipitation in convective as well as in stratiform rain.

a. Problems unique to rainfall estimation from space

The first task in the estimation of rain rate is detection of a radar return from the precipitation. In the case of the TRMM PR, the presence of precipitation is triggered if the signal plus noise level at three consecutive ranges exceeds by some threshold the noise-only measurements in rain-free range gates. Since returns from the surface are always present and generally much larger than the rain return, the two types of return must be separated. Mistaking the surface return for rain can occur when highly reflective surfaces contribute to radar returns along an antenna sidelobe. Conversely, detection of the surface return can be obscured in the presence of strong rain attenuation or by weak surface returns. At off-nadir incidence, the lowest portion of the rain will be masked by the surface clutter; the height over which the rain is obscured depends on the antenna beamwidth, the radar cross section of the surface, and the rain intensity. In the case of the TRMM radar, light rain is obscured below about 1.8 km at the scan edge at 17°. As in the case of ground-based radar, the inability to detect rain near the surface causes uncertainties in the estimate of surface rain rate. This poses particular challenges to the use of the surface reference technique in that the radar reflectivities are unavailable over a portion of the path even though the SRT provides an estimate of attenuation over the full path.

Once rain has been detected and isolated from the surface clutter, an algorithm is used to search for the presence of a bright band. This is closely related to the rain classification algorithm in that a bright band is taken as a sufficient condition for stratiform rain. In the absence of a detectable bright band, the horizontal structure of the reflectivity (dBZ) field is examined to classify the rain type into either stratiform or convective (Awaka et al. 1998). Rain detected below the 0°C isotherm is separately categorized. Brightband detection is used to distinguish regions of rain, snow, and partially melted precipitation. This information is not only useful in itself but is employed in the application of attenuation correction algorithms. Rain-type classification is used to select initial k–Z and Z–R relationships and serve as a first-order indicator of the vertical profile of latent heating in the atmosphere.

b. Design and algorithm issues

The design of spaceborne weather radar poses a different set of problems than those faced in the design of its ground-based counterpart. The spaceborne radar generally encounters ice and mixed-phase precipitation at the storm top and strong surface returns in the lowest part of the path. The large platform speed (7 km s^{-1}) places great demands on the pulse repetition frequency, beamwidth, and pointing angle accuracy if Doppler measurements are to be attempted (Amayenc et al. 1993). Perhaps the most fundamental requirement for spaceborne radar is the need for adequate resolution coupled with restrictions on the size and weight of the antenna. To resolve a typical convective cell requires that the horizontal resolution at the surface be better than about 5 km. Achieving this from an altitude of 400 km requires an antenna on the order of 100 wavelengths in diameter. For a typical ground-based weather radar, operating at a wavelength of 10 cm, an antenna diameter on the order of 10 m would be required. Although the use of inflatable antennas may ultimately achieve this, the technology is not feasible at present. By decreasing the wavelength to 2 cm, we are able to achieve the same resolution with a 2-m diameter antenna. It is well known, however, that at X-band frequencies and above the signals are strongly attenuated in moderate and heavy rainfall.

An independent measure of the path-integrated attenuation serves as a good reference to constrain spaceborne algorithms. This can be done by using a fixed target where the two-way path attenuation from the radar to the target is taken as the difference between the return powers measured in the absence and presence of rain (Harrold 1967; Atlas and Ulbrich 1977; Ihara et al. 1984; Ruf et al. 1996). To adapt the idea to spaceborne radar, we replace the fixed target with the surface itself and arrive at the surface reference technique (SRT) (Meneghini et al. 1983; Fujita 1983; Marzoug and Amayenc 1991, 1994; Iguchi and Meneghini 1994; Kozu et al. 1991; Durden and Haddad 1998; Meneghini et al. 2000). Of course, the surface return is not constant; nevertheless, at near-nadir incidence angles (0°–20°) over ocean, the scattering cross section within and outside the rain is often stable enough to provide a good estimate of the path-integrated attenuation.

An example of this type of estimate is shown in Fig. 9.6 where the upper panel shows the TRMM PR return from the precipitation, surface, and mirror image at a fixed incidence angle of 3.55° over a span of 200 scans or about 860 km. The change of the normalized surface cross section of the surface σ^0 (dB) over the same in-

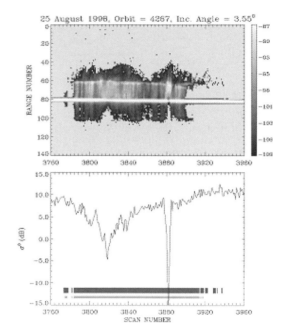

FIG. 9.6. (top) Range bin vs scan number of the TRMM PR radar return (dBm) at a fixed incidence angle of 3.55°; (bottom) normalized radar cross section of the surface (dB).

terval is shown in the lower panel. In the most recent version of the algorithm, the σ^0 data in rain-free areas both in the along-track and cross-track directions are used to determine a reference value, σ^0_{ref}. The estimate of the two-way path attenuation is taken to be the difference between σ^0_{ref}, and the apparent σ^0 at the raining location of interest. The fractional error of the estimate is approximated by the ratio of the estimated attenuation to the standard deviation of the rain-free measurements used to determine σ^0_{ref}. This information on the reliability of the estimate is used in the attenuation correction algorithm to determine the relative weighting between the Hitschfeld–Bordan and SRT estimates of path attenuation.

Even if an exact and independent measurement of path attenuation could be made, the rain rate would not, in general, be exact because of other error sources. These include variations and offsets in the radar calibration, changes in the drop size distribution, fluctuations in the radar return power resulting from a finite number of samples, and reflectivity gradients within the beam. Path attenuation, on the other hand, is a single constraint and allows us the liberty of adjusting only one parameter. However, if the radar is well calibrated, we can argue that the primary reason for the discrepancy between the Hitschfeld–Bordan and SRT estimates of the attenuation is an error in the initial γ–Z relationship. By adjusting α in the $\gamma = \alpha Z^\beta$ relationship to, say, $\alpha' = \alpha\varepsilon$, where ε is chosen so that the Hitschfeld–Bordan and SRT estimates of path attenuation are equal, we arrive at a γ–Z relationship independent of the initial selection, as shown in section 2. But this new γ–Z relationship also tells us something about the nature of the drop size distribution and, in particular, suggests that we should choose the coefficient a in the R–Z relationship $R = aZ^b$ to maximize the expectation $E(a \mid \alpha)$ using sets of possible drop size distributions (Iguchi et al. 2000; Ferreira et al. 2001). Although this approach is reasonable and part of the operational TRMM algorithm, it is not without drawbacks. Since there is only one scaling parameter, it is not possible to adjust separately the attenuation in the mixed-phase, ice, and rain regions. For example, in cases of strong brightband attenuation (e.g., Bellon et al. 1997), the attenuation in this region would be underestimated with a corresponding overestimation, and therefore overcorrection, in the rain region. The method also depends on the correct identification of the medium so that misidentifying hail as rain can result in a significant error. Although attenuation from cloud liquid water is usually small at 13.6 GHz, its contribution can be significant at 35 GHz, one of the frequencies proposed for the dual-frequency radar aboard the GPM satellite.

The idea of constraining possible solutions by an independent measure of path attenuation has been extended to ground radars. A somewhat similar approach is used to address the attenuation problem encountered at C-band radars in the estimate of K_{dp} (Testud et al. 2000; Bringi et al. 2001). In estimating the parameters of the drop size distribution from a spaceborne dual-wavelength radar we find that an iterative solution beginning at the storm top becomes unstable in the presence of attenuation. In analogy to the single-wavelength problem we require estimates of the path attenuation at both frequencies. These quantities can be obtained by means of a dual-wavelength SRT or the standard dual-wavelength method (Eccles and Mueller 1971; Eccles 1979) discussed in section 2, along with empirical laws that relate the differential attenuation to the single-wavelength attenuation. With this information, the iteration can proceed from the surface upward where it can be shown that this solution is more robust than is the initial value problem (Meneghini et al. 1992). In snow, the dual-wavelength problem is somewhat different. In the absence of a significant amount of cloud liquid water, the attenuation can usually be neglected and iteration is not required; that is, the solution at a particular range gate is independent of the DSD at all other gates (Matrosov 1992). In this case, however, the dual-wavelength equations require specification of snow density as well.

Although the Bayesian approach has not received much attention in the literature of radar meteorology, it is the basis of many microwave radiometric retrieval algorithms (Olson et al. 1999; Smith et al. 1994) and some combined radar–radiometer algorithms as well (Haddad et al. 1997). These algorithms are model based where the forward radiative transfer calculations are run over ensembles of storm models at various stages in their evolution (e.g., Tao et al. 1993). The inverse problem then consists of matching the measured brightness temperatures with the model-derived values, taking into account the a priori probability of a particular microphysical model. Most radar retrieval methods are, in a restricted sense, model based as well in that a set of assumed or measured drop size distributions and a model for the effective dielectric constant for mixed-phase hydrometeors are employed. Nevertheless, the typical model-based and physically based approaches are fundamentally different. Whether new methods can be devised to take advantage of the best features of each approach is an issue that is particularly important in the development of algorithms for the Global Precipitation Mission and airborne experiments where coincident radar and radiometer measurements will be available.

Statistical approaches to the estimation of rainfall over large space–time regions can be thought of as a search for regularity and pattern in a parameter—rainfall—that at high space–time resolutions seems so unpredictable. In a general sense, these methods can be thought of as constraint based. For example, the fact that rain rate is approximately lognormally distributed can be used to recover rain rates at the low and high end that cannot be directly measured (Lopez 1977; Wilheit et al. 1991). Probability matched methods discussed in section 3 have found application in the derivation of gauge-adjusted Z–R relations. Area–time integral methods rely on the fact that the fractional area above a threshold is well correlated with area-average rain rate (Donneaud et al. 1981, 1984; Chiu et al. 1993; Atlas et al. 1990; Atlas and Bell 1992; Rosenfeld et al. 1990; Sauvageot 1994). For low earth orbiting satellites, where the temporal resolution is poor, statistical methods offer the promise of deducing useful features of the rainfall when continuous observations over an area are not possible. Shown in Fig. 9.7 are scatterplots of the fractional "area" of rain rates above thresholds of 0.65 (top), 2.7 (center), and 11.5 mm h^{-1} (bottom) versus the monthly rainfall totals derived from TRMM PR data. Each point represents a 5° lat $\times$ 5° long $\times$ 1 month space–time box where the monthly rainfall total is computed from the individual high-resolution (4 km $\times$ 4 km) rain-rate estimates and the fractional area at R_0 is the ratio of the number of rain-rate counts above R_0 to the total number of observations (Meneghini et al. 2001). It is noteworthy that the best correlation between the fractional area and area averages occurs for thresholds at or near the conditional mean rain rate (about 3 mm h^{-1} for the TRMM "global" observations; Kedem and Pavlopolous 1991; Short et al. 1993). The high correlation evident between these parameters implies that the conditional rain-rate distribution over large space–timescales can be well characterized by a single parameter. It also implies that the mean and standard deviation of the rain-rate distribution are well correlated. This close relationship between the mean and standard deviation of rain rate has been observed using disdrometer data, rain gauge data, and the TRMM radar data. These characteristics of the large-scale rain field are not only useful constraints but

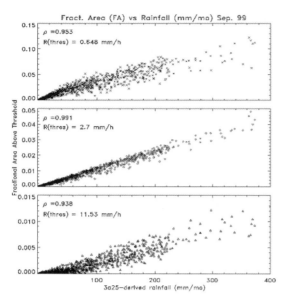

FIG. 9.7. Fractional area above rain-rate thresholds of 0.65 (top), 2.7 (center), and 11.5 mm h^{-1} (bottom) vs rainfall (mm month^{-1}) for 5° lat × 5° long × 1 month space–time regions for Sep 1999.

suggestive of a more general principle concerning the spatial and temporal distribution of rainfall. Whether they can be used as the basis of new retrieval methods and how the relationships change with different space–timescales are open issues.

c. Calibration and validation of spaceborne radars with ground-based radars

Tests of the calibration accuracy of spaceborne radars can be accomplished with a combination of internal monitoring of the system parameters of the radar along with a model that combines the measurements from the major radar subsystems into an overall system calibration (Kozu et al. 2001). This internal characterization must be supplemented with an external calibration procedure. The active radar calibrator (ARC) used in the TRMM PR has shown itself to be useful in demonstrating both the stability and accuracy of the radar calibration (Kumagai et al. 1995). The device can be operated in several modes including a receive-only mode to measure the transmit power and antenna gain of the PR, a transmit-only mode to check the antenna pattern-receiver gain chain, and a transponder mode to test the full transmit–receive chain of the PR. A second method of external calibration consists of a statistical comparison of the nadir backscattering surface cross sections over ocean as measured by the PR with those measured by a well-calibrated altimeter such as the Topography Experiment for Ocean Circulation (TOPEX).

Comparisons of the radar reflectivity factor and rain rate as derived from the PR and well-calibrated ground-based radars are an indispensable part of assessing the accuracy of spaceborne retrievals of rainfall (Bolen and

Chandrasekar 2000, 2003a; Schumacher and Houze 2000; Liao et al. 2001; Anagnostou et al. 2001). However, a simple comparison between spaceborne and ground-based weather radars raises the question as to which measurement is correct if the estimates differ. One approach is to adopt an indirect method. Based on an independent assessment of the calibration via the methods described above, we accept the accuracy of the PR estimate of radar reflectivity factor (dBZ) at high altitudes where attenuation effects are minimal. When good agreement is found between the PR and ground-based dBZ values at high altitudes (where attenuation of the PR signal is negligible), we can interpret differences between them near the surface to be caused primarily from errors in the PR attenuation correction algorithm. A comparison between the PR and Weather Surveillance Radar (WSR-88D) is shown in Fig. 9.8 for a TRMM overpass of the Melbourne, FL, WSR-88D site on 9 March 1998 (Liao et al. 2001). Despite the intense rain rate seen in the overpass, the attenuation-corrected radar reflectivities (center) match up well with those from the WSR-88D at the same height of 3 km (bottom). From the 45 overpasses of this site in 1998–99 we find that the mean reflectivity factor from the PR to be 0.5 dB larger than the WSR-88D at 6 km and about 1 dB larger at 1.5 km. Some of the discrepancy at the low altitude can be attributed to Mie scattering effects at the high (PR) frequency, which acts to increase the dBZ values. This good agreement is also seen in comparisons of area-averaged rainfall rates derived from the PR and WSR-88D where we find a correlation coefficient of 0.94 and an agreement in mean values to within 10% (Fig. 9.9, top). Comparisons of area-average rain rate from the PR and the Houston WSR-88D also show good agreement (Fig. 9.9, bottom). Bolen and Chandrasekar (2003b) used polarimetric radar data to compute direct estimates of attenuation to compare against the attenuation correction method of PR. They observed no systematic bias between the two estimates. It is important to add here that not all comparisons show so favorable an agreement as the papers just cited and issues regarding the PR attenuation correction and the choice of Z–R relationships are subjects that will continue to be studied.

6. Summary and conclusions

Quantitative precipitation estimation is a challenging problem that has been pursued for over four decades. This pursuit has led to improvements in radar hardware, processing and algorithm development, as well as the development of dual-polarization and spaceborne weather radars. Though polarimetric radars were originally proposed for advancing rainfall estimates, they have had tremendous impact on microphysical research as well. Technology has advanced significantly over time making the radar systems very stable. Engineers who maintain the WSR-88D radars devote a great deal of

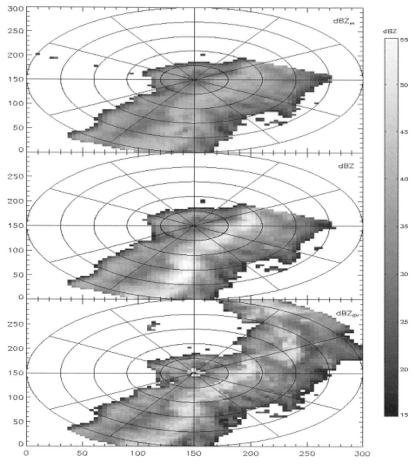

FIG. 9.8. TRMM radar reflectivity factors at a height of 3 km above surface without (top) and with (center) attenuation correction; (bottom) Corresponding WSR-88D radar reflectivity factors are shown.

effort on calibration because of an awareness of its impact on rainfall algorithms. We are also more cognizant of the error structure of precipitation estimates. Some of the wide variations in Z–R relations reported in the past may have come from instrumentation- and calibration-related problems and variations in analysis methods. Data quality has emerged as a significant problem that can affect the radar rainfall estimates as much as or more than errors arising from the variability in the drop size distribution. It is also increasingly evident that it is difficult to collect long-term high-quality gauge data over a network of gauges. The number of radars deployed over the world has increased multifold over the past two decades. Several auxiliary technologies such as Global Positioning System, electronic recording, and telemetering of gauge data have also advanced and become very affordable. With the assistance of technology, which has made it easier to collect high-quality coordinated datasets, we have considerably improved our understanding of rainfall processes and the associated radar observations. Moreover, as more radars are deployed, a greater variety of rain events are being observed. Regarding local precipitation estimation from

ground radars, it appears that for long-term averages over large areas, engineering or statistical solutions will yield sufficiently accurate results because adjustments derived from ground observations will ensure unbiased rainfall estimates. However, if the total rainfall from individual storms or if short-term rainfall estimation is important such as in flash flood or extreme rainfall events, then physically based algorithms are likely to be the best choice. This was observed during the Fort Collins flood event, which is one of the few well-documented cases with polarimetric radar measurements.

Regarding global-scale precipitation measurements, the space-borne weather radar is still in its infancy with many opportunities and challenges ahead. The problems that we face can be divided into instrument and science issues. Some of the questions go beyond the radar capabilities alone and are concerned with instrument combinations that will be used to improve the accuracy, detail, and sampling of rain, snow, and cloud. What can we expect to achieve, for example, from combining radar and radiometer and what retrieval techniques are available to analyze the data? Are Doppler or polarization diversity measurements feasible from space? Can

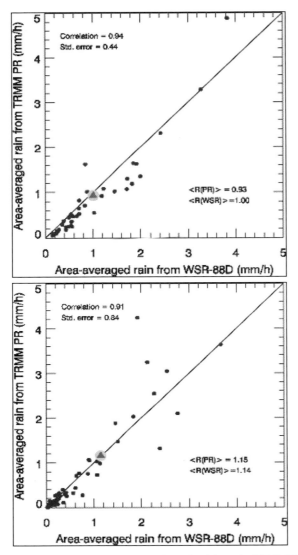

FIG. 9.9. Area-average rain rates determined from the TRMM PR vs that determined from the Melbourne, FL, WSR-88D radar (top) and Houston, TX, WSR-88D radar (bottom) for TRMM overpasses during 1998–99.

spaceborne weather radars be built inexpensively so that they will be deployed as one of the standard instruments aboard low earth orbiting weather satellites? Is a geostationary weather radar feasible and, if so, what technological problems must be solved, other than the construction of very large antenna? From a science perspective, how do the measurements from a particular instrument advance our understanding of the global distribution of rainfall and estimates of latent heating? In assimilating rainfall data from spaceborne radars and radiometers into general circulation models, can we ensure that the data are available in a timely fashion with adequate space and time resolution, and without significant biases?

In summary, it appears that in the near future, significant advancs will be made in the characterization of the space–time error covariance of radar rainfall estimates at the local scale. Polarimetric radars will not only contribute to advancing the characterization of precipitation microphysics, but will be critical in three areas, namely, (i) improving data quality, (ii) hydrometeor identification and quantification, and (iii) extreme event precipitation estimation. At the global scale, advances can be expected in the following areas: improved space–time sampling with the goal of 3-h revisit time over most of the globe; progress in dual wavelength methods for phase-state identification, and improved rain rate and DSD estimation; and development of affordable technologies for airborne and spaceborne weather radars. Global radar datasets also offer the opportunity for new validation strategies for both ground-based and spaceborne radars. These advances should serve the ultimate goal of accurate and timely precipitation estimates over much of the globe.

Acknowledgments. The authors acknowledge the funding from various agencies over the years that have led to the research progress described in this paper. National Science Foundation and the UMBC GEST center are acknowledged by VC whereas the NASA TRMM program is acknowledged by RM and VC. IZ acknowledges the support of the Meteorological Service of Canada and is grateful to Gyu Won Lee for preparing Figs. 9.2, 9.3, and 9.4 for inclusion in this paper.

REFERENCES

Amayenc, P., J. Testud, and M. Marzoug, 1993: Proposal for a spaceborne dual-beam rain radar with Doppler capability. *J. Atmos. Oceanic Technol.,* **10,** 262–276.

Anagnostou, E. N., C. A. Morales, and T. Dinku, 2001: On the use of TRMM precipitation radar observations in determining ground radar calibration biases. *J. Atmos. Oceanic Technol.,* **18,** 616–628.

Andsager, K., K. V. Beard, and N. F. Laird, 1999: Laboratory measurements of axis ratios for large raindrops. *J. Atmos. Sci.,* **56,** 2673–2683.

Atlas, D., 1954: The estimation of cloud parameters by radar. *J. Meteor.,* **11,** 309–317.

——, and F. H. Ludlum, 1961: Multi-wavelength radar reflectivity of hailstorms. *Quart. J. Roy. Meteor. Soc.,* **87,** 523–534.

——, and C. W. Ulbrich, 1977: Path- and area-integrated rainfall measurement by microwave attenuation in the 1–3 cm band. *J. Appl. Meteor.,* **16,** 1322–1331.

——, and T. L. Bell, 1992: The relation of radar to cloud area-time integrals and implications for rain measurements from space. *Mon. Wea. Rev.,* **120,** 1997–2008.

——, D. Rosenfeld, and D. A. Short, 1990: The estimation of convective rainfall by area integrals. 1: The theoretical and empirical basis. *J. Geophys. Res.,* **95,** 2153–2160.

——, C. W. Ulbrich, F. D. Marks, E. Amitai, and C. R. Williams, 1999: Systematic variation of drop size and radar-rainfall relations. *J. Geophys. Res.,* **104,** 6155–6169.

Austin, P. M., 1947: Measurement of approximate raindrop size distribution by microwave attenuation. *J. Atmos. Sci.,* **4,** 121–124.

Awaka, J., T. Iguchi, and K. Okamoto, 1998: Early results on rain type classification by the Tropical Rainfall Measuring Mission (TRMM) precipitation radar. *Proc. 8th URSI Commission F Open Symp.,* Aveiro, Portugal, URSI 143–146.

Beard, K. V., 1985: Simple altitude adjustments to raindrop velocities for Doppler radar analysis. *J. Atmos. Oceanic Technol.,* **2,** 468–471.

——, and C. Chuang, 1987: A new model for the equilibrium shape of raindrops. *J. Atmos. Sci.,* **44,** 1509–1524.

Bellon, A., I. Zawadzki, and F. Fabry, 1997: Measurements of melting layer attenuation at X-band frequencies. *Radio Sci.,* **32,** 943–955.

Berjulev, G. P., and V. V. Kostarev, 1974: Dual wavelength radar measurements of 8 mm radiowave attenuation by atmospheric precipitation and clouds. *J. Rech. Atmos.,* **8,** 358–363.

Bolen, S., and V. Chandrasekar, 2000: Quantitative cross validation of space-based and ground-based radar observations. *J. Appl. Meteor.,* **39,** 2071–2079.

——, and ——, 2003a: Methodology for aligning and comparing spaceborne radar and ground-based radar observations. *J. Atmos. Oceanic Technol.,* **20,** 647–659.

——, and ——, 2003b: Quantitative estimation of TRMM PR signals along the PR path from ground-based polarimetric radar observations. *Radio Sci.,* in press.

——, V. N. Bringi, and V. Chandrasekar, 1998: An optimal area approach to intercomparing polarimetric radar rainrate algorithms with gauge data. *J. Atmos. Oceanic Technol.,* **15,** 605–623.

Bringi, V. N., and V. Chandrasekar, 2001: *Polarimetric Doppler Weather Radar—Principles and Applications.* Cambridge University Press, 636 pp.

——, ——, N. Balakrishnan, and D. S. Zrnic, 1990: An examination of propagation effects in rainfall on radar measurements at microwave frequencies. *J. Atmos. Oceanic Technol.,* **7,** 829–840.

——, ——, and R. Xiao, 1998: Raindrop axis ratios and size distributions in Florida rainshafts: An assessment of multiparameter radar algorithms. *IEEE Trans. Geosci. Remote Sens.,* **36,** 703–715.

——, T. D. Keenan, and V. Chandrasekar, 2001: Correcting C-band radar reflectivity and differential reflectivity data for rain attenuation: A self-consent method with constraints. *IEEE Geosci. Remote Sens.,* **39,** 1906–1915.

——, T. Tang, and V. Chandrasekar, 2002: Evaluation of a polarimetrically tuned *Z–R* relation. *Proc. Second European Conf. on Radar Meteor.,* Delft, Netherlands, ERAD, 217–221.

——, V. Chandrasekar, J. Hubbert, E. Gorgucci, W. L. Randeu, and M. Schoenhuber, 2003: Raindrop size distribution in different climatic regimes from disdrometer and dual-polarized radar analysis. *J. Atmos. Sci.,* **60,** 354–365.

Calheiros, R. V., and I. Zawadzki, 1987: Reflectivity–rain rate relationships for hydrology in Brazil. *J. Climate Appl. Meteor.,* **26,** 118–132.

Chandrasekar, V., W. A. Cooper, and V. N. Bringi, 1988: Axis ratios and oscillations of raindrops. *J. Atmos. Sci.,* **45,** 1323–1333.

——, V. N. Bringi, N. Balakrishnan, and D. S. Zrnic, 1990: Error structure of multiparameter radar and surface measurements of rainfall. Part III: Specific differential phase. *J. Atmos. Oceanic Technol.,* **7,** 621–629.

Chiu, L. S., A. T. C. Chang, and J. Janowiak, 1993: Comparison of monthly rain rates derived from GPI and SSM/I using probability distribution functions. *J. Appl. Meteor.,* **32,** 323–334.

Donneaud, A. A., P. L. Smith, S. A. Dennis, and S. Sengupta, 1981: A simple method for estimating convective rain over an area. *Water Resour. Res.,* **17,** 1676–1682.

——, S. Ionescu-Niscov, D. L. Priegnitz, and P. L. Smith, 1984: The area–time integral as an indicator for convective rain volumes. *J. Climate Appl. Meteor.,* **23,** 555–561.

Durden, S. L., and Z. S. Haddad, 1998: Comparison of radar rainfall retrieval algorithms in convective rain during TOGA COARE. *J. Atmos. Oceanic Technol.,* **15,** 1091–1096.

Eccles, P. J., 1979: Comparison of remote measurements by single and dual wavelength meteorological radars. *IEEE Trans. Geosci. Remote Sens.,* **17,** 205–218.

——, and E. A. Mueller, 1971: X-band attenuation and liquid water content estimation by dual-wavelength radar. *J. Appl. Meteor.,* **10,** 1252–1259.

Fabry, F., and I. Zawadzki, 1995: Long term radar observations of the melting layer of precipitation and their interpretation. *J. Atmos. Sci.,* **52,** 838–851.

Ferreira, F., P. Amayenc, S. Oury, and J. Testud, 2001: Study and tests of improved rain estimates from the TRMM precipitation radar. *J. Appl. Meteor.,* **40,** 1878–1899.

Foote, G. B., and P. S. Du Toit, 1969: Terminal velocity of raindrops aloft. *J. Appl. Meteor.,* **8,** 249–253.

Fujita, M., 1983: An algorithm for estimating rain rate by a dual-frequency radar. *Radio Sci.,* **18,** 697–708.

Goldhirsh, J., and I. Katz, 1974: Estimation of rain drop size distribution using multiple wavelength radar systems. *Radio Sci.,* **9,** 439–446.

Gorgucci, E., V. Chandrasekar, and G. Scarchilli, 1995: Radar and surface measurement of rainfall CaPE. *J. Appl. Meteor.,* **34,** 1570–1577.

——, ——, ——, and ——, 1999: A procedure to calibrate multiparameter weather using properties of the rain medium. *IEEE Trans. Geosci. Remote Sens.,* **38,** 269–276.

——, ——, ——, and V. N. Bringi, 2000: Measurements of mean raindrop shape from polarimetric radar observations. *J. Atmos. Sci.,* **57,** 3406–3413.

——, ——, ——, and ——, 2001: Rainfall estimation from polarimetric radar measurements: Composite algorithms independent of raindrop shape-size relation. *J. Atmos. Oceanic Technol.,* **18,** 1773–1786.

Haddad, Z. S., and Coauthors, 1997: The TRMM "Day-1" radar–radiometer combined rain-profiling algorithm. *J. Meteor. Soc. Japan,* **75,** 799–808.

Harrold, T. W., 1967: The attenuation of 8.6 mm wavelength radiation in rain. *Proc. Inst. Electr. Eng. London,* **114,** 201–203.

Hildebrand, P., 1978: Iterative correction for attenuation of 5 cm radar in rain. *J. Appl. Meteor.,* **17,** 508–514.

Hitschfeld, W., and J. Bordan, 1954: Errors inherent in the radar measurement of rainfall at attenuating wavelengths. *J. Meteor.,* **11,** 58–67.

Holt, A. R., 1988: Extraction of differential propagation phase from data from S-band circularly polarized radars. *Electron. Lett.,* **24,** 1241–1242.

Houze, R. A., Jr., 1989: Observed structure of mesoscale convective systems and implications for large-scale heating. *Quart. J. Roy. Meteor. Soc.,* **115,** 425–461.

Iguchi, T., and R. Meneghini, 1994: Intercomparison of single-frequency methods for retrieving a vertical rain profile from airborne or spaceborne radar data. *J. Atmos. Oceanic Technol.,* **11,** 1507–1516.

——, and ——, 1995: Differential equations for dual-frequency radar returns. Preprints, *27th Conf. on Radar Meteorology,* Vail, CO, Amer. Meteor. Soc., 190–193.

——, T. Kozu, R. Meneghini, J. Awaka, and K. Okamoto, 2000: Rain profiling algorithm for the TRMM precipitation radar. *J. Appl. Meteor.,* **39,** 2038–2052.

Ihara, T., Y. Furuhara, and T. Manabe, 1984: Inference of raindrop size distribution from rain attenuation statistics at 12, 35, and 82 GHz. *Trans. IECE Japan,* **E67,** 211–217.

Jameson, A. R., 1985: Microphysical interpretation of multi-parameter radar measurements in rain. Part III: Interpretation and measurements of propagation differential phase shift between orthogonal linear polarizations. *J. Atmos. Sci.,* **42,** 607–614.

Kedem, B., and H. Pavlopoulos, 1991: On the threshold method for rainfall estimation: Choosing the optimal threshold level. *J. Amer. Stat. Assoc.,* **86,** 626–633.

——, R. Pfeiffer, and D. A. Short, 1997: Variability of space–time mean rain rate. *J. Appl. Meteor.,* **36,** 443–451.

Keenan, T. D., L. D. Carey, D. S. Zrnic, and P. T. May, 2001: Sensitivity of 5-cm wavelength polarimetric radar variables to raindrop axial ratio and drop size distribution. *J. Appl. Meteor.,* **40,** 526–545.

Klett, J. D., 1981: Stable analytical inversion for processing lidar returns. *Appl. Opt.*, **20**, 211–220.

Kollias, P., R. Lhermitte, and B. A. Albrecht, 1999: Vertical air motion and raindrop size distributions in convective systems using a 94 GHz radar. *Geophys. Res. Lett.*, **26**, 3109–3112.

Kozu, T., and K. Nakamura, 1991: Rainfall parameter estimation from dual radar measurements combining reflectivity profile and path-integrated attenuation. *J. Atmos. Oceanic Technol.*, **8**, 259–270.

——, ——, R. Meneghini, and W. C. Boncyk, 1991: Dual-parameter radar rainfall measurement from space: A test result from an aircraft experiment. *IEEE Trans. Geosci. Remote Sens.*, **29**, 690–703.

——, and Coauthors, 2001: Development of precipitation radar onboard the Tropical Rainfall Measuring Mission (TRMM) satellite. *IEEE Trans. Geosci. Remote Sens.*, **39**, 102–116.

Kumagai, H., T. Kozu, M. Satake, H. Hanado, and K. Okamoto, 1995: Development of an active radar calibrator for the TRMM precipitation radar. *IEEE Trans. Geosci. Remote Sens.*, **33**, 1316–1318.

Kummerow, C., W. Barnes, T. Kozu, J. Shiue, and J. Simpson, 1998: The Tropical Rainfall Measuring Mission (TRMM) sensor package. *J. Atmos. Oceanic Technol.*, **15**, 809–817.

L'Ecuyer, T. S., and G. L. Stephens, 2002: An estimation-based precipitation retrieval algorithm for attenuating radars. *J. Appl. Meteor.*, **41**, 272–285.

Lhermitte, R., 1990: Attenuation and scattering of millimeter wavelength radiation by clouds and precipitation. *J. Atmos. Oceanic Technol.*, **7**, 464–479.

Liao, L., R. Meneghini, and T. Iguchi, 2001: Comparisons of rain rate and reflectivity factor derived from the TRMM precipitation radar and the WSR-88D over the Melbourne, Florida, site. *J. Atmos. Oceanic Technol.*, **18**, 1959–1974.

Liu, H., V. Chandrasekar, and G. Xu, 2001: An adaptive neural network scheme for radar rainfall estimation from WSR-88D observations. *J. Appl. Meteor.*, **40**, 2038–2050.

Lopez, R. E., 1977: The lognormal distribution and cumulus cloud populations. *Mon. Wea. Rev.*, **105**, 865–872.

Marzoug, M., and P. Amayenc, 1991: Improved range-profiling algorithm of rainfall rate with path-integrated attenuation constraint. *IEEE Trans. Geosci. Remote Sens.*, **29**, 584–592.

——, and ——, 1994: A class of single- and dual-frequency algorithms for rain-rate profiling from a space-borne radar. Part I: Principle and tests from numerical simulations. *J. Atmos. Oceanic Technol.*, **11**, 1480–1506.

Matrosov, S. Y., 1992: Radar reflectivity in snowfall. *IEEE Trans. Geosci. Remote Sens.*, **30**, 454–461.

Meneghini, R., and D. Atlas, 1986: Simultaneous ocean cross section and rainfall measurements from space with a nadir-looking radar. *J. Atmos. Oceanic Technol.*, **3**, 400–413.

——, and K. Nakamura, 1990: Range profiling of the rain rate by an airborne weather radar. *Remote Sens. Environ.*, **31**, 193–209.

——, J. Eckerman, and D. Atlas, 1983: Determination of rain rate from space-borne radar using measurements of total attenuation. *IEEE Trans. Geosci. Remote Sens.*, **21**, 34–43.

——, T. Kozu, J. Kumagai, and W. C. Boncyk, 1992: A study of rain estimation methods from space using dual-wavelength radar measurements at near-nadir incidence over ocean. *J. Atmos. Oceanic Technol.*, **9**, 364–382.

——, H. Kumagai, J. R. Wang, T. Iguchi, and T. Kozu, 1997: Microphysical retrievals over stratiform rain using measurements from an airborne dual-wavelength radar–radiometer. *IEEE Trans. Geosci. Remote Sens.*, **35**, 487–506.

——, T. Iguchi, T. Kozu, L. Liao, K. Okamoto, J. A. Jones, and J. Kwiatkowski, 2000: Use of the surface reference technique for path attenuation estimates for the TRMM radar. *J. Appl. Meteor.*, **39**, 2053–2070.

——, J. A. Jones, T. Iguchi, K. Okamoto, and J. Kwiatkowski, 2001: Statistical methods of estimating average rainfall over large space–time scales using data from the TRMM precipitation radar. *J. Appl. Meteor.*, **40**, 568–585.

Olson, W. S., C. D. Kummerow, Y. Hong, and W.-K. Tao, 1999: Atmospheric latent heating distributions in the Tropics derived from passive microwave radiometer measurements. *J. Appl. Meteor.*, **38**, 633–644.

Pruppacher, H. R., and K. V. Beard, 1971: A semi-empirical determination of the shape of cloud and rain drops. *J. Atmos. Sci.*, **28**, 86–94.

Raghavan, R., and V. Chandrasekar, 1994: Multiparameter radar study of rainfall: Potential application to area time integral studies. *J. Appl. Meteor.*, **33**, 1636–1645.

Rinehart, R. E., and J. D. Tuttle, 1984: Dual-wavelength processing: Some effects of mismatched antenna beam patterns. *Radio Sci.*, **19**, 121–131.

Rosenfeld, D., and C. W. Ulbrich, 2003: Cloud microphysical properties, processes and rainfall estimation opportunities. *Radar and Atmospheric Science: A Collection of Essays in Honor of David Atlas, Meteor. Monogr.*, No. 52, Amer. Meteor. Soc., 237–258.

——, D. Atlas, and D. A. Short, 1990: The estimation of convective rainfall by area-integrals. 2: The height-area rainfall threshold (HART) method. *J. Geophys. Res.*, **95**, 2161–2176.

——, D. B. Wolff, and E. Amitai, 1994: The window probability matching method for rainfall measurements with radar. *J. Appl. Meteor.*, **33**, 682–693.

Ruf, C. S., K. Aydin, S. Mathur, and J. P. Bobak, 1996: 35 GHz dual-polarization propagation link for rain-rate estimation. *J. Atmos. Oceanic Technol.*, **13**, 419–425.

Ryzhkov, A. V., D. S. Zrnic, and R. Fulton, 2000: Areal rainfall estimates using differential phase. *J. Appl. Meteor.*, **39**, 263–268.

Sachidananda, M., and D. S. Zrnic, 1986: Differential propagation phase shift and rainfall rate estimation. *Radio Sci.*, **21**, 235–247.

Sauvageot, H., 1994: The probability density function of rain rate and the estimation of rainfall by area integrals. *J. Appl. Meteor.*, **33**, 1255–1262.

Scarchilli, G., V. Gorgucci, V. Chandrasekar, and A. Dobaie, 1996: Self-consistency of polarization diversity measurement of rainfall. *IEEE Trans. Geosci. Remote Sens.*, **34**, 22–26.

Schumacher, C., and R. A. Houze, 2000: Comparison of radar from the TRMM satellite and Kwajalein ocean validation site. *J. Appl. Meteor.*, **39**, 2151–2164.

Sekhon, R. S., and R. C. Srivastava, 1971: Doppler radar observations of drop size distributions in a thunderstorm. *J. Atmos. Sci.*, **28**, 983–994.

Seliga, T. A., and V. N. Bringi, 1976: Potential use of radar differential reflectivity measurements at orthogonal polarization for measuring precipitation. *J. Appl. Meteor.*, **15**, 69–76.

——, and ——, 1978: Differential reflectivity and differential phase shift—Applications in radar meteorology. *Radio Sci.*, **13**, 271–275.

Short, D. A., K. Shimizu, and B. Kedem, 1993: Optimal threshold for the estimation of area rain-rate moments by the threshold method. *J. Appl. Meteor.*, **32**, 182–192.

Simpson, J., C. Kummerow, W. K. Tao, and R. F. Adler, 1996: On the Tropical Rainfall Measuring Mission (TRMM). *Meteor. Atmos. Phys.*, **60**, 19–36.

Smith, E. A., X. Xiang, A. Mugnai, and G. J. Tripoli, 1994: Design of an inversion-based precipitation profile retrieval algorithm using explicit cloud model for initial guess microphysics. *Meteor. Atmos. Phys.*, **54**, 53–78.

Srivastava, R. C., and A. R. Jameson, 1977: Radar detection of hail. *Hail: A Review of Hail Science and Hail Suppression, Meteor. Monogr.*, No. 38, Amer. Meteor. Soc., 269–277.

Tao, W.-K., S. Lang, J. Simpson, and R. Adler, 1993: Retrieval algorithms for estimating the vertical profiles of latent heat release: Their applicaitons to TRMM. *J. Meteor. Soc. Japan*, **71**, 685–700.

Testud, J., and P. Amayenc, 1989: Stereoradar meteorology—A promising technique for observation from a mobile platform. *J. Atmos. Oceanic Technol.*, **6**, 89–108.

——, ——, and M. Marzoug, 1992: Rainfall-rate retrieval from a spaceborne radar: Comparison between single-frequency, dual-frequency, and dual-beam techniques. *J. Atmos. Oceanic Technol.,* **9,** 599–623.

——, E. LeBouar, E. Obligis, and M. Ali-Mehenni, 2000: The rain-profiling algorithm applied to polarimetric weather radar. *J. Atmos. Oceanic Technol.,* **17,** 322–356.

——, S. Oury, P. Amayenc, and R. A. Black, 2001: The concept of "normalized" distributions to describe raindrop spectra: A tool for cloud physics and cloud remote sensing. *J. Appl. Meteor.,* **40,** 1118–1140.

Ulbrich, C. W., 1983: Natural variations in the analytical form of the raindrop size distribution. *J. Climate Appl. Meteor.,* **22,** 1764–1775.

van de Hulst, H. C., 1981: *Light-Scattering by Small Particles.* Dover, 470 pp.

Wilheit, T. T., A. Chang, and L. Chiu, 1991: Retrieval of monthly rainfall indices from microwave radiometric measurements using probability distribution functions. *J. Atmos. Oceanic Technol.,* **8,** 118–136.

Willis, P. T., 1984: Functional fits to some observed drop size distributions and parametrization of rain. *J. Atmos. Sci.,* **41,** 1648–1661.

Xiao, R., and V. Chandrasekar, 1995: Multiparameter radar rainfall estimation using neural network techniques. Preprints, *27th Conf. on Radar Meteorology,* Vail, CO, Amer. Meteor. Soc., 199–204.

Zawadzki, I. I., 1975: On radar-raingage comparison. *J. Appl. Meteor.,* **14,** 1430–1436.

——, 1984: Factors affecting the precision of radar measurements of rain. Preprints, *22nd Conf. on Radar Meteorology,* Zurich, Switzerland, Amer. Meteor. Soc., 251–256.

Chapter 10

Cloud Microphysical Properties, Processes, and Rainfall Estimation Opportunities

Daniel Rosenfeld

Institute of Earth Sciences, Hebrew University of Jerusalem, Jerusalem, Israel

Carlton W. Ulbrich

Department of Physics and Astronomy, Clemson University, Clemson, South Carolina

Rosenfeld

1. Introduction

In this work the longstanding question of the connections between raindrop-size distributions (RDSDs) and radar reflectivity–rainfall rate (Z–R) relationships is revisited, this time from the combined approach of rain-forming physical processes that shape the RDSD, and a formulation of the RDSD into the simplest free parameters of the rain intensity R, rainwater content W, and median volume drop diameter D_0. This is accomplished through a theoretical analysis, using a gamma RDSD, of D_0–R and W–R relations implied by the coefficients and exponents in empirical Z–R relations. The results provide a means by which these Z–R relations can be classified. The most dramatic of these classifications involves the relation between D_0 and W, which shows a remarkable ordering with the rain types.

This work also summarizes the effects of various physical processes in modifying the RDSD in clouds. These individual processes are combined into conceptual models of the way different microphysical and dynamical rain-forming processes can build different kinds of RDSDs. Much of the physical insights that are at the heart of this study came from examining the evolution of the RDSD with respect to its ultimate mature state

of the equilibrium raindrop-size distribution described by Hu and Srivastava (1995).

Finally, the different components of the previous sections are combined in an examination of integral parameters deduced from the raindrop-size distributions associated with the host of RDSD-based Z–R relations found in the literature. Only those relations are used that could be associated to the cloud microstructure and dynamic context of the conceptual model. It is found that there exists a well-defined sequence in the transition from extreme continental to equatorial maritime for convective rainfall. In addition, similar behavior is found in tropical convective versus stratiform rainfall and in orographic rainfall as a function of altitude. These results offer promise for the development of algorithms for classification of the rainfall with respect to type in the remote measurement of rainfall either from satellite platforms or from ground-based radars.

Early attempts to explain the variability in Z–R relations are reviewed in section 2. Section 3 reviews the formulation of the RDSD and provides the tools to restore RDSD parameters from published Z–R power-law relations. Section 4 describes the way the different individual processes that modify the RDSD can be combined into conceptual models of the rain-forming pro-

TABLE 10.1. Microphysical and kinematic influences on Z–R relationships and the effect on radar rainfall estimates when no adjustment is applied (after Wilson and Brandes 1979).

Process	Change in $Z = AR^b$		Probable effect on radar rainfall if Z–R not adjusted	Possible region of max influence
	A	b		
Microphysical				
Evaporation (Atlas and Chmela 1957)	Increase	Decrease	Overestimate	Inflow regions, fringe areas
Accretion of cloud particles (Atlas and Chmela 1957; Rigby et al. 1954)	Decrease	Increase	Underestimate	Downdraft
Collision, coalescence (Srivastava 1971)	Increase	Decrease	Overestimate	Reflectivity core
Breakup (Srivastava 1971)	Decrease	Decrase	Underestimate	Reflectivity core
Kinematic				
Size sorting (Gunn and Marshall 1955; Atlas and Chmela 1957)	Increase	Decrease	Tendency to overestimate	Regions of strong inflow and outflow
Vertical motion				
Updraft	Increase	Decrease	Overestimate	
Downdraft	Decrease	Increase	Underestimate	

cesses. Section 5 applies the different components in the previous sections to deduce the Z–R classification scheme. Section 6 summarizes the results and offers suggestions for implementation of a dynamic Z–R classification method.

2. Early attempts to classify Z–R relations

It has long been recognized that wide range of values found for the coefficient A and exponent b in Z–R relations of the form $Z = AR^b$ is due to variations in the form of the RDSD. Chandrasekar et al. recognize this connection between RDSD variability and the values of A and b (chapter 9 in this monograph). They also point to the importance of separating Z–R relations according to the type of rainfall. One of the earliest studies to recognize these effects was that due to Atlas and Chmela (1957) who showed that RDSD sorting at the scale of the individual rain shaft could occur due to drop sorting by wind shear and updrafts. Beyond the scale of the individual rain shafts, the causes of variability in Z–R relations were sought in differences in rainfall types, atmospheric conditions, and geographical locations (Fujiwara 1965; Stout and Mueller 1968; Cataneo and Stout 1968). The rationale was that different conditions would prefer different rain processes, and the effects of these processes were summarized in the form of a table in Wilson and Brandes (1979), which is reproduced here as Table 10.1.

Wilson and Brandes (1979) provided this table with little discussion. Such a discussion is provided later in this work, with some explanations on the causes for the trends of the coefficient and exponent. That is done after the various analytical forms of the RDSD that have been employed in the past are introduced.

To depict the relationships between the various parameters of the RDSD, Atlas and Chmela (1957) produced a rain parameter diagram (RAPAD) with Z plotted versus R and on which isopleths of distribution param-

eters were displayed for an exponential distribution of the form

$$N(D) = N_0 \exp(-\Lambda D), \quad (10.1)$$

where $N(D)$ (m^{-3} cm^{-1}) is the number of drops per unit volume per unit size interval and N_0 and Λ are the RDSD parameters. In addition, as shown by Atlas (1955), $\Lambda = 3.67/D_0$, where D_0 is the median volume diameter. The Atlas–Chmela diagram is reproduced in Atlas (1964), but a more recent version is shown in Fig. 10.1 for an exponential distribution with isopleths of W, D_0, and N_0, where W(g m^{-3}) is the liquid rainwater concentration. At the time the Atlas–Chmela diagram was published, the use of Z (the single radar measurable then available) to measure R through the use of a Z–R relation was the focus of research in radar meteorology. As the field expanded the number of measurements of drop-size spectra (and Z–R relations derived from them) grew rapidly and it was discovered quickly that there was no unique relationship between Z and R; that is, there were no unique values for A and b. The advantage of the Atlas–Chmela Z–R rain parameter diagram was that for a given Z–R relation (*and* an exponential RDSD), it permitted the relationships between all of the drop-size distribution (DSD) integral parameters to be determined. That is, for a given Z–R relation the diagram implied corresponding relations between D_0–R, W–R, Z–W, D_0–W, etc.. The disadvantage was that a different diagram had to be produced for distributions different from exponential.

There have been many attempts to relate the large observed variations in the coefficient A and exponent b in the Z–R law to the meteorological conditions associated with the rainfall and to the parameters of the drop-size distribution. It is well known in radar meteorology that there is a great lack of consistency in the drop-size distribution for various meteorological conditions. Even when the conditions appear to be similar the size distributions can be widely different. This is apparent in

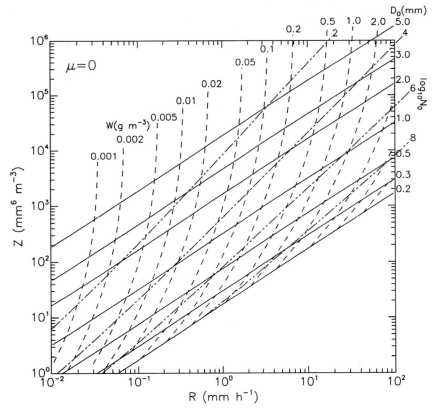

FIG. 10.1. Rain parameter diagram of Ulbrich and Atlas (1978) similar to that presented by Atlas and Chmela (1957) for an exponential RDSD. The solid, dashed, and dash–double-dotted lines are isopleths of median volume diameter D_0, liquid water concentration W, and exponential intercept parameter N_0. The isopleths are labeled with the values to which they correspond with the units of mm for D_0, g m^{-3} for W, and m^{-3} cm^{-1} for N_0.

the work of Fujiwara (1965) who uses results for A and b derived from analysis of data collected at the surface with a raindrop camera (Mueller 1965) together with radar data and National Weather Service reports to deduce those regions on a plot of A versus b, which correspond to a given rainfall type. Data were analyzed for four locations, namely, Florida, Illinois, Germany, and Japan. Fujiwara considers only three types of rainfall, that is, thunderstorms, showers, and continuous rain and, although there is much scatter on the A–b plot, he enumerates some general findings. He finds that large A (300–1000) and moderate b (1.25–1.65) are associated with thunderstorms, while both A and b are somewhat smaller and more variable for rain showers. For continuous rain the values of A are generally smaller than for either of the previous two types, but the range of b is large (1.0–2.0). The results found for thunderstorms in Illinois are in essential agreement with those found for the Florida data. He also attempts to relate A and b to the shapes of the drop-size distribution and to the characteristics of the radar echo. The distributions for thunderstorm rain were found to tend toward exponentiality with drops of large diameters and several peaks. For weak rain showers the distribution is sharply peaked at small diameters and concave downward on a plot of

log[$N(D)$] versus D. However, all drop-size distributions in Fujiwara's analysis are concave downward and exhibit considerable shortage of drops with diameters less than about 1 mm. This may be due to an inability of the drop camera to detect these drops and thus precludes definitive conclusions about the dependence of A and b on RDSD shape. In any event, the absence of small drops has little effect on the values of Z and R and, hence, on the values of A and b. For the three types of rainfall Fujiwara finds central values of $(A, b) = (450, 1.46)$ for thunderstorms, $(300, 1.37)$ for rain showers, and $(205, 1.48)$ for continuous rain. There is a great deal of scatter in Fujiwara's results for A and b when plotted on an A versus b diagram, but there is a weak suggestion of an inverse dependence of A on b; that is, large A corresponds to small b, etc.

Similar work of this nature has been conducted by investigators at the Illinois State Water Survey using data from the same instrument as that used by Fujiwara. Stout and Mueller (1968) report measurements from Florida, the Marshall Islands, and Oregon and classify their Z–R relations separately according to rainfall type (continuous, showers, or thunderstorms), synoptic situation (air mass, warm front, cold front, etc.), and thermodynamic instability. The results of classification by

TABLE 10.2. Values of A and b in $Z = AR^b$ for tropical stratiform and convective rainfall in TOGA COARE.

Source	Stratiform		Convective	
	A	b	A	b
Tokay et al. (1995)	335	1.37	175	1.37
Tokay and Short (1996)	367	1.30	139	1.43
Atlas et al. (2000)	224	1.28	129	1.38
Ulbrich and Atlas (2002)	203	1.46	120	1.43

thermodynamic instability were found not to be useful, but those found from the classification by rainfall type and synoptic situation displayed large systematic variations in A and b, indicating the importance of using a stratification technique for measurement of rainfall using Z–R relations. In spite of this finding, neither of the first two techniques was found to be superior in measuring rainfall amounts to the method that uses the Z–R relation given by Marshall et al. (1947). In fact, in several cases the latter was found to produce more accurate results than either stratification method.

Some limited progress in this area has been made recently for tropical rainfall. Well-defined differences in stratiform and convective rainfall in the Tropics have been found by several investigators during Tropical Oceans Global Atmosphere Coupled Ocean–Atmosphere Response Experiment (TOGA COARE). Some of the results found for A and b by various investigators are listed in Table 10.2. It must be recognized that these relations are based on long-term temporal and spatial averages of experimental RDSDs. For individual storms and for stages of such storms the Z–R relations can vary appreciably. For example, Atlas et al. (1999) show data for tropical squall lines that are segmented into convective (C), transition (T), and stratiform (S) stages. The variations in A and b between different storms for each of these stages are very large and can also vary appreciably between storms. In any event, it is clear from Table 10.2 that there is not much difference between these relations for convective rain when plotted on the rain parameter diagram of Atlas and Chmela (1957; Fig. 10.1). The differences between the stratiform relations lie mostly in the coefficients, which Atlas et al. (2000) attribute to the inclusion by Tokay and Short (1996) of transition rain in the convective category. Nevertheless, it may be concluded from the results shown above that the principal differences between stratiform and convective rain in the Tropics is that the coefficient A for stratiform rain is somewhat larger (at least 70%) than the coefficient for convective rain. Examination of these relations, when plotted on the RAPAD of Fig. 10.1, indicates that the larger coefficients A for stratiform rain are associated with larger Z values (for the same R) than convective rain and therefore also with larger values of D_0.

3. Formulations of the raindrop-size distribution

Raindrop-size distributions have been a subject of extensive investigation for nearly 100 years. The earliest carefully performed measurements of raindrop sizes were reported by Laws and Parsons (1943), Marshall and Palmer (1948), and Best (1950) and indicated that the distribution could be approximated well by an exponential function of the form of Eq. (10.1). (In the following the term "raindrop size" is used to mean raindrop diameter). This mathematical approximation to the raindrop-size distribution has been in widespread use for decades and is especially convenient because of its simplicity. However, even in the early experimental work just cited distinct deviations from exponentiality were noted. Since these deviations are reflective of the physics of rain formation in clouds it has been considered imperative that an accurate mathematical representation of the distribution be found.

To account for distribution shape effects Atlas (1955) introduced a "moment" G of the distribution, which related the reflectivity factor Z to the median volume diameter D_0 and the liquid water concentration M. They also showed Z to be related to the rainfall rate R and D_0 through the moment G. Joss and Gori (1978) also defined measures of distribution shape $S(PQ)$, where P and Q are any two integral parameters of the distribution. For distributions that have breadth narrower than, equal to, or broader than an exponential distribution, S is less than, equal to, or greater than 1, respectively. For the experimental distributions they investigate, Joss and Gori find that S is always less than 1, the more so the shorter the time interval used to average the data. Joss and Gori also found that considerable long-term averaging of disdrometer data is required for the distributions to approach exponentiality; the longer the averaging period the closer the approach to exponentiality. Periods as long as 256 min were required to find average distributions close to exponential, regardless of the type of rainfall. Their work further demonstrates the need for RDSDs of greater generality than the exponential distribution.

Other attempts to account for distribution shape have involved the use of specific mathematical forms different from exponential. One of the earliest of these was a lognormal function suggested by Levin (1954) of the form

$$N(D) = N_0 D^{-1} \exp(-c \ln^2[D/D_g]), \quad (10.2)$$

with N_0, c, and D_g as parameters. This form has been applied to the analysis of cloud droplet and raindrop distributions by many investigators including Mueller and Sims (1966), Bradley and Stow (1974), and Markowitz (1976). Although this function approximates drop-size distributions well, it does not allow for as broad a spectrum of RDSD shapes as other representations and does not reduce to the exponential function as a special case. An alternative function that has come into widespread use is the gamma function having the form

$$N(D) = N_0 D^\mu \exp(-\Lambda D), \qquad (10.3)$$

with N_0, μ, and Λ as parameters (Deirmendjian 1969; Willis 1984; Ulbrich 1983). The advantages of this distribution are that it reduces to the exponential distribution when $\mu = 0$ and it allows for distributions with a wide variety of shapes including those that are either concave upward or downward on a plot of $\log[N(D)]$ versus D. RDSD shapes of this type are very apparent in experimental spectra collected at the earth's surface using various sampling devices, such as drop cameras, disdrometers, 2D optical probes, and video recorders. An early example of an investigation that displays these effects is that of Dingle and Hardy (1962). More recent examples are very prevalent; one that includes extensive analysis of tropical raindrop spectra is that of Tokay and Short (1996). Such data usually consist of samples of short duration (e.g., 1 min). However, Levin et al. (1991) find such effects in disdrometer data even when averaged for periods as long as 2 h. It might also be stated that these effects may not be representative of RDSD shapes observed aloft with radar. However, shape effects similar to that found with surface instruments also exist in RDSDs aloft as is apparent from the early work of Rogers and Pilié (1962) and Caton (1966), who acquired Doppler radar spectra of rain at vertical incidence. They are also apparent in the analysis by Atlas et al. (2000) of 2D optical probe data acquired aloft during TOGA COARE by an National Center for Atmospheric Research (NCAR) *Electra* aircraft. The gamma distribution has properties that provide an accurate representation to be made of these shape effects. In addition, integral rainfall parameters generally are simple to calculate with the gamma distribution.

In spite of its advantages there are features that make this function troublesome. First, the coefficient N_0 no longer has the simple units as the equivalent coefficient in the exponential distribution and, in fact, includes the parameter μ. As a result N_0 and μ are strongly correlated as shown by Ulbrich (1983), but this correlation is demonstrated by Chandrasekar and Bringi (1987) not to imply any physical basis. To avoid this problem they rewrite the distribution in the form

$$N(D) = \frac{N_T}{\Gamma(\mu + 1)\Lambda^{\mu+1}} D^\mu \exp(-\Lambda D), \quad (10.4)$$

where N_T is the total concentration of raindrops, and recommend using N_T, μ, and Λ as the distribution parameters. Note that N_T can be written as

$$N_T = N_0 \Gamma(\mu + 1)/\Lambda^{\mu+1} \qquad (10.5)$$

so that this form requires that $\mu > -1$. Values of $\mu \le -1$ will produce results for N_T that are undefined. Willis (1984) normalized the distribution so that it assumed the form

$$N(D) = N_G(D/D_0)^\mu \exp(-\Lambda D/D_0), \qquad (10.6)$$

where the coefficient N_G is expressed in terms of the liquid water concentration W by

$$N_G = (6W\Lambda^{4+\mu})/(\pi \Gamma(4 + \mu)). \qquad (10.7)$$

In similar fashion, Testud et al. (2001) normalized the gamma distribution using W and the mean volume diameter D_m by writing it as

$$N(D) = N_0^* F_\mu(X), \qquad (10.8)$$

where $X = D/D_m$, $N_0^* = 4^4 W/(\pi D_m^4)$,

$$F_\mu(X) = C_\mu X^\mu \exp(-(4 + \mu)X), \qquad (10.9)$$

and $C_\mu = (\Gamma(4)(4 + \mu)^{4+\mu})/(\Gamma(4 + \mu)4^4)$. As long as values of N_T are not required, the latter two normalizations will yield useful results for $\mu > -3$.

Another analytical expression for the drop-size distribution that represents some types of experimental drop-size data fairly well is that of Imai (1964). It may be written

$$N(D) = 6\sqrt{\frac{h}{\pi^3}} W D^{-3} \exp[-h(D - D_0)^2], \quad (10.10)$$

where W is the liquid water concentration, D_0 is the median volume diameter, and h is a parameter. The distribution may be easily normalized such that it has the form

$$N^*(D) = C\Delta^{-3} \exp(-(\Delta - \Delta_0)^2), \qquad (10.11)$$

where $C = (6h^3 W)/\pi^{3/2}$ and $\Delta = \sqrt{h}D$. For $\sqrt{h}D_0 > \sqrt{6}$ the normalized curves show one maximum and one minimum and for $\sqrt{h}D_0 < \sqrt{6}$ the curves show no extrema but one inflection point. The form of the DSD with $\sqrt{h}D_0 < \sqrt{6}$ is similar to that displayed by the experimental data of Marshall and Palmer (1948), whereas the form for $\sqrt{h}D_0 > \sqrt{6}$ is similar to that of the equilibrium distributions of Hu and Srivastava (1995). The disadvantage of this distribution is that it implies that the mass is distributed normally with respect to diameter.

None of these "normalization" methods removes the strong correlation among the DSD parameters that is commonly observed in experimental data. These parameters cannot, therefore, be considered strictly independent, a property of distribution parameters that is highly desirable in remote sensing algorithms. In an investigation of methods to avoid this problem, Haddad et al. (1996) found parameters of the gamma distribution that are negligibly correlated and may therefore be considered independent. One of these is chosen to be the rainfall rate R and the other two, D' and s', are defined in terms of R, D_m (the mean volume diameter), and s_m (the relative standard deviation of the mass spectrum) as

$$D' = D_m R^{-0.155} \quad \text{and} \qquad (10.12)$$

$$s' = s_m D_m^{-0.2} R^{0.031} \exp(0.017 R^{0.74}). \qquad (10.13)$$

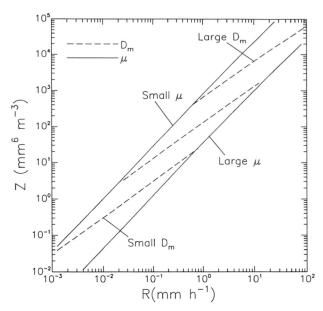

FIG. 10.2. The simplified rain parameter diagram for a gamma RDSD from Ulbrich and Atlas (1998). The solid lines are isopleths of the gamma distribution shape parameter μ, and the dashed lines are isopleths of the mean volume diameter D_m. The μ isopleths have slopes equal to that of the Z–R relation that applies to the dataset under consideration.

The values of the coefficients and exponents in these expressions were determined for a set of raindrop-size distribution data acquired by a Joss disdrometer near Darwin, Australia, in the summers of 1988/89 and 1989/90. It remains to be demonstrated that these expressions apply to raindrop distributions in general.

Ulbrich and Atlas (1998) have shown a version of the rain parameter diagram that employs the gamma RDSD and, as in the Atlas–Chmela diagram, consists of a logarithmic plot of Z versus R but with isopleths of μ and D_m (the mean volume diameter) for a gamma distribution. It is shown in Fig. 10.2. The μ isopleths are parallel to the Z–R relation, which applies to the dataset of interest, and the D_m isopleths have slopes equal to 1. For a set of experimental Z–R data, the diagram shows how the shape μ and breadth D_m vary within the dataset. It does not require the assumption of a specific distribution shape but does require knowledge of the empirical Z–R relation for the dataset in question.

Although all of the analytical expressions given above provide useful and usually accurate approximations to most observed drop-size distributions, not all experimental distributions behave as smoothly as these functions. When such behavior is evident it is difficult, perhaps impossible, to describe the distribution accurately using mathematical formulas similar to those given above. An example may be found in the work of Steiner and Waldvogel (1987) who found multiple distribution peaks at raindrop diameters of approximately 0.7, 1.0, 1.9, and (perhaps) 3.2 mm. Similar measurements by

de Beauville et al. (1988) in a maritime tropical environment find peaks at 0.6, 1.0, 1.8, and 3 mm. Each of these investigations support the predictions of earlier theoretical work by several investigators who predict trimodal distributions in equilibrium rainfall, that is, rainfall in which the actions of collisional breakup and coalescence produce equilibrium size distributions. Examples of such calculations are found in the work of Valdez and Young (1985), List et al. (1987), Hu and Srivastava (1995), and Brown (1989). Hu and Srivastava (1995) also find that when evaporation is included an approximate equilibrium RDSD is still possible. Valdez and Young (1985) find peaks at 0.268, 0.79, and 1.76 mm; List et al. (1987) find them at 0.24, 0.87, and 2.0 mm; and Hu and Srivastava (1995) determine peaks to be at 0.2, 0.9, and 1.5 mm. None of the experimental work has been able to resolve diameters as small as the first of these peaks, but the second and third are very similar to the experimentally determined values. Disagreements between the experimental and theoretical values may be due to insufficient time for the observed clouds to reach an equilibrium situation. For example, the clouds observed by de Beauville et al. (1988) were all showers of duration less than 10 min. In spite of the peakedness of these spectra, Steiner and Waldvogel (1987) show that they have little influence on Z–R relations used in the remote measurement of rainfall. Integral parameters such as Z and R are not affected very much by the presence of these peaks in the distribution so that one may have confidence that the analytical approximations described above will be adequate representations of the actual distribution except for the finescale details.

Some of the earliest work on equilibrium distributions was done by Srivastava (1971), who considered the evolution of drop-size distributions under the action of drop coalescence and spontaneous breakup. He found computed distributions with considerable shape tending toward downward concavity but with numbers of small diameter drops greater than could be described by a downward concave gamma distribution. The important characteristic feature of the distributions corresponding to different rainfall rates was that they were essentially parallel to one another. In other words, the distributions corresponding to different rainfall rates were all multiples of one another. This result has also been found by Donaldson (1984) and List et al. (1987) using a similar theoretical model that includes drop breakup and coalescence. List (1988) therefore writes the equilibrium distribution as a product of the rainfall rate R and a shape function, that is,

$$f_N(D, R) = R\psi_N(D), \qquad (10.14)$$

which means that all integral rainfall parameters can be written as the product of a constant and the rainfall rate. For example, the total number concentration is $N_T = C_N R$, the liquid water concentration is $W = C_M R$, the reflectivity factor is $Z = C_Z R$, etc., where C_N, C_M, and

TABLE 10.3. Definitions of various raindrop integral parameters.

Symbol	Parameter	p	a_P
Z	Reflectivity factor	6	10^6 mm^6 cm^{-6}
W	Liquid water concentration	3	0.524 g cm^{-3}
R	Rainfall rate	3.67	33.31 mm h^{-1} m^3 cm$^{-3.67}$
N_T	Total number concentration	0	1.0

C_Z are constants. Note that the latter relation implies that in equilibrium rainfall a linear relation exists between Z and R. Atlas and Ulbrich (2000) show that a direct proportionality between Z and R also implies that the median volume diameter D_0 must be constant in time during the rainfall event. Although this behavior has been observed infrequently in nature with surface disdrometer data, Atlas and Ulbrich (2000) show spectra acquired aloft for storms in TOGA COARE for which the constancy of D_0 and proportionality of Z and R are evident. These spectra closely resemble the equilibrium spectra of List et al. (1987) and Hu and Srivastava (1995) and display a tendency for a peak occurring near a diameter of 1.5 mm, in agreement with the theoretical predictions.

In this work the gamma RDSD will be employed to deduce the behavior of integral parameters implied by the values of the coefficient A and exponent b in the Z–R law. This is done in the manner described by Ulbrich (1983), which uses A and b to compute the gamma distribution parameters and proceeds in the following way. Any integral parameter P may be expressed in terms of the gamma distribution as

$$P = a_P \int_0^\infty D^p N(D) \, dD$$

$$= a_P \frac{\Gamma(p + \mu + 1)}{(3.67 + \mu)^{p+\mu+1}} N_0 D_0^{p+\mu+1}. \quad (10.15)$$

Note that the method assumes $D_{\min} = 0$, $D_{\max} \to \infty$. The effects on integral parameters and empirical relations of assuming values for $D_{\min}$ and $D_{\max}$ different from these values have been investigated by Ulbrich (1985, 1992, 1993). For parameters Z, W, R, and N_T the values of p and a_p are listed in Table 10.3.

In the above it has been assumed that the fall speeds of the drops in still air are given by a power law in terms of the diameter as given by Atlas and Ulbrich (1977), that is,

$$v(D) = 17.67 D^{0.67}, \quad (10.16)$$

where $v(D)$ is in meters per second and D is in centimeters. This is a good approximation to the raindrop fall speeds at sea level and is sufficiently accurate for the present purposes.

Consider a pair of integral parameters P and Q. The theoretical expression for Q is the same as that shown

TABLE 10.4. Relations between integral rainfall parameters found from Eqs. (10.18)–(10.20) assuming a gamma RDSD.

P–Q relation	Coefficient		Exponent
$Z = AR^b$	$A = \dfrac{10^6 \Gamma(7 + \mu) N_0^{-2.33/(4.67+\mu)}}{[33.31\Gamma(4.67 + 1)]^{(7+\mu)/(4.67+\mu)}}$		$b = \dfrac{7 + \mu}{4.67 + \mu}$
$D_0 = \varepsilon R^\delta$	$\varepsilon = \dfrac{3.67 + \mu}{[33.31 N_0 \Gamma(4.67 + 1)]^{1/(4.67+\mu)}}$		$\delta = \dfrac{1}{4.67 + \mu}$
$N_T = \xi R^\eta$	$\xi = \dfrac{\Gamma(1 + \mu) N_0^{3.67/(4.67+\mu)}}{[33.31\Gamma(4.67 + \mu)]^{(1+\mu)/(4.67+\mu)}}$		$\eta = \dfrac{1 + \mu}{4.67 + \mu}$
$W = \zeta R^\kappa$	$\zeta = \dfrac{\pi \Gamma(4 + \mu) N_0^{0.67/(4.67+\mu)}}{6[33.31\Gamma(4.67 + \mu)]^{(4+\mu)/(4.67+\mu)}}$		$\kappa = \dfrac{4 + \mu}{4.67 + \mu}$

above for P with a_Q and q substituted for a_p and p, respectively:

$$Q = a_Q \int_0^\infty D^q N(D) \, dD$$

$$= a_Q \frac{\Gamma(q + \mu + 1)}{(3.67 + \mu)^{q+\mu+1}} N_0 D_0^{q+\mu+1}. \quad (10.17)$$

Elimination of D_0 between P and Q results in the form

$$P = \alpha Q^\beta, \quad (10.18)$$

where

$$\beta = \frac{p + \mu + 1}{q + \mu + 1} \quad \text{and} \quad (10.19)$$

$$\alpha = \frac{a_P \Gamma(p + \mu + 1) N_0^{1-\beta}}{[a_Q \Gamma(q + \mu + 1)]^\beta}. \quad (10.20)$$

These equations can be inverted to obtain expressions for μ and N_0, that is,

$$\mu = \frac{p - \beta q}{\beta - 1} - 1 \quad \text{and} \quad (10.21)$$

$$N_0 = \left\{ \frac{\alpha \left[a_Q \Gamma\left(\dfrac{p - q}{\beta - 1}\right) \right]^\beta}{a_P \Gamma\left[\dfrac{\beta(p - q)}{\beta - 1}\right]} \right\}^{1/(1-\beta)}. \quad (10.22)$$

This approach assumes that the parameter N_0 is constant or at least slowly varying with R. In this work we consider the examples of P–Q relations of the form $Z = AR^b$, $D_0 = \epsilon R^\delta$, $N_T = \xi R^\eta$ and $W = \zeta R^\kappa$. The coefficient A and exponent b in an empirical Z–R relation are used to find μ and N_0 from which the coefficients and exponents for the remaining three relations are calculated. The coefficients and exponents for all these relations are listed in the Table 10.4.

The corresponding expressions for N_0 and μ in terms of A and b are

$$N_0 = \left\{ \frac{A\left[33.31\Gamma\left(\frac{2.33}{b-1}\right)\right]^b}{10^6\Gamma\left(\frac{2.33b}{b-1}\right)} \right\}^{1/(1-b)} \quad \text{and} \quad (10.23)$$

$$\mu = \frac{7 - 4.67b}{b - 1}. \quad (10.24)$$

It should be noted that there are a couple of instances in which this approach will not yield useful results. First of all, if $\mu \le -1$, then N_T is undefined. Second, if b is very close to 1 (as in the case of the equilibrium drop-size distribution), then μ becomes very large. In neither of these cases are the results found for the coefficients and exponents in the above table considered to be physically meaningful.

These results may now be used to illustrate the effects on the coefficient A and exponent b of each of the various physical process listed in Table 10.1. For the purposes of illustration the RDSD before modification is shown as an exponential distribution on most of the diagrams that follow and is represented by a straight line. However, it is clear that the RDSD could have any shape before modification. In describing the changes that take place, the equation for N_T in terms of N_0, μ, and D_0 is used. From Eq. (10.15) the expression for N_T may be shown to have the form

$$N_T = \frac{N_0 D_0^{1+\mu}\Gamma(1+\mu)}{(3.67 + \mu)^{1+\mu}}. \quad (10.25)$$

Wilson and Brandes (1979) provided qualitative microphysical and kinematic influences on Z–R relationships, presented in Table 10.1. Here, with the added benefit of the additional RDSD formulations, this can be expanded and tie the different physical processes to the parameters presented in Table 10.3 and the Z–R relation parameters. The discussion is presented for each factor acting alone, assuming everything else is held constant. Admittedly, it is rarely the case in reality; however, it serves the purpose of understanding the various processes that combine to form the actual RDSDs.

a. Coalescence (Fig. 10.3a)

Modification of the RDSD by coalescence alone decreases the numbers of small diameter drops and increases those of the larger drops. Consequently, D_0 must increase and the total number concentration of drops N_T must decrease. The process also increases μ; the amount by which it would change depends on the efficiency of the coalescence process. The result is a decrease in N_0 and a consequent increase in A and small decrease in b. There would be an approximate parallel shift in the Z–R relation on the RAPAD upward and to the left, perpendicular to the D_0 isopleths. It would be necessary

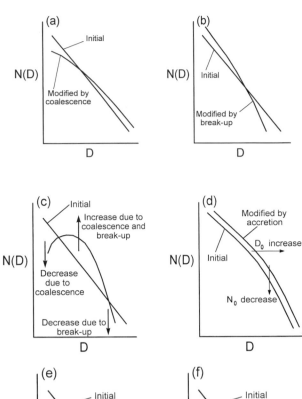

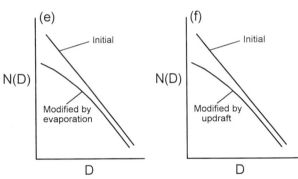

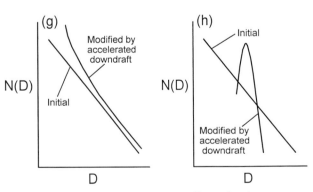

FIG. 10.3. Schematic depictions of the effects of various processes on the shape of the RDSD. The processes illustrated are (a) raindrop coalescence, (b) raindrop breakup, (c) coalescence and breakup acting simultaneously, (d) accretion of cloud droplets, (e) evaporation, (f) an updraft, (g) an accelerated downdraft, and (h) size sorting.

to use a RAPAD for a different value of μ before and after modification.

b. Breakup (Fig. 10.3b)

Modification of the RDSD by breakup alone increases the numbers of small diameter drops and decreases the numbers of large diameter drops. There must be a consequent decrease in D_0 and an increase in N_T. Accordingly, N_0 must increase. There is probably a small change in μ with a tendency toward a decrease. The end result is a decrease in A and a small increase (or perhaps no change) in b. Because of the small change in μ we may use the same RAPAD before and after modification. There is therefore an approximate shift in the Z–R relation on the RAPAD downward and to the right, approximately parallel to the D_0 isopleths.

c. Coalescence and breakup combined (Fig. 10.3c)

Breakup is more important at the larger sizes, coalescence more important at small sizes (insofar as numbers are concerned). Both processes acting together increase μ substantially. The degree to which μ changes depends on the relative strengths of the two processes in real cloud situations. This requires using a different RAPAD before and after modification; that is, the isopleths will shift. The apparent increase in μ will decrease b. What will happen to A depends on which of the two processes is predominant.

d. Accretion (Fig. 10.3d)

Since accretion of cloud particles by raindrops acts to increase the sizes of all particles without increasing their numbers, then N_T must remain unchanged. If all drops grow at the same rate, this implies a shift of the RDSD parallel to itself to larger diameters with a consequent increase in D_0. Since N_T remains constant, N_0 must decrease. The result is an increase in A and probably little change in b. Since μ is unchanged, the same RAPAD before and after modification can be used. Therefore, the Z–R law will be shifted parallel to itself upward on the RAPAD. In reality, larger raindrops have greater terminal velocity and therefore have larger growth rate by accretion. When also adding to consideration the creation of new small drops, the process mirrors the effect of evaporation, as discussed next.

e. Evaporation (Fig. 10.3e)

The presence of evaporation acting alone will result in a greater loss of the numbers of small diameter particles than large drops. Consequently, N_T is not constant and must decrease. There must also be a substantial change in the shape of the RDSD so that μ increases. Also, D_0 must increase. The result is a decrease in N_0 and an increase in A. In addition, since μ increases, b must decrease. Since there is a change in μ it is necessary to use a different RAPAD before and after modification.

f. Updraft (Fig. 10.3f)

Because of the retarding influence of gravity preferably on the larger drops, an updraft eliminates the smallest precipitation particles from the RDSD at the lower levels. This is especially true in thunderstorms, where the smallest particles are deposited in the anvil or are carried aloft to other regions where they fall out later. The effect on the RDSD is therefore the same as evaporation.

g. Downdraft (Fig. 10.3g)

A downdraft would increase the downward flux of particles of all diameters. A change in the shape of the RDSD is likely, but the details of the changes are not certain. One possible modification is shown in Fig. 10.3g. Should such changes in RDSD shape occur it would necessitate the use of a different RAPAD before and after modification.

h. Size sorting (Fig. 10.3h)

Size sorting tends to make the RDSD much narrower, which means μ must increase substantially. Therefore, b must decrease. Obviously N_T must decrease, but what happens to D_0 depends on which segment of the precipitation streamer is being observed. So A would either increase or decrease depending on what happens to D_0. It is likely that the decrease in N_T will dominate the change in D_0 so that A would increase, but this is not certain. Because of the dramatic change in μ a different RAPAD would have to be used before and after modification.

4. Conceptual model of rain formation

In the present work a review is presented of the physical considerations enumerated above and a test is made against the wealth of RDSD-based Z–R relations that have been reported during the last five decades. Our departure point is the equilibrium drop-size distribution (DSDe). In that regard the theoretical calculations of the DSDe of Hu and Srivastava (1995) are used in this work. We applied the method of moments [as described by Tokay and Short (1996)] to the DSDe data of Hu and Srivastava (1995) and found values of $\mu = 9$, $D_0 = 1.76$ mm, and a Z–R relation of the form $Z = 600\ R$. A RAPAD corresponding to this value of μ is shown in Fig. 10.4, in which a point corresponding to the DSDe is shown as the large filled circle. Also shown are the four Z–R relations depicted by Atlas and Chmela (1957) on their RAPAD. It is notable that, when extended to

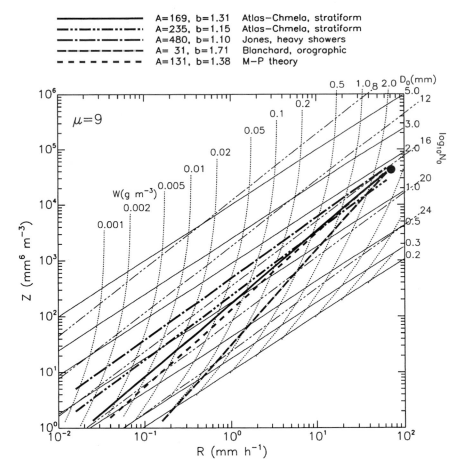

FIG. 10.4. Rain parameter diagram for a gamma distribution with $\mu = 9$. Isopleths are the same as in Fig. 10.1. Also shown are the four Z–R relations displayed by Atlas and Chmela (1957). The large filled circle corresponds to the theoretical equilibrium RDSD of Hu and Srivastava (1995) for which $\mu = 9$ and $D_0 = 1.76$ mm. It is notable that the four Z–R relations are close to converging to the point for the equilibrium distribution.

large rainfall rates, these four relations tend to converge to the DSDe.

a. Equilibrium DSD and Z–R relations

The DSDe is the drop-size distribution that would be developed in a rain shaft that falls a sufficiently long time for the rates of drop merging and breakup processes to reach equilibrium, assuming no gain or loss of raindrops to other processes. The time required for reaching DSDe is shorter for greater R, because of the greater W and respectively shorter time between drop interactions. Hu and Srivastava (1995) have calculated that reaching DSDe from initial Marshall–Palmer exponential RDSD would take about 10 min for $R = 90$ mm h^{-1}, but the main features of DSDe would appear already after 2.5 min. The time for reaching DSDe becomes longer linearly with R^{-1}. Hu and Srivastava (1995, p. 1768) stated that "for heavy rainfall rates, say 50 mm h^{-1}, approximate equilibrium between collisional processes may perhaps be expected within the usual lifetime of a con-

vective shower." Supporting observations in heavy tropical rain showers were reported by Zawadzki and de Agostinho Antonio (1988). Observations at extra-tropical rain showers showed considerable deviations from DSDe (Carbone and Nelson 1978; Sauvageot and Lacaux 1995).

Because DSDe is independent of R, D_0 is constant at 1.76 mm, and R is linear with N_T. This means that the exponent in the Z–R power-law relation is unity (List 1988), and the Z–R relation is simply $Z = 600 R$ (after Hu and Srivastava 1995). An exponent greater than 1 means that D_0 is increasing with R. During the growth phase of the precipitation before reaching DSDe, D_0 does increase with R. The conversion of cloud droplets to precipitation occurs by forming small precipitation particles that increase in size with the progress of their collection of the cloud water and so increasing R, until the drops become sufficiently large for breakup to compensate their additional growth. Because breakup does not occur in convective ice precipitation, that is, graupel and hail, no equilibrium is expected there and the equivalent melted hydrometeors would have ever-increasing D_0 with R. This is why rain formed from the melting of hail can reach extreme reflectivities, which translate to impossibly high R when applying to the hail Z–R relations for rainfall.

b. Evolution of warm rain

In a hypothetical rising cloud column with active co-alescence, the initial dominant process would be widening of the cloud drop-size distribution into large concentrations of drizzle drops; the drizzle continues to coalesce with other drizzle and cloud drops into raindrops, which will continue to grow asymptotically to D_{0e}. Therefore, during the growth phase of precipitation R increases with D_0, and this would increase the exponent b. Ideally, for rainfall with drops that fall from cloud top while growing, R would increase with the fall distance from the cloud top, mainly by growth of the falling drops due to accretion and coalescence, and to a lesser extent by addition of new small rain drops, until R becomes sufficiently large for breakup to become significant. Shallow orographic clouds can present conditions such as some distance below the tops of convective clouds. Therefore, similar evolution of R can be observed on a mountain slope, such as documented by Fujiwara (1965). Different values of R near cloud top or in shallow orographic clouds can come mainly from changing N_T, because the drop size is bounded by the limited vertical fall distance along which they can grow. This would cause orographic precipitation to have small coefficient A, and more so with shallower clouds and stronger orographic ascent, because the stronger rising component supplies more water for the production of many small raindrops not too far below cloud top, which are manifested as a larger R.

c. Evolution of cold rain

Microphysically "continental" clouds are characterized by narrow cloud drop-size distributions and, therefore, by having little drop coalescence and warm rain. Most raindrops originate from melting of ice hydrometeors that are typically graupel or hail in the convective elements, and snowflakes in the mature or stratiform clouds. Graupel and hail particles grow without breakup while falling through the supercooled portion of the cloud, and continue to grow by accretion in the warm part of the cloud, where they melt. Large melting hailstones shed the excess meltwater in the form of an RDSD about which little is known. The shedding stops when the melting particles approach the size of the largest stable raindrops, which are later subject to further breakup due to collisions with other raindrops. In fact, new raindrop formation is limited only to the breakup of pre-existing larger precipitation particles. Therefore, we should expect that in such clouds there would be, for a given R, a relative dearth of small drops and excess

of large drops compared to microphysically "maritime" clouds with active cloud drop coalescence. Deep continental convective clouds would therefore initiate the precipitation by forming large drops that, with maturing, approach DSDe from above. This is in contrast with the approach from below for maturing maritime RDSD.

Recent satellite studies (Rosenfeld and Lensky 1998) have shown that microphysically maritime clouds are associated typically with a "rain-out" zone; that is, the fast conversion of cloud water to precipitation causes the convective elements to lose water to precipitation while growing. This leaves less water carried upward to the supercooled zone, so that weaker ice precipitation can develop aloft. Williams et al. (2002) have recognized this as a potential cause to the much greater occurrence of lightning in continental compared to maritime clouds. Williams et al. (2002) noted that frequent lightning occurred also in very clean air during high atmospheric instability, probably because the strong updraft leaves little time to the formation of warm rain and carries the large raindrops that manage to form up to the supercooled levels of the clouds, where they freeze and participate in the cloud electrification processes.

This difference between continental and maritime clouds means that mostly warm rain would fall even from the very deep maritime convection, which reaches well above the freezing level, whereas precipitation from continental clouds would originate mainly in ice processes. Therefore, the expected difference in RDSD between microphysically maritime and continental clouds is expected to exist also for the deepest convective clouds that extend well into the subfreezing temperatures.

5. Proposed method of classifying Z–R relations

Equipped with this conceptual model, now we can turn our attention to actual measurements of RDSDs and their Z–R relations, which can be related to the precipitation-forming processes as discussed above. These Z–R relations are provided in Table 10.5 and are classified according to combinations of the categories of microphysically maritime and continental, convective, stratiform, and orographic. The values of A and b are those listed in the source of the Z–R relation, whereas the values of the RDSD parameters N_0 and μ were calculated from A and b using Eqs. (10.23) and (10.24), respectively. The coefficients and exponents (ε, δ) and (ζ, κ) in the D_0–R and W–R relations, respectively, were calculated using the expressions in Table 10.3. Also shown for reference in Table 10.5 are values of $D_0(10)$, $D_0(30)$, $W(10)$, and $W(30)$, the values of D_0 and W at $R = 10$ and 30 mm h^{-1}. It should be recognized that the D_0–R and W–R relations derived in this way are strictly theoretical and are only approximations to those that might be found from empirical analyses of the data from which the Z–R relations were found. However, in

TABLE 10.5. RDSD parameters μ and N_0 deduced from values of A and b in $Z = AR^b$ for the sources shown in the second column. Also shown are the theoretical values of the coefficients and exponents (ε, δ), and (ζ, κ) in the corresponding D_0–R and W–R relations of the form $Y = cX^a$, where D_0 is the median volume diameter (cm) and W is the liquid water concentration (g m^{-3}). The reference number in the first column refers to the description of the data given in part (b) of the table. Values of D_0 and W for $R = 10$ and 30 mm h^{-1} are listed in the columns labelled $D_0(10)$, $D_0(30)$, $W(10)$, and $W(30)$, respectively.

a) Parameters

Ref.	Source	A	b	μ	N_0	ε	δ	ζ	κ	$D_0(10)$	$D_0(30)$	$W(10)$	$W(30)$
1	Joss and Waldvogel (1970)	830	1.50	−0.010	0.634D+04	0.148	0.215	0.055	0.856	0.243	0.308	0.393	1.007
2	Foote (1966)	646	1.46	0.395	0.321D+05	0.137	0.197	0.058	0.868	0.216	0.269	0.426	1.106
3	Rinehart (2002)	429	1.59	−0.721	0.289D+04	0.104	0.215	0.069	0.830	0.187	0.246	0.466	1.160
4	Sims (1964)	446	1.43	0.749	0.201D+06	0.120	0.185	0.063	0.876	0.183	0.225	0.478	1.251
5	Petrocchi and Banis (1980)	316	1.36	1.802	0.112D+08	0.109	0.155	0.068	0.896	0.156	0.184	0.536	1.435
6	Sauvageot (1994)	425	1.29	3.364	0.458D+09	0.130	0.124	0.061	0.917	0.174	0.199	0.500	1.369
7	Ulbrich et al. (1999)	261	1.43	0.749	0.700D+06	0.095	0.185	0.074	0.876	0.146	0.178	0.557	1.459
8	Maki et al. (2001)	232	1.38	1.462	0.892D+07	0.094	0.163	0.075	0.891	0.137	0.164	0.583	1.551
9	Tokay et al. (1995)	175	1.37	1.627	0.324D+08	0.084	0.159	0.081	0.894	0.121	0.144	0.634	1.691
10	Tokay and Short (1996)	139	1.43	0.749	0.303D+07	0.073	0.185	0.089	0.876	0.111	0.136	0.668	1.749
11	Stout and Mueller (1968)	126	1.47	0.287	0.772D+06	0.068	0.202	0.093	0.865	0.108	0.134	0.681	1.761
12	Stout and Mueller (1968)	146	1.42	0.878	0.418D+07	0.075	0.180	0.087	0.879	0.113	0.138	0.660	1.734
13	Tokay et al. (1995)	335	1.37	1.627	0.560D+07	0.111	0.159	0.067	0.894	0.160	0.190	0.526	1.403
14	Tokay and Short (1996)	367	1.30	3.097	0.344D+09	0.122	0.129	0.063	0.914	0.163	0.188	0.520	1.420
15	Stout and Mueller (1968)	226	1.46	0.395	0.315D+06	0.088	0.197	0.078	0.868	0.138	0.171	0.577	1.496
16	Ulbrich and Atlas (2001)	120	1.43	0.749	0.426D+07	0.068	0.185	0.093	0.876	0.104	0.128	0.697	1.824
17	Ulbrich and Atlas (2001)	203	1.46	0.395	0.398D+06	0.084	0.197	0.081	0.868	0.132	0.164	0.595	1.543
18	Jorgensen and Willis (1982)	287	1.27	3.960	0.110D+11	0.112	0.116	0.067	0.922	0.146	0.166	0.562	1.549
19	Jorgensen and Willis (1982)	301	1.38	1.462	0.450D+07	0.105	0.163	0.070	0.891	0.153	0.183	0.541	1.439
20	Fujiwara and Yanase (1968)	240	1.48	0.184	0.141D+06	0.088	0.206	0.078	0.862	0.142	0.178	0.564	1.455
21	Fujiwara and Yanase (1968)	88	1.28	3.651	0.291D+12	0.067	0.120	0.095	0.919	0.088	0.101	0.788	2.164
22	Fujiwara and Yanase (1968)	48	1.11	16.512	0.436D+34	0.058	0.047	0.105	0.968	0.065	0.068	0.976	2.828
23	Blanchard (1953)	31	1.71	−1.388	0.159D+05	0.031	0.305	0.155	0.796	0.062	0.086	0.967	2.318

TABLE 10.5. (*Continued*)

(b) Source and notes.

Continental

1. Joss and Waldvogel (1970)	Thunderstorms. 25 days total, Locarno, Switzerland. Disdrometer data.
2. Foote (1966)	Mountain thunderstorm in AZ. Filter paper measurements. 62 spectra for 37 storms.
3. Rinehart (2002)	Grand Forks, ND during several autumn seasons. Filter paper measurements.
4. Sims (1964)	Thundershowers, 1963. ISWS drop camera data.

Moderate continental

5. Petrocchi and Banis (1980)	Thunderstorm, Norman, OK. Disdrometer data.

Tropical continental

6. Sauvageot (1994)	Tropical squall line. Congo. Disdrometer data.
7. Ulbrich et al. (1999)	Average of seven afternoon thunderstorms in Arecibo, PR. Disdrometer data.
8. Maki et al. (2001)	Darwin, Australia, 1997–98. 15 squall lines, all stages. Disdrometer data.

Tropical maritime

9. Tokay et al. (1995)	Tropical maritime, coastal, convective, Darwin, Australia. Disdrometer data.
10. Tokay and Short (1996)	Tropical convective, equatorial, maritime, TOGA COARE, Disdrometer data.
11. Stout and Mueller (1968)	Marshall Islands, trade wind cumulus, warm rain, maritime. ISWS drop camera data.
12. Stout and Mueller (1968)	Marshall Islands, showers, equatorial maritime. ISWS drop camera data.
13. Tokay et al. (1995)	Darwin, Australia. Stratiform, coastal, tropical maritime. Disdrometer data.
14. Tokay and Short (1996)	TOGA COARE, stratiform, coastal, equatorial maritime. Disdrometer data.
15. Stout and Mueller (1968)	Marshall Islands, continuous rain. Equatorial maritime. ISWS drp camera data.

Tropical maritime aloft

16. Ulbrich and Atlas (2001)	TOGA COARE Convective aloft, by updraft. PMM analysis. 2DP probe data.
17. Ulbrich and Atlas (2001)	TOGA COARE Stratiform aloft, by updraft. PMM analysis. 2DP probe data.

Hurricane

18. Jorgensen and Willis (1982)	Hurricane eye wall. Aircraft data aloft. 2DP probe data.
19. Jorgensen and Willis (1982)	Hurricane rain bands. Aircraft data aloft. 2DP probe data.

Orographic

20. Fujiwara and Yanese (1968)	Orographic rain, Mount Fuji at altitude of 1300 m. Filter paper.
21. Fujiwara and Yanese (1968)	Orographic rain, Mount Fuji at altitude of 2100 m. Filter paper.
22. Fujiwara and Yanese (1968)	Orographic rain, Mount Fuji at altitude of 3400 m. Filter paper.
23. Blanchard (1953)	Orographic rain, Hawaii, Mauna Loa, altitudes between 670–920 m. Filter paper.

the vast majority of the empirical Z–R relations investigated in this work there is no information available concerning the corresponding D_0–R and W–R relations. The method employed in this work finds theoretical D_0–R and W–R relations that are used for classifying the type of rainfall even though they may not be accurate representations of the actual empirical relations. Nevertheless, Atlas (1964) has shown that the D_0–R relations found from the Z–R relations he investigated are in good agreement with those found directly from empirical analysis of experimental drop-size spectra. In Figs. 10.6, 10.9, 10.11, and 10.12 that follow, theoretical D_0–W relations are plotted for values of $\mu = -2$ and 12 and for $R = 10$ and 30 mm h^{-1}. These relations are found from Eq. (10.15) using the definitions in Table 10.3 and have the form

$$\frac{W}{R} = 0.0157(3.67 + \mu)^{0.67}\frac{\Gamma(4 + \mu)}{\Gamma(4.67 + \mu)}D_0^{-0.67}. \quad (10.26)$$

The two curves in each figure represent the range of uncertainty (for given R) in nature associated with the theoretical relations. In all cases it is seen that the the uncertainty due to such variations is small, thus adding further credibility to the method of classifying the Z–R relations.

a. Maritime–continental classification

The most fundamental classification of rain clouds can be done into maritime and continental. Cloud physicists have traditionally designated clouds as maritime and continental based on their microstructure, where maritime clouds contain small concentrations (about 50–100 cm^{-3}) of large droplets, and continental clouds contain tenfold-larger concentrations of respectively smaller droplets. Maritime clouds precipitate easily by warm processes, whereas coalescence is often suppressed in continental clouds, which often have to grow to supercooled levels to precipitate by "cold" processes, that is, involving the ice phase. Some notable differences have been documented between maritime and continental convective clouds.

- There is distinctly less supercooled water and it is limited to warmer temperatures in the tropical maritime compared to the continental clouds (Zipser and LeMone 1980; Black and Hallett 1986).
- The updraft velocities in maritime clouds are characteristically limited to below the terminal fall velocity of the raindrops, whereas no such maximum for the updraft was noted in continental convection (Zip-

R (Z) Convective: Continental - Maritime

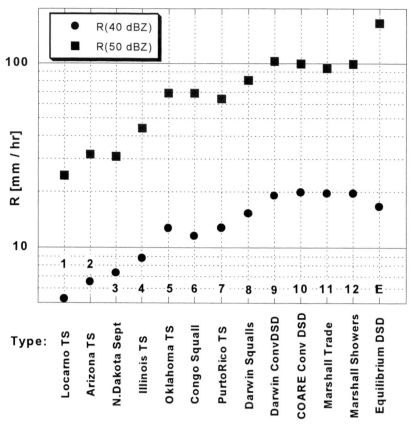

FIG. 10.5. The Z–R relations for rainfall from maritime and continental convective clouds. The rain intensities for 40 and 50 dBZ are plotted in the figure. Note the systematic increase of R for a given Z for the transition from continental to maritime clouds. The Z–R relations used in this figure are as follows:

		A	b
1.	Swiss Locarno thunderstorms, continental (Joss and Waldvogel 1970)	830	1.50
2.	Arizona mountain thunderstorms (Foote 1966)	646	1.46
3.	Grand Forks, North Dakota, in autumn (R. E. Rinehart 2002, personal communication)	429	1.59
4.	Illinois thunderstorms, continental (Sims 1964)	446	1.43
5.	Oklahoma thunderstorms, moderate continental (Petrocchi and Banis 1980)	316	1.36
6.	Congo squall line, tropical continental (Sauvageot 1994)	425	1.29
7.	Puerto Rico thunderstorms, coastal, moderate maritime (Ulbrich et al. 1999)	261	1.43
8.	Darwin Squalls, coastal, tropical maritime (Maki et al. 2001)	232	1.38
9.	Darwin convective DSD, coastal, tropical maritime (Tokay et al. 1995)	175	1.37
10.	COARE convective DSD, equatorial maritime (Tokay and Short 1996)	139	1.43
11.	Marshall trade wind cumulus, warm rain maritime (Stout and Mueller 1968)	126	1.47
12.	Marshall Showers, equatorial maritime (Stout and Mueller 1968)	146	1.42
E.	Equilibrium DSD	600	1.00

ser and LeMone 1980; Jorgensen and LeMone 1989; Zipser and Lutz 1994).
- The vertical profiles of radar reflectivity in the mixed-phase region are substantially stronger in continental than in maritime clouds. This was ascribed mainly to the greater updraft velocities in the more continental conditions (Williams et al. 1992; Rutledge et al. 1992; Zipser 1994; Zipser and Lutz 1994).

- All of these differences can potentially explain the dramatic contrast between the lightning over land and ocean that was revealed when observations of lightning from space became available (Orville and Henderson 1986).

Given the fundamental importance of the classification of clouds into maritime and continental, and in view of

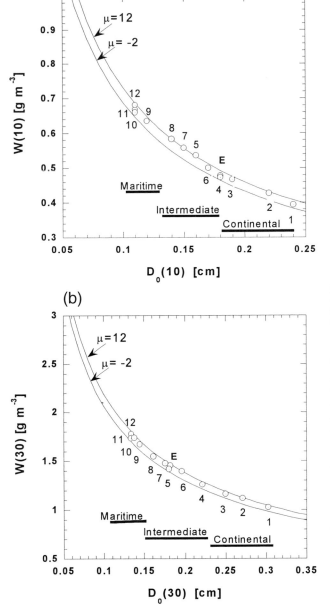

(a)

(b)

FIG. 10.6. (a) The liquid water content W vs median volume diameter D_0 for $R = 10$ mm h^{-1} of convective rainfall in maritime and continental regimes. The label of the points is according to Fig. 10.5. Note that D_0 decreases for more maritime clouds. The curves labeled with values of μ are theoretical D_0–W relations for $R = 10$ mm h^{-1}. The range of values of μ from -2 to 12 represent the range of observed variability of the D_0–W relations that would be expected in nature. Curves for values of μ greater than 12 are identical to that for $\mu = 12$. (b) The liquid water content W vs median volume diameter D_0 for $R = 30$ mm h^{-1} of convective rainfall in maritime and continental regimes. The label of the points is according to Fig. 10.5. Note that D_0 decreases for more maritime clouds. Also note that D_0 for maritime clouds is smaller than the equilibrium D_{0e} and vice versa for rainfall from continental clouds. Other details are the same as in (a).

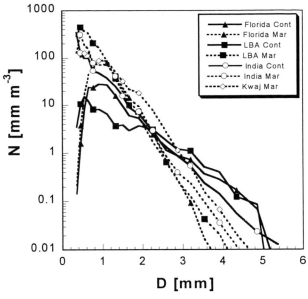

FIG. 10.7. Disdrometer-measured RDSDs of continental and maritime rainfall, as micophysically classified by VIRS overpass. The RDSD is averaged for the rainfall during ±18 h of the overpass time, and the concentrations are scaled to 1 mm h^{-1}. The disdrometers are in Florida (Teflun B), Amazon (LBA), India (Madras), and Kwajalein.

the implications to RDSD as alluded to in the section describing the conceptual model, it is expected that systematic differences in RDSD would also be ordered by this classification. The classification was done on all of the Z–R relations available to us that were based on RDSD measurements and could be related to a relative scale of estimated continentality–maritimity of the clouds. Figure 10.5 presents 12 such Z–Rs, ordered from extreme continental through moderate continental, tropical continental, moderate maritime, to equatorial maritime. According to Fig. 10.5, R increases for a given Z with increasing maritimity of the clouds, by a factor of more than 3. This substantial factor is the manifestation of major differences in the RDSD for continental and maritime clouds. Figures 10.6a,b illustrate the dependence of the drop sizes and number concentrations on the continentality of the clouds, based on the RDSD parameters that are calculated in Table 10.5. According to Fig. 10.6, D_0 increases systematically from maritime to continental clouds, reaching the greatest value in the most extreme continental clouds in Switzerland and in Arizona. At $R = 30$ mm h^{-1} (Fig. 10.6b) the drops are larger than the equilibrium DSD ($D_0 > D_{0e}$) for continental clouds, and $D_0 < D_{0e}$ for maritime clouds, in agreement with the conceptual model. This explains the substantial decrease of R for the same Z in more continental clouds, as shown in Fig. 10.5.

Is it the microstructure of the cloud or the impact of the surface properties (land or ocean) that makes the clouds continental or maritime and so produces such vastly different RDSDs? There is no clear separation between these two alternatives, because both aerosol

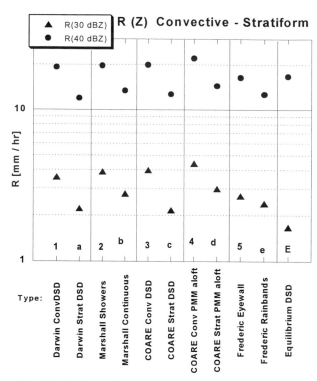

FIG. 10.8. The Z–R relations for rainfall from convective and strat-iform tropical rainfall and from a hurricane. The rain intensities for 30 and 40 dBZ are plotted in the figure. Note the systematic increase of R for a given Z for the transition from stratiform to convective clouds. Apparently hurricane rainfall is more similar to stratiform, even in the eyewall. The Z–R relations used in this figure, numbered in digits for convective and characters for stratiform, are as follows:

		A	b
1.	Darwin convective DSD, coastal, tropical maritime (Tokay et al. 1995)	175	1.37
a.	Darwin stratiform DSD, coastal, tropical maritime (Tokay et al. 1995)	335	1.37
2.	Marshall showers, equatorial maritime (Stout and Mueller 1968)	146	1.42
b.	Marshall continuous, equatorial maritime (Stout and Mueller 1968)	226	1.46
3.	COARE convective DSD, equatorial maritime (Tokay and Short 1996)	139	1.43
c.	COARE stratiform DSD, equatorial maritime (Tokay and Short 1996)	367	1.30
4.	COARE convective aloft, by updraft (Ulbrich and Atlas 2002)	120	1.43
d.	COARE stratiform aloft, by updraft (Ulbrich and Atlas 2002)	203	1.46
5.	Hurricane eyewall (Jorgensen and Willis 1982)	287	1.27
e.	Hurricane rainbands (Jorgensen and Willis 1982)	301	1.38
E.	Equilibrium DSD	600	1.00

content and updraft velocities determine cloud micro-structure. Larger aerosol concentrations make the coa-lescence slower, and greater updrafts leave less time for the progress of the coalescence, so that the product of these two factors ultimately determines the "continen-tality" of the clouds, as manifested by the evolution of cloud drop-size distribution with height in the growing

convective elements. The ultimate test for the role of cloud microstructure is comparing the RDSD of clouds at the same location, but at different times, when they possess maritime or continental microstructure.

That is exactly what is done in Fig. 10.7. The visible and infrared scanner onboard (VIRs) the Tropical Rain-fall Measuring Mission (TRMM) satellite was used to retrieve the microstructure of rain clouds over disdrome-ter sites. The clouds were classified into continental, intermediate, and maritime, using the methodology of Rosenfeld and Lensky (1998). The DSDs from the con-tinental and maritime classes during the overpass time ±18 h were lumped together and plotted in Fig. 10.7. Indeed, the continental and maritime DSDs are well separated in Fig. 10.7, with the continental clouds pro-ducing greater concentrations of large drops and smaller concentrations of small drops. A comparison between the directly measured disdrometer rainfall and the cal-culated accumulation by applying the TRMM Z–R re-lations (Iguchi et al. 2000) to the disdrometer measured Z resulted in a relative overestimate by more than a factor of 2 of the rainfall from the microphysically con-tinental clouds compared to the maritime clouds.

The evidence shows that it is mainly the cloud mi-crostructure that is responsible to the large systematic difference in the RDSD and Z–R relations between mar-itime and continental clouds. There are several possible causes for these differences, as described in the follow-ing sections, all working at the same direction.

1) EXTENT OF COALESCENCE

The cloud drop coalescence in highly maritime clouds is so fast that rainfall is developed low in the growing convective elements and precipitates while the clouds are still growing. The large concentrations of raindrops that form low in the cloud typically fall before they have the time to grow and reach equilibrium RDSD, thereby creating the rain-out zone (Rosenfeld and Len-sky 1998) less than 2 km above cloud-base height. Therefore, D_0 remains much smaller than D_{0e}, as can be seen in Fig. 10.6b.

In microphysically continental clouds with sup-pressed coalescence the cloud has to grow into large depth before it will start precipitating, by either warm or cold processes. The raindrops that fall through the lower part of the cloud grow by accretion of small cloud drops, so that they tend to break up much less than drops that grow mainly by collisions with other raindrops, as is the case for maritime clouds. This process allows D_0 to exceed D_{0e} in the growing stages of the precipitation and later to approach it from above when the raindrop collisions become more frequent with the intensification of the rainfall.

2) WARM VERSUS COLD PRECIPITATION PROCESSES

The rain-out of the maritime clouds (Rosenfeld and Lensky 1988) depletes the cloud water before reaching

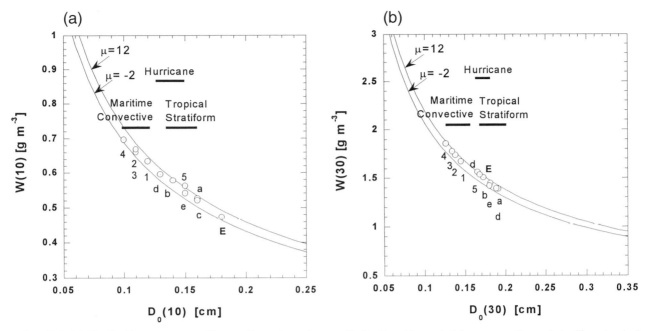

FIG. 10.9. (a) The liquid water content W vs median volume diameter D_0 for $R = 10$ mm h^{-1} from convective and stratiform tropical rainfall, and from a hurricane. The labels of the points are according to Fig. 10.8. Other details are the same as in Fig. 10.6a. (b) The liquid water content W vs median volume diameter D_0 for $R = 30$ mm h^{-1} from convective and stratiform tropical rainfall and from a hurricane. The labels of the points are according to Fig. 10.8. Note the distinct separation into convective and stratiform groups, where the convective rainfall has much smaller D_0. The hurricane rainfall fits into the stratiform group. Other details are the same as in Fig. 10.6a.

the supercooled levels (Zipser and LeMone 1980; Black and Hallett 1986), so that mixed-phase precipitation would be much less developed in the maritime clouds compared to the continental. This is manifested in the smaller reflectivity aloft in the maritime clouds (Zipser and Lutz 1994), which is a manifestation of the smaller hydrometeors that form there (Zipser 1994). In contrast, the suppressed coalescence in continental clouds leaves most of the cloud water available for growth of ice hydrometeors aloft, typically in the form of graupel and hail. These ice hydrometeors can grow indefinitely without breakup, until they fall into the warm part of the cloud and melt. The melted hydrometeors continue to grow by accretion of cloud droplets, until they exceed the size of spontaneous breakup or collide with other raindrops. Therefore, convective rainfall that originates as ice hydrometeors would have $D_0 > D_{0e}$ and would approach D_{0e} from above with maturing of the RDSD.

3) STRENGTH OF THE UPDRAFTS

Updrafts are typically stronger in more continental clouds and therefore contribute to more microphysically continental clouds and less warm rain processes, as discussed already above. In addition, stronger updrafts allow drops with greater minimal size to fall through them. In addition, stronger updrafts leave less time for forming of warm rain and rain-out, and advect more cloud water to the supercooled zone. Therefore, due to the reasons already discussed in sections 5a(1) and

5a(2), the stronger updrafts are likely to lead to precipitation with greater D_0 and smaller R for the same Z.

4) EVAPORATION

More continental environments have typically higher cloud base and lower relative humidity at the subcloud layer. Evaporation depletes preferentially the smaller raindrops and works to increase D_0.

b. Convective–stratiform classification

The mature elements of organized deep convective cloud systems often merge into widespread light to moderate rainfall area, which is called "stratiform," although it is eventually generated by convection. The classification is obvious in typical squall lines, which have a simple structure with three characteristic regions: convective, stratiform, and transition (Houze 1989). The rainfall in the convective and transition regions is formed as warm rain and graupel melt. Typically, there is more warm rain falling through the updraft, and more graupel melt falling with the downdraft toward the transition zone. The stratiform precipitation is composed typically of ice particles that were advected from the convective portion of the storm, and from aggregation of newly formed ice crystals in the moderate mesoscale updraft that develops in the merged anvils over the stratiform rainfall area. Waldvogel (1974) has shown that the onset of the stratiform precipitation is associated

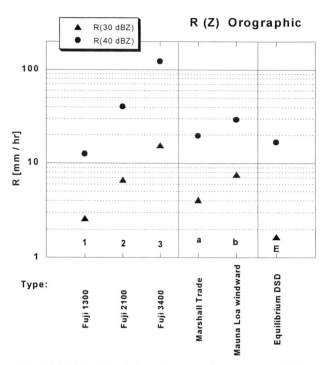

FIG. 10.10. The *Z–R* relations for warm rain over orographic barriers. The rain intensities for 30 and 40 dBZ are plotted in the figure. Note the systematic increase of *R* for a given *Z* for a greater height of the sampling location. The *Z–R* relations used in this figure, numbered in digits for Mount Fuji and characters for Hawaii, are as follows:

		A	b
1.	Mount Fuji, at height of 1300 m (Fujiwara and Yanase 1968)	240	1.48
2.	Mount Fuji, at height of 2100 m (Fujiwara and Yanase 1968)	88	1.28
3.	Mount Fuji, at height of 3400 m (Fujiwara and Yanase 1968)	48	1.11
a.	Marshall trade wind cumulus, warm rain maritime (Stout and Mueller 1968)	126	1.47
b.	Hawaii, windward side of Mauna Loa, at 670–920 m (Blanchard 1953)	31	1.71
E.	Equilibrium DSD	600	1.00

with sharp decrease of N_0, leading to greater Z for the same R. This was ascribed to the aggregation of the ice particles (Stewart et al. 1984), as evident by the bright band that is typically associated with the stratiform rainfall with large drops (Huggel et al. 1996). Substantial fraction of the stratiform rainfall can evaporate during its long fall from the melting level, so that the smaller drops evaporate preferentially, with further decrease in N_0 and increase in D_0.

The differences between convective and stratiform DSD are probably the best recognized and documented, mainly in the context of tropical rainfall. The DSD-based *Z–R*s available to us were compiled in Fig. 10.8. According to the figure, the same Z translates to R greater by a factor of 1.5–2 in maritime convective compared to stratiform rainfall.

The convective and stratiform rainfall is nicely sep-

arated in Fig. 10.9 by the values of D_0, where, as expected, D_0 is much greater for stratiform than convective rainfall. Stratiform rainfall usually occurs at $R < 15$ mm h^{-1}. The extrapolation of stratiform rain to 30 mm h^{-1} in Fig. 10.9b is rarely achieved, except for in hurricanes. The eyewall and rainbands in hurricanes are not considered normally as stratiform because R often exceeds 15 mm h^{-1}. However, according to Figs. 10.8 and 10.9, at least the *Z–R* and D_0 of hurricane Frederic had stratiform values. The eyewall value of D_0 was somewhat smaller and thus more convective than D_0 of the rainbands, but still closer to the stratiform than the convective range of values.

As expected for such a mature and deep rain system, the values of D_0 in the hurricane are not far from D_{0e}, especially for the larger rain intensities. For regular stratiform rain, D_0 is still much smaller than D_{0e}, suggesting that the raindrop coalescence and breakup do not play a major role in shaping the stratiform RDSD.

c. Orographic rainfall classification

The few available RDSDs for orographic rain are compiled in Figs. 10.10 and 10.11. According to the figures, orographic lifting of maritime air can supply a large amount of condensates, which create a large numer of small raindrops that fall to the mountain slope. This highly immature RDSD has extremely small D_0, which leads to very small Z for a given R. According to Fig. 10.10, the low-level (i.e., close to the ground) orographic enhancement can cause an underestimate of R by up to a factor of 10, when using the same *Z–R* as for the upwind rainfall. This low-level enhancement depends on effective cloud drop coalescence. Therefore, the low-level enhancement probably would be weaker in more microphysically continental clouds, with respectively less radar underestimate of its magnitude.

6. Summary

The longstanding question of RDSD and *Z–R* relationships has been revisited in this work, this time from the combined approach of rain-forming physical processes that shape the RDSD, and a formulation of the RDSD into the simplest free parameters of *R*, *W*, and D_0. It was found that the major processes that shape RDSD are, by order of practical importance as viewed by the authors,

a) cloud microstructure, with the two end members being (i) microphysically continental with small cloud drops, suppressed cloud drop coalescence and warm rain processes and with strong updrafts; and (ii) microphysically maritime with large cloud drops, active coalescence and warm rain, and weak updrafts;

b) cloud dynamics, with the two end members being (i) "convective," where precipitation elements grow in convective updrafts by coalescence and accretion

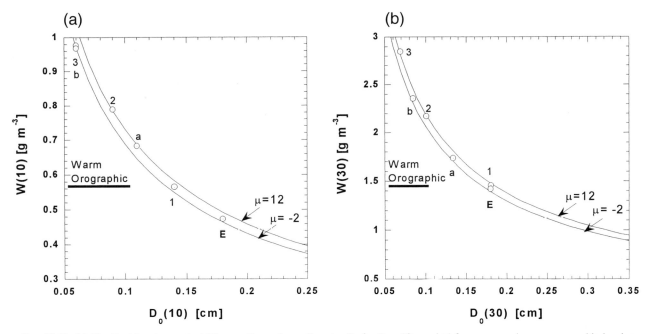

FIG. 10.11. (a) The liquid water content W vs median volume diameter D_0 for $R = 10$ mm h^{-1} from warm rain over orographic barriers. The labels of the points are according to Fig. 10.10 Other details are the same as Fig. 10.6a. (b) The liquid water content W vs median volume diameter D_0 for $R = 30$ mm h^{-1} from warm rain over orographic barriers. The labels of the points are according to Fig. 10.10. Note the systematic decrease of D_0 for increasing height. Other details are the same as in Fig. 10.6a.

of cloud droplets, both on water and ice hydrometeors; and (ii) "stratiform," where precipitation forms in updrafts <1 m s^{-1} mainly as ice crystals that aggregate into snowflakes and melt into rainfall in a radar "bright band;" and

c) orography, where low-level orographic lifting in

clouds with active cloud drop coalescence can produce extremely small D_0 compared to all other types of rainfall and, hence, a gross radar underestimate of R.

Figures 10.12a,b illustrates that this classification orders

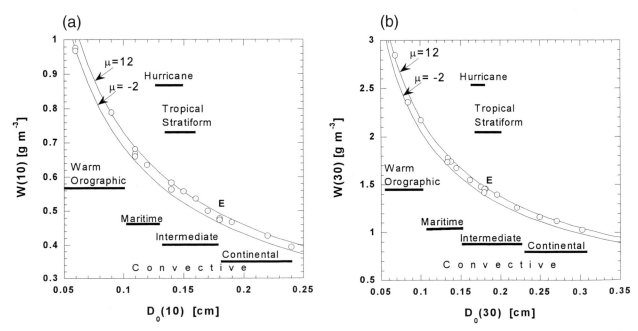

FIG. 10.12. (a) The liquid water content W for $R = 10$ mm h^{-1} for all the sampled rain types. The horizontal bars show the range of D_0 for the various rain types; E denotes the value of equilibrium RDSD. Other details are the same as in Fig. 10.6a. (b) The same as in (a) but for $R = 30$ mm h^{-1}.

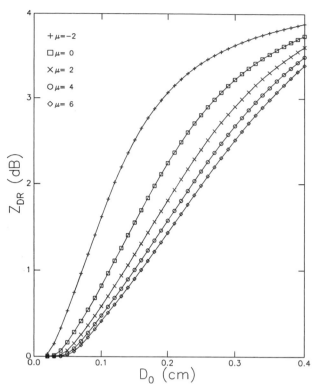

FIG. 10.13. The differential reflectivity factor Z_{DR} as a function of the median volume diameter D_0. The calculations assume that the axial ratio–diameter relation is that defined in Keenan et al. (2001) and that the raindrops are distributed according to a gamma distribution having shape parameter μ. Results are shown for values of μ = −2, 0, 2, 4, and 6. The radar wave length is assumed to be 10.7 cm and the maximum raindrop diameter D_{max} = 8 mm.

in a physically meaningful way the large range of variability of the Z–R and D_0. This classification scheme can explain a variability of R for a given Z by a factor of 1.5–2 for convective–stratiform classification, a factor of more than 3 for continental–maritime classification, and up to a factor of 10 in orographic precipitation.

The classification scheme reveals the potential for significant improvements in radar rainfall estimates by application of a dynamic Z–R relation, based on the microphysical, topographical, and dynamical context of the rain clouds while following known practices, such as

- determination of cloud microstructure by analyzing the dependence of the cloud drop effective radius on cloud-top temperature of growing convective elements, as done by Rosenfeld and Lensky (1998);
- determination of convective–stratiform separation from existence of bright band or from the horizontal structure of the reflectivity field (e.g., Churchill and Houze 1984);
- synoptic analysis of the low-level moisture orographic uplifting in clouds.

There are probably many more ways that can lead to this classification. This can serve as the foundation for

a new generation of combined cloud physics–radar algorithms that will produce variable Z–R, which will hopefully lead to improvements in the rainfall measurements, not only when using reflectivity-only data, but also with polarimetric radars. The application of polarimetric radar data for the measurement of rainfall parameters and characterization of rainfall by type is covered in detail by Bringi and Chandrasekar (2001). In this work a discussion is presented only of the differential reflectivity technique. These radars possess the capability of acquiring simultaneous estimates of the two parameters of an exponential raindrop-size distribution through measurement of the backscattered reflectivity factors at horizontal and vertical polarization, Z_H and Z_V, respectively (Seliga and Bringi 1976). One of these parameters, the median volume diameter D_0, can be determined directly from the differential reflectivity Z_{DR}, defined as

$$Z_{DR} = 10 \log_{10}\left(\frac{Z_H}{Z_V}\right),$$

which is a function of only D_0, assuming that the RDSD is a function of only two parameters, as in the case of the exponential distribution, or, for the gamma distribution, that the shape of the RDSD is known. To illustrate the latter point, the differential reflectivity Z_{DR} is plotted in Fig. 10.13 versus D_0 for several values of the parameter μ in the gamma distribution. These results were calculated for a radar wavelength in the S band and assume that the maximum diameter of the RDSD is D_{max} = 8 mm. It is evident that for D_0 values as small as about 0.5 mm Z_{DR} has values at least as large as 0.2 dB. The latter value is considered to be well within the accuracy with which Z_{DR} can be measured so it may be concluded that it can be determined with high accuracy for all the situations discussed in this work. A method has been developed by Zhang et al. (2001) that enables polarimetric radar measurements to determine all three of the parameters of a gamma RDSD using only the reflectivity factor at horizontal polarization Z_{HH} and Z_{DR}. The method involves an empirical relation between μ and Λ and estimates raindrop median size with good accuracy, at least for storms in central Florida. It may therefore be concluded that Z_{DR} can be used as a means of classifying rainfall by type even for median volume diameters as small as 0.5 mm as in the case of orographic rainfall and even in the case where the RDSD shape is unknown. Another recent method for determination of the RDSD parameters N_0, D_m, and μ is described by Bringi et al. (2003). They show that these RDSD parameters change systematically between microphysically continental and maritime clouds along the same lines that were documented and physically explained in the current study.

Spaceborne radars, however, cannot employ dual-polarization measurements, because a falling drop remains perfectly round at the horizontal cross section regardless of its size. Other radar methods, such as dual wavelength, will have to be developed for obtaining D_0. In the meantime, the classification by cloud microstructure, dynamical structure (convective–stratiform), and orographic component will have to play a major role in improving spaceborne rainfall measurements from satellites such as the Tropical Rainfall Measuring Mission.

Acknowledgments. A portion of the participation of CWU in this work was sponsored by Contract NAS 5-7782 with the National Aeronautics and Space Administration (NASA) Goddard Space Flight Center. DR's participation was funded by EURAINSAT, a shared-cost project (Contract EVG1-2000-00030) cofunded by the Research DG of the European Commission within the RTD activities of a generic nature of the Environment and Sustainable Development subprogram (5th Framework Programme).

REFERENCES

Atlas, D., 1955: The radar measurement of precipitation growth. Ph.D. thesis, Massachusetts Institute of Technology, 239 pp.

——, 1964: Advances in radar meteorology. *Advances in Geophysics,* Vol. 10, Academic Press, 317–478.

——, and A. C. Chmela, 1957: Physical–synoptic variations of drop-size parameters. *Proc. Sixth Weather Radar Conf.,* Cambridge, MA, Amer. Meteor. Soc., 21–29.

——, and C. W. Ulbrich, 1977: Path- and area-integrated rainfall measurement by microwave attenuation in the 1–3 cm band. *J. Appl. Meteor.,* **16,** 1322–1331.

——, and ——, 2000: An observationally based conceptual model of warm oceanic convective rain in the Tropics. *J. Appl. Meteor.,* **39,** 2165–2181.

——, ——, F. D. Marks Jr., E. Amitai, and C. R. Williams, 1999: Systematic variation of drop size and radar–rainfall relations. *J. Geophys. Res.,* **104,** 6155–6169.

——, ——, R. A. Black, E. Amitai, P. T. Willis, and C. E. Samsury, 2000: Partitioning tropical oceanic and stratiform rains by draft strength. *J. Geophys. Res.,* **105,** 2259–2267.

Best, A. C., 1950: The size distribution of raindrops. *Quart. J. Roy. Meteor. Soc.,* **76,** 16–36.

Black, R. A., and J. Hallett, 1986: Observations of the distribution of ice in hurricanes. *J. Atmos. Sci.,* **43,** 802–822.

Blanchard, D. C., 1953: Raindrop size-distribution in Hawaiian rains. *J. Meteor.,* **10,** 457–473.

Bradley, S. G., and C. D. Stow, 1974: The measurement of charge and size of raindrops: Part II. Results and analysis at ground level. *J. Appl. Meteor.,* **13,** 131–147.

Bringi, V. N., and V. Chandrasekar, 2001: *Polarimetric Doppler Weather Radar.* Cambridge University Press, 636 pp.

——, ——, J. Hubbert, E. Gorgucci, W. L. Randeau, and M. Schoenhuber, 2003: Raindrop size distribution in different climatic regimes from disdrometer and dual-polarized radar analysis. *J. Atmos. Sci.,* **60,** 354–365.

Brown, P. S., Jr., 1989: Coalescence and breakup-induced oscillations in the evolution of the raindrop size distribution. *J. Atmos. Sci.,* **46,** 1186–1192.

Carbone, R. E., and L. D. Nelson, 1978: The evolution of raindrop spectra in warm-based convective storms as observed and numerically modeled. *J. Atmos. Sci.,* **35,** 2302–2314.

Cataneo, R., and G. E. Stout, 1968: Raindrop-size distributions in humid continental climates, and associated rainfall rate–radar reflectivity relationships. *J. Appl. Meteor.,* **7,** 901–907.

Caton, P. G. F., 1966: A study of raindrop size distributions in the free atmosphere. *Quart. J. Roy. Meteor. Soc.,* **92,** 15–30.

Chandrasekar, V., and V. N. Bringi, 1987: Simulation of radar reflectivity and surface measurements of rainfall. *J. Atmos. Oceanic Technol.,* **4,** 464–478.

——, R. Meneghini, and I. Zawadzki, 2003: Global and local precipitation measurements by radar. *Radar and Atmospheric Science: A Collection of Essays in Honor of David Atlas,Meteor. Monogr.,* No.52, Amer. Meteor. Soc., 215–236.

Churchill, D. D., and R. A. Houze Jr., 1984: Development and structure of winter monsoon cloud clusters on 10 December 1978. *J. Atmos. Sci.,* **41,** 933–960.

de Beauville, C. A., R. H. Petit, G. Marion, and J. P. Lacaux, 1988: Evolution of peaks in the spectral distribution of raindrops from warm isolated maritime clouds. *J. Atmos. Sci.,* **45,** 3320–3332.

Deirmendjian, D., 1969: *Electromagnetic Scattering on Spherical Polydispersions.* Elsevier, 290 pp.

Dingle, A. N., and K. R. Hardy, 1962: The description of rain by means of sequential rain-drop size distributions. *Quart. J. Roy. Meteor. Soc.,* **88,** 301–314.

Donaldson, N. R., 1984: Raindrop evolution with collisional breakup: Theory and models. Ph.D. thesis, University of Toronto, 181 pp.

Foote, G. B., 1966: A *Z–R* relation for mountain thunderstorms. *J. Appl. Meteor.,* **5,** 229–231.

Fujiwara, M., 1965: Raindrop-size distribution from individual storms. *J. Atmos. Sci.,* **22,** 585–591.

——, and T. Yanase, 1968: Raindrop *Z–R* relationships in different altitudes. *Proc. 13th Radar Meteorology Conf.,* Montreal, QC, Canada, Amer. Meteor. Soc., 380–383.

Haddad, Z. S., S. L. Durden, and E. Im, 1996: Parameterizing the raindrop size distribution. *J. Appl. Meteor.,* **35,** 3–13.

Houze, R. A., Jr., 1989: Observed structure of mesoscale convective systems and implications for large scale heating. *Quart. J. Roy. Meteor. Soc.,* **115,** 425–461.

Hu, Z., and R. Srivastava, 1995: Evolution of the raindrop size distribution by coalescence, breakup, and evaporation: Theory and observations. *J. Atmos. Sci.,* **52,** 1761–1783.

Huggel, A., W. Schmid, and A. Waldvogel, 1996: Raindrop size distributions and the radar bright band. *J. Appl. Meteor.,* **35,** 1688–1701.

Iguchi, T., T. Kozu, R. Meneghini, J. Awaka, and K. Okamoto, 2000: Rain-profiling algorithm for the TRMM precipitation radar. *J. Appl. Meteor.,* **39,** 2038–2052.

Imai, I., 1964: A fitting equation for raindrop-size distributions in various weather situations. *Proc. World Conf. on Radio Meteorology and 11th Weather Radar Conf.,* Boulder, CO, Amer. Meteor. Soc., 149A–149D.

Jorgensen, D. P., and P. T. Willis, 1982: A Z–R relationship for hurricanes. *J. Appl. Meteor.,* **21,** 356–366.

——, and M. A. LeMone, 1989: Vertical velocity characteristics of oceanic convection. *J. Atmos. Sci.,* **46,** 621–640.

Joss, J., and A. Waldvogel, 1970: A method to improve the accuracy of radar measured amounts of precipitation. Preprints, *14th Radar Meteorology Conf.,* Tucson, AZ, Amer. Meteor. Soc., 237–238.

——, and E. G. Gori, 1978: Shapes of raindrop size distributions. *J. Appl. Meteor.,* **17,** 1054–1061.

Keenan, T. D., L. D. Carey, D. S. Zrnic, and P. T. May, 2001: Sensitivity of 5-cm wavelength polarimetric radar variables to raindrop axial ratio and drop size distribution. *J. Appl. Meteor.,* **40,** 526–545.

Laws, J. O., and D. A. Parsons, 1943: The relation of raindrop-size to intensity. *Trans. Amer. Geophys. Union,* **24,** 452–460.

Levin, L. M., 1954: On the size distribution function for cloud droplets and rain drops. *Dokl. Akad. Nauk SSSR,* **94,** 1045–1053.

Levin, Z., G. Feingold, S. Tzivion, and A. Waldvogel, 1991: The evolution of raindrop spectra: Comparisons between modeled and observed spectra along a mountain slope in Switzerland. *J. Appl. Meteor.,* **30,** 893–900.

List, R., 1988: A linear radar reflectivity–rain-rate relationship for steady tropical rain. *J. Atmos. Sci.,* **45,** 3564–3572.

——, T. B. Low, N. Donaldson, E. Freire, and J. R. Gillespie, 1987:

Temporal evolution of drop spectra to collisional equilibrium in steady and pulsating rain. *J. Atmos. Sci.,* **44,** 362–372.

Maki, M., T. D. Keenan, Y. Sasaki, and K. Nakamura, 2001: Characteristics of the raindrop size distribution in tropical continental squall lines observed in Darwin, Australia. *J. Appl. Meteor.,* **40,** 1393–1412.

Markowitz, A. H., 1976: Raindrop size distribution expressions. *J. Appl. Meteor.,* **15,** 1029–1031.

Marshall, J. S., and W. McK. Palmer, 1948: The distribution of raindrops with size. *J. Meteor.,* **5,** 165–166.

——, R. C. Langille, and W. McK. Palmer, 1947: Measurement of rainfall by radar. *J. Meteor.,* **4,** 186–192.

Mueller, E. A., 1965: Radar rainfall studies. Ph.D. dissertation, University of Illinois at Urbana–Champaign, 89 pp.

——, and A. L. Sims, 1966: Radar cross sections from drop size spectra. Tech. Rep. ECOM-00032-F, Contract DA-28-043 AMC-00032(E), Illinois State Water Survey, Urbana, Illinois, 110 pp.

Orville, R. E., and R. W. Henderson, 1986: Global distribution of midnight lightning: September 1977 to August 1978. *Mon. Wea. Rev.,* **114,** 2640–2653.

Petrocchi, P. J., and K. J. Banis, 1980: Computer analysis of raindrop disdrometers' spectral data acquired during the 1979 SESAME project. Preprints, *19th Conf. on Radar Meteorology,* Miami Beach, FL, Amer. Meteor. Soc., 490–492.

Rogers, R. R., and R. J. Pilié, 1962: Radar measurements of drop size distribution. *J. Atmos. Sci.,* **19,** 503–506.

Rosenfeld, D., and I. M. Lensky, 1998: Spaceborne sensed insights into precipitation formation processes in continental and maritime clouds. *Bull. Amer. Meteor. Soc.,* **79,** 2457–2476.

Rutledge, S. A., E. R. Williams, and T. D. Keenan, 1992: The Down Under Doppler and Electricity Experiment (DUNDEE): Overview and preliminary results. *Bull. Amer. Meteor. Soc.,* **73,** 3–16.

Sauvageot, H., 1994: Rainfall measurement by radar: A review. *Atmos. Res.,* **35,** 27–54.

——, and J.-P. Lacaux, 1995: The shape of averaged drop size distributions. *J. Atmos. Sci.,* **52,** 1070–1083.

Seliga, T. A., and V. N. Bringi, 1976: Potential use of radar differential reflectivity measurements at orthogonal polarizations for measuring precipitation. *J. Appl. Meteor.,* **15,** 69–76.

Sims, A. L., 1964: Case studies of the areal variations in raindrop size distributions. *Proc. World Conf. on Radio Meteorology and 11th Weather Radar Conf.,* Boulder, CO, Amer. Meteor. Soc., 162–165.

Srivastava, R. C., 1971: Size distribution of raindrops generated by their breakup and coalescence. *J. Atmos. Sci.,* **28,** 410–415.

Steiner, M., and A. Waldvogel, 1987: Peaks in raindrop size distributions. *J. Atmos. Sci.,* **44,** 3127–3133.

Stewart, R. E., J. D. Marwitz, J. C. Pace, and R. E. Carbone, 1984: Characteristics through the melting layer of stratiform clouds. *J. Atmos. Sci.,* **41,** 3227–3237.

Stout, G. E., and E. A. Mueller, 1968: Survey of relationships between rainfall rate and radar reflectivity in the measurement of precipitation. *J. Appl. Meteor.,* **7,** 465–474.

Testud, J., S. Oury, R. A. Black, P. Amayenc, and X. Dou, 2001: The concept of "normalized" distributions to describe raindrop spectra: A tool for cloud physics and cloud remote sensing. *J. Appl. Meteor.,* **40,** 1118–1140.

Tokay, A., and D. A. Short, 1996: Evidence from tropical raindrop spectra of the origin of rain from stratiform and convective clouds. *J. Appl. Meteor.,* **35,** 355–371.

——, ——, and B. Fisher, 1995: Convective vs. stratiform precipitation classification from surface measured drop size distributions at Darwin, Australia and Kapingamarangi atoll. Preprints, *27th Conf. on Radar Meteorology,* Vail, CO, Amer. Meteor. Soc., 690–693.

Ulbrich, C. W., 1983: Natural variations in the analytical form of the raindrop size distribution. *J. Climate Appl. Meteor.,* **22,** 1764–1775.

——, 1985: The effects of drop size distribution truncation on rainfall integral parameters and empirical relations. *J. Climate Appl. Meteor.,* **24,** 580–590.

——, 1992: Effects of drop size distribution truncation on computer simulations of dual-measurement radar methods. *J. Appl. Meteor.,* **31,** 689–699.

——, 1993: Corrections to empirical relations derived from rainfall disdrometer data for effects due to drop size distribution truncation. *Atmos. Res.,* **34,** 207–215.

——, and D. Atlas, 1978: The rain parameter diagram: Methods and applications. *J. Geophys. Res.,* **83C** 1319–1325.

——, and ——, 1998: Rainfall microphysics and radar properties: Analysis methods for drop size spectra. *J. Appl. Meteor.,* **37,** 912–923.

——, and ——, 2002: On the separation of tropical convective and stratiform rains. *J. Appl. Meteor.,* **41,** 188–195.

——, M. Petitdidier, and E. F. Campos, 1999: Radar properties of tropical rain found from disdrometer data at Arecibo, PR. Preprints, *29th Int. Conf. on Radar Meteorology,* Montreal, QC, Canada, Amer. Meteor. Soc., 676–679.

Valdez, M. P., and K. C. Young, 1985: Number fluxes in equilibrium raindrop populations: A Markov chain analysis. *J. Atmos. Sci.,* **42,** 1024–1036.

Waldvogel, A., 1974: The N_0 jump of raindrop spectra. *J. Atmos. Sci.,* **31,** 1067–1078.

Williams, E. R., S. A. Rutledge, S. G. Geotis, N. Renno, E. Rasmussen, and T. Rickenbach, 1992: A radar and electrical study of tropical "hot towers." *J. Atmos. Sci.,* **49,** 1386–1395.

——, and Coauthors, 2002: Contrasting convective regimes over the Amazon: Implications for cloud electrification. *J. Geophys. Res.,* **107,** doi:10.1029/2001JD000380.

Willis, P. T., 1984: Functional fits to some observed drop size distributions and parameterization of rain. *J. Atmos. Sci.,* **41,** 1648–1661.

Wilson, J. W., and E. A. Brandes, 1979: Radar measurement of rainfall—A summary. *Bull. Amer. Meteor. Soc.,* **60,** 1048–1058.

Zawadzki, I., and M. de Agostinho Antonio, 1988: Equilibrium raindrop size distributions in tropical rain. *J. Atmos. Sci.,* **45,** 3452–3459.

Zhang, G., J. Vivekanandan, and E. Brandes, 2001: A method for estimating rain rate and drop size distribution from polarimetric radar measurements. *IEEE Trans. Geosci. Remote Sens.,* **39,** 830–841.

Zipser, E. J., 1994: Deep cumulonimbus cloud systems in the Tropics with and without lightning. *Mon. Wea. Rev.,* **122,** 1837–1851.

——, and M. A. LeMone, 1980: Cumulonimbus vertical velocity events in GATE. Part II: Synthesis and model core structure. *J. Atmos. Sci.,* **37,** 2458–2469.

——, and K. Lutz, 1994: The vertical profile of radar reflectivity of convective cells: A strong indicator of storm intensity and lightning probability. *Mon. Wea. Rev.,* **122,** 1751–1759.

Chapter 11

Some Educational Innovations in Radar Meteorology

STEVEN A. RUTLEDGE

Department of Atmospheric Science, Colorado State University, Fort Collins, Colorado

V. CHANDRASEKAR

Department of Electrical and Computer Engineering, Colorado State University, Fort Collins, Colorado

Rutledge   **Chandrasekar**

1. Introduction

This study is aimed at describing past, current, and future thrusts associated with educating radar meteorologists. The study is geared toward graduate educational activities in academia and is purposely limited to the mid-1950s onward. The review is influenced by the personal experiences of the authors, in particular, through their experiences at their graduate institutions (S. Rutledge, University of Washington; V. Chandrasekar, Colorado State University) and via their faculty appointments at Colorado State University (CSU). The methods used to educate radar meteorologists have continually evolved with the arrival of new radar hardware, a rather expected occurrence, since this educational discipline is obviously hardware intensive. Today, many institutions have their own radar systems, which form focal points for their in-house educational activities. Still, other universities do not have ready access to radar systems, leaving their students to gain hands-on experience via field programs that employ radars, or by other means [e.g., viewing Next Generation Weather Radar (NEXRAD) data over the Internet]. To address this deficiency in the future, it is hoped that students can acquire "virtual" hands-on experience with radar systems through imaginative uses of the Internet. As will be discussed below in more detail, mobile radar systems can also fill this niche by moving around to various universities, providing valuable hands-on experience for students, even just for limited periods of time.

2. Brief historical perspective

A survey was sent to some 20 radar meteorologists asking, among other things, to describe their early memories as to how they were initially trained or exposed to radar measurements. The survey included the following questions.

1) State briefly what exposure you had to radar equipment as a graduate student, including the when/where/what. What were some of the most memorable aspects of your graduate-level education in radar meteorology?
2) Briefly describe the radar educational aspects of your position at present. Do you regularly use and have access to radar equipment? What use do you make of the Internet in radar education?

Selected responses to question 1 include the following.

• "In 1960, assembling a AN/APS-15 antenna, AN/

APS-3 radar, and a Navy VE-2 PPI display, and later working with a CPS-9 radar at the Illinois State Water Survey under the guidance of Gene Mueller." (Prof. R. Rinehart)
- "Having daily contact with Speed Geotis and Pauline Austin. Using tracing paper and colored pencils to track echoes observed by the MIT [Massachusetts Institute of Technology] radar. Using filter paper to measure drop size distributions by hand. Hearing Juerg Joss talk about early disdrometers." (Prof. R. A. Houze)
- "Memorable discussions with Dave Atlas, Bernard Vonnegut, Lou Battan, and Roger Lhermitte on the use of radar to assess the idea that gravitational power of settling precipitation particles provides energy for electrical breakdown. Assembling an X-band radar with Roger Lhermitte provided a valuable hands-on experience." (Dr. E. Williams)
- "My interest in radar began at the University of Texas where an 'old boat radar' was located on top of the engineering building. My interest was further heightened by working with the NCAR [National Center for Atmospheric Research] C bands and the NOAA [National Oceanic and Atmospheric Administration] airborne radars under Bob Houze's guidance at the U. of Washington." (Prof. M. Biggerstaff)
- "I was exposed to CP-3, CP-4, and CHILL as a second-year graduate student at the University of Chicago during the NIMROD experiment. Seeing the first-ever color displays of reflectivity and velocity." (Prof. R. Wakimoto)
- "At McGill, I worked to assemble an X-band vertically pointing radar, wrote the data acquisition code, and analyzed the data. Long hours were spent at the radar observatory where I was exposed to radar maintenance issues and could spend hours looking at displays of all sorts of weather." (Prof. F. Fabry)

Based on responses to question 1, it was evident that during the late 1950s to early 1960s, there was relatively widespread use of U.S. Army Signal Corps surplus radars, as well as emerging weather radars, for example, the CPS-9. Numerous universities and institutions had access to weather radars in this period, including MIT, Texas A&M University, the University of Chicago, and the Illinois State Water Survey, among others. The early education efforts that centered on these facilities emphasized the measurement of surface rainfall and the study of precipitation processes. Formal courses in radar meteorology were now being taught at institutions such as MIT (P. Austin), the University of Chicago (D. Atlas), and the University, of Arizona (L. Battan). Indeed, the first edition of Battan's popular text (Battan 1959) became available during this period of time. By the late 1960s and early 1970s, many radar systems (with both Doppler and polarimetric capabilities) were being built or designed at various universities and laboratories. The CP-2, CP-3, and CP-4 radars built at NCAR (Fig. 11.1);

the CHILL radar, assembled by the University of Chicago and Illinois State Water Survey; X-band radar systems developed at the University of Miami and NOAA/ETL; and the Norman and Cimmaron S-band Doppler radars fielded by NOAA/NSSL, among others, were all instrumental in education efforts during this period, largely through their use in various field projects such as the National Hail Research Experiment (NHRE) and many others. Conception, design, and fabrication of the CHILL S-band radar system was led by D. Atlas (University of Chicago) and E. Mueller (Illinois State Water Survey). This system was originally designed with the capability of computing (via fast Fourier transform processing) and recording Doppler spectra in 32 range gates. A second, X-band, frequency was soon added, as were polarimetric measurements at S band (using a ferrite polarization switch). In this same period, R. Lhermitte was building an X-band Doppler radar using pulse-pair processing to obtain the mean Doppler velocity in all range gates. Dual-Doppler observations soon followed from Dr. Lhermitte's contributions. The NCAR CP-2 radar was developed when Dr. Dave Atlas relocated to NCAR in the early 1970s. Soon after that, the NCAR C-band workhorses (CP-3 and CP-4) were developed, spearheaded by Dr. R. Serafin, director emeritus of NCAR, then the manager of the Field Observing Facility. Many fledgling radar meteorologists in this period received their "training" by reading Battan's book *Radar Meteorology* (1959) and going to the field with an NCAR radar (the first author included)! This was not all that bad actually, but far from where we are today.

There are some common, important threads that can be distilled from our survey. First, hands-on experience was, and remains today, a vital component of the education process. There is no replacement for direct experience when it comes to radar meteorological education. These experiences allow students to gain knowledge of the hardware systems and calibrations techniques, which allows them to better judge data quality and adapt the radar system for a multitude of advanced applications. It is also clear that students best learn when the radar is the focal point of simple experiments, for example, comparing radar-derived rain rates against observations, or calculating reflectivity from measurements of the drop-size distribution and comparing against measurements.

It was only after the emergence of these more complex radar systems that university curricula began to broadly include formal courses on radar meteorology. Early curricula were focused on applications of radar meteorology and the use of radar data (as part of instrumentation classes) and did not initially include topics on radar engineering and signal processing. Only recently have university curricula begun to include, in a serious way, radar hardware, signal processing, antenna theory, and other related engineering topics. Indeed, with the complexity of today's radar systems, ra-

(a)

(c)

(b)

FIG. 11.1. NCAR radars in the 1970s and 1980s era: (a) CP-3 5-cm Doppler radar, vintage 1975; (b) CP-4 5-cm Doppler radar, shown deployed in 1982 during the Cyclonic Extratropical Storms (CYCLES) project conducted by the University of Washington at Ocean Shores, Washington; (c) CP-2 dual-wavelength antenna system. Photos courtesy of D. Parsons, NCAR/ATD.

dar meteorologists must have a full appreciation for all technical aspects of the measurements. Today, a rigorous course on graduate-level radar meteorology may begin with Maxwell's equations and electromagnetic wave propagation, investigate Rayleigh and Mie scattering processes, discuss hardware design, antenna theory, and signal-processing algorithms, develop Doppler and polarimetric measurements, and conclude with highly advanced meteorological applications using fixed, ground-mobile, airborne, shipborne, and spaceborne radar systems at various wavelengths from S- to W-band. In addition to the formal university-based courses, online radar educational materials have been developed by the UCAR/COMET program (information available online at www.comet.ucar.edu). These modules are typically used as stand-alone educational materials for undergraduate programs, and to supplement graduate-level courses. Also, many universities have benefited by access to the NEXRAD datastream via the RIDDS/BDDS system, which is extremely useful for illustrating various weather phenomena. A listing of relevant books and references (with our apologies to authors we missed) includes *Radar Observations of the Atmosphere* (Battan 1973), *Introduction to Radar Systems* (Skolnik 1980), *Spaceborne Weather Radar* (Meneghini and Kozu 1990), *Radar in Meteorology* (Atlas 1990), *Radar Meteorology* (H. Sauvageot 1991), *Doppler Radar and Weather Observations* (Doviak and Zrnic 1993), *Radar for Meteorologists* (Rinehart 1997), *Polarimetric Doppler Weather Radar* (Bringi and Chandrasekar 2001), and

(a)

(c)

(b)

FIG. 11.2. Selected mobile radar systems. (a) The SMART-R radar, shown deployed for CAMEX-4 during summer 2001 in Key West, Florida; photo courtesy of M. Biggerstaff, Texas A&M University. (b) The DOW radar system at the University of Oklahoma; photo courtesy of J. Wurman, University of Oklahoma. (c) The UMASS–OU mobile 95-GHz Doppler radar; photo courtesy of H. Bluestein, University of Oklahoma.

Weather Radar Technology Beyond NEXRAD (National Research Council 2002).

3. Current perspectives

Responses to question 2 above returned a general vision that many universities have access to radar systems at their home institutions. It was also evident that other front-line institutions focused on graduate education lack hardware, despite their best efforts to acquire such hardware. Perhaps a partial solution to these deficits can be filled by the mobile X-band radars in use at the University of Oklahoma (OU) and National Severe Storms Laboratory (Doppler on Wheels systems); the two mobile C-band Shared Mobile Atmospheric Research and Teaching Radars (SMART-Rs) developed by NSSL, Texas A&M University, Texas Tech University, and OU, the University of Massachusetts–University of Oklahoma mobile 95-GHz Doppler radar; the University of Miami portable 94-GHz Doppler radar (Fig. 11.2);

and other systems. Yet another similar opportunity exists with the Florida State University mobile C-band system, known as the "Seminole Hurricane Hunter." Indeed, future plans call for some of these mobile systems to be upgraded to dual-polarization capabilities. These mobile facilities could move around to various undergraduate institutions as a way of exposing undergraduates to the exciting opportunities in radar meteorology. Longer visits could perhaps be made to graduate-based institutions so that the mobile facility could be operated during a portion of a quarter-or semester-long course (Bluestein et al. 2001).

One of the focal points for modern day educational activities is the CSU-CHILL National Radar Facility (Fig. 11.3), sponsored by the National Science Foundation and Colorado State University. As such, it serves a broad element of the community. A detailed description of the radar can be found in Brunkow et al. (2000; see also www.chill.colostate.edu). The CSU-CHILL system has been used in many field projects [Winter

(a)

(b)

FIG. 11.3. The CSU-CHILL National Radar Facility: (a) The CSU-CHILL site near Greeley, Colorado; (b) the CSU-CHILL antenna system; photos courtesy of P. Kennedy, Colorado State University.

Icing and Storms Project (WISP), Program for Regional Observing and Forecasting Services Operational Work Station (POWS), Stratosphere–Troposphere Experiments: Radiation, Aerosoles, Ozone (STERAO-A), Severe Thunderstorm, Electrification, and Precipitation Study (STEPS), etc.], thus providing access for students. Furthermore, the system is operated during a large por-

tion of the storm season in order to build a data archive on a wide variety of weather situations. These data form the bases for class projects at CSU (Rutledge et al. 1993) and are also made available via the web for use in classrooms well beyond those at CSU. The radar has also been the focal point for Research Experience for Undergraduates (REU) projects under the direction of V.

VCHILL Concept

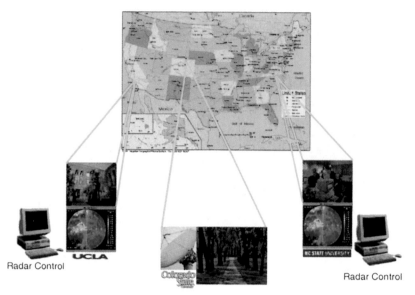

FIG. 11.4. Schematic of the VCHILL system. See text for details.

Chandrasekar. These REU projects have focused on bringing approximately 10 undergraduate engineering students from universities in the western United States to the facility for 6–8-week-long visits, providing hands-on experience in radar engineering and radar meteorology. Students return to their home institutions to complete senior theses on data they collected during their visit. The CSU-CHILL facility regularly hosts visits from a wide range of student groups, as well as playing a major role in an outreach project, the Colorado Cooperative Rain and Hail Studies (CoCoRaHS). In this project, hundreds of public citizen volunteers have been trained to make quantitative rain and hail measurements and post these on a web site. Such data are also used to provide validation data for the CSU-CHILL polarimetric data. An Internet course largely based upon the CSU-CHILL radar, entitled Introduction to Meteorological Systems, will be offered through Colorado State University's education outreach program starting in Fall 2003.

In addition to CHILL, other radars also provide valuable educational experiences via field projects and REU projects. Most notable is the role that the portable S-pol 10-cm polarimetric radar operated by NCAR now plays in the educational arena. The S-pol has also been used in many field projects, providing excellent opportunities to expose students to the field of radar. NOAA/ETL radars also reach students via field projects. Yet another opportunity has recently come on line with the NASA N-pol radar, a highly portable, 10-cm Doppler polarimetric radar.

4. A glimpse at the future

Given the recent advances in Internet technology, it is now possible to essentially "bring" radar systems into classrooms. CSU has embarked on a major initiative called Virtual CHILL (VCHILL), with the objective of bringing the research and educational mission of CSU-CHILL closer to research and educational institutions, in a virtual sense. The VCHILL initiative enables real-time operation of the radar to be distributed over the Internet, which is a paradigm shift from conventional mode of operation where all the radar users assemble at the radar facility. Figure 11.4 shows the VCHILL concept, where students from remote universities will participate in conducting radar experiments as part of their educational experience. Full control of the radar (i.e., adjustment of antenna scanning parameters) is possible from these remote classrooms. The access that is being developed is far beyond what is presently available. One of the fundamental requirements assumed, from an educational perspective, is that the user must see the rotating radar beam in real time, just as one would see on the radar display. The CSU-CHILL also has the capability to transmit high-bandwidth digitized radar samples to conduct experiments on radar signals. Currently, six universities are using or planning to use the VCHILL system as an integral part of their meteorology courses. We expect many more institutions to utilize VCHILL in the near future. The VCHILL system is also being developed for various network protocols. There are other online capabilities that are used for supporting teaching radars at various institutions. Two-way video conferencing capability over the Internet enables researchers to conduct online experiments. For example, in a radar system class, the concept of noise temperature at different points on the receiver chain can be demonstrated over a video link. Calibration pointing to the sun can also be done over the Internet.

As part of VCHILL, CSU-CHILL staff have developed a complete set of Internet tools to facilitate the process of analyzing archived radar data, such as radar data browsers, hydrometeor classification systems, and detailed analyzing systems on the Internet. Staff also implemented a Virtual Private Network (VPN), such that students from multiple institutions can participate in a field experiment without physically having to leave their home institution. VCHILL was recently demonstrated at the Atlas Symposium during the 2002 American Meteorological Society (AMS) Annual Meeting. At the time of this writing, CHILL personnel are developing ways to inform the broader community about VCHILL and describe how VCHILL may be accessed.

5. Conclusions and recommendations

The future of radar meteorology (with this field including radar engineering as well) is bright. Many universities have ready access to radar systems. Others operate their own systems, including maintaining national radar facilities. Portable radars such as the Doppler on Wheels (DOWs) and the SMART-R systems are playing an emerging role in radar education by visiting sites that do not have regular hands-on access to radar systems. We are also starting to make good use of distance education offerings in radar meteorology via the Internet, especially couched at the undergraduate level. In this way, we can sustain a good influx of graduate students into this discipline, as there is a national need for such a pool given the ubiquitous presence of weather radars. With the increased capabilities of the Internet and data dissemination, fixed radar systems, such as the CSU-CHILL radar, can simulate "hands-on" access to any classroom setting equipped with Internet capabilities. Indeed, capabilities now exist to pipe live audio and video from this radar into the classroom, and to permit real-time control of the radar from the remote classroom locations. The future of radar meteorology will continue to make use of these virtual radar systems via the Internet. These efforts can especially benefit those institutions worldwide that do not have ready access to radar systems.

Since meteorological radars are typically embedded with an observational network of other sensors to study convective and mesoscale processes, we are constantly challenged to create learning environments that expose students not only to radars, but also to profilers, lidars, disdrometers, and other instruments operating in an integrated fashion. It is likely that such instrument suites are only viable for specific field projects. Therefore, these venues must continue to serve as hands-on learning environments. Our community should continue to make broad use of programs such as the National Science Foundation (NSF) Research Experience for Undergraduates opportunity for getting students to the field. Efforts should be made to create similar exposure for graduate students, beyond just the students of the principal investigators for the field project.

Finally, there is a vast resource of radar systems, as well as science and engineering expertise, in laboratories at NCAR, NOAA, and the National Aeronautics and Space Administration (NASA), as well as in industry. These resources should be tapped through an organized effort to allow an expanded role in radar meteorology and radar engineering education. Furthermore, efforts to offer joint courses between atmospheric science and electrical engineering at universities should continue. Such training is vital for future generations of radar meteorologists that must deal with ever increasing complexity in radar systems.

Today, radar meteorologists regularly use ground-based radars, both fixed and mobile systems. Many of these systems have been mentioned in this article. Obviously, such systems are only one component of the broad radar arsenal used in atmospheric research. Airborne systems (such as the NCAR ELDORA dual-Doppler radar and the NOAA P-3 tail Doppler and lower-fuselage radars), spaceborne radars [such as the NASA Tropical Rainfall Measuring Mission (TRMM) precipitation radar], and shipborne weather radars, (such as operated on the R/V *Ron Brown*) are also at our disposal. Therefore, these types of systems need to be standard components of radar meteorological training as well. For example, the R/V *Brown* could play a significant role in radar meteorological training by dedicating cruises for education. The advantage of such an activity is that the radar observations on the R/V *Brown* are nested within atmospheric and oceanographic measurements, thus allowing an integrated educational approach. Indeed, this is how practicing radar meteorologists use radars.

Acknowledgments. This work was supported by the National Science Foundation, through the Cooperative Agreement for the CSU-CHILL National Radar Facility, ATM-0118021, and NSF ITR Grant ATM-0121546. Colorado State University also supports the CSU-CHILL radar facility.

REFERENCES

Atlas, D., Ed., 1990: *Radar in Meteorology.* Amer. Meteor. Soc., 806 pp.

Battan, L., 1959: *Radar Meteorology.* University of Chicago Press, 161 pp.

——, 1973: *Radar Observations of the Atmosphere.* 2d ed. University of Chicago Press, 324 pp.

Bluestein, H. B., B. A. Albrecht, R. M. Hardesty, W. D. Rust, D. Parsons, R. Wakimoto, and R. M. Rauber, 2001: Ground-based mobile instrument workshop summary, 23–24 February 2000, Boulder, Colorado. *Bull. Amer. Meteor. Soc.,* **82,** 681–694.

Bringi, V. N., and V. Chandrasekar, 2001: *Polarimetric Doppler Weather Radar.* Cambridge University Press, 636 pp.

Brunkow, D. A., V. N. Bringi, P. C. Kennedy, S. A. Rutledge, V. Chandrasekar, E. A. Mueller, and R. K. Bowie, 2000: A description of the CSU–CHILL National Radar Facility. *J. Atmos. Oceanic Technol.,* **17,** 1596–1608.

Doviak, R., and D. Zrnic, 1993: *Doppler Radar and Weather Observations.* 2d ed. Academic Press, 562 pp.

Meneghini, R., and T. Kozu, 1990: *Spaceborne Weather Radar.* Artech House, 199 pp.

National Research Council, 2002: *Weather Radar Technology Beyond NEXRAD.* National Academy Press, 81 pp.

Rinehart, R., 1997: *Radar for Meteorologists.* 3d ed. Rinehart, 428 pp.

Rutledge, S. A., P. C. Kennedy, and D. A. Brunkow, 1993: Use of the CSU–CHILL radar in radar meteorology education at Colorado State University. *Bull. Amer. Meteor. Soc.,* **74,** 25–31.

Sauvageot, H., 1992: *Radar Meteorology.* Artech House, 366 pp.

Skolnik, M. I., 1980: *Introduction to Radar Systems.* 2d ed. McGraw-Hill, 581 pp.

Evening Banquet at The Rosen Plaza Hotel in Orlando, Florida

Lucille and Dave Atlas

Jenny Sun, Frédéric Fabry, and Howie Bluestein

Vonnie & Richard Johnson, and Steve Rutledge

Eugene Rasmusson and Don Johnson

Paul Smith and Roddy Rogers

Roger Lhermitte and Frank Marks

Ramesh Srivastava and Keith Browning

Betsy Serafin and Rit Carbone

Dave & Lucille Atlas

Frank Marks and Dave Atlas

Vonnie and Richard Johnson

Linda Carbone, Dave Atlas, and Keith Browning

Keith Browning, Roger Lhermitte,
Dave Atlas, and Peter Black

Bob Serafin, Dave Atlas, and Rit Carbone

Betsy & Bob Serafin, Lucille & Dave Atlas,
and Rit & Linda Carbone

Ramesh Srivastava

Dave Atlas

Dave & Lucille Atlas

Eunice & Paul Twitchell, and Dave Atlas

Bob Serafin, Dave & Lucille Atlas,
and V. Chandrasekar

Dave & Lucille Atlas